LEPTONS, HADRONS AND NUCLEI

LEPTONS, HADRONS and NUCLEI

Florian SCHECK

Johannes Gutenberg-Universität, Mainz, Germany

1983

NORTH-HOLLAND PHYSICS PUBLISHING
AMSTERDAM · NEW YORK · OXFORD

ISBN: 0 444 86719 8

Published by:
North-Holland Physics Publishing
a division of
Elsevier Science Publishers B.V.
P.O. Box 103
1000 AC Amsterdam
The Netherlands

Sole distributors for the U.S.A. and Canada:
Elsevier Science Publishing Company, Inc.
52 Vanderbilt Avenue
New York, N.Y. 10017
U.S.A.

Library of Congress Cataloging in Publication Data

Scheck, Florian, 1936-
Leptons, hadrons, and nuclei.

Bibliography: p.
Includes index.
1. Leptons. 2. Hadrons. 3. Weak interactions
(Nuclear physics) 4. Electromagnetic interactions.
I. Title.
QC793.5.L428S33 1984 539.7'2 83-23631
ISBN 0-444-86719-8 (U.S.)

Printed in the Netherlands

Res Jost in Freundschaft
und Verehrung zugeeignet

PREFACE

In the early days, at the time of their founders, the theory of elementary particles and the physics of nuclei used to be just two facets of one and the same field whose primary motivation was the investigation of the nature of the elementary constituents of matter and the interactions between them. As is well known, under the impetus of the great discoveries in particle physics during the 1950's and the early 1960's on the one hand, and under the impression of the specific complexity of the nuclear many-body problem after the invention of the shell model on the other hand, the two fields drifted apart and they both became respectable, complex and multiply ramified disciplines of their own. Nevertheless, thanks to the vision and encouragement of a generation of physicists between the two worlds such as Amos de-Shalit and Hans Jensen and some others who are still with us, there has been a continuing effort all over the years to save at least something of the original unity and to bridge the now independent fields, to the benefit of both. These efforts materialized in an ever increasing and important branch of research at the "meson factories", at the electron accelerators in the range of up to several GeV, and at some of the particle accelerators at high energies—a set of research activities which is often referred to (rather inaccurately) as Intermediate Energy Physics and which is regularly covered by the well-known series of conferences on High-Energy Physics and Nuclear Structure.

This book emanated from a series of lectures on general Particle Physics, Electroweak Interactions, and Intermediate Energy Nuclear Physics that I gave at the Eidgenössische Technische Hochschule in Zurich, at the 3^{e} cycle de la Physique en Suisse Romande, and at the University of Mainz, at the level of *advanced quantum mechanics*. The material of the book can be covered in two semesters (a one-year course), but, depending on the time available, it may seem necessary to apply some cuts. The book is written on the assumption that the reader is familiar with the general principles of quantum mechanics as well as with the basics of the theory of special relativity. Although some knowledge of the basic notions of quantum field theory is certainly useful this is not a prerequisite. For instance, the theory of relativistic spinor fields including their quantization, is dealt with in Chap. III in rather great detail. The occasional use of Feynman rules in formulating simple, lowest-order amplitudes does not mean that I presuppose detailed knowledge of quantum electrodynamics. For the sake of reference, the rules are summarized in Appendix C so that the reader who is not acquainted with covariant perturbation theory can understand and reconstruct any particular ansatz in the text. Apart from this the book is self-contained to a large extent. All important definitions and basic notions are explained at the appropriate place in the text and many examples are given. In addition, I have collected a set of exercises at the end of each chapter which may help to test one's understanding of the material and to further illustrate the content of the corresponding chapter.

For those who are no longer students, who, consequently, have less time for reading and who feel discouraged by long textbooks, I wish to add the following "guide": The book can be read in several sections and ways. The first and trivial option ("not at all"), I hope, will not be chosen. The reader who is primarily interested in meson–nucleus physics at intermediate energies may concentrate, in a first reading, on Chaps. I and II. If electroweak interactions are at the heart of interest, the combinations III, IV, V or III, IV (purely electromagnetic interactions), or III, V (mostly weak and electroweak interactions) may be chosen. The book was conceived in such a way that the aforementioned combinations form coherent but largely self-contained blocks. Chap. III, in particular, deals with the relativistic theory of fermion fields and forms the basis for Chaps. IV and V. For this reason these two chapters contain many cross-references to topics and results of Chap. III. Dirac theory is formulated both in its natural language of Van der Waerden spinors, i.e. of "dotted" and "undotted" two-component spinors, and in the conventional four-component notation. As explained in and illustrated by Chaps. III and V, the two-component formalism is more adequate in analyzing basic properties of the theory (see its importance in modern supersymmetries!) and for certain questions in electroweak interactions which it renders more transparent, whereas the conventional notation is certainly more useful in practical calculations. Unfortunately, neither the notation nor the phase conventions are standardized in this matter. For this reason we spend some space in Chap. III at explaining our conventions in some detail and at illustrating them by explicit examples. For the sake of clarity we write out the spinor indices in many cases even when a component-free notation would have been more elegant.

The times are over where a book like Schweber's Relativistic Quantum Field Theory could give a complete survey of the literature at the time of writing. The literature on particle physics has been growing at a pace that even a list of the "characteristic fossils" would be exorbitantly long. Therefore, and in order to avoid the delicate problem of giving balanced credit to the workers in the field, some authors deliberately choose to give no reference to the original literature at all, letting the reader find his own way to the sources of wisdom. In this book I have adopted the following, admittedly imperfect compromise: At the end of the book there is a general bibliography of textbooks, handbooks and monographs to which I refer throughout all chapters (by the symbols R*i*). These references are useful for refreshment of knowledge in other fields, for complementary reading on specific topics, or as sources for specific mathematical properties needed in the text. In addition, each chapter contains a shorter list of articles (referred to by names of first author and year of publication) from which I learnt a lot about the subject in question or which I chose for the sake of illustration and as a useful starting point for further reading. Clearly, this choice is neither representative nor balanced nor objective. However, this incompleteness is easily repaired by going first to the review papers that we quote (including my own) which do give a fair account of the existing literature.

After a long hesitation I decided to omit the description of bag models which form indeed a topic of ever increasing importance in nuclear and particle physics. At the time of this writing this field is still in an exploratory stage and very few of its results are generally accepted by the different schools of thought. A detailed discussion of this topic would soon be out of date, if included in the book at this time, but I will certainly reconsider the question for a later edition. Also omitted are the more detailed and rather sophisticated theories of meson–nucleus scattering (beyond what is discussed in Chap. I), as well as the topics of muon capture, pion radiative capture on nuclei, and exchange current effects in nuclei on which there already exist a number of excellent treatises.

Regarding the spirit in which this book is written I may say this: The technical aspects of a field are much easier to master if one has a clear guiding line of the topic under consideration, if one has some *Anschauung* and a feeling for the relevant features of it. This is the reason why I try to motivate and to explain in great detail the basic notions and leading ideas before embarking in the unavoidable technicalities. Experimental data are quoted in many occasions. Especially in the discussion of electroweak interactions in the framework of the standard model and beyond it, I discuss a set of key experiments (most of which are fairly recent) that illustrate the state of knowledge and the limitations in testing our theoretical ideas. I also point out open ends and future possibilities where progress is to be expected in the years to come.

Because of lack of philological training my approach to the english language is a rather intuitive one. I can only hope that the text is intelligible and that the educated reader will not suffer too much. For the rest I wish I could say this:

"Zum Nutzen und Gebrauch der Lehr-begierigen
*Physicali*schen Jugend, als auch derer in diesem *studio*
schon *habil* seyenden besonderem ZeitVertreib auffgesetzet
und verfertiget..."

(with slight modification, from: Das Wohl*temperierte Clavier* by Johann Sebastian Bach, Köthen 1722).

I wish to thank my colleagues, collaborators and students here at Mainz, at SIN and at Zurich for numerous discussions on many topics dealt with in this book and for constructive criticism at very many occasions both in teaching and in active research. In particular, I am grateful to Jörg Hüfner who encouraged me to write this book at various stages of the project. I acknowledge the stimulating response of my students whom I confronted with some of the more original presentations of the material and whose questions were of great help in improving the form, the content and the style. Special thanks go to Nikos Papadopoulos, to Alfons Heil, John Missimer and Gerhard Rufa who read major portions of the manuscript and made numerous suggestions for improvement and pointed out errors and inaccuracies to me. G. Rufa kindly provided the calculations and the drawings for the examples in Glauber theory shown in figs. (I.6) and the cross sections for electron scattering shown in fig. (IV.5).

Last but not least I owe special thanks to Maraike zur Hausen who patiently typed various versions of the book and who took great care in preparing an excellent original from a complex manuscript.

Mainz, 1 May 1983 Florian Scheck

NOTATION, CONVENTIONS, SYMBOLS

To a large extent, the notation is explained in the text. Nevertheless it may be useful to go first through this short section on notations and conventions, or to return to it when one is not absolutely sure about a symbol or definition that is used in the text.

(i) *Units*. We use natural units

$$\hbar = c = 1$$

throughout the book. As is well-known this convention is not sufficient to fix the units completely. What is needed, in addition, is a unit of energy (or mass, or time). Following standard practice, we use the energy units

1 eV	$1\ \text{meV} = 10^{-3}\ \text{eV}$	$1\ \text{keV} = 10^{3}\ \text{eV}$
$1\ \text{MeV} = 10^{6}\ \text{eV}$	$1\ \text{GeV} = 10^{9}\ \text{eV}$	$1\ \text{TeV} = 10^{12}\ \text{eV}$

It is easy to translate length l, cross section σ, and time t from natural units back to conventional units. Let $\mathring{l}$, $\mathring{\sigma}$, $\mathring{t}$ be such quantities expressed in natural units, l, σ, t the same quantities in standard units. Then

$$l[\text{fm}] \triangleq \hbar c \cdot \mathring{l}[\text{MeV}^{-1}],$$

$$t[\text{sec}] \triangleq \frac{\hbar c}{c} \cdot \mathring{t}[\text{MeV}^{-1}],$$

$$\sigma[\text{fm}^2] \triangleq (\hbar c)^2 \cdot \mathring{\sigma}[\text{MeV}^{-2}],$$

with $\hbar c = 197.3286\ \text{MeV} \cdot \text{fm}$, $1\ \text{fm} = 10^{-13}\ \text{cm}$, $c = 2.9979 \times 10^{10}\ \text{cm/sec}$. Cross sections are usually expressed in units of $1\ \text{b} = 10^{-24}\ \text{cm}^2 = 10^{2}\ \text{fm}^2$. Thus, if $\mathring{\sigma}$ is found in GeV^{-2}, for example, then it follows that

$$1\ \text{GeV}^{-2} \triangleq 0.3894\ \text{mb}.$$

Momenta p are expressed in energy units if $\hbar$ and c are set equal to one, and are given in (energy unit)/c when conventional units are used.

(ii) *Experimental results and errors* are generally quoted with the error of the last digits in parentheses. For example,

0.777 3(13) means 0.7773 ± 0.0013,

0.511 26(5) means 0.51126 ± 0.00005.

The abbreviation ppm stands for "parts per million". For example, a measurement giving the result 0.511 003 4(14) is a "2.7 ppm measurement".

(iii) *Metric and normalization*. The metric is explained in more detail in App. A. We use the form $g^{00}=+1$, $g^{ii}=-1$ $(i=1,2,3)$ for the diagonal metric tensor. A contravariant vector is denoted by $x^{\mu}=(x^0, \boldsymbol{x})$ so that $x_{\mu}=(x^0, -\boldsymbol{x})$. One-particle states appear with the covariant normalization

$$\langle \boldsymbol{p}'|\boldsymbol{p}\rangle = 2E_p\delta(\boldsymbol{p}-\boldsymbol{p}')$$

for both bosons and fermions. In case of particles with spin there is an additional Kronecker δ-symbol for the spin indices.

(iv) *Some symbols*. T denotes the scattering matrix. Note, however, that in Chap. I, which is based on a nonrelativistic formalism, this is the noncovariant T-matrix of nonrelativistic scattering theory. In Chaps. II and V, on the other hand, we use the covariant T-matrix as defined in App. B.

$\overleftrightarrow{\nabla}$ and $\overleftrightarrow{\partial}_{\mu}$ are short-hand notations for antisymmetric derivatives

$$f(\boldsymbol{x})\overleftrightarrow{\nabla}g(\boldsymbol{x})=f(\boldsymbol{x})(\nabla g(\boldsymbol{x}))-(\nabla f(\boldsymbol{x}))g(\boldsymbol{x}),$$
$$f(x)\overleftrightarrow{\partial}_{\mu}g(x)=f(x)(\partial_{\mu}g(x))-(\partial_{\mu}f(x))g(x).$$

(v) *Rotation matrices*. For the representation coefficients of the rotation group we use the definitions of ref. R4, that is

$$D^{(j)}_{KM}(\psi,\theta,\phi)=\mathrm{e}^{\mathrm{i}K\psi}d^{(j)}_{KM}(\theta)\mathrm{e}^{\mathrm{i}M\phi}$$

with Euler angles as defined in Fig. III.1 (p. 94). As explained in more detail in sec. III.2, these are the matrices which transform the expansion coefficients. Basis functions then transform according to D^*.

(vi) *Reference to equations, tables, figures and subsections*. Equations are identified by their number, e.g. (21), when referred to from within the same chapter in which they appear. If referred to from another chapter, they are identified by the number of their chapter and their equation number. For example, in Chap. V the text may refer to eq. (II.21), i.e. eq. (21) of Chap. II. An analogous system is adopted for reference to tables and to figures. Similarly, reference to sec. (*i*. *j*.) means a section or subsection of the same chapter, whereas reference to sec. (III.*i*. *j*.) means the corresponding section in Chap. III.

Aber sage nur niemand, daß uns das Schicksal trenne! Wir sinds, wir! Wir haben unsere Lust daran, uns in die Nacht des Unbekannten, in die kalte Fremde irgend einer andern Welt zu stürzen, und wär' es möglich, wir verließen der Sonne Gebiet und stürmten über des Irrsterns Grenzen hinaus.

(Friedrich Hölderlin, Hyperion, 1. Band, 1. Buch)

CONTENTS

Chapter I

SCATTERING OF STRONGLY INTERACTING PARTICLES ON NUCLEI

This chapter deals with the interaction of strongly interacting particles (hadrons) with nuclei, at low and intermediate energies. The hadronic projectiles are used as probes of nuclear properties and of the hadron–nucleon interaction in the nuclear medium, and they are selected accordingly. We discuss mainly the case of pions, to some extent also kaons, because of the specific properties of the pion–(kaon)–nucleon system. Some of the methods and results can also be applied to nucleons if these are chosen to be projectiles, but we do not go into the full complexity of the spin analysis of the nucleon–nucleon system. In this sense this field is distinct from the conventional topic of *nuclear reactions* for which there are many excellent treatises in the literature on Nuclear Physics. Conventional nuclear reactions are mostly studied in their own right with the primary aim of understanding and formulating the reaction mechanism itself. In intermediate energy physics the reaction mechanism is believed, or assumed, to be known and the scattering process is analyzed in terms of properties of the nucleus and of the elementary projectile–nucleon system.*)

After a reminder of some results of general scattering theory we discuss three specific situations: (i) the domain of scattering at "high" energies where eikonal methods are adequate tools to describe the reaction; (ii) scattering at very low energies which can be analyzed in terms of an optical potential; and (iii) pion–nucleus scattering around the $\Delta(1232)$ resonance. The application of the optical potential to bound states in hadronic atoms is treated in Chap. II.

This field is still in rapid evolution. It would be pointless to present in detail specific models and methods of analysis advocated by the different schools in the field. Too much of the more detailed analyses would be out of date very soon. Instead, we concentrate on setting a general frame of results that is likely to remain valid for some time to come and that might remain the starting point for more specific and detailed studies.

1. Selected topics in potential scattering theory

The theory of the scattering of hadrons on nuclei at low and intermediate energies is generally formulated on the basis of potential scattering theory. Even though some of the results obtained from nonrelativistic wave mechanics can be generalized to the case of relativistic wave equations (Klein–Gordon and Dirac equations), a fully relativistic treatment of multiple scattering of hadrons on composite targets is not

*) The separation is never as clear as said here. A serious analysis of any hadron–nucleus process must rely, to some extent, on the now classical theory of nuclear reactions.

available. In other words, such a treatment can be generalized such as to take into account relativistic *kinematics* but its *dynamical* basis is essentially nonrelativistic. In this section we collect those results of nonrelativistic scattering theory, in a very condensed form, which are relevant for the subsequent sections on multiple scattering and optical potentials. These sections (1.1 to 1.5) are meant as a reminder of some basic results of quantum scattering theory, and can by no means replace a detailed introduction to this important topic. The formulation of scattering and decay processes in the frame of general S-matrix theory, needed in later chapters of this book, is summarized in Appendix B.

1.1. Scattering amplitude for scalar wave equations

Consider the scattering of a spinless particle of mass m on a potential $V(\boldsymbol{r})$ of finite range R. The wave function describing the scattering process for given energy E and given asymptotic momentum $\boldsymbol{k}$ satisfies the stationary equation

$$(\Delta + \boldsymbol{k}^2)\psi_{\boldsymbol{k}}(\boldsymbol{r}) = 2mV(\boldsymbol{r})\psi_{\boldsymbol{k}}(\boldsymbol{r}). \tag{1}$$

In the nonrelativistic limit the energy E is related to the momentum $\boldsymbol{k}$ by $E = \boldsymbol{k}^2/2m$, and eq. (1) is the Schrödinger equation. For relativistic motion

$$E^2 = \boldsymbol{k}^2 + m^2. \tag{2}$$

In this case eq. (1) is a Klein–Gordon equation with $V(\boldsymbol{r})$ an external potential. Note that here there is an ambiguity: the way it is introduced into eq. (1) the potential takes the place of a Lorentz *scalar* quantity. It could as well appear in the form

$$E^2 \to [E - V(\boldsymbol{r})]^2$$

in eq. (1), in which case it would look like the fourth component of a Lorentz *vector*. In the corresponding nonrelativistic situation this ambiguity disappears (see exercise 1). In either case the boundary condition on ψ is

$$\psi_{\boldsymbol{k}}(\boldsymbol{r}) \underset{r\to\infty}{\sim} e^{i\boldsymbol{k}\cdot\boldsymbol{r}} + f(\theta, \phi)\frac{e^{ikr}}{r}. \tag{3}$$

The formal solution of the problem is obtained by means of suitably chosen Green functions $g(\boldsymbol{r}, \boldsymbol{r}')$, viz.

$$\psi_{k}(\boldsymbol{r}) = e^{i\boldsymbol{k}\cdot\boldsymbol{r}} + \int d^3r'\, g(\boldsymbol{r}, \boldsymbol{r}')V(\boldsymbol{r}')\psi_{\boldsymbol{k}}(\boldsymbol{r}'), \tag{4}$$

where g satisfies

$$(\Delta + \boldsymbol{k}^2)g(\boldsymbol{r}, \boldsymbol{r}') = 2m\delta(\boldsymbol{r} - \boldsymbol{r}') \tag{5}$$

and is given by

$$g(\boldsymbol{r}, \boldsymbol{r}') = -\frac{m}{2\pi}\frac{1}{|\boldsymbol{r} - \boldsymbol{r}'|}\{a e^{ik|\boldsymbol{r}-\boldsymbol{r}'|} + b e^{-ik|\boldsymbol{r}-\boldsymbol{r}'|}\}. \tag{6}$$

Here $k \equiv |\boldsymbol{k}|$ while the parameters a and b satisfy the condition $a + b = 1$. The

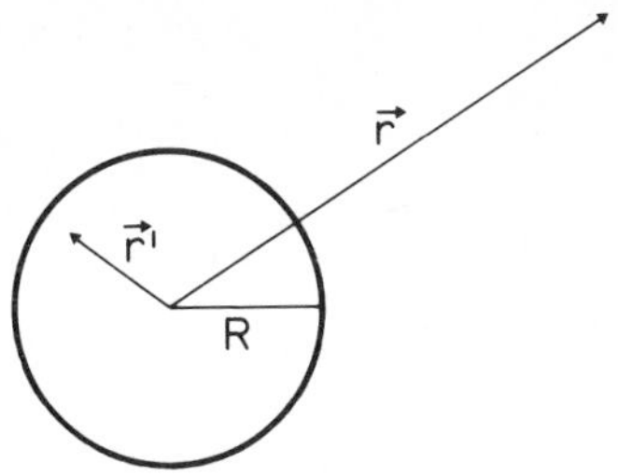

Fig. 1. Coordinates $\boldsymbol{r}$ of the scattered particle and integration variable $\boldsymbol{r}'$ that covers the domain over which the potential is different from zero.

boundary condition (3) is obtained by the choice $a=1$, $b=0$. Indeed, in this case, letting $r \equiv |\boldsymbol{r}| \to \infty$,

$$|\boldsymbol{r}-\boldsymbol{r}'| \simeq r - \frac{1}{r}\boldsymbol{r}'\cdot\boldsymbol{r}$$

(see Fig. 1) and

$$\psi_k(\boldsymbol{r}) \underset{r\to\infty}{\sim} \mathrm{e}^{\mathrm{i}\boldsymbol{k}\cdot\boldsymbol{r}} - \frac{m}{2\pi}\frac{\mathrm{e}^{\mathrm{i}kr}}{r}\int \mathrm{e}^{-\mathrm{i}k(\boldsymbol{r}/r)\cdot\boldsymbol{r}'}V(\boldsymbol{r}')\psi_k(\boldsymbol{r}')\,\mathrm{d}^3r'.$$

Defining $k\boldsymbol{r}/r =: \boldsymbol{k}'$, this yields the following expression for the scattering amplitude:

$$f(\theta,\phi) = -\frac{m}{2\pi}\int \mathrm{e}^{-\mathrm{i}\boldsymbol{k}'\cdot\boldsymbol{r}}V(\boldsymbol{r})\psi_k(\boldsymbol{r})\,\mathrm{d}^3r. \tag{7}$$

Note the following: This expression is useful only if we already know $\psi_k(\boldsymbol{r})$ or if we have a suitable approximation for it. However, as $V(\boldsymbol{r})$ has the finite range R, we need to know ψ_k only within the sphere of radius R, i.e. in a limited domain of space. We shall make use of this fact below. If $V(\boldsymbol{r})$ is symmetric with respect to rotations about the axis $\hat{\boldsymbol{k}}$, then f depends on θ, not on the azimuth ϕ. This will be assumed for what follows.

1.2. Partial wave decomposition of scalar scattering amplitude

The scattering amplitude $f(\theta)$ for a scalar particle can be expanded in terms of Legendre polynomials,

$$f(\theta) = \sum_{l=0}^{\infty}(2l+1)f_l(k)P_l(\cos\theta). \tag{8}$$

The integrated *elastic* cross section is

$$\sigma_{\mathrm{el}} = \int \mathrm{d}\Omega\,|f(\theta)|^2$$

$$= 4\pi\sum_{l=0}^{\infty}(2l+1)|f_l(k)|^2. \tag{9}$$

If the potential $V(\boldsymbol{r})$ is complex, i.e. if there is also scattering into inelastic channels, this is distinct from the *total* cross section

$$\sigma_{\text{tot}} = \sigma_{\text{el}} + \sigma_{\text{abs}}, \tag{10}$$

which, by the optical theorem, is given by the imaginary part of the forward scattering amplitude,

$$\sigma_{\text{tot}} = \frac{4\pi}{k} \operatorname{Im} f(\theta = 0) = \frac{4\pi}{k} \sum_l (2l+1) \operatorname{Im} f_l(k). \tag{11}$$

As $\sigma_{\text{tot}} \geq \sigma_{\text{el}}$ and as this inequality holds for each partial wave separately, one obtains the positivity condition

$$\operatorname{Im} f_l(k) \geq k|f_l(k)|^2. \tag{12}$$

This condition is met if $f_l(k)$ has the general form

$$f_l(k) = \frac{1}{2ik}\left(e^{2i\delta_l(k)} - 1\right), \tag{13}$$

where $\delta_l(k)$ are complex phases which depend on the orbital angular momentum l and on the momentum k. The total cross section, the integrated elastic, and the absorption cross section are then written as follows,

$$\sigma_x = \sum_{l=0}^{\infty} \sigma_x^l, \tag{14a}$$

with

$$\sigma_{\text{tot}}^l = \frac{2\pi}{k^2}(2l+1)\left[1 - \operatorname{Re}(e^{2i\delta_l})\right], \tag{14b}$$

$$\sigma_{\text{el}}^l = \frac{\pi}{k^2}(2l+1)\left|e^{2i\delta_l} - 1\right|^2, \tag{14c}$$

$$\sigma_{\text{abs}}^l = \sigma_{\text{tot}}^l - \sigma_{\text{el}}^l = \frac{\pi}{k^2}(2l+1)\left[1 - \left|e^{2i\delta_l}\right|^2\right]. \tag{14d}$$

It is convenient to write the factor $e^{2i\delta_l}$ in terms of its modulus and a real phase viz.

$$e^{2i\delta_l} = \eta_l e^{2i\varepsilon_l},$$

where

$$\varepsilon_l = \operatorname{Re} \delta_l$$

is the *real* scattering phase,

$$\eta_l = e^{-2 \operatorname{Im} \delta_l}$$

is the *inelasticity* and describes the amount of extinction, in a given partial wave, due to the disappearance of the projectile into inelastic channels. The inelasticity η_l varies between 1 and 0.

We distinguish the two extreme cases

(i) $\eta=1$: no absorption in the partial wave amplitude l. From eqs. (14) we see that

$$\sigma^l_{\rm abs}=0, \qquad \sigma^l_{\rm tot}=\sigma^l_{\rm el}. \tag{15a}$$

As the phase δ_l is real, we can write the amplitude (13) in the alternative form

$$f_l(k)=\frac{1}{k}\,{\rm e}^{{\rm i}\delta_l}\sin\delta_l. \tag{15b}$$

(ii) $\eta_l=0$: the absorption in the partial wave amplitude is maximal. We find from eqs. (14)

$$\sigma^l_{\rm abs}=\sigma^l_{\rm el}=\tfrac{1}{2}\sigma^l_{\rm tot}. \tag{16a}$$

The absorption cross section is equal to the integrated elastic and is equal to half the total cross section. The amplitude (13) becomes pure imaginary,

$$f_l(k)={\rm i}/2k. \tag{16b}$$

Comparing eqs. (14c) and (14d) one obtains the inequality

$$\sigma^l_{\rm abs}\le\sigma^l_{\rm el},$$

the equal sign being realized for maximal absorption. The same inequality holds for the cross sections summed over l.

1.3. Green functions and scattering equation

Consider scattering solutions of the wave equation

$$\{H_0-(E-V(\boldsymbol{r}))\}\psi(\boldsymbol{r})=0. \tag{17}$$

In the nonrelativistic case $H_0=-(1/2m)\Delta$ is the operator of kinetic energy and $E=\boldsymbol{k}^2/2m$.

Introduce the following notation:

$\psi_{\rm out}(\boldsymbol{k},\boldsymbol{r})$: scattering solution for momentum $\boldsymbol{k}$, containing an *out*going spherical wave,

$\psi_{\rm in}(\boldsymbol{k},\boldsymbol{r})$: scattering solution with *in*coming spherical wave,

$\phi(\boldsymbol{k},\boldsymbol{r})\equiv{\rm e}^{{\rm i}\boldsymbol{kr}}$: plane wave for energy $E=\boldsymbol{k}^2/2m$.

The functions ϕ are solutions of the equation

$$(E-H_0)\phi(\boldsymbol{k},\boldsymbol{r})=0, \tag{18}$$

and satisfy the orthogonality and completeness relations

$$\int {\rm d}^3r\,\phi^*(\boldsymbol{k}',\boldsymbol{r})\phi(\boldsymbol{k},\boldsymbol{r})=(2\pi)^3\delta(\boldsymbol{k}'-\boldsymbol{k}), \tag{19a}$$

$$\int {\rm d}^3k\,\phi^*(\boldsymbol{k},\boldsymbol{r}')\phi(\boldsymbol{k},\boldsymbol{r})=(2\pi)^3\delta(\boldsymbol{r}'-\boldsymbol{r}). \tag{19b}$$

Expanding the function $\psi_{out/in}$ in terms of the plane wave solutions,

$$\psi(\boldsymbol{k}, \boldsymbol{r}) = \int a(\boldsymbol{k}')\phi(\boldsymbol{k}', \boldsymbol{r})\, \mathrm{d}^3k', \tag{20}$$

and making use of the obvious equation

$$(E - H_0)\phi(\boldsymbol{k}', \boldsymbol{r}) = (E - E')\phi(\boldsymbol{k}', \boldsymbol{r}), \tag{18'}$$

we can write eq. (17) as follows:

$$\begin{aligned}(E - H_0)\psi(\boldsymbol{k}, \boldsymbol{r}) &= \int \mathrm{d}^3k''\, a(\boldsymbol{k}'')(E - E'')\phi(\boldsymbol{k}'', \boldsymbol{r}) \\ &= V(\boldsymbol{r})\psi(\boldsymbol{k}, \boldsymbol{r}).\end{aligned}$$

If we multiply this equation by $\phi(\boldsymbol{k}', \boldsymbol{r})$ from the left and integrate over $\boldsymbol{r}$, we obtain by means of eq. (19a)

$$a(\boldsymbol{k}') = \frac{1}{(2\pi)^3(E - E')} \int \mathrm{d}^3r'\, \phi^*(\boldsymbol{k}', \boldsymbol{r}')V(\boldsymbol{r}')\psi(\boldsymbol{k}, \boldsymbol{r}').$$

Thus,

$$\psi(\boldsymbol{k}, \boldsymbol{r}) = \frac{1}{(2\pi)^3} \int \mathrm{d}^3k' \int \mathrm{d}^3r'\, \frac{\phi(\boldsymbol{k}', \boldsymbol{r})\phi^*(\boldsymbol{k}', \boldsymbol{r}')}{E - E'} V(\boldsymbol{r}')\psi(\boldsymbol{k}, \boldsymbol{r}').$$

The functions $\psi_{out/in}$ are obtained from this by choosing the integration path in the plane of the integration variable $\boldsymbol{k}'^2$ so as to obtain the proper asymptotic spherical wave, and by adding a solution of the corresponding homogeneous equation to it:

$$\psi_{out/in}(\boldsymbol{k}, \boldsymbol{r}) = \phi(\boldsymbol{k}, \boldsymbol{r}) + \int \mathrm{d}^3r'\, G_0^{out/in}(\boldsymbol{r}, \boldsymbol{r}')V(\boldsymbol{r}')\psi_{out/in}(\boldsymbol{k}, \boldsymbol{r}'), \tag{21}$$

where

$$\begin{aligned}G_0^{out/in}(\boldsymbol{r}, \boldsymbol{r}') &= \frac{2m}{(2\pi)^3} \int \mathrm{d}^3k'\, \frac{\mathrm{e}^{\mathrm{i}\boldsymbol{k}'\cdot(\boldsymbol{r}-\boldsymbol{r}')}}{k^2 - k'^2 +/- \mathrm{i}\varepsilon} \\ &= -\frac{m}{2\pi}\frac{\mathrm{e}^{+/-\mathrm{i}k|\boldsymbol{r}-\boldsymbol{r}'|}}{|\boldsymbol{r}-\boldsymbol{r}'|}.\end{aligned} \tag{22}$$

Eq. (21) is an integral equation for $\psi_{out/in}$ which is equivalent to the differential equation (17), but supplemented by the required asymptotic conditions. It is convenient to reformulate and generalize (21) in two respects: First, to write this equation in operator form; second, to write it in a form independent of a specific representation. Let $(E - H_0)^{-1}$ be the inverse of the operator $E - H_0$. Then, from eq. (18′), we find

$$(E - H_0)^{-1}\phi(\boldsymbol{k}', \boldsymbol{r}) = \frac{1}{E - E'}\phi(\boldsymbol{k}', \boldsymbol{r}).$$

In other words, $\phi(\boldsymbol{k}', \boldsymbol{r})$ is an eigenfunction of $(E - H_0)^{-1}$ with eigenvalue $1/(E - E')$, with $E' > 0$ and $E' \neq E$.

Similarly, from eq. (17),

$$\psi(\boldsymbol{k}, \boldsymbol{r}) = (E - H_0)^{-1} V(\boldsymbol{r}) \psi(\boldsymbol{k}, \boldsymbol{r}).$$

Hence, eq. (21) can be written as follows:

$$\psi_{\text{out/in}}(\boldsymbol{k}, \boldsymbol{r}) = \phi(\boldsymbol{k}, \boldsymbol{r}) + (E - H_0 +/- \mathrm{i}\varepsilon)^{-1} V(\boldsymbol{r}) \psi_{\text{out/in}}(\boldsymbol{k}, \boldsymbol{r}). \tag{21'}$$

One says that $G_0^{\text{out/in}}(\boldsymbol{r}, \boldsymbol{r}')$ is the Green function of the operator $O^{\text{out/in}} \equiv (E - H_0 +/- \mathrm{i}\varepsilon)^{-1}$. Indeed, if we write (21′) in the form

$$\psi_{\text{out/in}} = \phi + \int \mathrm{d}^3 r' \, O_{\text{out/in}} \delta(\boldsymbol{r}' - \boldsymbol{r}) V(\boldsymbol{r}') \psi_{\text{out/in}}(\boldsymbol{k}, \boldsymbol{r}'),$$

replace then the δ-distribution by means of eq. (19b), and use

$$O_{\text{out/in}} \phi(\boldsymbol{k}, \boldsymbol{r}) = (E - E' +/- \mathrm{i}\varepsilon)^{-1} \phi(\boldsymbol{k}, \boldsymbol{r}),$$

we recover eq. (21) in its original form.

In a next step we formulate these equations in a more abstract form. As they are written they are formulated in coordinate space. Therefore let us introduce abstract operators G_0^{out} and G_0^{in}, and a representation-free notation $|\psi_{\text{out/in}}\rangle$ for the scattering states, so that

$$G_0^{\text{out/in}}(\boldsymbol{r}, \boldsymbol{r}') = \langle \boldsymbol{r} | G_0^{\text{out/in}} | \boldsymbol{r}' \rangle \tag{23}$$

and

$$\psi_{\text{out/in}}(\boldsymbol{k}, \boldsymbol{r}) = \langle \boldsymbol{r} | \psi_{\text{out/in}} \rangle.$$

The action of the operators G_0 on the states $|\psi\rangle$ in coordinate representation is defined by

$$\langle \boldsymbol{r} | G_0^{\text{out/in}} | \psi^{\text{out/in}} \rangle = \int \mathrm{d}^3 r' \, G_0^{\text{out/in}}(\boldsymbol{r}, \boldsymbol{r}') \psi^{\text{out/in}}(\boldsymbol{k}, \boldsymbol{r}').$$

Clearly, $G_0^{\text{in}} = (G_0^{\text{out}})^\dagger$. Eq. (21) can then be written as

$$|\psi^{\text{out/in}}\rangle = |\phi\rangle + G_0^{\text{out/in}} V |\psi^{\text{out/in}}\rangle, \tag{24}$$

where

$$G_0^{\text{out/in}} := (E - H_0 +/- \mathrm{i}\varepsilon)^{-1}. \tag{25}$$

This is the Lippmann–Schwinger equation.

Instead of G_0, the Green functions of $(E - H_0 \pm \mathrm{i}\varepsilon)^{-1}$, one can also introduce the corresponding Green functions of the inhomogeneous equation (17), with H_0 replaced by $H = H_0 + V(\boldsymbol{r})$:

$$G^{\text{out/in}} := (E - H +/- \mathrm{i}\varepsilon)^{-1}. \tag{26}$$

There are several ways of relating G_0 and G by making use of the operator identity

$$A^{-1} - B^{-1} = B^{-1}(B - A) A^{-1}.$$

(i) Choose $A \equiv E - H \pm \mathrm{i}\varepsilon$, $B = E - H_0 \pm \mathrm{i}\varepsilon$, to find

$$G^{\text{out/in}} = G_0^{\text{out/in}} + G_0^{\text{out/in}} V G^{\text{out/in}}. \tag{27}$$

(ii) Choose $A = E - H_0 + \mathrm{i}\varepsilon$, $B = E - H \pm \mathrm{i}\varepsilon$, to find

$$G_0^{\text{out/in}} = G^{\text{out/in}} - G^{\text{out/in}} V G_0^{\text{out/in}}. \tag{27'}$$

Introducing (27′) into eq. (24), and dropping the "ket" notation, we have

$$\psi^{\text{out/in}} = \phi + G^{\text{out/in}} V \left[\psi^{\text{out/in}} - G_0^{\text{out/in}} V \psi^{\text{out/in}} \right] = \phi + G^{\text{out/in}} V \phi. \tag{28}$$

1.4. *T-matrix and T-operator*

Equation (28) can be read as follows: ψ is the result that is obtained if either one of the operators

$$\Omega^{\text{out(in)}} := \mathbb{1} + G^{\text{out(in)}} V \tag{29}$$

is applied to the free solution ϕ, viz.

$$\psi^{\text{out/in}} = \Omega^{\text{out/in}} \phi. \tag{28'}$$

The operators Ω are called *Møller operators*.

In eq. (7) we have found a general expression for the scattering amplitude. In the notation of sec. 1.3 it is

$$f(\theta, \phi) = -\frac{m}{2\pi} \langle \phi | V | \psi^{\text{out}} \rangle.$$

The matrix elements on the right-hand side of this equation form a matrix, called T-matrix,

$$T_{\text{fi}} := \langle \phi | V | \psi^{\text{out}} \rangle = \langle \phi | V \Omega^{\text{out}} | \phi \rangle. \tag{30}$$

The connection between T-matrix element and scattering amplitude is, therefore,

$$f(\theta, \phi) = -\frac{m}{2\pi} T_{\text{fi}}. \tag{31}$$

The definition (30) suggests to define the *T-operator*,

$$T := V \Omega^{\text{out}}. \tag{30'}$$

It is not difficult to derive the following integral equation for T (again in abstract, representation-free form),

$$T = V + V G_0^{\text{out}} T. \tag{32}$$

Proof:

$$\begin{aligned} T = V\Omega^{\text{out}} &= V(\mathbb{1} + G^{\text{out}} V) \\ &= V + V G_0^{\text{out}} V \{ \mathbb{1} + G^{\text{out}} V \} \\ &= V + V G_0^{\text{out}} T. \end{aligned}$$

In the first line we have used the definitions (30′) and (29); in the second we have made use of eq. (27).

Finally we can derive another useful equation relating the wave function ψ^{out}, the Green function G_0 and the T-matrix. Introducing the definition (28′) into eq. (24) we have

$$\psi^{\text{out}} = \phi + G_0^{\text{out}} V \psi^{\text{out}} = \phi + G_0^{\text{out}} V \Omega^{\text{out}} \phi.$$

$V\Omega^{\text{out}}$ is the T-matrix, by virtue of the definition (30′).

We obtain, therefore, the important equation

$$\psi^{\text{out}} = \phi + G_0^{\text{out}} T \phi. \tag{33}$$

Remarks and conclusions: Eq. (33) will be the starting point for a formulation of multiple scattering of a hadronic projectile on a composite target. T is the scattering operator pertaining to the potential V. The potential V may be an external (real or complex) potential, as was tacitly assumed above. It may as well be the potential $V(\boldsymbol{r}_\text{p} - \boldsymbol{r}_\text{t})$ describing the interaction between the projectile (p) and a target particle (t) which then depends on the relative coordinate $\boldsymbol{r}_\text{p} - \boldsymbol{r}_\text{t}$. In this case, the appropriate transformation to center-of-mass and relative coordinates must be performed, and the mass m in eq. (31) must be replaced with the reduced mass. If the target is a composite target then V will be the sum of individual projectile–target particle interactions,

$$V = \sum_{n=1}^{A} v_n(\boldsymbol{r}_\text{p} - \boldsymbol{r}_r). \tag{34}$$

As the A target particles are themselves part of a more complicated, composite quantum mechanical system, this interaction will be responsible for all kinds of elastic and inelastic scattering processes of the projectile off that composite system. The only assumption is that the interaction of the projectile with any individual target particle can be represented by a potential. This assumption will limit the use of these methods to low and medium energies.

1.5. Multiple scattering: some definitions

In the following sections 2, 3 we develop different approaches to the description of multiple scattering of a hadronic projectile on a composite target (atom or nucleus). The methods and approximations are chosen according to the energy range for which they are supposed to hold. However, all these methods are based on the theory of potential scattering, that is, the interaction is represented by the sum (34) over individual projectile–particle potentials v_n. It is then useful to introduce the following definitions:

(i) By *scattering amplitude* we mean the amplitude that describes the scattering of projectile p on the composite target as a whole.

(ii) The system of the projectile (p) and a given one of the target constituents (n), regarded in isolation from the rest of the target, is called *elementary system*. It is described by the potential v_n and the corresponding scattering operator t_n, which satisfies an equation of type (32):

$$t_n = v_n + v_n \frac{1}{E - H_0 + \mathrm{i}\varepsilon} t_n . \tag{35}$$

If the particle n is free, H_0 is the sum of the kinetic energy of projectile p and particle n. Solving eq. (35) for this case yields the free elementary scattering matrix t_n and the corresponding scattering amplitude. However, as particle n is embedded in the composite target,

$$H_0 = T_{\mathrm{kin}} + H_{\mathrm{target}} , \tag{36}$$

where H_{target} is the Hamiltonian describing the target. τ_n, the solution of eq. (35), is then the elementary scattering matrix *in medium*, which is different from t_n, the free scattering matrix.

2. Potential scattering at "high" energies: eikonal approximation

When there are many partial waves contributing to the scattering amplitude (8) it is useful to express $f(\theta)$ in terms of an integral over impact parameter. This representation allows for eikonal approximations in a rather transparent manner. In the following section we derive a simple expression for the scattering amplitude and discuss the nature of the approximations on which it is based. This formula is then applied to the case of multiple scattering at high energies. The expression obtained for the scattering amplitude on a composite target can be expressed in terms of the elementary scattering amplitudes, thereby getting rid of the potentials themselves.

2.1. Impact parameter representation of scattering amplitude and eikonal approximation

In classical mechanics the asymptotic linear momentum $\boldsymbol{k}$ and the orbital angular momentum $\boldsymbol{l}$ with respect to the scattering center are related by $\boldsymbol{l} = \boldsymbol{r} \wedge \boldsymbol{k}$.

The modulus l, in particular, is related to the impact parameter b, as defined in Fig. 2, by $l = k \cdot b$. In quantum mechanics a partial wave state is an eigenstate of $\boldsymbol{l}$

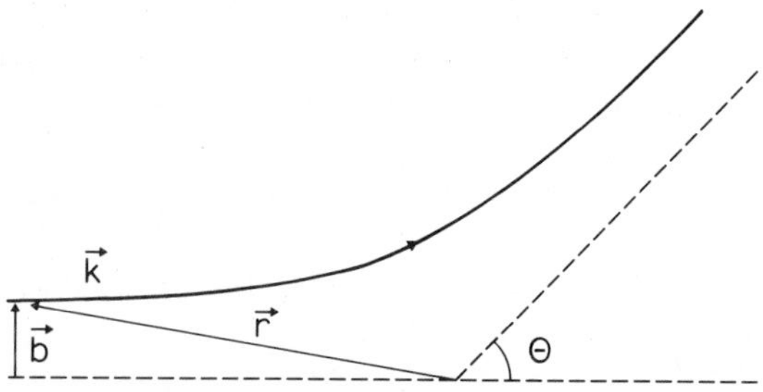

Fig. 2. Definition of impact parameter $\boldsymbol{b}$ for a classical scattering trajectory. The asymptote of the outgoing branch defines the scattering angle θ.

and is characterized by its energy or, equivalently, by $k=|\boldsymbol{k}|$, the modulus of its asymptotic momentum. It does not correspond, however, to a definite asymptotic *direction* $\hat{\boldsymbol{k}}$. Nevertheless, one can always introduce the notion of an impact parameter, instead of l, by a relation of the type (36), or some other definition such as $\sqrt{l(l+1)}=k\cdot b$ or

$$l=kb-\tfrac{1}{2}. \tag{37}$$

Similarly, instead of the partial wave series (8), one may as well introduce an impact parameter representation of the scattering amplitude (Wallace 1973), viz.

$$f(\boldsymbol{q})=\frac{\mathrm{i}k}{2\pi}\int \mathrm{d}^2b\,\mathrm{e}^{\mathrm{i}\boldsymbol{q}\cdot\boldsymbol{b}}\Gamma(\boldsymbol{b}). \tag{38}$$

Here $\boldsymbol{q}$ is the momentum transfer, k the modulus of the initial momentum; the vector $\boldsymbol{b}$ lies in the plane perpendicular to $\boldsymbol{q}$. $\Gamma(\boldsymbol{b})$ is called the *profile function* and plays the role of the partial wave amplitude f_l in the decomposition (8).

If the scattering problem possesses axial or spherical symmetry the profile function depends on $b=|\boldsymbol{b}|$ only and the formula (38) can be simplified further. Introducing polar coordinates

$$\mathrm{d}^2b=b\,\mathrm{d}b\,\mathrm{d}\phi$$

and writing $\boldsymbol{q}\cdot\boldsymbol{b}=qb\cos\phi$, we can make use of the well-known relation*⁾

$$\int_0^{\pi}\mathrm{d}\phi\,\mathrm{e}^{\mathrm{i}qb\cos\phi}=\pi J_0(qb), \tag{39}$$

where $J_0(z)$ is the Bessel function of order zero. Thus, we obtain

$$f(\boldsymbol{q})=\mathrm{i}k\int b\,\mathrm{d}b\,\Gamma(b)J_0(qb). \tag{40}$$

Thus far eq. (38), or (40), is fairly general and is no more than another parametrization of the scattering amplitude. The relation of the profile function to the partial wave amplitudes f_l is not a simple one and one may wonder whether a representation in terms of impact parameters is at all useful. For instance, the unitarity constraint on the scattering amplitude is simple only when expressed in terms of partial wave amplitudes (see eqs. (11) and (12) above), but is much less transparent in terms of profile functions. Guided by the analogy to the classical situation one might expect that the impact parameter is a useful concept whenever the motion of the projectile is such that it follows approximately a classical trajectory. As Glauber has shown (Glauber 1959), such a situation is realized in potential scattering at high energies. This is the case we now wish to discuss. Let $\boldsymbol{k}$ ($\boldsymbol{k}'$) be the wave vector of the incident (outgoing) projectile, and let $\boldsymbol{q}$ be the momentum transfer. As we deal with elastic scattering

$$\boldsymbol{q}=\boldsymbol{k}-\boldsymbol{k}',\qquad |\boldsymbol{k}|=|\boldsymbol{k}'|\equiv k,$$
$$q\equiv|\boldsymbol{q}|=2k\sin\theta/2.$$

*⁾See Ref. R1.

Let R be the radius of a sphere within which the potential is sizeably different from zero. We make the following assumptions:

(I) The energy of the incoming state is large so that

$$kR \gg 1.$$

This means, in particular, that a large number of partial waves with $l \lesssim kR$ will contribute to the scattering amplitude.

(II) Scattering into small angles is predominant.

(III) The scattering phases $\delta_l(k)$ in eq. (13) are smooth functions of both l and k such that we can make l a continuous variable and approximate eq. (8) by an integral over l:

$$f(\theta) \simeq \frac{1}{2ik}\int_0^\infty dl(2l+1)(e^{2i\delta_l(k)}-1)P_l(\cos\theta).$$

(IV) For a fourth assumption in applying the formalism to multiple scattering see below.

We substitute l by b, by means of eq. (37) and obtain

$$f(\theta) \simeq -ik\int_{1/2k}^\infty db\, b[e^{2i\chi(b)}-1]P_{kb-1/2}(\cos\theta)$$

where

$$\chi(b) := \delta_{kb-1/2}(k).$$

In the limit $k \to \infty$ we can replace the Legendre function with the Bessel function of order zero by virtue of the relation*)

$$\lim_{n\to\infty} P_n(1 - z^2/2n^2) = J_0(z).$$

Indeed, if we take $z = 2kb\sin\theta/2$ and $n = kb - 1/2$

$$1 - \frac{z^2}{2n^2} = \left(\cos\theta - \frac{1}{kb} + \frac{1}{4k^2b^2}\right)\Bigg/\left(1 - \frac{1}{kb} + \frac{1}{4k^2b^2}\right)$$

$$\simeq \cos\theta + (\cos\theta - 1)\left[\frac{1}{kb} + \frac{3}{4k^2b^2}\right]$$

$$\xrightarrow[k\to\infty]{} \cos\theta.$$

Thus,

$$f(\theta) \simeq ik\int_0^\infty db\, b[1 - e^{2i\chi(b)}]J_0(2kb\sin\theta/2). \tag{41}$$

The same result (41) can be derived in the frame of the theory of potential scattering. At the same time this yields an explicit representation for the "phase" function $\chi(b)$ in terms of the potential.

*) See e.g. Ref. R1., eq. (9.1.71).

In order to see this, we write the wave equation (1) in the form

$$\Delta\psi + n^2(\boldsymbol{r})k^2\psi = 0, \tag{42}$$

with

$$n(\boldsymbol{r})k := \sqrt{k^2 - 2mV(\boldsymbol{r})}\,.$$

For calculating the scattering amplitude (7) we need to know ψ only in the domain $r \lesssim R$. For large k and small scattering angle (assumption II), ψ will be essentially the plane wave $e^{i\boldsymbol{k}\boldsymbol{r}}$ times a slowly varying function of $\boldsymbol{r}$. Therefore, if we write

$$\psi(\boldsymbol{r}) = e^{i\phi(\boldsymbol{r})}$$

we can neglect second and higher derivatives of $\phi(\boldsymbol{r})$:

$$\begin{aligned}\Delta\psi(\boldsymbol{r}) &= ie^{i\phi}\Delta\phi - e^{i\phi}(\nabla\phi)^2\\ &\simeq -e^{i\phi}(\nabla\phi)^2.\end{aligned}$$

Eq. (42) then becomes

$$(\nabla\phi)^2 \simeq n^2(\boldsymbol{r})k^2, \tag{42'}$$

whose solution is given by the path integral

$$\phi(\boldsymbol{r}) = \boldsymbol{k}\cdot\boldsymbol{r}_0 + k\int_{\boldsymbol{r}_0}^{\boldsymbol{r}} n(\boldsymbol{r}')\,ds'.$$

$\boldsymbol{r}_0$ is an arbitrary point outside the sphere with radius R about the origin. For the sequel it is convenient to choose the z-axis along the direction of the vector $\boldsymbol{K} = \boldsymbol{k} + \boldsymbol{k}'$, as done in Fig. 3. As scattering to small angles is assumed to dominate the integrals along the directions perpendicular to the z-axis are small and we can

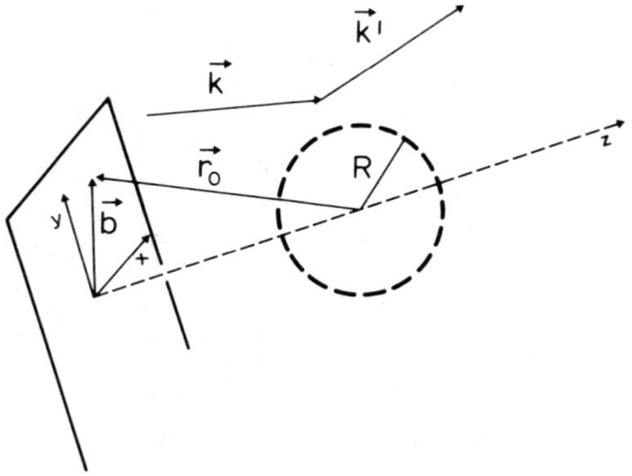

Fig. 3. Definition of impact parameter plane (x, y) in scattering process at high energies. The z-axis is chosen parallel to the sum $\boldsymbol{k} + \boldsymbol{k}'$ of incoming and outcoming momenta.

write approximately

$$\phi(\boldsymbol{r}) = \boldsymbol{k}\cdot\boldsymbol{r} + k\int_{\boldsymbol{r}_0}^{\boldsymbol{r}} [n(\boldsymbol{r}') - 1]\,\mathrm{d}s'$$

$$\simeq \boldsymbol{k}\cdot\boldsymbol{r} + k\int_{z_0}^{z} [n(\boldsymbol{r}') - 1]\,\mathrm{d}z'.$$

The point z_0 can be taken to minus infinity, without loss of generality. Furthermore, as k goes to infinity,

$$k[n(\boldsymbol{r}) - 1] \simeq -V(\boldsymbol{r})/(k/m) = -V(\boldsymbol{r})/v,$$

v being the modulus of the particle's initial velocity. This leads to the approximate representation for the scattering wave function

$$\psi(\boldsymbol{r}) \simeq \exp\left\{ \mathrm{i}\boldsymbol{k}\cdot\boldsymbol{r} - \frac{\mathrm{i}}{v}\int_{-\infty}^{z} \mathrm{d}z'\, V(\boldsymbol{b}, z') \right\}. \tag{43}$$

The derivation of eq. (43) stresses the analogy to the familiar eikonal approximation in classical optics. Clearly, eq. (43) cannot be an acceptable scattering function in the whole space, $-\infty < z < +\infty$: it is derived on the basis of the analogue of ray optics and does not contain the full diffraction pattern of the exact solution. In particular, $\psi(\boldsymbol{r})$ from eq. (43) does not contain the asymptotic outgoing spherical wave. However, eq. (43) is a perfectly acceptable approximation to the exact solution for values of the argument within the range R of the potential V, i.e. for $r \lesssim R$. This is all we need for calculating the scattering amplitude from the general formula (7). Inserting eq. (43) into eq. (7) we find

$$f(\theta) \simeq -\frac{m}{2\pi}\int \mathrm{d}^3 r\, \mathrm{e}^{-\mathrm{i}\boldsymbol{k}'\cdot\boldsymbol{r}} V(\boldsymbol{r}) \exp\left\{ \mathrm{i}\boldsymbol{k}\cdot\boldsymbol{r} - \frac{\mathrm{i}}{v}\int_{-\infty}^{z} \mathrm{d}z'\, V(b, z') \right\}$$

$$= -\frac{m}{2\pi}\int \mathrm{d}^2 b \int_{-\infty}^{+\infty} \mathrm{d}z\, \mathrm{e}^{\mathrm{i}\boldsymbol{q}\cdot\boldsymbol{b}} V(\boldsymbol{b} + \hat{\boldsymbol{K}}z)$$

$$\cdot \exp\left\{ -\frac{\mathrm{i}}{v}\int_{-\infty}^{z} \mathrm{d}z'\, V(\boldsymbol{b} + \hat{\boldsymbol{K}}z') \right\}. \tag{44}$$

In the second line of eq. (44) we have decomposed the coordinate vector $\boldsymbol{r} = \hat{\boldsymbol{K}}z + \boldsymbol{b}$ in its components along the z-axis and perpendicular to it, see fig. 3. As $\boldsymbol{q}$ is perpendicular to $\boldsymbol{K}$, the component of $\boldsymbol{r}$ along $\boldsymbol{K}$ does not contribute in the exponential $\exp(\mathrm{i}(\boldsymbol{k} - \boldsymbol{k}')\cdot\boldsymbol{r})$*). The integral over z can be done analytically using

$$V(\boldsymbol{b} + \hat{\boldsymbol{K}}z) \exp\left\{ -\frac{\mathrm{i}}{v}\int_{-\infty}^{z} \mathrm{d}z'\, V(\boldsymbol{b} + \hat{\boldsymbol{K}}z') \right\}$$

$$= \mathrm{i}v\frac{\mathrm{d}}{\mathrm{d}z} \exp\left\{ -\frac{\mathrm{i}}{v}\int_{-\infty}^{z} \mathrm{d}z'\, V(\boldsymbol{b} + \hat{\boldsymbol{K}}z') \right\}.$$

This leaves us with the two dimensional integration over $\boldsymbol{b}$ and, indeed, yields an

*) This is strictly true only for elastic scattering, $|\boldsymbol{k}'| = |\boldsymbol{k}|$. In other cases $\exp(\mathrm{i}\boldsymbol{q}\cdot\hat{\boldsymbol{K}}z) \simeq 1$ holds approximately, on account of assumption II, and the same procedure goes through.

impact parameter representation of the scattering amplitude. As a short-hand notation write

$$\chi(\boldsymbol{b}) =: -\frac{1}{2v}\int_{-\infty}^{+\infty} V(\boldsymbol{b}+\hat{\boldsymbol{K}}z)\,\mathrm{d}z. \tag{45}$$

The scattering amplitude then takes the form

$$f(\theta) \simeq \frac{\mathrm{i}k}{2\pi}\int \mathrm{d}^2b\,\mathrm{e}^{\mathrm{i}\boldsymbol{q}\cdot\boldsymbol{b}}\{1-\mathrm{e}^{2\mathrm{i}\chi(\boldsymbol{b})}\}. \tag{46}$$

Comparing this to the general form (38) we see that in the eikonal approximation the profile function is given by

$$\Gamma(\boldsymbol{b}) = 1-\mathrm{e}^{2\mathrm{i}\chi(\boldsymbol{b})}. \tag{47}$$

If the potential $V(\boldsymbol{r})$ is spherically symmetric the phase function χ depends only on $b=|\boldsymbol{b}|$, so that eq. (46) simplifies to

$$f(\theta) \simeq \mathrm{i}k\int_0^{\infty} \mathrm{d}b\, b\,\Gamma(b)\,J_0(2kb\sin\theta/2). \tag{48}$$

2.2. *Application to multiple scattering*

In applying the results (45) and (46) to the scattering of hadrons on a composite target we take advantage of two remarkable properties of these equations.

(i) If the target is represented by a sum of potentials,

$$V(\boldsymbol{r}) = \sum_{n=1}^{N} v_n(\boldsymbol{r}-\boldsymbol{r}_n),$$

the eikonal phase χ of the resultant potential $V(\boldsymbol{r})$ is equal to the sum of the phases χ_n pertaining to the individual potentials v_n.

For instance for $N=2$

$$V(\boldsymbol{r}) = v_1(\boldsymbol{r}-\boldsymbol{r}_1)+v_2(\boldsymbol{r}-\boldsymbol{r}_2)$$

and

$$\begin{aligned}\chi(\boldsymbol{b}) &= -\frac{1}{2v}\int_{-\infty}^{+\infty} \mathrm{d}z\,[v_1(\boldsymbol{b}-\boldsymbol{s}_1,z)+v_2(\boldsymbol{b}-\boldsymbol{s}_2,z)]\\ &= \chi_1(\boldsymbol{b}-\boldsymbol{s}_1)+\chi_2(\boldsymbol{b}-\boldsymbol{s}_2).\end{aligned} \tag{49}$$

This is illustrated by Fig. 4. The vectors $\boldsymbol{s}_i$ are the projections of the vectors $\boldsymbol{r}_i$ onto the (x, y)-plane, i.e. onto the plane of the impact parameter $\boldsymbol{b}$. Thus, the eikonal phases are *additive*. (Note that this is not true, in general, for the scattering phases δ_l in eq. (13).)

(ii) The scattering amplitude $f(\theta)$ and the profile function $\Gamma(\boldsymbol{b})$ are related by a Fourier transformation in two dimensions. Therefore, if one knows the scattering amplitude from some other source the profile function can be calculated from it by a simple Fourier transform. In particular, it is not necessary to know, or to reconstruct the corresponding potential.

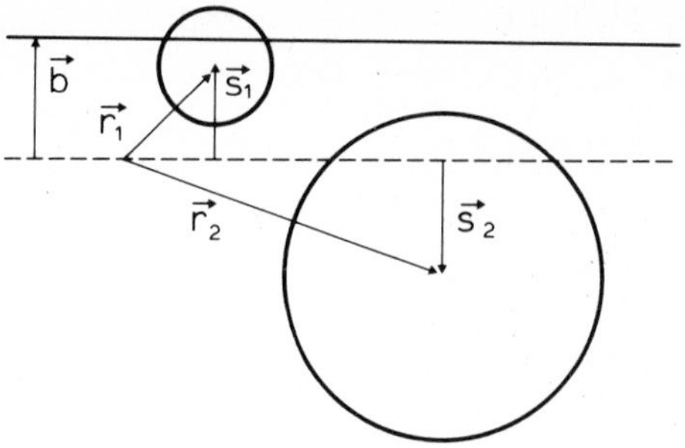

Fig. 4. Geometry of scattering process for two potentials of finite range.

Let us now apply these results to the approximate description of hadron scattering on a nucleus or, for that purpose, any other composite target. Suppose that the elementary (i.e. hadron–nucleon) amplitudes are known. By inverting the Fourier transform (46) one first constructs the corresponding profile functions $\Gamma_n(\boldsymbol{b}-\boldsymbol{s}_n)$. In a second step, the full scattering amplitude for the case of a composite target is calculated from eq. (46) by assuming that the full eikonal phase χ is the sum of elementary eikonal phases $\chi_n(\boldsymbol{b}-\boldsymbol{s}_n)$, the coordinates $\boldsymbol{s}_n$ being defined as shown in Fig. 5.

In this way one obtains an expression for the hadron–nucleus scattering amplitude which depends only on the elementary amplitudes and on the geometrical distribution of the target constituents (nucleons). At no point is it necessary to explicitly introduce potentials.

We illustrate the procedure that we described here by means of a particularly simple situation.

Scattering of a spinless particle on a nucleus. We consider the scattering of a pion, or any other spinless object, on the ground state $|i\rangle$ of a nucleus leading to the final state $|f\rangle$. The final state can be the same as the initial state (elastic scattering), but it may also be some excited state of the nucleus. The elementary amplitudes on a proton (p) or a neutron (n) can always be written in terms of an integral over the

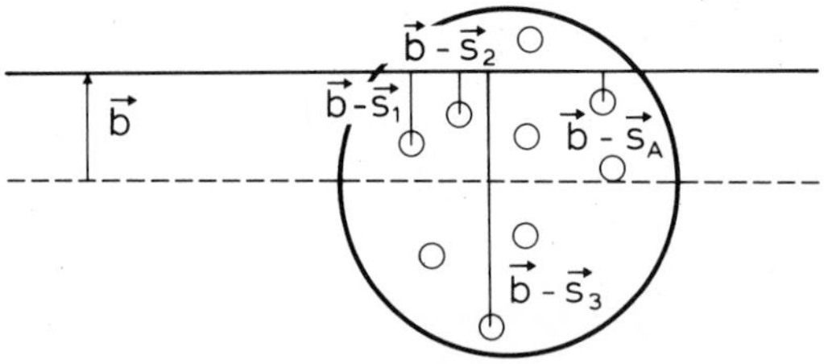

Fig. 5. Geometry of scattering process on a composite target made up of A potentials at fixed positions and having finite ranges.

impact parameter vector, viz.

$$f_{\mathrm{p(n)}}(\boldsymbol{q}) = \frac{\mathrm{i}k}{2\pi}\int \mathrm{d}^2 b\, \mathrm{e}^{\mathrm{i}\boldsymbol{q}\cdot\boldsymbol{b}}[1 - \mathrm{e}^{2\mathrm{i}\chi_{\mathrm{p(n)}}(\boldsymbol{b})}]. \tag{50a}$$

From this we obtain

$$\mathrm{e}^{2\mathrm{i}\chi_{\mathrm{p(n)}}(\boldsymbol{b})} = 1 - \frac{1}{2\pi\mathrm{i}k}\int_S \mathrm{d}^2 q\, f_{\mathrm{p(n)}}(\boldsymbol{q})\mathrm{e}^{-\mathrm{i}\boldsymbol{q}\cdot\boldsymbol{b}}. \tag{50b}$$

Here, the domain of integration S is the surface spanned by the momentum transfer $\boldsymbol{q}$, for fixed k (i.e. fixed energy). If scattering into small angles predominates, (i.e. if $f_{\mathrm{p(n)}}(\boldsymbol{q})$ drop rapidly with increasing $\boldsymbol{q}^2$) and if the energy is large, the surface S is approximately a plane. The eikonal phase for scattering off a nucleus is the sum

$$\chi_A(\boldsymbol{b}) = \sum_{n=1}^{A} \chi_n(\boldsymbol{b} - \boldsymbol{s}_n),$$

so that the amplitude becomes

$$\begin{aligned} F(\boldsymbol{q}) &= \frac{\mathrm{i}k}{2\pi}\int \mathrm{d}^2 b\, \mathrm{e}^{\mathrm{i}\boldsymbol{q}\cdot\boldsymbol{b}}[1 - \mathrm{e}^{2\mathrm{i}\chi_A(\boldsymbol{b})}] \\ &= \frac{\mathrm{i}k}{2\pi}\int \mathrm{d}^2 b\, \mathrm{e}^{\mathrm{i}\boldsymbol{q}\cdot\boldsymbol{b}}\left[1 - \prod_{j=1}^{A} \mathrm{e}^{2\mathrm{i}\chi_j(\boldsymbol{b}-\boldsymbol{s}_j)}\right]. \end{aligned} \tag{51}$$

A specific transition leading from the initial state $|\mathrm{i}\rangle$ to the final state $|\mathrm{f}\rangle$ is described by the matrix element

$$\begin{aligned} F_{\mathrm{fi}}(\boldsymbol{q}) &= \frac{\mathrm{i}k}{2\pi}\int \mathrm{d}^2 b\, \mathrm{e}^{\mathrm{i}\boldsymbol{q}\cdot\boldsymbol{b}}\langle \mathrm{f}|\left[1 - \prod_{1}^{A} \mathrm{e}^{2\mathrm{i}\chi_j(\boldsymbol{b}-\boldsymbol{s}_j)}\right]|\mathrm{i}\rangle \\ &= \frac{\mathrm{i}k}{2\pi}\int \mathrm{d}^2 b\, \mathrm{e}^{\mathrm{i}\boldsymbol{q}\cdot\boldsymbol{b}}\phi_{\mathrm{f}}^*(\boldsymbol{r}_1 \cdots \boldsymbol{r}_A)\left[1 - \prod_{1}^{A} \mathrm{e}^{2\mathrm{i}\chi_j(\boldsymbol{b}-\boldsymbol{s}_j)}\right]\phi_{\mathrm{i}}(\boldsymbol{r}_1 \cdots \boldsymbol{r}_A) \\ &\quad \times \delta\left(\frac{1}{A}(\boldsymbol{r}_1 + \boldsymbol{r}_2 + \cdots + \boldsymbol{r}_A)\right)\mathrm{d}^3 r_1 \cdots \mathrm{d}^3 r_A. \end{aligned} \tag{52}$$

The δ-distribution has been introduced in order to keep track of the requirement that the motion of the center-of-mass must be separated off and that the integration must be done over a set of $(A-1)$ relative coordinates only. Inserting the inversion formula (50b) we obtain the final result

$$\begin{aligned} F_{\mathrm{fi}}(\boldsymbol{q}) &= \frac{\mathrm{i}k}{2\pi}\int \mathrm{d}^2 b\, \mathrm{e}^{\mathrm{i}\boldsymbol{q}\cdot\boldsymbol{b}}\int \mathrm{d}^3 r_1 \cdots \mathrm{d}^3 r_A\, \phi_{\mathrm{f}}^*\delta\left(\frac{1}{A}\sum \boldsymbol{r}_n\right) \\ &\quad \times \left\{1 - \prod_{1}^{A}\left[1 - \frac{1}{2\pi\mathrm{i}k}\int \mathrm{d}^2 q'\, \mathrm{e}^{-\mathrm{i}\boldsymbol{q}'\cdot(\boldsymbol{b}-\boldsymbol{s}_j)} f_j(\boldsymbol{q}')\right]\right\}\phi_{\mathrm{i}}. \end{aligned} \tag{53}$$

Let us now understand the physics content of this result. We do this by analyzing first the structure of the multiple scattering series (53). Then, in a second step, we consider a model which allows to evaluate eq. (53) analytically and which has the advantage of being simple and transparent.

(i) By expanding the product in the curly brackets we see that the scattering amplitude F_{fi}, eq. (53), is a polynomial of degree A in the elementary amplitudes f_i. The first nonvanishing term contains only one such amplitude f_j and corresponds to single scattering of the projectile on nucleon (constituent) j of the target. The next term contains products $f_k(\boldsymbol{q}')f_l(\boldsymbol{q}'')$, with $k \neq l$, and corresponds to double scattering off two different nucleons, etc. The last term is given by products of exactly A elementary amplitudes,

$$f_{j1}(\boldsymbol{q}_1)f_{j2}(\boldsymbol{q}_2)\cdots f_{jA}(\boldsymbol{q}_A), \quad j1 \neq j2 \neq \cdots \neq jA.$$

Obviously, the projectile is scattered off any specific nucleon at most once. This is a consequence of the approximations made above: If the projectile were to scatter twice or more times on the *same* nucleon this would mean that at least one elementary scattering would have to involve a deflection to large scattering angles. As the eikonal approximation assumes scattering into small angles to be predominant (assumption II) this cannot happen.

(ii) Another important assumption which is implicit in the derivation of the expressions (52) and (53) is this: In these equations the nucleons are assigned fixed positions and all multiple scattering terms are evaluated at the same momentum transfer. This means that we have neglected the Fermi motion of the nucleons in their bound states. This assumption is in accord with assumption (I) above: during the short time that the projectile needs to cross the target the nucleons seem frozen at their instantaneous position. The average over the nuclear initial and final states is taken only afterwards, by evaluating the corresponding matrix element $F_{\mathrm{fi}}(\boldsymbol{q})$ between these states. This is called the *frozen nucleus assumption* (assumption IV).

(iii) *A soluble model* (Locher et al. 1971). As an example we consider proton scattering off nuclei at high energies. Except for spin effects which we neglect for simplicity, the differential cross section for nucleon–nucleon scattering is known to decrease exponentially with the momentum transfer squared. Thus, the proton–proton and proton–neutron scattering amplitudes may be parametrized as follows:

$$f_{\mathrm{N}}(q) = \frac{\mathrm{i}k}{4\pi}\sigma_{\mathrm{N}}(1 - \mathrm{i}\rho_{\mathrm{N}})\mathrm{e}^{-\beta_{\mathrm{N}}^2 q^2/2}, \tag{54a}$$

where q^2 is the invariant momentum transfer and where $\mathrm{N} = \mathrm{p}$ or n; σ_{N} is the total cross section,

$$\sigma_{\mathrm{N}} = \frac{4\pi}{k}\,\mathrm{Im}\, f_{\mathrm{N}} \quad (q^2 = 0), \tag{54b}$$

and ρ_{N} is the ratio of real to imaginary part of the forward scattering amplitude,

$$\rho_{\mathrm{N}} = (\mathrm{Re}\, f_{\mathrm{N}}/\mathrm{Im}\, f_{\mathrm{N}})_{q^2=0} = 4\pi\frac{\mathrm{Re}\, f_{\mathrm{N}}(0)}{k\sigma_{\mathrm{N}}}. \tag{54c}$$

We assume this ratio to be independent of q^2.

It is easy to compute the elementary profile function from eq. (50b) and from the ansatz (54), viz.

$$\Gamma_{\mathrm{N}}(b) = 1 - \mathrm{e}^{2\mathrm{i}\chi_{\mathrm{N}}(b)} = \frac{\sigma_{\mathrm{N}}(1-\mathrm{i}\rho_{\mathrm{N}})}{4\pi\beta_{\mathrm{N}}^2}\mathrm{e}^{-b^2/2\beta_{\mathrm{N}}^2}.$$

Suppose further that the nuclear ground state can be described by a potential of a harmonic oscillator, with oscillator constant $\alpha = \sqrt{m\omega}$, and that the A nucleons all occupy the 1s state (for $A > 4$ disregarding the Pauli principle). The ground state wave function then is a product of A identical functions,

$$\phi(\boldsymbol{r}_1 \cdots \boldsymbol{r}_A) = \prod_{j=1}^{A} \varphi(\boldsymbol{r}_j), \quad \varphi(\boldsymbol{r}_j) = \left(\frac{\alpha^2}{\pi}\right)^{3/4} \mathrm{e}^{-\alpha^2 r_j^2/2}. \tag{55}$$

The elastic scattering amplitude is

$$F_{\mathrm{ii}}(q) = \frac{\mathrm{i}k}{2\pi}\int \mathrm{d}^2 b\, \mathrm{e}^{\mathrm{i}\boldsymbol{q}\cdot\boldsymbol{b}} \int \mathrm{d}^3 r_1 \cdots \mathrm{d}^3 r_A \prod_1^A \left[\left(\frac{\alpha^2}{\pi}\right)^{3/2} \mathrm{e}^{-\alpha^2 r_j^2}\right] \delta\left(\frac{1}{A}\sum_1^A \boldsymbol{r}_n\right)$$

$$\times \left\{1 - \prod_{k=1}^{A}\left(1 - \frac{\sigma_k(1-\mathrm{i}\rho_k)}{4\pi\beta_{\mathrm{N}}^2}\exp\left[-\frac{(\boldsymbol{b}-\boldsymbol{s}_k)^2}{2\beta_{\mathrm{N}}^2}\right]\right)\right\}.$$

Both the harmonic potential $\frac{1}{2}m\omega^2 \sum_n r_n^2$ and the ground state wave function (55) are separable in center-of-mass and relative coordinates. In this case one can show that the center-of-mass condition expressed by the δ-distribution can be removed provided the whole amplitude is multiplied by the factor $\exp\{q^2/4A\alpha^2\}$ (see exercise 2). This leads to

$$F_{\mathrm{ii}}(q) = \frac{\mathrm{i}k}{2\pi}\mathrm{e}^{q^2/4A\alpha^2}\int \mathrm{d}^2 b\, \mathrm{e}^{\mathrm{i}\boldsymbol{q}\cdot\boldsymbol{b}}\left\{1 - \prod_1^A\left(1 - \frac{\sigma_k(1-\mathrm{i}\rho_k)}{4\pi\beta_{\mathrm{N}}^2}\left(\frac{\alpha^2}{\pi}\right)^{3/2}\right.\right.$$

$$\left.\left.\times \int \mathrm{d}^3 r_k\, \mathrm{e}^{-\alpha^2 r_k^2} \exp\left[-\frac{(\boldsymbol{b}-\boldsymbol{s}_k)^2}{2\beta_{\mathrm{N}}^2}\right]\right)\right\}.$$

The integral over $\mathrm{d}^3 r_k$ can be performed analytically and turns out to be independent of k, the nucleon index. Assuming the parameters σ_{N}, ρ_{N}, β_{N} to be the same for proton–proton and proton–neutron scattering and setting

$$x := \frac{\alpha^2\sigma(1-\mathrm{i}\rho)}{2\pi(1+2\alpha^2\beta^2)}, \tag{56}$$

we find

$$F_{\mathrm{ii}} = \frac{\mathrm{i}k}{2\pi}\mathrm{e}^{q^2/4A\alpha^2}\int \mathrm{d}^2 b\, \mathrm{e}^{\mathrm{i}\boldsymbol{q}\cdot\boldsymbol{b}}\left\{1 - \left[1 - x\exp\left(-\frac{\alpha^2 b^2}{1+2\alpha^2\beta^2}\right)\right]^A\right\}.$$

Finally, the integral over the impact parameter can also be performed, viz.

$$F_{\mathrm{ii}}(q)=\frac{1}{2}\mathrm{i}k\frac{1+2\alpha^2\beta^2}{\alpha^2}\mathrm{e}^{q^2/4A\alpha^2}$$
$$\times\sum_{m=1}^{A}\binom{A}{m}\frac{(-)^{m+1}}{m}x^m\exp\left[-\frac{q^2(1+2\alpha^2\beta^2)}{4m\alpha^2}\right]. \tag{57}$$

Even though the model is oversimplified the result (57) is instructive and transparent. The scattering amplitude appears as a finite sum of exponentials $\exp\{-\text{const.}\, q^2/m\}$ describing m-fold scattering of the projectile in the target. The larger m the slower the fall-off with q^2. The relative strength of the various multiple scattering terms is governed by the dimensionless quantity x, eq. (56).

The absolute square of the amplitude (57) and its decomposition into single scattering, single plus double scattering, etc. is illustrated in Figs. 6 for the case of pion scattering on ^{4}He, for two laboratory momenta: $k=280$ MeV$/c$ (Fig. 6a), and $k=500$ MeV$/c$ (Fig. 6b).

It is also possible, within this model, to evaluate the forward scattering amplitude and, thereby, the total cross section. Indeed, from eq. (57)

$$F_{\mathrm{ii}}(0)=\frac{1}{2}\mathrm{i}k\frac{1+2\alpha^2\beta^2}{\alpha^2}S(A),$$

where

$$S(A)=\sum_{m=1}^{A}\binom{A}{m}\frac{(-)^{m+1}}{m}x^m.$$

This latter function can be written (see exercise 3)

$$S(A)=\sum_{1}^{A}\frac{1}{m}+\ln x+\int_0^{1-x}\mathrm{d}y\,\frac{y^A}{1-y}.$$

The integral can be estimated and turns out to be small for $|x-1|<1$; one finds

$$\left|\int_0^{1-x}\mathrm{d}y\,\frac{y^A}{1-y}\right|<\frac{2|1-x|^{A+1}}{(A+1)|x|},$$

so that

$$F_{\mathrm{ii}}(0)=\frac{\mathrm{i}k}{2\alpha^2}(1+2\alpha^2\beta^2)\left[\sum_{1}^{A}\frac{1}{m}+\ln x\right]$$
$$+\frac{1}{|x|}\mathrm{O}\left(\frac{|1-x|^{A+1}}{A+1}\right). \tag{58}$$

2.3. *Comments and addenda*

As is always the case with models and theories of strong interactions it is difficult to indicate the domain of validity of the eikonal approximation. Assumptions (I) to

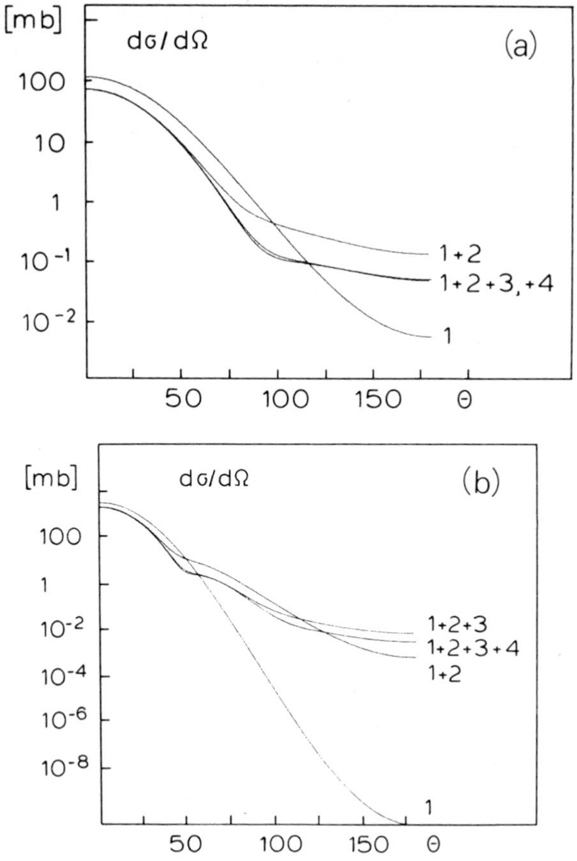

Fig. 6. Differential cross section for scattering of pions on ^4He at laboratory momenta 280 MeV/c (a), and 500 MeV/c (b), as a function of scattering angle θ. Curve marked 1 shows the single scattering term $m = 1$ of eq. (57), curve marked 1 + 2 contains single and double scattering terms, etc. In part (a) the approximations 1 + 2 + 3 and 1 + 2 + 3 + 4 are practically indistinguishable.

(III) are relatively easy to check. In particular, whether or not small angle scattering is predominant depends on the elementary amplitude: if the projectile–nucleon amplitude has a behaviour of the type indicated in eq. (54a) it is certainly true that scattering on any given nucleon more than once is very unlikely. In applying the eikonal approximation to multiple scattering it is more difficult to define, in any precise manner, what "high energy" is. The puzzling observation is that the approximation seems to be acceptable down to intermediate energies where some or all of the assumptions (I)–(III) break down. Perhaps this is due to the conjecture that even though the approximation is derived in the framework of potential theory it holds on more general grounds.

Regarding the treatment of the target in this formalism several comments should be made. The frozen nucleus assumption is good as long as the nucleon motion is adiabatically slow as compared to the motion of the projectile across the nucleus. Of course, this is a question about the dynamics of the target. It is not too difficult to judge, in any specific situation, whether this "snapshot" picture of the scattering process is a good approximation to reality.

The multiple scattering series (53) is derived on the assumption that each elementary scattering process can be described by the *free*, *on-shell* amplitude $f_j(\boldsymbol{q})$. This is equivalent to saying that between two successive scatterings the projectile propagates freely, on its mass shell. When is this assumption justified? Within the frame of potential theory this question can be answered uniquely by means of a theorem first proven by Bég.*) The theorem says this: Given a sum of elementary potentials $v_n(r - r_n)$ which have strictly finite range and which do not overlap, the scattering amplitude for the sum $V(r) = \Sigma_n v_n(r - r_n)$ is a function only of the elementary scattering amplitudes *on shell* and the geometrical disposition of the elementary potentials. It does not depend on any off-shell properties of the elementary scattering amplitudes. In applying this no-overlap theorem to a multiple scattering series like eq. (53) one has to check whether it is possible to ascribe a finite range λ to the elementary amplitude, and whether this range (or the maximum of them if there are several) is indeed smaller than the mean distance of two neighbouring scatterers in the target.

In spite of its limitations and in spite of its somewhat unclear theoretical basis, the eikonal approximation is a very useful tool in many scattering problems.

There are many extensions and generalizations of the eikonal method, some of which can be found already in Glauber's original work (Glauber 1959, Wu 1957, Gottfried 1971, Sugar et al. 1969). The approximation also appears in the framework of more general theories of high-energy scattering (Kerman et al. 1959, Foldy et al. 1969). Also in the more recent original literature on multiple scattering the reader will find many more applications of this conceptually simple and intuitively appealing approximation.

3. Scattering at low energies: optical potential

3.1. The basic equations

For the sake of simplicity we consider the scattering of a spinless particle (pion, kaon) on a nuclear ground state of spin zero. The scattering energy shall be low so that only the first two partial waves, s- and p-wave, of the elementary pion–nucleon amplitude need to be taken into account. In a first step we neglect the dependence of the elementary amplitudes on the nucleon spin and on the isospin degrees of freedom. This serves to make the formalism as simple and transparent as possible. As before we assume that the projectile–nucleon interaction can be described by means of *potentials*. Like in the previous section we will try to eliminate these

*) The original form is given in Bég (1961). Applications and generalizations are studied e.g. in Agassi et al. (1973), Hüfner (1973) and Scheck et al. (1972).

potentials at the end, as far as possible, and to replace them with the corresponding elementary *scattering amplitudes*. Nevertheless, it should be kept in mind that the description of multiple scattering is based on the theory of nonrelativistic potential scattering and, therefore, is limited to low energies. Let us introduce the following definitions:

t_n: T-matrix describing the scattering of the projectile on a *free* nucleon (labelled n),

τ_n: T-matrix describing the scattering of the projectile on nucleon n which is *bound* in the target nucleus,

$T_{\mathrm{p(N)}}$: Kinetic energy of the projectile p (nucleon N),

H_A: Hamiltonian of target nucleus describing the motion of the target nucleons,

T_A: T-matrix describing the scattering of the projectile on the *nucleus* as a whole.

From eq. (35) we see that t_n and τ_n satisfy the Lippmann–Schwinger (L.S.) equations

$$t_n(E) = v_n + v_n \frac{1}{E - T_\mathrm{p} - T_n + \mathrm{i}\varepsilon} t_n(E), \tag{59}$$

$$\tau_n(E) = v_n + v_n \frac{1}{E - T_\mathrm{p} - H_A + \mathrm{i}\varepsilon} \tau_n(E), \tag{60}$$

while $T_A(E)$ satisfies the equation

$$T_A(E) = \sum_{n=1}^{A} v_n + \sum_{n=1}^{A} v_n \frac{1}{E - T_\mathrm{p} - H_A + \mathrm{i}\varepsilon} T_A(E). \tag{61}$$

It is convenient to define the auxiliary quantities

$$T_A^n(E) = v_n + v_n \frac{1}{E - T_\mathrm{p} - H_A + \mathrm{i}\varepsilon} T_A(E), \tag{62}$$

in terms of which

$$T_A(E) = \sum_{n=1}^{A} T_A^n(E).$$

The basic L.S. equation (61) can be recast into a set of coupled equations for the auxiliary T-matrices T_A^n: From eq. (62)

$$\left(1 - v_n \frac{1}{E - T_\mathrm{p} - H_A + \mathrm{i}\varepsilon}\right) T_A^n = v_n \left(1 + \frac{1}{E - T_\mathrm{p} - H_A + \mathrm{i}\varepsilon}(T_A - T_A^n)\right).$$

Making use of eq. (60) this implies

$$T_A^n(E) = \tau_n(E)\left[1 + \frac{1}{E - T_\mathrm{p} - H_A + \mathrm{i}\varepsilon} \sum_{m \neq n} T_A^m(E)\right]. \tag{63}$$

These equations give an exact description of the multiple scattering of projectile p in the nuclear target, within the general frame of potential scattering. In this general form, they are not very useful, however, because they contain by far too much information and, obviously, cannot be solved in any exact manner. Indeed, we must keep in mind that the T-matrices t_n, τ_n, and T_A contain all projectile and all nucleon variables and thus describe all types of elastic and inelastic processes between the incident particle and the target. Depending on the specific process we shall consider, depending on the energy at which it all happens and depending on the specific dynamics of the target, we shall introduce further assumptions, thereby simplifying the multiple scattering equations (59) to (62).

As a first step, we may write down a formal iterative solution of eq. (63), viz.

$$T_A = \sum T_A^n$$
$$= \sum_n \tau_n + \sum_{m \neq n} \tau_n G^{\text{out}} \tau_m + \sum_{n \neq m,\, m \neq l} \tau_n G^{\text{out}} \tau_m G^{\text{out}} \tau_l + \cdots, \tag{64}$$

where $G^{\text{out}} = 1/(E - T_{\text{p}} - H_A + \mathrm{i}\varepsilon)$. This series expresses the projectile–nucleus T-matrix in terms of single, double,..., multiple scattering of the projectile off the nucleons bound in the nucleous. As such, it is a good starting point for approximations which help to reduce the full complexity of eqs. (59) to (62) to a simpler and, hopefully, solvable situation.

3.2. *Coherent approximation and optical potential to second order*

Let us assume that the nucleus remains in its ground state whenever the projectile has interacted with any one of the nucleons (coherent approximation). In other words, the nucleus is not excited at any stage of the multiple scattering process. Thus, when we take the expectation value of series (64) over the nuclear ground state $|\Omega\rangle$, we have approximately

$$\langle \Omega | \tau_n G^{\text{out}} \tau_m G^{\text{out}} \cdots \tau_r | \Omega \rangle$$
$$\simeq \langle \Omega | \tau_n | \Omega \rangle G_\Omega^{\text{out}} \langle \Omega | \tau_m | \Omega \rangle G_\Omega^{\text{out}} \cdots \langle \Omega | \tau_r | \Omega \rangle, \tag{65}$$

with $G_\Omega^{\text{out}} = 1/(\tilde{E} - T_{\text{p}} + \mathrm{i}\varepsilon)$. Inserting this into series (64) we have

$$T_A^0 := \langle \Omega | T_A | \Omega \rangle$$
$$\simeq A\langle \tau \rangle + A\langle \tau \rangle G_\Omega^{\text{out}} (A-1)\langle \tau \rangle$$
$$+ A\langle \tau \rangle G_\Omega^{\text{out}} (A-1)\langle \tau \rangle G_\Omega^{\text{out}} (A-1)\langle \tau \rangle + \cdots. \tag{66}$$

Here the antisymmetry of the state $|\Omega\rangle$ in the variables of the A nucleons has been used. The symbol $\langle \tau \rangle$ implies the average over the nuclear ground state wave function and, if such degrees of freedom had been taken into account, over the spins and isospins involved in the process. Therefore, $\langle \tau \rangle$ depends only on the coordinates of the projectile and we may define an *optical potential* for that particle by

$$U_A := (A-1)\langle \tau \rangle, \tag{67}$$

by means of which eq. (66) becomes equivalent to the one-body L.S. equation

$$\frac{A-1}{A} T_A^0 = U_A + U_A G_\Omega^{\text{out}} \left(\frac{A-1}{A} T_A^0 \right). \tag{68}$$

In this approximation the many-body multiple scattering problem is effectively reduced to a one-body scattering problem. Note, however, that the optical potential (67) contains the *bound* scattering amplitude which may not be known a priori.

In a first approximation, we might replace τ_n by the *free* scattering matrix t_n. This approximation may be justified if the nuclear density is low and if the elementary amplitude shows no rapid variation around the energy considered. For pure s-wave scattering, the optical potential in coordinate space is (using rel. (31)),

$$U_A(\boldsymbol{r}) \simeq U_A^{(1)}(\boldsymbol{r}) = -\frac{2\pi}{\bar{m}}\bar{f}\frac{A-1}{A}\rho(\boldsymbol{r}). \tag{69}$$

Here $\bar{f}$ is the averaged s-wave scattering amplitude, $\bar{m}$ the reduced mass of the projectile–nucleon system. The term $U_A^{(1)}(\boldsymbol{r})$, eq. (69), is the first term in the density expansion of the optical potential in nuclear matter.

As an example of how to proceed beyond the simple approximations (69) or (68), we discuss the derivation of the optical potential to second order in the density. Let

- ϕ: incident wave function describing the projectile in the absence of its interactions with the nucleons;
- ϕ_n: fractional wave that is incident on nucleon number n;
- ψ: wave function describing the projectile's state in the nuclear medium.

Note that these "wave functions" are in fact operators which depend on the projectile and on the nucleon coordinates. In order to obtain the true wave function of the projectile in the case of, say, elastic scattering the projection onto the nuclear ground state must be taken. For instance, in coordinate representation this true wave function is

$$\psi(\boldsymbol{r}) = \langle \boldsymbol{r}, \Omega | \psi \rangle.$$

The L.S. equations relating the operators ϕ, ϕ_n and ψ are obtained by direct application of eq. (33),

$$\psi = \phi + \frac{1}{E - T_\mathrm{p} - H_A + \mathrm{i}\varepsilon}\sum_1^A \tau_n \phi_n, \tag{70}$$

$$\phi_n = \phi + \frac{1}{E - T_\mathrm{p} - H_A + \mathrm{i}\varepsilon}\sum_{m \neq n}^{A} \tau_m \phi_m. \tag{71}$$

The physics of the multiple scattering process can be read off these equations: The wave in the medium is built up by the incident free wave and the contribution of all scattered waves emanating from the A nucleons in the medium (eq. (70)). The fractional wave ϕ_n incident on nucleon number n, in turn, is determined by the incident free wave and the scattered waves from all other nucleons $m \neq n$, eq. (71).

These equations express simple physical relations but, clearly, they cannot be solved in full generality. In order to transform them to a tractable form let us introduce some further approximations without worrying too much, for the moment at least, under which conditions these are justified. Suppose we can neglect the dynamics of the nucleus so that we can replace the propagator

$$\left(E - T_\mathrm{p} - H_A + \mathrm{i}\varepsilon\right)^{-1} \quad \text{by} \quad \left(\tilde{E} - T_\mathrm{p} + \mathrm{i}\varepsilon\right)^{-1}. \tag{72}$$

This would apply, for instance, in the approximation of fixed scatterers. When projected onto the nuclear ground state Ω, the multiple scattering equations (70) and (71) simplify considerably. Let us introduce the following notations and definitions:

$\boldsymbol{r}$: projectile coordinates,

$\{\boldsymbol{x}\} \equiv \{\boldsymbol{x}_1, \boldsymbol{x}_2, \ldots, \boldsymbol{x}_A\}$: nucleon coordinates,

$\rho(\boldsymbol{x}) = \dfrac{1}{A} \langle \Omega | \sum_{n=1}^{A} \delta(\boldsymbol{x}_n - \boldsymbol{x}) | \Omega \rangle$: nuclear one-body density,

$\rho(\boldsymbol{x}, \boldsymbol{y}) = \dfrac{1}{A(A-1)} \langle \Omega | \sum_{n \neq m}^{A} \delta(\boldsymbol{x}_n - \boldsymbol{x}) \delta(\boldsymbol{x}_m - \boldsymbol{y}) | \Omega \rangle$: nuclear two-body density.

Besides the two-body density $\rho(\boldsymbol{x}, \boldsymbol{y})$ it is also useful to introduce the two-body correlation function $C(\boldsymbol{x}, \boldsymbol{y})$ which we define by the equation (cf. exercise 4)

$$A(A-1)\rho(\boldsymbol{x}, \boldsymbol{y}) = A^2[1 + C(\boldsymbol{x}, \boldsymbol{y})]\rho(\boldsymbol{x})\rho(\boldsymbol{y}). \tag{73}$$

Finally, we define the uncorrelated and correlated projectile waves $\phi(\boldsymbol{r}; \boldsymbol{x}_i)$ and $\phi(\boldsymbol{r}; \boldsymbol{x}_i, \boldsymbol{x}_j)$ by

$$A\rho(\boldsymbol{x})\phi(\boldsymbol{r}; \boldsymbol{x}) := \langle \boldsymbol{r}, \Omega | \sum_{n=1}^{A} \delta(\boldsymbol{x}_n - \boldsymbol{x}) \phi_n | \Omega \rangle, \tag{74a}$$

$$A(A-1)\rho(\boldsymbol{x}, \boldsymbol{y})\phi(\boldsymbol{r}; \boldsymbol{x}, \boldsymbol{y}) := \langle \boldsymbol{r}, \Omega | \sum_{n \neq m}^{A} \delta(\boldsymbol{x}_n - \boldsymbol{x}) \delta(\boldsymbol{x}_m - \boldsymbol{y}) \phi_n | \Omega \rangle. \tag{74b}$$

Neglecting, for consistency, the recoil of the nucleons, the in-medium T-matrix is

$$\langle \boldsymbol{r}'; \{\boldsymbol{x}'\} | \tau_n | \boldsymbol{r}; \{\boldsymbol{x}\} \rangle \simeq \prod_{m=1}^{A} \delta(\boldsymbol{x}'_m - \boldsymbol{x}_m) \langle \boldsymbol{r}' - \boldsymbol{x}_n | \tau_n | \boldsymbol{r} - \boldsymbol{x}_n \rangle.$$

We replace the T-matrix by the scattering amplitude by means of rel. (31),

$$\langle \boldsymbol{r}' - \boldsymbol{x}_n | \tau_n | \boldsymbol{r} - \boldsymbol{x}_n \rangle = -\frac{2\pi}{\overline{m}} F(\boldsymbol{r}' - \boldsymbol{x}_n; \boldsymbol{r} - \boldsymbol{x}_n).$$

Eqs. (70) and (71) then simplify to the following:

$$\psi(\boldsymbol{r}) = \phi(\boldsymbol{r}) + A\int \mathrm{d}^3r'\, g(\boldsymbol{r} - \boldsymbol{r}') \int \mathrm{d}^3x\, \rho(\boldsymbol{x}) \int \mathrm{d}^3r''\, F(\boldsymbol{r}' - \boldsymbol{x}; \boldsymbol{r}'' - \boldsymbol{x}) \psi(\boldsymbol{r}''; \boldsymbol{x}), \tag{75a}$$

$$\psi(\boldsymbol{r}; \boldsymbol{x}) = \phi(\boldsymbol{r}) + A\int \mathrm{d}^3r'\, g(\boldsymbol{r} - \boldsymbol{r}') \int \mathrm{d}^3y\, \rho(\boldsymbol{y})$$
$$\times [1 + C(\boldsymbol{x}, \boldsymbol{y})] \int \mathrm{d}^3r''\, F(\boldsymbol{r}' - \boldsymbol{y}; \boldsymbol{r}'' - \boldsymbol{y}) \psi(\boldsymbol{r}''; \boldsymbol{x}, \boldsymbol{y}). \tag{75b}$$

Here the function $g(\boldsymbol{r} - \boldsymbol{r}')$ in coordinate space, according to eq. (22), is

$$g(\boldsymbol{r} - \boldsymbol{r}') = \mathrm{e}^{\mathrm{i}q|\boldsymbol{r} - \boldsymbol{r}'|}/|\boldsymbol{r} - \boldsymbol{r}'|.$$

Again, the reduced eqs. (75) can be interpreted rather easily: The wave function $\psi(\boldsymbol{r})$

which describes the motion of the projectile in the nucleus, through eq. (75a), is given by the incoming wave $\phi(\boldsymbol{r})$ plus the sum of all scattered waves originating from the scatterers in the nucleus. The strength of these scattered waves is proportional to the elementary scattering amplitude F and the nucleon density $\rho(\boldsymbol{x})$; $g(\boldsymbol{r}-\boldsymbol{r}')$ describes the propagation of the projectile in the medium. The scattering of the projectile on a nucleon cannot be independent of whether this nucleon is free or is embedded in the nuclear medium. For instance, if a proton sits at position $\boldsymbol{x}$ there can be no other proton at the same site (i.e. the correlation function $C(\boldsymbol{x}, \boldsymbol{y})$ must go to -1 for $|\boldsymbol{x}-\boldsymbol{y}| \to 0$). This kind of information is expressed by eq. (75b) which yields the wave $\psi(\boldsymbol{r}; \boldsymbol{x})$ incident on a nucleon at the site $\boldsymbol{x}$ in terms of the correlated wave $\psi(\boldsymbol{r}; \boldsymbol{x}; \boldsymbol{y})$ incident on another nucleon at the site $\boldsymbol{y}$.

The full dynamics of the nucleus, by assumption, has disappeared from these equations. The nucleus has been replaced by a medium characterized by its density, its two, three,... body densities or correlation functions. Thereby the full complexity of eqs. (70) and (71) is reduced to the problem of multiple scattering of waves (Lax 1951).

Unfortunately, even the simplified eqs. (75) do not close and are not soluble in a general manner. The quantity $\psi(\boldsymbol{r}; \boldsymbol{x}; \boldsymbol{y})$, for instance, is given by a third equation which contains also the three-body density or three-body correlation function, and so on. Eqs. (75a) and (75b) are only the first two of a (finite) system of coupled integral equations. They still contain the full elementary scattering amplitude $F(\boldsymbol{r}'-\boldsymbol{x}; \boldsymbol{r}-\boldsymbol{x})$ including its range and possible nonlocalities. Whether or not such finite range effects and nonlocalities are important depends on the specific dynamics of the projectile that one considers. Also the question as to whether this system of coupled equations can be truncated at some point depends on the properties of the nuclear medium and of the projectile–nucleon scattering amplitude (Fäldt 1973). Here we shall discuss the simplest cases for which eqs. (75) can be still solved with elementary techniques of quantum mechanics.

(i) *Approximation of homogeneous medium.* If one neglects all nucleon correlations the nucleus appears as a fully homogeneous classical medium of density $\rho(\boldsymbol{x})$. In this approximation

$$\psi(\boldsymbol{r}; \boldsymbol{x}) \simeq \psi(\boldsymbol{r}) \tag{76}$$

and eq. (75a) closes. For an incident plane wave $\phi(\boldsymbol{r}) = e^{i\boldsymbol{q}\cdot\boldsymbol{r}}$ and using the relations

$$(\Delta + \boldsymbol{q}^2) e^{i\boldsymbol{q}\cdot\boldsymbol{r}} = 0, \tag{77a}$$

$$(\Delta + \boldsymbol{q}^2) g(\boldsymbol{r}-\boldsymbol{r}') = -4\pi\delta(\boldsymbol{r}-\boldsymbol{r}'), \tag{77b}$$

eq. (75a) becomes

$$(\Delta + \boldsymbol{q}^2)\psi(\boldsymbol{r}) = -4\pi A \int d^3x\, \rho(\boldsymbol{x}) \int d^3r''\, F(\boldsymbol{r}-\boldsymbol{x}; \boldsymbol{r}''-\boldsymbol{x})\psi(\boldsymbol{r}''). \tag{78}$$

As it stands, eq. (78) is a wave equation containing an optical potential which, in general, is nonlocal. In order to make the physics of the approximation (76) and of eq. (78) as transparent as possible let us assume that the elementary scattering amplitude contains only s- and p-waves. Thus, in momentum space,

$$\tilde{F}(\boldsymbol{q}',\boldsymbol{q}) \simeq b_0 + c_0 \boldsymbol{q}\cdot\boldsymbol{q}', \tag{79}$$

where $\boldsymbol{q}$ ($\boldsymbol{q}'$) denotes the initial (final) three-momentum in the projectile–nucleon center-of-mass system. All spin and isospin degrees of freedom are neglected or averaged over. By double Fourier transform

$$F(z',z) = \frac{1}{(2\pi)^6}\int \mathrm{d}^3q' \int \mathrm{d}^3q\, \mathrm{e}^{-\mathrm{i}\boldsymbol{q}'\cdot z'} F(\boldsymbol{q}'\cdot\boldsymbol{q})\mathrm{e}^{\mathrm{i}\boldsymbol{q}\cdot z}, \tag{80}$$

the amplitude in coordinate space is then given by

$$F(\boldsymbol{r}'-\boldsymbol{x};\boldsymbol{r}-\boldsymbol{x}) \simeq b_0\delta(\boldsymbol{r}'-\boldsymbol{x})\delta(\boldsymbol{r}-\boldsymbol{x}) + c_0(\nabla_{r'}\delta(\boldsymbol{r}'-\boldsymbol{x}))\cdot(\nabla_r\delta(\boldsymbol{r}-\boldsymbol{x})). \tag{79'}$$

Inserting this into the wave equation (78) we obtain

$$(\Delta + \boldsymbol{q}^2)\psi(\boldsymbol{r}) = 2\overline{m}U^{(1)}(\boldsymbol{r})\psi(\boldsymbol{r}), \tag{81}$$

with the optical potential (Kisslinger 1955)

$$U^{(1)}(\boldsymbol{r}) = -\frac{4\pi A}{2\overline{m}}\{b_0\rho(\boldsymbol{r}) - c_0\nabla\rho(\boldsymbol{r})\nabla\}. \tag{82}$$

The squared momentum on the left-hand side of eq. (81) is given by

$$\boldsymbol{q}^2 = 2\overline{m}E \quad \text{for nonrelativistic motion,} \tag{83a}$$

$$\boldsymbol{q}^2 = E^2 - \overline{m}^2 \quad \text{for relativistic motion.} \tag{83b}$$

(ii) *Effect of two-body correlations.* In a next step beyond the approximation of a homogeneous medium we may set approximately

$$\psi(\boldsymbol{r}'';\boldsymbol{x};\boldsymbol{y}) \simeq \psi(\boldsymbol{r}'';\boldsymbol{y}) \tag{84}$$

on the right-hand side of eq. (75b).

In this case the system of two equations (75a) and (75b) closes. The approximation (84) now includes two-body correlations but no higher correlations. In other words, we take into account the granular structure of the nuclear medium in the sense that scattering on an individual nucleon is no more independent of the state of other nucleons around the struck nucleon.

Inserting eq. (75a) into eq. (75b) and introducing the approximation (84) one obtains

$$\psi(\boldsymbol{r};\boldsymbol{x}) \simeq \psi(\boldsymbol{r}) + A\int \mathrm{d}^3r'\, g(\boldsymbol{r}-\boldsymbol{r}')\int \mathrm{d}^3y\,\rho(\boldsymbol{y})C(\boldsymbol{x},\boldsymbol{y}) \times \int \mathrm{d}^3r''\, F(\boldsymbol{r}'-\boldsymbol{y};\boldsymbol{r}''-\boldsymbol{y})\psi(\boldsymbol{r}'',\boldsymbol{y}). \tag{85}$$

This approximate expression can be inserted under the integral on the right-hand side of eq. (75a). As we are working in second order, we must replace $\psi(\boldsymbol{r},\boldsymbol{y})$ by

$\psi(\boldsymbol{r})$ in the double scattering term, for consistency. Thus

$$\begin{aligned}\psi(\boldsymbol{r}) \simeq \phi(\boldsymbol{r}) &+ A\int d^3r''\, g(\boldsymbol{r}-\boldsymbol{r}'')\int d^3x\, \rho(\boldsymbol{x})\int d^3r'\, F(\boldsymbol{r}''-\boldsymbol{x};\boldsymbol{r}'-\boldsymbol{x})\psi(\boldsymbol{r}')\\ &+A^2\int d^3r''\, g(\boldsymbol{r}-\boldsymbol{r}'')\int d^3x\int d^3y\, \rho(\boldsymbol{x})\rho(\boldsymbol{y})C(\boldsymbol{x},\boldsymbol{y})\\ &\times\int d^3r'\, F(\boldsymbol{r}''-\boldsymbol{x};\boldsymbol{r}'-\boldsymbol{x})\int d^3s'\, g(\boldsymbol{r}'-\boldsymbol{s}')\\ &\times\int d^3s''\, F(\boldsymbol{s}'-\boldsymbol{y};\boldsymbol{s}''-\boldsymbol{y})\psi(\boldsymbol{s}''). \end{aligned} \tag{86}$$

Applying the operator $(\Delta + \boldsymbol{q}^2)$ to this equation and using the relations (77) leads to the second-order optical potential

$$\begin{aligned}(U^{(1)}+U^{(2)})\psi = -\frac{4\pi A}{2\overline{m}}\Big\{&\int d^3x\, \rho(\boldsymbol{x})\int d^3r'\, F(\boldsymbol{r}-\boldsymbol{x};\boldsymbol{r}'-\boldsymbol{x})\psi(\boldsymbol{r}')\\ &+A\int d^3x\int d^3y\, \rho(\boldsymbol{x})\rho(\boldsymbol{y})C(\boldsymbol{x},\boldsymbol{y})\\ &\times\int d^3r'\int d^3s'\int d^3s''\, F(\boldsymbol{r}-\boldsymbol{x};\boldsymbol{r}'-\boldsymbol{x})g(\boldsymbol{r}'-\boldsymbol{s}')\\ &\times F(\boldsymbol{s}'-\boldsymbol{y};\boldsymbol{s}''-\boldsymbol{y})\psi(\boldsymbol{s}'')\Big\}. \end{aligned} \tag{87}$$

As an example, let us work out the optical potential for the case of the amplitudes (79′). The first term in the curly brackets in eq. (87) gives the expression (82) above. The second-order potential is worked out for the s-wave and p-wave parts separately,

$$U^{(2)}(\boldsymbol{r}) \equiv U_s^{(2)}(\boldsymbol{r}) + U_p^{(2)}(\boldsymbol{r}). \tag{88a}$$

(i) The *s-wave* optical potential

$$U_s^{(2)}\psi = -\frac{4\pi A}{2\overline{m}} A b_0^2 \rho(\boldsymbol{r})\int d^3y\, \rho(\boldsymbol{y})C(\boldsymbol{r},\boldsymbol{y})g(\boldsymbol{r}-\boldsymbol{y})\psi(\boldsymbol{y}). \tag{88b}$$

The integral in eq. (88b) can only be evaluated if the correlation function $C(\boldsymbol{x}, \boldsymbol{y})$ is known. For example, suppose that C is a spherically symmetric function of the distance $r = |\boldsymbol{x} - \boldsymbol{y}|$ such that

$$\begin{aligned} C_{\rm HC}(r) &= -1 \quad \text{for } r \le r_{\rm c},\\ C_{\rm HC}(r) &= 0 \quad\;\; \text{for } r > r_{\rm c}. \end{aligned} \tag{89}$$

This means that the nucleons move independently as long as their separation is larger than the radius $r_{\rm c}$ but that they can never come closer to each other than this distance. This correlation acts as if the nucleon–nucleon interaction became infinitely repulsive at the relative distance $r = r_{\rm c}$. For this reason $r_{\rm c}$ is called "hard core", and the function (89) is called the hard core correlation function. In a practical situation such as pion scattering on nuclei at low energies, both $\rho(y)$ and $\psi(y)$ vary slowly over the distance r_c, so that

$$A\int \mathrm{d}^3y\,\rho(y)C_{\mathrm{HC}}(\boldsymbol{r},y)g(\boldsymbol{r}-y)\psi(y)$$

$$\simeq A\rho(\boldsymbol{r})\psi(\boldsymbol{r})\int \mathrm{d}^3z\,C_{\mathrm{HC}}(|z|)\frac{\mathrm{e}^{\mathrm{i}q|z|}}{|z|}$$

$$\simeq -A\rho(\boldsymbol{r})\psi(\boldsymbol{r})4\pi\frac{r_c^2}{2}.$$

In the last step we have assumed $qr_c \ll 1$. In a slightly more general situation, this same expression would be approximately equal to

$$A\int \mathrm{d}^3y\,\rho(y)C(\boldsymbol{r},y)g(|\boldsymbol{r}-y|)\psi(\boldsymbol{r}) =: \langle 1/r\rangle_{\mathrm{corr}}\cdot\psi(\boldsymbol{r}), \tag{90}$$

so that the s-wave potential, of first and second order, is given by

$$U_s^{(1+2)}(r) \simeq -\frac{4\pi A}{2\bar{m}}b_0\rho(\boldsymbol{r})\{1+b_0\langle 1/r\rangle_{\mathrm{corr}}\}. \tag{91}$$

The quantity $\langle 1/r\rangle_{\mathrm{corr}}$, defined through eq. (90), characterizes, roughly, the inverse correlation length of any two scatterers in the target nucleus and, therefore, leads to a simple physical understanding of the result (91). The convergence of the density expansion of the optical potential depends primarily on the dimensionless parameter $b_0\langle 1/r\rangle_{\mathrm{corr}}$.

We also note the following: Instead of the systematic expansion (86) that leads to the result (91) we could have utilized eq. (85) to obtain a simple relation between $\psi(\boldsymbol{r};\boldsymbol{x})$ and $\psi(\boldsymbol{r})$. Indeed, within the same approximations as before, eq. (85) gives

$$\psi(\boldsymbol{r};\boldsymbol{x}) \simeq \frac{1}{1-b_0\langle 1/r\rangle_{\mathrm{corr}}}\psi(\boldsymbol{r}). \tag{92}$$

Inserting this into eq. (75a), and applying the operator $(\Delta+\boldsymbol{q}^2)$ to the resulting equation, one obtains

$$(\Delta+\boldsymbol{q}^2)\psi(\boldsymbol{r}) = 2\bar{m}\,U_s^{\mathrm{eff}}(\boldsymbol{r})\psi(\boldsymbol{r}),$$

with

$$U_s^{\mathrm{eff}} = -\frac{4\pi A}{2\bar{m}}\frac{b_0}{1-b_0\langle 1/r\rangle_{\mathrm{corr}}}\rho(\boldsymbol{r}). \tag{93}$$

Thus, the free scattering length b_0 is renormalized to an effective scattering length in the medium

$$b_0^{\mathrm{eff}} = b_0/(1-b_0\langle 1/r\rangle_{\mathrm{corr}}). \tag{94}$$

Although this is an interesting and intuitively appealing result it must be viewed with some caution. In contrast to eq. (91) the expression (93) contains also terms of third order (and higher) in the density. However, such terms will also be generated by the remaining equations beyond eqs. (75) for the quantity $\psi(\boldsymbol{r};\boldsymbol{x},\boldsymbol{y})$ etc. which depend on three-body (and higher) correlations. Since we have neglected these terms, it appears more consistent to truncate (93) at the term quadratic in the density, i.e. to use eq. (91) instead. The point is not entirely trivial, however, insofar as the three-body (and higher) correlations, in general, may be characterized by other dimensioned quantities. So the expansion parameters might not be the same.

(ii) The *p-wave* potential. From eqs. (87) and (79′)

$$U_{\mathrm{p}}^{(2)}\psi = -\frac{4\pi A}{2\bar{m}} A c_0^2 \int \mathrm{d}^3x\,\mathrm{d}^3y\,\rho(\boldsymbol{x})\rho(\boldsymbol{y})C(\boldsymbol{x},\boldsymbol{y})$$
$$\times \int \mathrm{d}^3r'\,\mathrm{d}^3s'\,\mathrm{d}^3s''\left(\nabla_r^i\delta(\boldsymbol{r}-\boldsymbol{x})\right)\left(\nabla_{r'}^i\delta(\boldsymbol{r}'-\boldsymbol{x})\right)g(\boldsymbol{r}'-\boldsymbol{s}')$$
$$\times\left(\nabla_{s'}^j\delta(\boldsymbol{s}'-\boldsymbol{y})\right)\left(\nabla_{s''}^j\delta(\boldsymbol{s}''-\boldsymbol{y})\right)\psi(\boldsymbol{s}'').$$

The gradient ∇_r can be replaced with $-\nabla_x$, $\nabla_{s'}$ with $-\nabla_y$. This allows us to apply partial integrations to the integrals over $\boldsymbol{x}$ and over $\boldsymbol{y}$. $\nabla_{r'}$, by partial integration, is shifted to the Green function; similarly $\nabla_{s''}$ is shifted over to the wave function $\psi(s'')$. This gives

$$U_{\mathrm{p}}^{(2)}\psi = -\frac{4\pi A}{2\bar{m}} A c_0^2 \int \mathrm{d}^3y\left[\nabla_r^i\nabla_y^j\rho(\boldsymbol{r})\rho(\boldsymbol{y})C(\boldsymbol{r},\boldsymbol{y})\right]\nabla_r^i g(\boldsymbol{r}-\boldsymbol{y})\,\nabla_y^j\psi(\boldsymbol{y}). \tag{88c}$$

Again, this last integral can only be evaluated if the two-body correlation function $C(\boldsymbol{x},\boldsymbol{y})$ is known. However, as for the s-wave part, we can obtain an approximate expression for the case of correlations of very short range. In this case

$$U_{\mathrm{p}}^{(2)}\psi \simeq -\frac{4\pi A}{2\bar{m}} A c_0^2\,\nabla_r^i\rho(\boldsymbol{r})J^{ji}\,\nabla_r^j\psi(\boldsymbol{r}),$$

with

$$J^{ji} = \int \mathrm{d}^3y\left[\nabla_y^j\rho(\boldsymbol{y})C(|\boldsymbol{r}-\boldsymbol{y}|)\right]\nabla_r^i g(|\boldsymbol{r}-\boldsymbol{y}|)$$
$$= -\int \mathrm{d}^3z\left[\nabla_z^j\rho(\boldsymbol{r}-\boldsymbol{z})C(|\boldsymbol{z}|)\right]\nabla_z^i g(|\boldsymbol{z}|).$$

In the limit $qr_c \ll 1$

$$\nabla_z^i g(|\boldsymbol{z}|) \simeq -z^i/|\boldsymbol{z}|^3.$$

Since C is different from zero essentially only at $|\boldsymbol{z}|=0$, the first term in J^{ji} (with the gradient acting on ρ) vanishes. The remainder can be evaluated as follows:

$$J^{ji} \simeq \rho(\boldsymbol{r})\int_0^\infty \mathrm{d}|\boldsymbol{z}|\,\frac{\mathrm{d}C(|\boldsymbol{z}|)}{\mathrm{d}|\boldsymbol{z}|}\int \mathrm{d}\Omega_z\,\frac{z^j z^i}{|\boldsymbol{z}|^2} = -\frac{4\pi}{3}C(0)\,\delta^{ji}\rho(\boldsymbol{r})$$

(see exercise 5). The p-wave potential is then given by the approximate expression

$$U_{\mathrm{p}}^{(1+2)}(\boldsymbol{r}) \simeq \frac{4\pi A}{2\bar{m}}\nabla\left\{c_0\rho(\boldsymbol{r}) + \frac{4\pi}{3}c_0^2 C(0)A\rho^2(\boldsymbol{r})\right\}\nabla. \tag{95}$$

For repulsive correlations (such as the Pauli correlations for identical particles) $C(0) = -1$.

Additions and remarks.

(i) If one takes into account a finite range of the elementary interaction (79) and if one uses a somewhat more realistic functional form for the two-body correlation function these second order effects will be changed to some extent. Let λ be the characteristic range of the interaction, r_c the correlation length. Then the second order term in eq. (95) is multiplied with a parameter ξ which is equal to 1 for $r_c \gg \lambda$ but decreases rapidly when r_c becomes comparable to, or even smaller than, λ. Thus, eq. (95) is to be replaced by (Fäldt 1973, Eisenberg et al. 1973),

$$U_{\mathrm{p}}^{(1+2)}(\boldsymbol{r}) \simeq \frac{4\pi A}{2\overline{m}} \nabla c_0 \rho(\boldsymbol{r}) \left\{1 - \frac{4\pi}{3} \xi c_0 A \rho(\boldsymbol{r})\right\} \nabla . \tag{95'}$$

(ii) As for the s-wave potential we could have solved eq. (85) for $\psi(\boldsymbol{r}, \boldsymbol{x})$, on the basis of the approximations introduced above. This would have led us to the effective p-wave potential

$$U_{\mathrm{p}}^{\mathrm{eff}}(\boldsymbol{r}) = \frac{1}{2\overline{m}} \nabla \frac{4\pi c_0 A \rho(\boldsymbol{r})}{1 + \frac{1}{3}\xi 4\pi c_0 A \rho(\boldsymbol{r})} \nabla , \tag{96}$$

of which eq. (95′) contains the first two terms in an expansion in powers of the density. This form of the p-wave potential was first derived by Ericson and Ericson. Because of its analogy to the well-known nonlinear density dependence of the index of refraction in a polarizable medium they named this renormalization of the p-wave potential the Lorentz–Lorenz effect (Ericson 1966).

(iii) The Lorentz–Lorenz–Ericson form of the potential (96) must be viewed with the same caution as in the case of the s-wave potential. If genuine three-body effects are important, these must also be taken into account. In any event, it might be safer to stop at the second order term (95′).

(iv) As it stands, the optical potential (95) or (96) depends only on the elementary scattering amplitudes taken *on-shell* and on static properties of the nucleus. This is a reflection of Bég's theorem which says roughly that the particle–nucleus scattering should only depend on the on-shell elementary scattering amplitudes as long as the elementary interactions do not overlap.

One can show (Scheck et al. 1972) that the Lorentz–Lorenz correction can be understood as a consequence of Bég's theorem. For example if instead of the amplitude (79) one starts from the equivalent form

$$\tilde{F}(\boldsymbol{q}', \boldsymbol{q}) \simeq b_0 + c_0\left[2\overline{m}E - \tfrac{1}{2}(\boldsymbol{q} - \boldsymbol{q}')^2\right], \tag{97}$$

then the following optical potential is obtained in first order,

$$V_{\mathrm{p}}^{(1)}(\boldsymbol{r}) \simeq -\frac{4\pi A}{2\overline{m}} \left\{ b_0 \rho(\boldsymbol{r}) + c_0 \left(2\overline{m}E + \tfrac{1}{2}\Delta\right) \rho(\boldsymbol{r}) \right\} \tag{98}$$

(the so-called "local" potential). Even though the amplitude (97) is the same as the amplitude (79) as long as $2\overline{m}E = \boldsymbol{q}^2 = \boldsymbol{q}'^2$, the optical potential (98) is very different from the previous form (82)—in conflict with Bég's theorem. The answer to this puzzle is that also the local potential (98) must be supplemented by its own Lorentz–Lorenz effect (which has a different form in this case). After this is done, the differences disappear and the potentials become the same, to the order that one considers—in accord with the theorem.

3.3. Formal theory of the optical potential

The derivation of the optical potential given in the preceding section 3.2 stresses its physical content and the nature of the approximations on which it is based. This section deals with a more formal construction of the optical potential from general (nonrelativistic) scattering theory. Practical applications will be encountered below in the context of bound states in hadronic atoms.

We start from eq. (61) for the projectile–composite target scattering operator $T_A(E)$,

$$T_A(E) = V + VG(E)T_A(E), \tag{99}$$

with

$$V = \sum_{n=1}^{A} v_n, \qquad G(E) = \frac{1}{E - T_{\mathrm{p}} - H_A + \mathrm{i}\varepsilon}. \tag{100}$$

As before, T_{p} denotes the kinetic energy of the projectile, H_A the Hamiltonian describing the target. $G(E)$, as well as all other Green functions of this section, is chosen to match the outgoing-state asymptotic condition. The superscript "out" is dropped, for simplicity. Recall that the scattering operator T_A describes all possible elastic and inelastic transitions of the projectile–target system, and, therefore, contains by far too much information for the problem at hand: scattering of the projectile off the nuclear ground state. Therefore, it is appropriate to divide the space of all possible states into those where the projectile moves in a plane wave, while the target is in its ground state, and the remainder. This can be done by means of projection operators,

$$P \coloneqq \int \frac{\mathrm{d}^3 q}{(2\pi)^3} |\boldsymbol{q};\, \Omega\rangle\langle\Omega; \boldsymbol{q}|, \tag{101a}$$

and

$$Q \coloneqq \mathbb{1} - P. \tag{101b}$$

By means of these operators write eq. (99) as a pair of equations,

$$PT_A(E)P = PVP + (PVP)G(E)(PT_A(E)P) + (PVQ)G(E)(QT_A(E)P), \tag{102a}$$

$$QT_A(E)P = QVP + (QVQ)G(E)(QT_A(E)P) + (QVP)G(E)(PT_A(E)P). \tag{102b}$$

For the case of elastic scattering we need an equation for the projected amplitude

$(PT_A(E)P)$ alone. This can be obtained from eqs. (102) by eliminating the operator (QT_AP). From eq. (102b) we find

$$QT_AP = [1-(QVQ)G]^{-1}(QVP)[1+G(PT_AP)].$$

Inserting this into eq. (102a) we have

$$PT_AP = PVP + (PVQ)G[1-(QVQ)G]^{-1}(QVP) \\ + \{PVP + (PVQ)G[1-(QVQ)G]^{-1}(QVP)\}G(PT_AP).$$

Now use the identity

$$A[1+BA]^{-1} = [(1+BA)A^{-1}]^{-1} = [A^{-1}+B]^{-1}$$

to obtain

$$PT_AP = (PVP + (PVQ)[G^{-1}-(QVQ)]^{-1}(QVP))(1+G(PT_AP)).$$

If we define

$$W(E) := V + VQ[G^{-1}(E) - QVQ]^{-1}QV, \tag{103}$$

the last equation assumes the form of a Lippmann–Schwinger equation for a single particle, viz.

$$(PT_AP) = (PW(E)P) + (PW(E)P)G(E)(PT_A(E)P). \tag{104}$$

Either $W(E)$ or, as done in the previous section, its projection

$$U(E) := PW(E)P, \tag{103'}$$

may be called optical potential. Loosely speaking, it sums up the scattering from the initial state (projectile in plane wave state, target in its ground state) to all intermediate inelastic states.

We shall now reexpress the optical potential up to second order in terms of the elementary scattering amplitude. Let

$$g_n(E') := \frac{1}{E' - T_\mathrm{p} - T_\mathrm{n} + \mathrm{i}\varepsilon},$$

$$g_0(E) := \frac{P}{E - T_\mathrm{p} - T_A + \mathrm{i}\varepsilon},$$

where T_A denotes the kinetic energy of the nucleus as a whole. From eqs. (59) and (60) one obtains τ_n, the elementary scattering operator in medium, in terms of the free scattering operator t_n,

$$\tau_n(E) = t_n(E')[1 + (g_n(E') - G(E))t_n(E')]^{-1}. \tag{106}$$

The energy E' at which the free amplitude is taken, is arbitrary. One may in fact make use of this arbitrariness, in approximations for the optical potential, such that these approximations become optimal. In practice, one usually identifies E' with E, the energy of the relative motion projectile–nucleon.

Expanding eq. (106) to second order,

$$\tau_n(E) \simeq t_n(E') - t_n(E')[g_n(E') - G(E)]\,t_n(E'), \tag{106'}$$

one obtains from eq. (104)

$$\begin{aligned} W(E) &= T_A(E)[1 + PG(E)PT_A]^{-1} \\ &= T_A(E)[1 + g_0(E)T_A(E)]^{-1} \\ &\simeq T_A(E) - T_A(E)g_0(E)T_A(E). \end{aligned} \tag{104'}$$

Finally, we recall eq. (63) which gives

$$T_A(E) = \sum_n T_A^n(E) = \sum_n \tau_n(E)\left[1 + G(E)\sum_{m\neq n} T_A^m(E)\right]. \tag{107}$$

Inserting eqs. (106′) and (107) into the expansion (104′) we obtain, in second order in t_n,

$$\begin{aligned} W(E) \simeq{}& \sum_n t_n(E') + \sum_{m\neq n} t_n(E')[G(E) - g_0(E)]\,t_m(E') \\ &+ \sum_n t_n(E')[G(E) - g_0(E) - g_n(E')]\,t_n(E'). \end{aligned} \tag{108}$$

As shown in exercise 12, this expression can also be derived from eqs. (64) and (106′).

The optical potential, eq. (103′), is obtained from eq. (108) by projection onto the ground state of the target. From here on one derives approximate expressions for the first order and second order potentials, on the basis of additional assumptions which must be adapted to the specific situation under consideration. Clearly, for the case of scattering of spin zero projectiles on nuclei at low energies the same expressions as above are obtained. However, the result (108) is much more general and can lead to local as well as truly nonlocal effective optical potentials (Kerman et al. 1959, Moniz 1977, 1981).

4. Pion–nucleus scattering near the $\Delta(I = J = 3/2)$ resonance

The eikonal approximation for multiple scattering (treated in sec. 2) is applicable to a situation were *many* partial waves contribute to the elementary amplitude and where none of these partial waves exhibits any anomalously rapid energy variation. The static optical potential (95) of sec. 2 was derived from the assumption (among others) that the elementary interaction is short-ranged and only weakly energy dependent.

Scattering of pions on nuclei at *intermediate* energies, i.e. at kinetic energies between 100 and 300 MeV meets neither of these conditions: In this energy range the elementary pion–nucleon amplitude is completely dominated by a resonance with isospin $I = \frac{3}{2}$ which appears in the relative p-wave and has spin $J = \frac{3}{2}$. The pion–nucleon p-wave amplitude exhibits a strong energy dependence around this resonance. The finite width of the resonance introduces important nonlocalities. As a consequence the assumptions on which the on-shell optical potential was based are not justified in this energy range and a different approach to the optical potential must be found.

This field is still in rapid evolution. Different theoretical frameworks for the treatment of multiple scattering in the presence of a dominating resonance were proposed and have been tested in various applications to pion–nuclear scattering at intermediate energies. We do not go into these theories in any detail since none of them has reached its final form and since they are all rather technical as to the details.

Instead, in this section we collect the main properties of the pion–nucleus scattering amplitude around the $\Delta(\frac{3}{2}, \frac{3}{2})$ resonance and provide the theoretical tools for the so-called isobar models. We also discuss, in a more qualitative way, the implications of this resonance for the propagation of pions in the nucleus.

4.1. The $\Delta(\frac{3}{2}, \frac{3}{2})$ resonance in pion–nucleon scattering

In Fig. 7 we show the measured total cross sections $\sigma_{tot}(\pi^+ p)$ and $\sigma_{tot}(\pi^- p)$ as a function of the kinetic energy of the pion in the laboratory system. Fig. 8 shows the

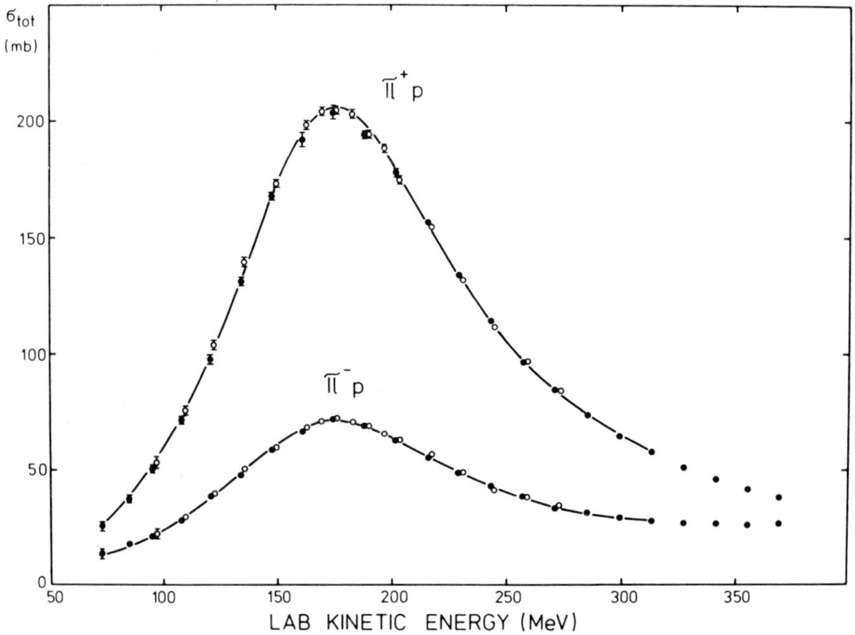

Fig. 7. Total cross sections for πp scattering around the resonance. Taken from Pedroni et al., 1978, Nucl. Phys. A300, 321.

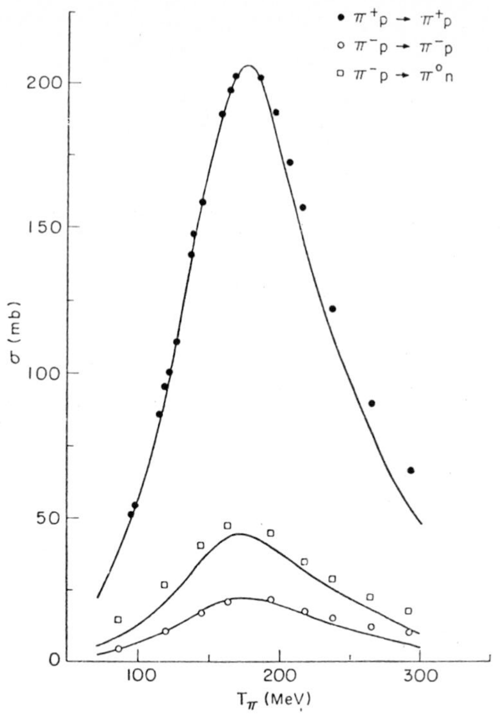

Fig. 8. The pion–nucleon cross sections around the resonance (open and full circles and squares). The solid lines show the contributions from the resonant p-wave channel with spin and isospin 3/2. Taken from Moniz (1981).

experimental, integrated, elastic and charge exchange cross sections $\sigma(\pi^{\pm}p \to \pi^{\pm}p)$, $\sigma(\pi^- p \to \pi^0 n)$, around the Δ resonance. This resonance is known to have spin and parity $J = \frac{3}{2}^+$ and isospin $I = \frac{3}{2}$. Its mass is $m(\Delta) \cong 1232$ MeV, its width $\Gamma \simeq 115$ MeV. It is not difficult to calculate that the resonance position occurs at the following value of the pion beam momentum in the laboratory system: $p_\pi = 298$ MeV/c. The pion kinetic energy is then $T_\pi^{\text{lab}} = 189$ MeV (see exercise 6). The corresponding three-momentum in the centre-of-mass system is $k_{\text{res}} = 227$ MeV/c.

Isospin invariance yields relations between the scattering amplitudes and the total cross sections (see below, sec. II.1). In particular, the total cross sections can be expressed in terms of such cross sections for pion–nucleon states with definite isospin, viz.

$$\sigma_{\text{tot}}(\pi^+ p) = \sigma_{\text{tot}}\left(I = \tfrac{3}{2}\right), \tag{109}$$

$$\sigma_{\text{tot}}(\pi^- p) = \tfrac{2}{3}\sigma_{\text{tot}}\left(I = \tfrac{1}{2}\right) + \tfrac{1}{3}\sigma_{\text{tot}}\left(I = \tfrac{3}{2}\right).$$

As $\sigma_{\text{tot}}(\pi^- p)$ from Fig. 7 is about one third of $\sigma_{\text{tot}}(\pi^+ p)$, the isospin-3/2 channel completely dominates pion–nucleon scattering at the resonance. The isospin-1/2 channel has a very small cross section. This is also borne out by the results shown in

Fig. 8. The solid lines show the contribution from the $I = J = \frac{3}{2}$ p-wave channel. At the resonance, the total cross section satisfies the unitarity bound

$$\sigma_{\text{tot}}(k_{\text{res}}) = \frac{4\pi}{k_{\text{res}}^2}\left(J + \tfrac{1}{2}\right). \tag{110}$$

(This follows from generalizing eqs. (14) to the case with spin, see below, sec. II.1.) Inserting $J = \frac{3}{2}$ and $k_{\text{res}} = 227$ MeV$/c \triangleq 1.15$ fm^{-1} gives $\sigma_{\text{tot}}(k_{\text{res}}) \simeq 190$ mb, which indeed is of the order of $\sigma_{\text{tot}}(\pi^+\text{p})$ at the resonance, see Fig. 7.

Of course, the detailed properties of the pion–nucleon scattering amplitudes around this resonance can only be obtained from a phase shift analysis of all spin–isospin partial waves. In particular, the mass and width of the $\Delta(\frac{3}{2}, \frac{3}{2})$ depend upon the shape of the resonance form that is used in parametrizing the data. It may be more useful to give the pole position corresponding to this resonance, viz. $M - \mathrm{i}\Gamma/2 \simeq (1211 - \mathrm{i}\ 51)$ MeV [R2], which is much less dependent on the resonance shape. This remark should be kept in mind in constructing explicit forms for the resonating p-wave amplitude.

The presence of a strongly dominating, resonating partial wave in the elementary pion–nucleon system has a number of interesting consequences for the behaviour of pions in nuclei which we now discuss in a sketchy and qualitative way.

4.2. *Non-static optical potential in presence of a resonating amplitude*

We start from eq. (60) for the scattering matrix τ_n of the pion on the *bound* nucleon number n. We decompose the nuclear Hamiltonian H_A in its fraction describing nucleon n,

$$T_n + W_n = \frac{\boldsymbol{K}^2}{2m_{\text{N}}} + \sum_{m \neq n} V(\boldsymbol{x}_m - \boldsymbol{x}_n), \tag{111}$$

and the Hamiltonian H_{A-1} describing the residual nucleus:

$$H_A = H_{A-1} + T_n + W_n. \tag{112}$$

The kinetic energies of pion and nucleon n are decomposed into the kinetic energies of the center-of-mass and of the relative motion,

$$T_\pi + T_n = T_{\text{c.m.}} + T_{\text{rel.}} = \frac{\boldsymbol{P}^2}{2M} + \frac{\boldsymbol{\kappa}^2}{2\mu}, \tag{113}$$

with $M = m_\pi + m_{\text{N}}$, $\mu = m_\pi m_{\text{N}}/M$, $\boldsymbol{P} = \boldsymbol{k} + \boldsymbol{K}$: total momentum, $\boldsymbol{\kappa} = (1/M) \times (m_{\text{N}}\boldsymbol{k} - m_\pi \boldsymbol{K})$: relative momentum.

The corresponding center-of-mass and relative coordinates are

$$\boldsymbol{R} = \frac{1}{M}(m_\pi \boldsymbol{r} + m_{\text{N}}\boldsymbol{x}_n), \qquad \boldsymbol{\rho} = \boldsymbol{r} - \boldsymbol{x}_n.$$

The separation into relative and center-of-mass motion is essential since the resonance appears in the relative motion of the pion and the nucleon. In particular, the *free* amplitude $t_n(E')$, described by eq. (59), must be written as a function of the relative kinetic energy. Thus, if we compare it to the *bound* amplitude $\tau_n(E)$ at energy E, this means that it should be taken at $E' = E - H_{A-1} - T_{\text{c.m.}}$. The relation of t_n and τ_n is given by eq. (106). In the limit of weak binding, $\omega_n \simeq 0$, the difference of the Green functions vanishes and we have from eq. (106)

$$\tau_n(E) \simeq t_n(E' = E - H_{A-1} - T_{\text{c.m.}}). \tag{114}$$

When taken between momentum eigenstates, this gives

$$\langle \boldsymbol{k}', \boldsymbol{K}' | \tau_n | \boldsymbol{k}, \boldsymbol{K} \rangle = \delta^{(3)}(\boldsymbol{k} + \boldsymbol{K} - \boldsymbol{k}' - \boldsymbol{K}') t_n \left(E - H_{A-1} - \frac{(\boldsymbol{k} + \boldsymbol{K})^2}{2M}; \boldsymbol{\kappa}', \boldsymbol{\kappa} \right), \tag{114'}$$

where $\boldsymbol{\kappa}$ and $\boldsymbol{\kappa}'$ are the relative momenta before and after the collision.

Suppose that the elementary amplitude is dominated by the resonant subamplitude with spin and isospin $\frac{3}{2}$,

$$t_{33}(E, \boldsymbol{\kappa}', \boldsymbol{\kappa}) = t_{33}^0 \frac{h(\boldsymbol{\kappa}') h(\boldsymbol{\kappa})}{D(E)}. \tag{115}$$

Here $D(E)$ is the resonance denominator and $h(\boldsymbol{\kappa})$, $h(\boldsymbol{\kappa}')$ denote form factors for the case where t_n has to be extrapolated off-shell. The term (115) appears in the p-wave part of the pion–nucleon scattering matrix $T_{\beta\alpha}(E, \boldsymbol{\kappa}', \boldsymbol{\kappa})$ (α and β denoting the pion's isospin component before and after the scattering). Therefore, it comes with the usual factor $P_{l=1}(\cos\theta) = \cos\theta = \hat{\boldsymbol{\kappa}}' \cdot \hat{\boldsymbol{\kappa}}$. If the pion is scattered on a nucleus with spin and isospin zero, α must equal β, and the corresponding average of $T_{\beta\alpha}$ over the spin and isospin must be taken, viz.

$$\langle T_{\beta\alpha} \rangle = c t_{33}(E, \boldsymbol{\kappa}', \boldsymbol{\kappa}) \hat{\boldsymbol{\kappa}}' \cdot \hat{\boldsymbol{\kappa}} \delta_{\alpha\beta},$$

where c is a numerical factor from the average. (One can show that $c = \frac{4}{9}$, see below, sec. II.1.1.)

Spherical nuclei are well described by the single-particle shell model. In the simplest case the shell model describes the nuclear ground state (with spin zero) by a Slater determinant of single-particle wave functions $|\psi_n\rangle$. A single particle state described by the wave function $|\psi_n\rangle$ carries the energy ε_n, the binding energy of the nucleon in that state.

The optical potential of first order is then given by, in momentum space,

$$\begin{aligned} \tilde{U}(E, \boldsymbol{k}', \boldsymbol{k}) &= \sum_{n=1}^{A} \langle \Omega, \boldsymbol{k}' | \tau_n(E) | \Omega, \boldsymbol{k} \rangle \\ &\simeq c t_{33}^0 \sum_{n=1}^{A} \int \mathrm{d}^3K \int \mathrm{d}^3K' \, \tilde{\psi}_n^*(\boldsymbol{K}') \tilde{\psi}_n(\boldsymbol{K}) \delta^{(3)}(\boldsymbol{k}' + \boldsymbol{K}' - \boldsymbol{k} - \boldsymbol{K}) \\ &\quad \times (\hat{\boldsymbol{\kappa}}' \cdot \hat{\boldsymbol{\kappa}}) h(\boldsymbol{\kappa}') h(\boldsymbol{\kappa}) D^{-1}\left(E + \varepsilon_n - (\boldsymbol{k} + \boldsymbol{K})^2/2M\right). \end{aligned} \tag{116}$$

This result contains a number of new effects which we have neglected hitherto, viz.

(i) transformation to center-of-mass and relative motion is taken into account properly;

(ii) the Fermi motion of the bound nucleons is included by way of their shell model wave functions. At least some off-shell behaviour is taken care of through the form factor $h(\boldsymbol{\kappa})$;

(iii) the energy dominator D contains the correct recoil term.

As noted above, the first effect is important whenever the energy (or, equivalently the momentum) dependence of the partial waves is not smooth, and if one partial wave dominates the amplitude. Point (ii) is typical for this same situation, at low or intermediate energies. The contribution of any individual nucleon to the optical potential depends on its state and cannot be approximated by the average nucleon density $\rho(\boldsymbol{r})$. The recoil term, finally, gives rise to a genuine nonlocality of the optical potential.

In order to exhibit these new features in a simple and transparent manner, let us assume the $I = J = \frac{3}{2}$ amplitude to be given by

$$F = F_0 \frac{-\Gamma/2}{E - R + \mathrm{i}\Gamma/2}, \tag{117}$$

where E denotes the kinetic energy of the relative motion, $R = m(\Delta) - m_\pi - m_\mathrm{N} = 154$ MeV, $\Gamma/2 = 51$ MeV. The factor F_0 is derived from the total cross section $\langle\sigma\rangle$ at the resonance, averaged over spin and isospin, through the optical theorem (11)

$$F_0 = \frac{1}{4\pi} p_\mathrm{c}\langle\sigma\rangle,$$

with p_c the pion's momentum in the c.m. system.

By virtue of this ansatz eq. (116) becomes

$$\tilde{U} \simeq u_0 \sum_{n=1}^{A} \int \mathrm{d}^3K\, \tilde{\psi}_n^*(\boldsymbol{K} - \boldsymbol{k}' + \boldsymbol{k}) \times \frac{-\Gamma/2}{E + \varepsilon_n - R - (1/2M)(\boldsymbol{k} + \boldsymbol{K})^2 + \mathrm{i}\Gamma/2} (\hat{\boldsymbol{\kappa}}' \cdot \hat{\boldsymbol{\kappa}}) \tilde{\psi}_n(\boldsymbol{K}),$$

with

$$u_0 = -F_0 p / (4\pi^2 p_\mathrm{c} E)$$

(p being the pion's momentum in the laboratory system). This expression may be integrated approximately if one neglects the factor $(\hat{\boldsymbol{\kappa}}' \cdot \hat{\boldsymbol{\kappa}})$. We set $\boldsymbol{k} + \boldsymbol{K} =: \boldsymbol{Q}$ and use the relation

$$\int \mathrm{d}^3Q \frac{\mathrm{e}^{-\mathrm{i}\boldsymbol{Q}\cdot\boldsymbol{z}}}{\kappa^2 - \boldsymbol{Q}^2} = -2\pi^2 \frac{\mathrm{e}^{\mathrm{i}\kappa|\boldsymbol{z}|}}{|\boldsymbol{z}|}, \tag{118}$$

the definition of the wave function in coordinate space,

$$\psi_n(\boldsymbol{r}) = \frac{1}{(2\pi)^3} \int \mathrm{d}^3k\, \mathrm{e}^{\mathrm{i}\boldsymbol{k}\cdot\boldsymbol{r}} \tilde{\psi}_n(\boldsymbol{k}), \tag{119}$$

and the transformation formula from momentum space to coordinate space,

$$U(E,\boldsymbol{r}',\boldsymbol{r})=\frac{1}{(2\pi)^3}\int \mathrm{d}^3k'\int \mathrm{d}^3k\, \mathrm{e}^{-\mathrm{i}\boldsymbol{k}'\cdot\boldsymbol{r}'}\tilde{U}(E,\boldsymbol{k}',\boldsymbol{k})\mathrm{e}^{\mathrm{i}\boldsymbol{k}\cdot\boldsymbol{r}}. \tag{120}$$

This gives after some calculation

$$U(E,\boldsymbol{r}',\boldsymbol{r})\simeq -4\pi^2 M u_0 \sum_{n=1}^{A} \psi_n^*(\boldsymbol{r}')\frac{\mathrm{e}^{\mathrm{i}\kappa_n|\boldsymbol{r}'-\boldsymbol{r}|}}{|\boldsymbol{r}'-\boldsymbol{r}|}\psi_n(\boldsymbol{r}), \tag{121}$$

with

$$\kappa_n=\sqrt{2M(E+\varepsilon_n-R+\mathrm{i}\Gamma/2)}\,. \tag{122}$$

We shall not pursue this somewhat technical analysis any further and shall restrict ourselves to a few remarks on the result (121) and on further consequences deriving from it. The simplified but still realistic expression (121) exhibits clearly the origin of the nonlocality: The resonating pion-nucleon system propagates from $\boldsymbol{r}$ to $\boldsymbol{r}'$ with the characteristic wave number κ_n, as given by eq. (122). Note that in a fixed-scatterer approximation we would have neglected this intermediate propagation, that is we would have replaced

$$\frac{\mathrm{e}^{\mathrm{i}\kappa_n|\boldsymbol{r}'-\boldsymbol{r}|}}{|\boldsymbol{r}'-\boldsymbol{r}|} \quad \text{by} \quad \delta(\boldsymbol{r}'-\boldsymbol{r}).$$

In this case we would recover the local potential of first order, proportional to the density

$$\rho(\boldsymbol{r})=\sum_{n=1}^{A}\psi_n^*(\boldsymbol{r})\psi_n(\boldsymbol{r}).$$

It is instructive to compare the "resonating-amplitude" potential (116) or (121) with the "smooth-amplitude" potentials considered in the preceding sections: In the latter case the nuclear target enters only via its ground state density (for elastic scattering), or via some transition density (for inelastic scattering). In either case the pion propagates, in a single mode, in a dispersive medium.

In the former case, to the contrary, the nuclear dynamics as well as the pion–nucleon dynamics enter in a complex manner. It is now the single particle density matrix that determines the scattering, not the ground state (or transition) density. Furthermore it can be shown that the inclusion of the Δ–nucleon dynamics leads to a complex multi-mode propagation of the pion in the nuclear medium. Basically, this comes about because the pion can be absorbed, in an intermediate step, by a bound nucleon forming a Δ-resonance. This leaves the combined pion–nuclear system in a state with a nucleon hole and a Δ within the nucleus. This Δ-hole state can decay back to a single pion state, as illustrated by Fig. 9. Clearly, diagrams of this type may appear nested in more complex configurations. At the same time it becomes clear that the behaviour of such an intermediate Δ within the

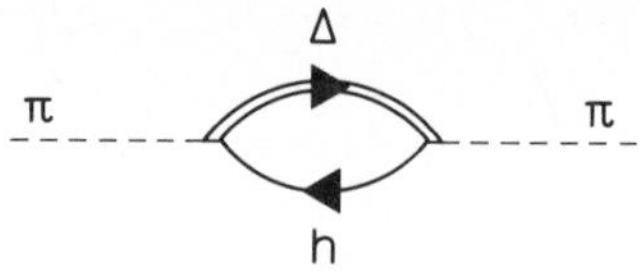

Fig. 9. Δ-hole intermediate state in pion propagation through nuclear matter.

nucleus becomes relevant. For instance, in the spirit of the nuclear shell model, it is conceivable that the Δ experiences itself effects like binding, spin–orbit coupling, quenching or enhancement of its decay width due to its interaction with the nuclear medium (Lenz 1975, Hirata et al. 1979).

The introduction of the resonance dynamics into pion–nucleus physics at intermediate energies has opened up a rich and diversified field of research which is currently being explored theoretically and experimentally.

4.3. *Elements of the isobar model*

In this section we wish to provide the tools which are indispensable for the understanding of current work in the isobar model.

Let the pion–nucleon vertex be described by a coupling of the form

$$\frac{\mathrm{i}f_{\pi\mathrm{NN}}}{m_\pi}\boldsymbol{q}\cdot\boldsymbol{\sigma}\,\tau_i . \tag{123}$$

$\boldsymbol{q}$ is the momentum transfer, the operators $\boldsymbol{\sigma}$ and τ_i act on nucleon states. The nucleons are assumed to be very heavy (static approximation) and, therefore, are treated nonrelativistically. $f_{\pi\mathrm{NN}}$ is the pion–nucleon coupling constant whose experimental value is

$$f^2_{\pi\mathrm{NN}} = 0.08.$$

The index i on τ_i, finally, refers to the charge state of the pion.

The interaction potential which arises from the exchange of a pion between two (static) nucleons is found to be [R3]

$$V_{\pi\mathrm{NN}}(\boldsymbol{r}) = \frac{f^2_{\pi\mathrm{NN}}}{m^2_\pi}\boldsymbol{\tau}^{(1)}\cdot\boldsymbol{\tau}^{(2)}\,\boldsymbol{\sigma}^{(1)}\cdot\nabla\,\boldsymbol{\sigma}^{(2)}\cdot\nabla\,\frac{\mathrm{e}^{-m_\pi r}}{r} \tag{124}$$

and is illustrated by Fig. 10a. $r = |\boldsymbol{r}_1 - \boldsymbol{r}_2|$ is the distance of two nucleons. In analogy to (123) we introduce a coupling term for the transition $\pi + \mathrm{N} \to \Delta$, viz.

$$\frac{\mathrm{i}f_{\pi\mathrm{N}\Delta}}{m_\pi}\boldsymbol{q}\cdot\boldsymbol{S}\,T_i . \tag{125}$$

Here, $\boldsymbol{S}$ and $\boldsymbol{T}$ are operators describing the spin and isospin pieces, respectively, of the transition between a nucleon state ($I = J = \frac{1}{2}$) and a Δ-resonance state ($I = J = \frac{3}{2}$), again in a static picture. The transition potential describing the transition $\mathrm{NN} \to \Delta\mathrm{N}$

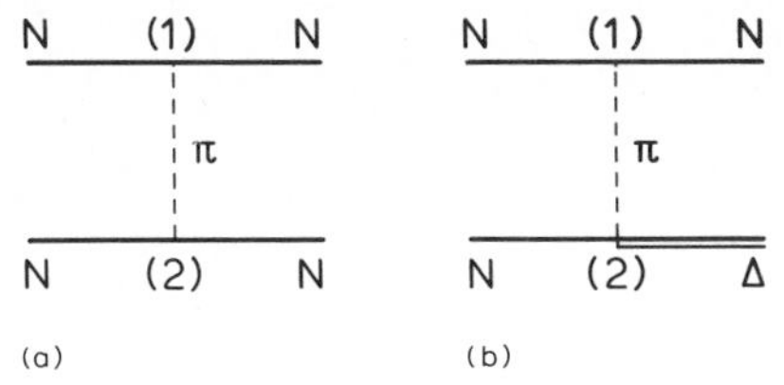

Fig. 10. (a) One-pion exchange in NN → NN scattering. (b) One-pion exchange in the process NN → NΔ.

by one pion exchange (Fig. 10b) is then found to be

$$V(r) = \frac{f_{\pi N\Delta} f_{\pi NN}}{m_\pi^2} \boldsymbol{\tau}^{(1)} \cdot \boldsymbol{T}^{(2)} \boldsymbol{\sigma}^{(1)} \cdot \nabla \boldsymbol{S}^{(2)} \cdot \nabla \frac{e^{-m_\pi r}}{r}. \tag{126}$$

It is not difficult to construct the transition operators $\boldsymbol{S}$ and $\boldsymbol{T}$. We do this explicitly for $\boldsymbol{S}$, the result being the same for $\boldsymbol{T}$. Denoting the nucleon Pauli spinor by $|\frac{1}{2}m\rangle$, the Δ-spinor by $|\frac{3}{2}M\rangle$, the μ-th spherical component of $\boldsymbol{S}$ is given by

$$S_\mu = \sum_{m,M} |\tfrac{3}{2}M\rangle(1\mu, \tfrac{1}{2}m|\tfrac{3}{2}M)\langle\tfrac{1}{2}m|. \tag{127}$$

This formula can be understood as follows. Suppose we want to construct a transition operator between states of definite angular momentum $|jm\rangle$ and $|JM\rangle$ respectively and that we want this operator to be a spherical tensor of degree λ. Then its components must be given by a superposition of product transition operators $|JM\rangle\langle jm|$ such that J and j are coupled to λ, by means of Clebsch–Gordan coefficients, viz.

$$P_\mu^{(\lambda)} = \alpha \sum_{m,M} (-)^{j-m}(JM, j-m|\lambda\mu)|JM\rangle\langle jm|. \tag{128}$$

The factor α is a constant to be chosen according to the needs. (The phase $(-)^{j-m}$ and the sign change on m in the coupling coefficient are necessary in order to make the vector $|jm\rangle$ *cogredient* to the vector $|JM\rangle$ [R4].) We may interchange (JM) and $(\lambda\mu)$ by making use of the symmetry relation of the Clebsch–Gordan coefficients,

$$(j_1m_1, j_2m_2|j_3m_3) = (-)^{j_1-j_3+m_2}\sqrt{\frac{2j_3+1}{2j_1+1}}\,(j_3m_3, j_2-m_2|j_1m_1).$$

This gives

$$P_\mu^{(\lambda)} = \alpha\sqrt{\frac{2\lambda+1}{2J+1}}\,(-)^{J-j-\lambda} \sum_{m,M} (\lambda\mu, jm|JM)|JM\rangle\langle jm|. \tag{128$'$}$$

It remains to fix the normalization factor α. In the specific situation at hand, $|jm\rangle$ is an elementary spin state while $|JM\rangle$ is a resonance state which is the result of coupling the spin j and an orbital angular momentum λ (p-wave, π–N amplitude and spin-1/2 of nucleon coupled to the spin-3/2 of the Δ-resonance). In this case it is convenient to choose α such that $P_\mu^{(\lambda)}$ projects onto the *normalized* intermediate state $|JM\rangle$. By inspecting eq. (128′) one finds that this condition is met by the choice

$$\alpha=\sqrt{\frac{2J+1}{2\lambda+1}}\,. \tag{129}$$

This can also be seen explicitly by considering the product

$$P_{\mu}^{(\lambda)\dagger}P_{\bar{\mu}}^{(\lambda)}=\sum_{m,\bar{m},M}(\lambda\mu,\,jm|JM)(\lambda\bar{\mu},\,j\bar{m}|JM)|jm\rangle\langle j\bar{m}|, \tag{130}$$

which is an operator in the space of the states $|jm\rangle$ only.

The special case $j=\frac{1}{2}$, $\lambda=1$, $J=\frac{3}{2}$ gives indeed the result shown in eq. (127), with $S_\mu\equiv P_\mu^{(1)}$. In this case the expression (130) can be worked out more explicitly, viz.

$$S_\mu^\dagger S_{\bar{\mu}}=\frac{4}{3}\sum_{m,\bar{m},M}(-)^{m-\bar{m}}(\tfrac{3}{2}M,\tfrac{1}{2}-m|1\mu)(\tfrac{3}{2}M,\tfrac{1}{2}-\bar{m}|1\bar{\mu})|\tfrac{1}{2}m\rangle\langle\tfrac{1}{2}\bar{m}| \tag{130a}$$

$$=\sum_{m,\bar{m},M}(1\mu,\tfrac{1}{2}m|\tfrac{3}{2}M)(1\bar{\mu},\tfrac{1}{2}\bar{m}|\tfrac{3}{2}M)|\tfrac{1}{2}m\rangle\langle\tfrac{1}{2}\bar{m}|. \tag{130b}$$

Let us introduce the following spherical tensor operators

$$O_0^{(0)}:=\mathbb{1},$$

$$O_{\pm1}^{(1)}:=\mp\frac{1}{\sqrt{2}}(\sigma^{(1)}\pm i\sigma^{(2)}),\qquad O_0^{(1)}:=\sigma^{(3)}, \tag{131}$$

where $\sigma^{(i)}$ are the Pauli matrices. The direct product $|\frac{1}{2}m\rangle\langle\frac{1}{2}\bar{m}|$ which is an operator in the nucleon's spin space, can always be expressed in terms of the four linearly independent matrices (131). On the basis of its behavior under rotations it must be given by the inverse Clebsch–Gordan series

$$|\tfrac{1}{2}m\rangle\langle\tfrac{1}{2}\bar{m}|=\sum_{\lambda=0}^{1}\beta_\lambda(-)^{1/2-\bar{m}}(\tfrac{1}{2}m,\tfrac{1}{2}-\bar{m}|\lambda\nu)O_\nu^{(\lambda)}, \tag{132}$$

with $\nu=m-\bar{m}$. It is not difficult to prove that the factors β_λ must be $\beta_0=\beta_1=1/\sqrt{2}$ if the states $|\frac{1}{2}m\rangle$ are normalized to 1 (see exercise 7).

For the same reasons the operator $S_\mu^\dagger S_{\bar{\mu}}$ must be given by an expression of the form

$$S_\mu^\dagger S_{\bar{\mu}}=\sum_{\lambda=0}^{1}\gamma_\lambda(-)^{1-\mu}(1-\mu,1\bar{\mu}|\lambda\nu)O_\nu^{(\lambda)}. \tag{133}$$

The constants γ_λ are determined as follows. Consider first

$$\sum_m\langle\tfrac{1}{2}m|S_\mu^\dagger S_\mu|\tfrac{1}{2}m\rangle=2\gamma_0(-)^{1-\mu}(1-\mu,1\mu|00)=\frac{2}{\sqrt{3}}\gamma_0.$$

On the other hand, using eq. (130a) and the completeness relation of the Clebsch–Gordan coefficients this is equal to 4/3. Thus,

$$\gamma_0=\frac{2}{\sqrt{3}}\,. \tag{134a}$$

Similarly, taking the difference of the expectation values of, say, $S_1^\dagger S_1$ in the states $m=\frac{1}{2}$ and $m=-\frac{1}{2}$ projects out the vector term in eq. (133), $\lambda=1$:

$$\langle \tfrac{1}{2}\tfrac{1}{2}|S_{+1}^\dagger S_{+1}|\tfrac{1}{2}\tfrac{1}{2}\rangle - \langle \tfrac{1}{2}-\tfrac{1}{2}|S_{+1}^\dagger S_{+1}|\tfrac{1}{2}-\tfrac{1}{2}\rangle = 2\gamma_1(1-1,11|10) = -\sqrt{2}\,\gamma_1 .$$

From eq. (130a) this same expression is equal to

$$\tfrac{4}{3}\left\{(\tfrac{3}{2}\tfrac{3}{2},\tfrac{1}{2}-\tfrac{1}{2}|11)^2 - (\tfrac{3}{2}\tfrac{1}{2},\tfrac{1}{2}\tfrac{1}{2}|11)^2\right\} = \tfrac{2}{3}.$$

Thus,

$$\gamma_1 = -\frac{\sqrt{2}}{3}. \tag{134b}$$

An equivalent representation of eqs. (133) and (134) is this: Instead of the spherical components S_μ introduce Cartesian coordinates,

$$S_1 = \frac{1}{\sqrt{2}}(S_{-1} - S_{+1}), \qquad S_2 = \frac{\mathrm{i}}{\sqrt{2}}(S_{-1} + S_{+1}), \qquad S_3 = S_0 .$$

Then one obtains the relation

$$S_i^\dagger S_j = \tfrac{2}{3}\delta_{ij}\mathbb{1} - \frac{\mathrm{i}}{3}\varepsilon_{ijk}\sigma^{(k)} \quad \text{(Cartesian coord.)} \tag{135}$$

(see exercise 8).

Finally, we write out the matrix representation of the operators S_μ in the space of the nucleon spinors:

$$S_{\mu=+1} = \begin{pmatrix} -1 & 0 \\ 0 & -\dfrac{1}{\sqrt{3}} \\ 0 & 0 \\ 0 & 0 \end{pmatrix}, \qquad S_{\mu=0} = \begin{pmatrix} 0 & 0 \\ \sqrt{\frac{2}{3}} & 0 \\ 0 & \sqrt{\frac{2}{3}} \\ 0 & 0 \end{pmatrix}, \qquad S_{\mu=-1} = \begin{pmatrix} 0 & 0 \\ 0 & 0 \\ \dfrac{1}{\sqrt{3}} & 0 \\ 0 & 1 \end{pmatrix}. \tag{136}$$

Clearly, the analogous isospin transition operators T_μ are determined by essentially the same formulae in isospin space.

As an example for the practical application of these results we consider pion–nucleon scattering via the Δ-resonance. For the sake of simplifying the notation we suppress the isospin degrees of freedom. The reader should have no difficulty in inserting them at will.

In the isobar approximation pion–nucleon scattering is described by the Lippmann–Schwinger equation

$$T(E) = V + V\frac{1}{E - T_\pi - T_\mathrm{N} + \mathrm{i}\varepsilon}T(E), \tag{137}$$

where V is constructed from the diagram of Fig. 11 and the vertex coupling (125). In

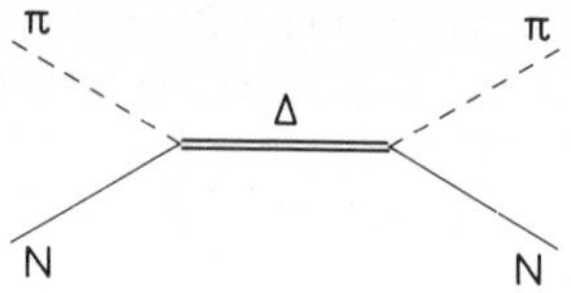

Fig. 11. Intermediate resonance state in pion–nucleon scattering.

momentum space we have

$$\langle \boldsymbol{\kappa}'|V|\boldsymbol{\kappa}\rangle = \frac{f^2_{\pi N\Delta}}{m^2_\pi}\frac{h(\kappa'^2)h(\kappa^2)(\boldsymbol{\kappa}'\cdot \boldsymbol{S}^\dagger)(\boldsymbol{\kappa}\cdot \boldsymbol{S})}{E-m(\Delta)}, \tag{138}$$

where the vertex function $h(\kappa^2)$ is introduced by hand, with the condition $h(0)=1$. Eq. (137) can be solved analytically and one finds

$$\langle \boldsymbol{\kappa}'|T(E)|\boldsymbol{\kappa}\rangle = \frac{f^2_{\pi N\Delta}}{m^2_\pi}h(\kappa'^2)h(\kappa^2)\left[\frac{2}{3}\boldsymbol{\kappa}'\cdot\boldsymbol{\kappa} - \frac{\mathrm{i}}{3}\boldsymbol{\sigma}\cdot(\boldsymbol{\kappa}'\wedge\boldsymbol{\kappa})\right]\Big/D(E), \tag{139}$$

where

$$D(E) = E - m(\Delta) - \frac{f^2_{\pi N\Delta}}{3m^2_\pi}\int\frac{\mathrm{d}^3q}{(2\pi)^3}\frac{q^2h(q^2)}{E-q^2/2\mu+\mathrm{i}\varepsilon}, \tag{140}$$

μ being the reduced mass. $D^{-1}(E)$ is the dressed propagator of the Δ-resonance.

References

Agassi, D. and A. Gal, 1973, *Ann. Phys.* (N.Y.) 75, 56.
Bég, M.A.B., 1961, *Ann. Phys.* (N.Y.) 13, 110.
Eisenberg, J.M., J. Hüfner, and E. Moniz, 1973, *Phys. Lett.* 47B, 381.
Ericson, M. and T.E.O. Ericson, 1966, *Ann. Phys.* (N.Y.) 36, 323.
Fäldt, G., 1973, *Nucl. Phys.* A206, 176.
Foldy, L.L. and J.D. Walecka, 1969, *Ann. Phys.* (N.Y.) 54, 447.
Glauber, R.J., 1959, *Boulder Lectures in Theoretical Physics*, Vol. 1 (Interscience, New York) p. 315.
Gottfried, K., 1971, *Ann. Phys.* 66, 868.
Hirata, M., J. Koch, F. Lenz, and E. Moniz, 1979, *Ann. Phys.* (N.Y.) 120, 205.
Hüfner, J., 1973, *Nucl. Phys.* B58, 55.
Kerman, A.K., H. McManus, and R.M. Thaler, 1959, *Ann. Phys.* (N.Y.) 8, 551.
Kisslinger, L., 1965, *Phys. Rev.* 98, 761.
Lax, M., 1951, *Rev. Mod. Phys.* 23, 287.
Lenz, F., 1975, *Ann. Phys.* (N.Y.) 95, 348.
Locher, M., O. Steinmann, and N. Straumann, 1971, *Nucl. Phys.* B27, 598.
Moniz, E., 1981, *Nucl. Phys.* A354, 535.
Moniz, E., 1977, *Lectures at Summer School on Nuclear Physics, Les Houches* (North-Holland, Amsterdam).
Scheck, F. and C. Wilkin, 1972, *Nucl. Phys.* B49, 541.
Sugar, R. L. and R. Blankenbecler, 1969, *Phys. Rev.* 183, 1387.
Wallace, S., 1973, *Phys. Rev.* D8, 1846.
Wu, T.T., 1957, *Phys. Rev.* 108, 466.

Exercises

1. Consider the stationary Klein–Gordon equation with a Lorentz scalar potential V and another potential W_0 which is the fourth component of a vector,

$$\left\{\Delta+(E-W_0)^2-m^2-2mV\right\}\psi(\boldsymbol{r})=0.$$

Show that in the nonrelativistic limit this goes over into the Schrödinger equation with potential $V+W_0$.

2. In the s-state wave function of a two-particle system perform the separation into c.m. and relative coordinates and show that the center-of-mass condition leads to the factor $\exp\{q^2/8\alpha^2\}$ in the elastic amplitude F_{ii}. Do the same for A particles, making use of Jacobi coordinates.

3. In the model of sec. 2.2.(iii) show that the function

$$S(A,x)=\sum_{m=1}^{A}\binom{A}{m}\frac{(-)^{m+1}x^m}{m}$$

can be written as an integral,

$$S=\int_0^x\frac{\mathrm{d}x'}{x'}\left[1-(1-x')^A\right].$$

Substituting $y=1-x'$ and splitting the integral over y into $\int_0^1\mathrm{d}y-\int_0^{1-x}\mathrm{d}y$ leads to the representation given in the text.

4. Calculate and discuss the two-body correlation function $C(\boldsymbol{x},\boldsymbol{y})$ for a two-particle state which contains only the correlations due to the Pauli principle,

$$\psi(\boldsymbol{x},\boldsymbol{y})=\frac{1}{\sqrt{2}}\{\phi_1(\boldsymbol{x})\phi_2(\boldsymbol{y})-\phi_1(\boldsymbol{y})\phi_2(\boldsymbol{x})\}.$$

5. In the derivation leading to eq. (95) calculate the integral J^{ji} in the approximation of $C(|\boldsymbol{z}|)$ being very short-ranged.

6. Study the kinematics of pion–nucleon scattering at the resonance. In particular, calculate momentum and kinetic energy of the pion, in the laboratory system, at which the Δ-resonance occurs.

7. Derive the factors β_λ of eq. (132).

8. Derive eq. (135) from the results (133) and (134).

9. Consider the amplitude $t_l(k):=kf_l(k)$ with $f_l(k)=(2\mathrm{i}k)^{-1}(\eta_l\mathrm{e}^{2\mathrm{i}\varepsilon_l}-1)$ as defined in sec. 1.2. Show that the complex quantity t_l lies either on or inside the *unitarity circle* $(\mathrm{Re}\,t_l)^2+(\mathrm{Im}\,t_l-\frac{1}{2})^2=\frac{1}{4}$. For a partial wave with $l>0$, $\delta_l(k=0)$ vanishes. Assume that $\eta_l(0)=1$, that the partial wave amplitude t_l has a resonance at some energy E_r, i.e. $\delta_l(k_\mathrm{r})=\pi/2$ and that in the neighbourhood of this resonance the amplitude has the form

$$t_l(k)=\frac{\Gamma_{\mathrm{el}}/2}{E-E_\mathrm{r}+\mathrm{i}\Gamma/2}$$

(Γ_{el}: elastic width, Γ: total width). Express Γ_{el} in terms of Γ and η. Write the partial cross sections (14) in terms of $x:=\Gamma_{\mathrm{el}}/\Gamma$ and $\varepsilon:=2(E-E_\mathrm{r})/\Gamma$ and discuss their behaviour in the neighbourhood of the resonance. In particular draw $\sigma_{\mathrm{tot}}^l(E)$ around the point $E=E_\mathrm{r}$ and identify the width of the resonance at half the maximum of the peak.

10. If $\pi^+ p$ scattering at the Δ-resonance proceeds exclusively via the two-step process

$$\pi^+ p \to \Delta^{++} \to \pi^+ p,$$

show that the differential cross section in the c.m. system is symmetric with respect to 90°. *Hints*: Use parity conservation and consider the two spin components (along the initial momentum) of the incoming proton separately.

11. Show: if parity is conserved the excited 1^+ level of ^{20}Ne cannot decay to the ground state of ^{16}O by emitting an α-particle.

12. Rederive eq. (108) by starting from eqs. (64) and (108′).

Chapter II

HADRONIC ATOMS

Any long-lived particle with negative charge can be captured in the Coulomb field of an atomic nucleus thus forming a hydrogenlike *exotic atom*. The particle which initially may be trapped in a state with high principal quantum number n runs through its atomic cascade, primarily by emitting electrons of the host atom through the Auger effect, or by emitting atomic X-rays, depending on the transition energy between two successive atomic levels. One can estimate that this cascade takes between $\tau_c = 10^{-15}$ and 10^{-19} seconds, depending on the nuclear charge number Z.*) Therefore, the lifetime τ of the trapped particle must be long as compared to the cascade time τ_c. As far as hadrons are concerned, these must be particles which are either stable or which decay via weak interactions. Some candidates are listed in table 1.

All these particles have masses m_i which are large as compared to m_e, the electron mass. As the Bohr radius is inversely proportional to the mass, the spatial size of exotic atoms is scaled down by the mass ratio m_e/m_i. Their orbits quickly fall inside the innermost electron shell of the host atom. For instance, for a circular orbit ($l = n - 1$), one has

$$\sqrt{r^2} \simeq n^2 a_B^{(i)}.$$

For this to lie inside the electronic 1s-shell, we must have

$$n^2 a_B^{(i)} \lesssim a_B^{(e)} \quad \text{or} \quad n \lesssim \sqrt{m_i/m_e}\,. \tag{1}$$

For a pion, for example, this means that the pionic orbits with $n \leq 16$ are localized in the interior of the K-shell of the host atom and, therefore, are hydrogenlike to a good approximation.

Yet, the energy levels of these exotic atoms**) deviate from the energy levels in a point-like Coulomb field for two reasons:

(i) They are shifted upwards (i.e. towards lesser binding) because the nuclear charge distribution $\rho(\boldsymbol{r})$ is not point-like.

(ii) The levels experience an additional shift, either attractive or repulsive, due to the strong interaction of the bound hadron with the nucleons. At the same time the atomic levels are broadened by all those processes which lead to disappearance of the trapped particle when it interacts strongly with the nucleus.

In all practical situations the level shift due to strong interaction is sizeably larger than the shift due to the finite size of the nuclear charge distribution.

*) The time scales of an exotic atom are estimated in more detail in the context of muonic atoms, see below, sec. 6.2 of Chapter IV.

**) This term was first introduced in a lecture at the SIN summer school in Leysin, in 1969, and quickly became the generic name for muonic, mesonic and hadronic atoms.

Table 1
Quantum numbers (isospin I and strangeness S), lifetimes, masses and Bohr radii of some negatively charged, quasi-stable, hadrons that can form exotic atoms in matter. The masses are in units of $m_e = 0.511\,003\,4(14)$ keV.

Particle	I	S	τ [sec]	m/m_e	a_∞ [fm]
π^-	1	0	2.603×10^{-8}	273.12	$193.8/Z$
K^-	$\frac{1}{2}$	-1	1.237×10^{-8}	966.08	$54.77/Z$
$\bar{p}$	$\frac{1}{2}$	0	$(> 8 \times 10^{30}$ yr)	1836.15	$28.82/Z$
Σ^-	1	-1	1.482×10^{-10}	2343.11	$22.58/Z$
Ξ^-	$\frac{1}{2}$	-2	1.64×10^{-10}	2585.7	$20.47/Z$
Ω^-	0	-3	0.82×10^{-10}	3272.4	$16.17/Z$

Regarding the level broadening we can argue in a qualitative manner as follows: As the strong interactions are of short range the absorption is proportional to the overlap of the atomic probability density $|\psi_{nl}|^2$ and the nuclear mass density $\rho_A(\boldsymbol{r})$. High-lying orbits which have very little overlap with the nucleus are not affected by the strong interaction and their width is dominated by the electromagnetic decay channels (Auger and X-ray transitions). As n and l decrease the overlap of $|\psi_{nl}|^2$ and $\rho_A(\boldsymbol{r})$ increases rapidly, up to a point where the strong interaction width becomes comparable and then even larger (as we lower n and l) than the electromagnetic width. Where this crossing happens depends primarily on Z, the nuclear charge number, and on the particle's mass m_i. As a consequence, the atomic 1s-state is not observable but in the lightest atoms. Similarly, the 2p-state is observable only in light and medium-weight systems. Its strong interaction width increases rapidly with Z and beyond a critical value of Z the state becomes too broad to be detectable; etc. (Examples will be given below.)

The shifts and absorption widths of atomic levels with hadrons are the quantities of primary interest in this chapter. These effects reflect the strong interactions of the trapped particle with the nucleons and are derived from the multiple scattering of the particle within the nuclear medium. Multiple scattering in this case has a number of unique features.

(i) The trapped hadron is bound in a hydrogenlike state; thus, the multiple scattering occurs practically at zero energy or, more precisely, somewhat *below* the threshold for free particle–nucleon scattering.

(ii) In the exotic atom the Coulomb field is the dominant feature which sets the scale and the pattern of the atomic spectrum. The strong interaction effects are measured with reference to that Coulomb pattern.

(iii) In contrast to scattering at high energies only the first few partial waves of the elementary amplitude are important.

(iv) The hadron moves in a state of definite angular momentum l (with respect to the center-of-mass of the total system particle plus nucleus). As a consequence, atomic states whose angular momentum differs, probe the elementary partial wave amplitudes with different relative weights. Furthermore, if the nucleus has nonvanishing multipole moments such states exhibit hyperfine structure—another source of information on the particle–nucleus interaction. We study primarily the case of pionic and kaonic atoms. The cases of antiprotons, Σ^-, etc., bound in an atom are touched upon only briefly. As spin and isospin are important in this context we start by analyzing the spin–isospin structure of the elementary amplitude. After a discussion of the hydrogen spectrum in the framework of the Klein–Gordon equation we analyze the strong interaction shifts and widths by means of multiple scattering theory and optical potentials as developed in Chapter I. The theory is extended to the case of exotic atoms whose nuclei have nonvanishing multipole moments and the case of the strong interaction quadrupole hyperfine structure is analyzed in some detail.

1. The elementary pion–nucleon and kaon–nucleon amplitudes at low energies

1.1. Isospin analysis of pion–nucleon amplitudes

The nucleons carry isospin $I^{\mathrm{N}} = 1/2$, the 3-component being $I_3^{\mathrm{N}} = +1/2$ for the proton, and $I_3^{\mathrm{N}} = -1/2$ for the neutron. The pion carries isospin $I^{\pi} = 1$ and it occurs in three charge states π^+, π^0, π^- which have $I_3^{\pi} = +1, 0, -1$, respectively. Suppose we wish to compare different charge combinations of one pion $(1, I_3^{\pi})$ and one nucleon $(\frac{1}{2}, I_3^{\mathrm{N}})$ in one and the same dynamical state. Irrespective of what that (spin and orbital) state is we can always expand a given combination of charge states in terms of eigenstates of the coupled isospin $\boldsymbol{I} = \boldsymbol{I}^{\pi} + \boldsymbol{I}^{\mathrm{N}}$ and its 3-component I_3,

$$|1 I_3^{\pi}; \tfrac{1}{2} I_3^{\mathrm{N}}\rangle = \sum_{I} \left(1 I_3^{\pi}, \tfrac{1}{2} I_3^{\mathrm{N}} | I I_3\right) | I I_3\rangle, \tag{2}$$

with $I_3 = I_3^{\pi} + I_3^{\mathrm{N}}$ and where I is $\frac{1}{2}$ or $\frac{3}{2}$. Inserting the Clebsch–Gordan coefficients in eq. (2), we have more explicitly,

$$|\pi^- \mathrm{p}\rangle = -\sqrt{\tfrac{2}{3}}\, |\tfrac{1}{2}\ -\tfrac{1}{2}\rangle + \sqrt{\tfrac{1}{3}}\, |\tfrac{3}{2}\ -\tfrac{1}{2}\rangle, \tag{3a}$$

$$|\pi^- \mathrm{n}\rangle = |\tfrac{3}{2}\ -\tfrac{3}{2}\rangle, \tag{3b}$$

$$|\pi^0 \mathrm{n}\rangle = \sqrt{\tfrac{1}{3}}\, |\tfrac{1}{2}\ -\tfrac{1}{2}\rangle + \sqrt{\tfrac{2}{3}}\, |\tfrac{3}{2}\ -\tfrac{1}{2}\rangle, \tag{3c}$$

and similar expressions for $|\pi^0 \mathrm{p}\rangle$ and the states involving positive pions. So far this is no more than expressing one complete basis in terms of another. As the states $|I I_3\rangle$ cannot be prepared in the laboratory (in contrast to coupled spin states), this expansion is a formal one. The coupled basis becomes important and useful, however, if the pion–nucleon interaction is invariant under unitary transformations in isospin space. Consider, for example, an elastic or charge exchange scattering amplitude at given and fixed kinematical conditions (including spin),

$$\langle \pi_1 N_1 | T | \pi_2 N_2 \rangle,$$

where T is the scattering matrix, and let us assume that T is a scalar under rotations in isospin space. This assumption says that any matrix element of T, when expressed in the coupled basis, must be diagonal not only in I_3 (charge conservation) but also in the eigenvalue of I

$$\langle I' I_3' | T | I I_3 \rangle \propto \delta_{II'} \delta_{I_3 I_3'}.$$

As I can only be $\frac{1}{2}$ or $\frac{3}{2}$, there are two amplitudes of this kind. For example, the amplitude with a π^- in the initial state can be expressed in terms of these, using the expansions (3), as follows

$$\langle \pi^- p | T | \pi^- p \rangle = \tfrac{2}{3} \left(\tfrac{1}{2} | T | \tfrac{1}{2} \right) + \tfrac{1}{3} \left(\tfrac{3}{2} | T | \tfrac{3}{2} \right).$$

$$\langle \pi^0 n | T | \pi^- p \rangle = -\frac{\sqrt{2}}{3} \left(\tfrac{1}{2} | T | \tfrac{1}{2} \right) + \frac{\sqrt{2}}{3} \left(\tfrac{3}{2} | T | \tfrac{3}{2} \right),$$

$$\langle \pi^- n | T | \pi^- n \rangle = \left(\tfrac{3}{2} | T | \tfrac{3}{2} \right).$$

The three physical amplitudes are expressed in terms of two independent amplitudes with good I and, therefore, there must exist a relation between them. This is

$$\langle \pi^- n | T | \pi^- n \rangle - \langle \pi^- p | T | \pi^- p \rangle = \sqrt{2} \langle \pi^0 n | T | \pi^- p \rangle. \tag{4}$$

This relation holds provided the dynamic, spin and orbital, state of the pion and the nucleon is the same in the three amplitudes. For example, we may apply it (as we shall do below) to a partial wave amplitude of given angular momentum J and at a given energy E. Clearly, a relation of this type can only hold, at best, for the piece of the amplitude due to the strong interactions. It must break down if the Coulomb force between the pion and the nucleon and the differences between the masses of the various charge states are taken into account.

1.2. Isospin analysis of kaon–nucleon amplitudes

For the case of atoms with a K^--meson which carries isospin $I^K = \frac{1}{2}$ and strangeness $S = -1$ only the states $|K^- N\rangle$ and $|\overline{K^0} N\rangle$ matter. The isospin decomposition of these states is

$$|K^- p\rangle = \sum_I \left(\tfrac{1}{2}\ -\tfrac{1}{2}, \tfrac{1}{2} \tfrac{1}{2} | I0 \right) | I0 \rangle = -\frac{1}{\sqrt{2}} |00\rangle + \frac{1}{\sqrt{2}} |10\rangle, \tag{5a}$$

$$|K^- n\rangle = |1\ -1\rangle, \tag{5b}$$

$$|\overline{K^0} n\rangle = \frac{1}{\sqrt{2}} |00\rangle + \frac{1}{\sqrt{2}} |10\rangle. \tag{5c}$$

[The same formulae hold for the states $|K^+ N\rangle$ and $|K^0 N\rangle$, however, with strangeness $S = +1$.] As before, the recoupling of the $\boldsymbol{I}^K$ and $\boldsymbol{I}^N$ to $\boldsymbol{I}$ is formal. It is useful, provided the scattering matrix is an isospin scalar, for deriving relations between different charge states of a given dynamical amplitude. For example, the elastic and charge exchange amplitudes of the $\overline{\mathrm{K}}$N-system are expressed in terms of the two independent amplitudes with good I, by the relations

$$\langle \mathrm{K^-p}|T|\mathrm{K^-p}\rangle = \tfrac{1}{2}\{(0|T|0) + (1|T|1)\},$$

$$\langle \mathrm{K^-n}|T|\mathrm{K^-n}\rangle = \langle \overline{\mathrm{K}^0}\mathrm{p}|T|\overline{\mathrm{K}^0}\mathrm{p}\rangle = (1|T|1),$$

$$\langle \overline{\mathrm{K}^0}\mathrm{n}|T|\mathrm{K^-p}\rangle = -\tfrac{1}{2}\{(0|T|0) - (1|T|1)\}.$$

T being an isoscalar the nondiagonal amplitude $(0|T|1)$ vanishes. Therefore there is a relation between the three physical amplitudes which reads

$$\langle \mathrm{K^-n}|T|\mathrm{K^-n}\rangle - \langle \mathrm{K^-p}|T|\mathrm{K^-p}\rangle = \langle \overline{\mathrm{K}^0}\mathrm{n}|T|\mathrm{K^-p}\rangle. \tag{6}$$

*1.3. Unitarity and optical theorem**)

In a general covariant theory of scattering the S-matrix is written (symbolically) as the sum

$$S = \mathbb{1} + R \tag{7}$$

of the identity $\mathbb{1}$ and the reaction matrix R. The identity stands for the case of "no scattering", i.e. a situation where the projectile would pass by the target without interacting at all. The reaction matrix R describes the nontrivial part of the scattering process and, therefore, is the dynamic quantity from which the cross section is derived. Suppose we wish to describe the transition from an initial asymptotic state i to some final asymptotic state f by means of the matrix element R_{fi}. This process conserves energy and momentum so that we can factor out a δ-distribution taking account of the requirement that the total four-momenta P_{i} and P_{f} must be equal,

$$R_{\mathrm{fi}} = \mathrm{i}(2\pi)^4\delta^{(4)}(P_{\mathrm{f}} - P_{\mathrm{i}})T_{\mathrm{fi}}. \tag{8}$$

With the factors i and $(2\pi)^4$, which are a matter of convention, this equation *defines* the T-matrix. Note that in a Lorentz covariant theory and employing the covariant normalization of single particle states,

$$\langle \boldsymbol{p}'|\boldsymbol{p}\rangle = 2E_p\delta^{(3)}(\boldsymbol{p}' - \boldsymbol{p}), \tag{9}$$

the T-matrix is invariant under Lorentz transformations. [It is *not* identical with the nonrelativistic scattering matrix introduced in Chap. I. It is related to it, however, in the center-of-mass frame.]

If the basis of asymptotic free states is complete, the S-matrix is unitary, $S^\dagger S = \mathbb{1}$. Unitarity implies the following conditions for R and T, respectively:

$$R + R^\dagger = -R^\dagger R, \tag{10a}$$

$$\mathrm{i}(T^\dagger - T) = (2\pi)^4 T^\dagger T. \tag{10b}$$

*) The reader who is familiar with the optical theorem and other consequences of the unitary of the S-matrix, may wish to go directly to eqs. (14), (15) and (17) which are relevant for the following sections.

For example, for the problem of elastic scattering of two particles with momenta p and q, the relation (10) reads more explicitly

$$\begin{aligned}
&\mathrm{i}\{T^*(p,q;p',q')-T(p',q';p,q)\}\\
&\quad=(2\pi)^4\sum_n\sum_{\text{spins}}\Pi^{(n)}\delta(p+q-k_1-k_2\ldots-k_n)\\
&\qquad\times T^*(k_1,\ldots,k_n;p'q')T(k_1,\ldots,k_n;pq),
\end{aligned}\tag{11}$$

where the symbol $\Pi^{(n)}$ stands for the $3n$-fold integral

$$\Pi^{(n)}\equiv\int\frac{\mathrm{d}^3k_1}{2E_1}\int\frac{\mathrm{d}^3k_2}{2E_2}\cdots\int\frac{\mathrm{d}^3k_n}{2E_n}\tag{12}$$

over the phase space of n particles in the intermediate state. Relation (11) is obtained by taking eqs. (10) between the states $|\mathrm{i}\rangle=|p,q\rangle$ and $|f\rangle=|p',q'\rangle$ and by inserting a complete set of asymptotic states (i.e. *on-shell* states) between $R^\dagger$ and R. The sum over n runs over all intermediate states which are kinematically allowed for given values of total energy and momentum in the initial state. For example, in pion–nucleon scattering below the first inelastic channels, the only allowed intermediate states are again two-particle states of one nucleon and one pion $|p'',q''\rangle$ with $p''+q''=p+q$. In this case eq. (11) reduces to the relation of so-called *elastic unitarity*, viz.

$$\begin{aligned}
&\mathrm{i}\{T^*(p,q;p',q')-T(p',q';p,q)\}\\
&\quad=(2\pi)^4\sum_{\text{spins}}\int\frac{\mathrm{d}^3p''}{2E_{p''}}\int\frac{\mathrm{d}^3q''}{2E_{q''}}\delta^{(4)}(p+q-p''-q'')\\
&\qquad\times T^*(p'',q'';p',q')T(p'',q'';p,q).
\end{aligned}\tag{13}$$

Here only the elastic channel (including charge exchange) is open. The relation (13) can be transformed and simplified if it is evaluated in the center-of-mass system where $\boldsymbol{p}+\boldsymbol{q}=0$. Let $\boldsymbol{\kappa}$, $\boldsymbol{\kappa}'$ be the three-momenta before and after the collision, κ their magnitude, so that $\kappa=|\boldsymbol{q}|=|\boldsymbol{p}|=|\boldsymbol{q}'|=|\boldsymbol{p}'|$. We substitute

$$\begin{aligned}
&\boldsymbol{Q}:=\tfrac{1}{2}(\boldsymbol{q}''-\boldsymbol{p}''),\\
&\boldsymbol{P}:=\boldsymbol{p}''+\boldsymbol{q}'',\\
&\mathrm{d}^3p''\,\mathrm{d}^3q''=\mathrm{d}^3Q\,\mathrm{d}^3P,
\end{aligned}$$

and perform first the integral over d^3P. The spatial part of the delta distribution gives the condition $\boldsymbol{P}=0$ or, equivalently, $\boldsymbol{p}''=-\boldsymbol{q}''$ and $\boldsymbol{Q}=\boldsymbol{q}''$. Let $x:=|\boldsymbol{Q}|$. We are left with the integral

$$\begin{aligned}
&\int x^2\,\mathrm{d}x\int\mathrm{d}\Omega_Q\frac{1}{4\sqrt{x^2+m_\pi^2}\sqrt{x^2+m_\mathrm{N}^2}}\\
&\quad\times\delta\left(\sqrt{x^2+m_\pi^2}+\sqrt{x^2+m_\mathrm{N}^2}-\sqrt{\kappa^2+m_\pi^2}-\sqrt{\kappa^2+m_\mathrm{N}^2}\right),
\end{aligned}$$

which gives

$$\int \mathrm{d}\Omega_Q \frac{\kappa}{4\left[\sqrt{\kappa^2+m_\pi^2}+\sqrt{\kappa^2+m_\mathrm{N}^2}\right]} \times (\text{integrand taken at } x=\kappa).$$

Eq. (13) then reads

$$2\,\mathrm{Im}\,T(\boldsymbol{\kappa}',\boldsymbol{\kappa}) = (2\pi)^4 \frac{\kappa}{4E} \sum_{\text{spins}} \int \mathrm{d}\Omega_{\kappa''} T^*(\boldsymbol{\kappa}',\boldsymbol{\kappa}'') T(\boldsymbol{\kappa}'',\boldsymbol{\kappa}), \tag{14a}$$

where E is the total energy in the c.m. system and where the sum runs over the spin orientations of the intermediate state. This same elastic unitarity relation, when expressed in terms of the c.m. scattering amplitude $f(\boldsymbol{\kappa}',\boldsymbol{\kappa}) \equiv f(E,\theta)$, reads

$$\mathrm{Im}\, f(\boldsymbol{\kappa}',\boldsymbol{\kappa}) = \frac{\kappa}{4\pi} \sum_{\text{spins}} \int \mathrm{d}\Omega_{\kappa''} f^*(\boldsymbol{\kappa}',\boldsymbol{\kappa}'') f(\boldsymbol{\kappa}'',\boldsymbol{\kappa}). \tag{14b}$$

This follows, indeed, from the expression (I.9) for σ_{el}, the integrated elastic cross section, and the form (I.11) of the optical theorem: In the kinematic domain where elastic unitarity applies, $\sigma_{\text{tot}} = \sigma_{\text{el}}$. It suffices then to take eq. (14b) in the forward direction where $\boldsymbol{\kappa}' = \boldsymbol{\kappa}$.

By comparing eqs. (14a) and (14b) we deduce the following relation between the scattering amplitude in the c.m. system and the corresponding T-matrix element:

$$f(\boldsymbol{\kappa}',\boldsymbol{\kappa}) \equiv f(E,\theta) = \frac{8\pi^5}{E} T(\boldsymbol{\kappa}',\boldsymbol{\kappa}). \tag{15}$$

In kaon–nucleon scattering there are inelastic channels open right at threshold. For instance,

$$\mathrm{K}^-\mathrm{p} \to \mathrm{Y}^0\mathrm{n},$$

where Y^0 is a hyperon at mass 1405 MeV with strangeness -1. Here, very much like in pion–nucleon scattering above the first threshold for inelastic channels, the unitarity relation must be applied in its general form (11). A particularly important special case is obtained by applying eq. (11) to the forward scattering amplitude, $q' = q$, $p' = p$. In this case the right-hand side of eq. (11) is proportional to the total cross section. From eq. (B.3) of App. B

$$\begin{aligned} &\sigma(p+q\to n) \\ &= \frac{(2\pi)^{10}}{4\kappa E} \int \frac{\mathrm{d}^3k_1}{2E_1} \cdots \int \frac{\mathrm{d}^3k_n}{2E_n} \\ &\quad \times \sum_{\text{spins}} \delta(k_1+k_2+\cdots+k_n-p-q)|T(k_1,\ldots,k_n;\,p,q)|^2. \end{aligned} \tag{16}$$

The factor κE stems from the Møller factor, evaluated in the c.m. frame. As before, E is the total energy, κ the modulus of the three-momentum, in the c.m. frame. Upon comparison with eq. (11) we find

$$\mathrm{Im}\,T(p,q;\,p,q) = \frac{2\kappa E}{(2\pi)^6} \sum_n \sigma(p+q\to n) = \frac{2\kappa E}{(2\pi)^6}\sigma_{\text{tot}}. \tag{17a}$$

Finally, inserting rel. (15), we obtain the familiar form for the optical theorem

$$\operatorname{Im} f(E, \theta=0)=\frac{\kappa}{4\pi}\sigma_{\text{tot}}(E) \tag{17b}$$

that we have used in Chap. I, eq. (I.11).

Notice that σ_{tot} and the T-matrix elements are Lorentz scalars. Therefore, the optical theorem, in its form (17a), should be written in a manifestly invariant form. The energy E is just the square root of the invariant $s=(p+q)^2$. The c.m. momentum κ is given by

$$\kappa=\frac{1}{2\sqrt{s}}\sqrt{\left(s-m_1^2-m_2^2\right)^2-4m_1^2m_2^2},$$

where m_1 and m_2 are the masses of the incoming particles. Thus, eq. (17) reads

$$\operatorname{Im} T(p, q; p, q)=\frac{1}{(2\pi)^6}\sqrt{\left(s-m_1^2-m_2^2\right)^2-4m_1^2m_2^2}\,\sigma_{\text{tot}}. \tag{17c}$$

Clearly, $f(E,\theta)$ refers to a specific Lorentz frame and cannot be a Lorentz scalar.

Addendum. Let us return to the relation (14b) of elastic unitarity and let us assume that $f(\boldsymbol{\kappa}', \boldsymbol{\kappa})$ describes the scattering of spinless particles. In this case the amplitude can be expanded in terms of Legendre polynomials, cf. eq. (I.8). Making use of the addition theorem

$$P_l(\hat{\boldsymbol{\kappa}}'\cdot\hat{\boldsymbol{\kappa}}'')=\frac{4\pi}{2l+1}\sum_{m=-l}^{+l} Y_{lm}^*(\hat{\boldsymbol{\kappa}}')Y_{lm}(\hat{\boldsymbol{\kappa}}'') \tag{18}$$

and using the orthogonality of spherical harmonics, eq. (14b) gives

$$\begin{aligned}
&\sum_l (2l+1)\operatorname{Im} f_l(\kappa)P_l(\hat{\boldsymbol{\kappa}}'\cdot\hat{\boldsymbol{\kappa}})\\
&\quad=4\pi\kappa\sum_{lm}\sum_{l'm'} f_l^*(\kappa)f_{l'}(\kappa)Y_{lm}^*(\hat{\boldsymbol{\kappa}}')Y_{l'm'}(\hat{\boldsymbol{\kappa}})\delta_{ll'}\delta_{mm'}\\
&\quad=\kappa\sum_l (2l+1)|f_l(\kappa)|^2P_l(\hat{\boldsymbol{\kappa}}'\cdot\hat{\boldsymbol{\kappa}}).
\end{aligned}$$

This relation holds for every l independently. Thus,

$$\operatorname{Im} f_l(\kappa)=\kappa|f_l(\kappa)|^2. \tag{19}$$

Similarly, it is not difficult to see that in general unitarity implies the conditions (for the spinless case)

$$\operatorname{Im} f_l(\kappa)\geq\kappa|f_l(\kappa)|^2. \tag{20}$$

This is rel. (I.12) of the first Chapter. The equality sign holds for elastic unitarity.

1.4. Partial wave expansion of meson–nucleon amplitude

Let $f(E,\theta)$ be the meson–nucleon scattering amplitude in the c.m. frame. Its expansion in terms of Legendre polynomials cannot have the simple form (I.8) because, in general, the partial wave amplitudes depend on the spin of the nucleon. A given relative orbital angular momentum l can be coupled to either $j_1 = l + \frac{1}{2}$ or $j_2 = l - \frac{1}{2}$. Accordingly, the partial wave amplitudes must be classified according to j, the total angular momentum. We use the notation

$$f_{l+}(E) := f_{j_1}(E) \quad \text{for } j_1 = l + \tfrac{1}{2},$$

$$f_{l-}(E) := f_{j_2}(E) \quad \text{for } j_2 = l - \tfrac{1}{2},$$

and introduce projection operators onto the states of good j (see exercise 15),

$$\Pi_{j_1} = \frac{1}{2l+1}(l + 1 + \boldsymbol{\sigma} \cdot \boldsymbol{l}), \tag{21a}$$

$$\Pi_{j_2} = \frac{1}{2l+1}(l - \boldsymbol{\sigma} \cdot \boldsymbol{l}). \tag{21b}$$

If we understand that the scattering amplitude be taken between Pauli spinors for the nucleon, in the c.m. frame, we may write

$$f(E, \theta) = \sum_{l=0}^{\infty} (2l+1)\{ f_{l+}(E)\Pi_{j_1} + f_{l-}(E)\Pi_{j_2} \} P_l(\cos\theta). \tag{22}$$

Inserting the explicit expressions (21) for the projection operators Π_{j_i} and using the relation

$$\boldsymbol{\sigma} \cdot \boldsymbol{l} P_l(\hat{\boldsymbol{\kappa}}' \cdot \hat{\boldsymbol{\kappa}}) = \boldsymbol{\sigma} \cdot (-\mathrm{i}\hat{\boldsymbol{\kappa}}' \wedge \nabla_{\kappa'}) P_l(\hat{\boldsymbol{\kappa}}' \cdot \hat{\boldsymbol{\kappa}})$$
$$= -\mathrm{i}\boldsymbol{\sigma} \cdot \hat{\boldsymbol{\kappa}}' \wedge \hat{\boldsymbol{\kappa}} P_l'(\hat{\boldsymbol{\kappa}}' \cdot \hat{\boldsymbol{\kappa}}), \tag{23}$$

where $P_l'(z) = (\mathrm{d}/\mathrm{d}z)P_l(z)$, we can rewrite eq. (22) as follows,

$$f(E, \theta) = \sum_{l=0}^{\infty} \{ l f_{l-}(E) + (l+1) f_{l+}(E)\} P_l(\cos\theta)$$
$$- i\boldsymbol{\sigma} \cdot \hat{\boldsymbol{\kappa}}' \wedge \hat{\boldsymbol{\kappa}} \sum_{l=0}^{\infty} \{ f_{l+}(E) - f_{l-}(E)\} P_l'(\cos\theta). \tag{24}$$

We note that in the case where the amplitudes do not depend on the spin, $f_{l+} = f_{l-}$, the expansion (20) reduces to the expression (I.8) for the spinless case. It is not difficult to apply the elastic unitarity relation (14b) to this case. We start from the representation (22) of the scattering amplitude, taken between two Pauli spinors χ_α and χ_β, and use again the addition theorem (18). Upon using

$$\int \mathrm{d}\Omega_{\kappa''} Y_{lm}(\hat{\boldsymbol{\kappa}}'') Y_{lm}(\hat{\boldsymbol{\kappa}}'') = \delta_{ll'}\delta_{mm'},$$

we have

$$4\pi\big(\chi_\alpha, Y_{lm}^*(\hat{\boldsymbol{\kappa}}')\{\mathrm{Im}(f_{l+})\Pi_{j_1} + \mathrm{Im}(f_{l-})\Pi_{j_2}\} Y_{lm}(\hat{\boldsymbol{\kappa}})\chi_\beta\big)$$
$$= 4\pi\kappa\big(\chi_\alpha, Y_{lm}^*(\hat{\boldsymbol{\kappa}}')\big[f_{l+}^*\Pi_{j_1} + f_{l-}^*\Pi_{j_2}\big]\big[f_{l+}\Pi_{j_1} + f_{l-}\Pi_{j_2}\big] Y_{lm}(\hat{\boldsymbol{\kappa}})\chi_\beta\big).$$

Using

$$\Pi_{j_1}\Pi_{j_2} = 0 = \Pi_{j_2}\Pi_{j_1}, \qquad \Pi_{j_1}^2 = \Pi_{j_1}, \qquad \Pi_{j_2}^2 = \Pi_{j_2},$$

we deduce the conditions

$$\mathrm{Im}\, f_{l\pm} = \kappa |f_{l\pm}|^2. \tag{25a}$$

Thus, we recover the relations (19) of the spinless case, for each partial wave $j_1 = l + \frac{1}{2}$ and $j_2 = l - \frac{1}{2}$ separately. Clearly, beyond the first inelastic threshold the equal sign is replaced by a "greater than" sign so that, in general, we have

$$\mathrm{Im}\, f_{l\pm} \geq \kappa |f_{l\pm}|^2. \tag{25b}$$

As in the spinless case this implies that the amplitudes f_{l+} and f_{l-} can be parametrized by real or complex scattering phases, cf. eq. (I.13):

$$f_{l\pm} = \frac{1}{2\mathrm{i}\kappa}\left(\mathrm{e}^{2\mathrm{i}\delta_{l\pm}(\kappa)} - 1\right). \tag{26}$$

In the case of purely elastic scattering to which eq. (25a) applies, these phases are real and the parametrization (26) simplifies to

$$f_{l\pm} = \frac{1}{\kappa}\mathrm{e}^{\mathrm{i}\delta_{l\pm}(\kappa)} \sin\delta_{l\pm}(\kappa). \tag{27}$$

1.5. *The elementary amplitudes near threshold*

The interaction of pions and kaons with nucleons, in bound states of exotic atoms, involves energies close to threshold and very low momenta. In this situation it may be sufficient to restrict the expansion (24) to s- and p-waves. With $P_1(z) = z$, $P_1'(z) = 1$, and $|\hat{\boldsymbol{\kappa}}' \wedge \hat{\boldsymbol{\kappa}}| = \sin\theta$, we have

$$f(E,\theta) \simeq f_{0+} + (f_{1-} + 2f_{1+})\cos\theta + \mathrm{i}\boldsymbol{\sigma}\cdot\boldsymbol{n}(f_{1-} - f_{1+})\sin\theta, \tag{28}$$

where $\boldsymbol{n}$ is a unit vector normal to the scattering plane,

$$\boldsymbol{n} = \hat{\boldsymbol{\kappa}}' \wedge \hat{\boldsymbol{\kappa}} / |\hat{\boldsymbol{\kappa}}' \wedge \hat{\boldsymbol{\kappa}}|.$$

Furthermore, it may be sufficient to represent the partial wave amplitudes by their limiting expressions valid close to threshold, i.e. by scattering lengths or volumes and effective ranges. For the moment we restrict our attention to the limit of zero energy.

Using dispersion relations and the Froissart–Gribov representation of partial wave amplitudes one can show that the partial wave amplitude $f_{l\pm}$ has the threshold behaviour $f_{l\pm} \propto \kappa^{2l}$, well-known from potential theory. As the scattering phases $\delta_{l\pm}(\kappa)$ vanish at threshold, $\delta_{l\pm}(0) = 0$, and since they are continuous functions of κ, we have

$$\lim_{\kappa\to 0}\left(\frac{f_{l\pm}(\kappa)}{\kappa^{2l}}\right) = \lim_{\kappa\to 0}\left(\frac{\delta_{l\pm}(\kappa)}{\kappa^{2l+1}}\right) =: a_{l\pm}. \tag{29}$$

This limit defines the constants $a_{l\pm}$, the scattering amplitudes at threshold. $a_{l\pm}$ carries the dimension (length)$^{2l+1}$. Thus, for $l = 0$, it is a length—the so-called *s-wave scattering length*. Likewise, for $l=1$, it is a volume—*the p-wave scattering volume*, etc. (see exercises). In the study of pionic atoms and kaonic atoms we replace, from now on, the scattering amplitudes by their limits $\kappa \to 0$. In doing so we must also take care of the isospin indices that we had suppressed so far.

For *pion–nucleon scattering* the conventional notation is the following. Let I denote the total isospin, j, as before, the total angular momentum.

In the s-wave: a_{2I}, so that

$$a_{l=0,\, j=1/2}(\pi^- \mathrm{p}) \equiv a_{0+}(\pi^- \mathrm{p}) = \tfrac{1}{3}(2a_1 + a_3),$$

$$a_{l=0,\, j=1/2}(\pi^- \mathrm{n}) \equiv a_{0+}(\pi^- \mathrm{n}) = a_3. \tag{30}$$

In the p-wave: $a_{2I,2j}$, so that, for example,

$$a_{l=1,\, j}(\pi^- \mathrm{p}) = \tfrac{1}{3}(2a_{1,2j} + a_{3,2j}),$$

$$a_{l=1,\, j}(\pi^- \mathrm{n}) = a_{3,2j}. \tag{31}$$

In the limit $\kappa \to 0$ we have $\lim_{\kappa \to 0} f_{1j} = \kappa^2 a_{1j}$, so that

$$\lim_{\kappa \to 0} \hat{\boldsymbol{\kappa}}' \cdot \hat{\boldsymbol{\kappa}} f_{1j} = a_{1j} \boldsymbol{\kappa}' \cdot \boldsymbol{\kappa},$$

$$\lim_{\kappa \to 0} \boldsymbol{n} f_{1j} \sin\theta = a_{1j} \boldsymbol{\kappa}' \wedge \boldsymbol{\kappa}.$$

This means that in this limit the amplitude (28) for arbitrary charge states assumes the form

$$f(E,\theta) \simeq b_0 + b_1 \boldsymbol{t} \cdot \boldsymbol{\tau} + [c_0 + c_1 \boldsymbol{t} \cdot \boldsymbol{\tau}] \boldsymbol{\kappa}' \cdot \boldsymbol{\kappa} + \mathrm{i} \boldsymbol{\sigma} \cdot (\boldsymbol{\kappa}' \wedge \boldsymbol{\kappa})[d_0 + d_1 \boldsymbol{t} \cdot \boldsymbol{\tau}], \tag{32}$$

where $\boldsymbol{t}$ and $\boldsymbol{\tau}$ denote the isospin operators of the pion and of the nucleon, respectively. It remains to express the constants b_i and c_j, d_j in terms of the scattering lengths (30) and scattering volumes (31), respectively. This is done by means of the identities

$$\boldsymbol{t} \cdot \boldsymbol{\tau} = \boldsymbol{I}^2 - \boldsymbol{t}^2 - \tfrac{1}{4}\boldsymbol{\tau}^2,$$

$$\text{eigenvalues: } I(I+1) - 2 - \tfrac{3}{4},$$

$$\boldsymbol{\sigma} \cdot \boldsymbol{l} = \boldsymbol{j}^2 - \boldsymbol{l}^2 - \tfrac{1}{4}\boldsymbol{\sigma}^2,$$

$$\text{eigenvalues: } j(j+1) - l(l+1) - \tfrac{3}{4}.$$

So, for example, $b_0 + b_1 \boldsymbol{t} \cdot \boldsymbol{\tau}$, when applied to an eigenstate of total isospin, gives $b_0 + b_1(I(I+1) - \frac{11}{4})$. Thus

$$\text{for } I = \tfrac{1}{2}: \quad a_1 = b_0 - 2b_1,$$

$$\text{for } I = \tfrac{3}{2}: \quad a_3 = b_0 + b_1,$$

which gives

$$b_0 = \tfrac{1}{3}(a_1 + 2a_3), \tag{33a}$$

$$b_1 = \tfrac{1}{3}(a_3 - a_1). \tag{33b}$$

For the case of the p-wave we make use again of the relation (23), viz.

$$i\boldsymbol{\sigma}\cdot(\hat{\boldsymbol{\kappa}}'\wedge\hat{\boldsymbol{\kappa}})P_l' = -\boldsymbol{\sigma}\cdot\boldsymbol{l}P_l(\hat{\boldsymbol{\kappa}}'\cdot\hat{\boldsymbol{\kappa}}).$$

The p-wave part of eq. (32), when applied to a state of good I and good j, gives

$$\{[c_0+c_1(I(I+1)-\tfrac{11}{4})]-[d_0+d_1(I(I+1)-\tfrac{11}{4})][j(j+1)-\tfrac{11}{4}]\}\boldsymbol{\kappa}'\cdot\boldsymbol{\kappa},$$

with $\boldsymbol{\kappa}'\cdot\boldsymbol{\kappa}=\kappa^2\cos\theta$.
Upon comparison with the p-wave terms of eq. (28) in the limit $\kappa\to 0$,

$$[(a_{2I,1}+2a_{2I,3})-\boldsymbol{\sigma}\cdot\boldsymbol{l}(a_{2I,1}-a_{2I,3})]\kappa^2\cos\theta$$
$$\triangleq[(a_{2I,1}+2a_{2I,3})-(j(j+1)-\tfrac{11}{4})(a_{2I,1}-a_{2I,3})]\kappa^2\cos\theta,$$

the following relations are found:

$$I=\tfrac{1}{2},j=\tfrac{1}{2}:\quad 3a_{11}=c_0-2c_1+2(d_0-2d_1),$$
$$I=\tfrac{1}{2},j=\tfrac{3}{2}:\quad 3a_{13}=c_0-2c_1-(d_0-2d_1),$$
$$I=\tfrac{3}{2},j=\tfrac{1}{2}:\quad 3a_{31}=c_0+c_1+2(d_0+d_1),$$
$$I=\tfrac{3}{2},j=\tfrac{3}{2}:\quad 3a_{33}=c_0+c_1-(d_0+d_1).$$

From these we deduce

$$c_0=\tfrac{1}{3}\{(a_{11}+2a_{31})+2(a_{13}+2a_{33})\},\tag{34a}$$
$$c_1=\tfrac{1}{3}\{(a_{31}-a_{11})+2(a_{33}-a_{13})\},\tag{34b}$$
$$d_0=\tfrac{1}{3}\{(a_{11}+2a_{31})-(a_{13}+2a_{33})\},\tag{34c}$$
$$d_1=\tfrac{1}{3}\{(a_{31}-a_{11})-(a_{33}-a_{13})\}.\tag{34d}$$

In fact, we could have guessed the results (34) by looking at eqs. (33) and noticing the symmetry between spin and isospin: The combination (x_1+2x_3) is the symmetric combination in spin or isospin, (x_3-x_1) is the antisymmetric combination. This pattern is indeed reflected by the results (34).

2. Bound states of spin zero particles in the field of a point charge

Pions and kaons are particles with spin zero and, therefore, must be described by the Klein–Gordon equation,

$$(\partial_\mu\partial^\mu+m^2)\phi(x)=0,\tag{35}$$

where

$$\partial_\mu=\frac{\partial}{\partial x^\mu},\qquad \partial_\mu\partial^\mu\equiv\Box=\frac{\partial^2}{\partial x^{02}}-\Delta,$$

and where m is the mass. Stationary solutions with positive frequency, for the case of the Coulomb field of a static external point charge Ze fulfill the equation

$$\{\Delta+(E-V_c)^2-m^2\}\psi(\boldsymbol{r})=0,\tag{36}$$

with

$$V_c(r) = -Z\alpha/r. \tag{37}$$

Here E is the total energy which includes the rest mass of the meson and its binding energy. If the external charge stems from a nuclear partner of mass M, the rest mass m of the meson must be replaced by the reduced mass

$$\overline{m} = \frac{mM}{m+M}. \tag{38}$$

The ansatz, familiar from the analogous problem for the Schrödinger equation, by which

$$\psi(\boldsymbol{r}) = \frac{1}{r} y(r) Y_{lm},$$

leads to the following differential equation for the radial wave function:

$$\frac{\mathrm{d}^2 y(r)}{\mathrm{d}r^2} - \left[\frac{l(l+1)}{r^2} + m^2 - \left(E + \frac{Z\alpha}{r}\right)^2\right] y(r) = 0. \tag{36'}$$

We wish to determine bound state solutions of this equation, i.e. solutions with energy $E < m$. Therefore, it is convenient to define the quantity

$$k := \sqrt{m^2 - E^2}, \tag{39}$$

which is the analogue to the momentum variable for energies above m. It is useful to introduce the dimensionless variable

$$x := 2kr \tag{40}$$

and the quantity

$$\eta := \frac{Z\alpha E}{k}. \tag{41}$$

As one verifies easily, the radial equation then assumes the dimensionless form

$$\frac{\mathrm{d}^2 y(x)}{\mathrm{d}x^2} - \left[\tfrac{1}{4} + \frac{l(l+1) - (Z\alpha)^2}{x^2} - \frac{\eta}{x}\right] y(x) = 0. \tag{42}$$

There are several methods, well-known from nonrelativistic quantum mechanics, to solve equations of this type analytically. One of them is the following.

First, one notes that the solutions have a characteristic power behaviour at the origin,

$$y(x) = x^{s+1} \sum_{n=0}^{\infty} a_n x^n \quad (a_0 \neq 0).$$

The characteristic exponent s follows from the equation

$$s(s+1) - \left[l(l+1) - (Z\alpha)^2\right] = 0, \tag{43}$$

whose solutions are

$$s_+ = -\tfrac{1}{2} + \sqrt{(l+\tfrac{1}{2})^2 - (Z\alpha)^2}, \tag{44a}$$

$$s_- = -\tfrac{1}{2} - \sqrt{(l+\tfrac{1}{2})^2 - (Z\alpha)^2}. \tag{44b}$$

Only the first of these is acceptable because the bound state solutions must be regular at the origin.

Secondly, for asymptotically large values of x we expect the bound state solutions to decrease exponentially to zero. Indeed, from eq. (42)

$$\frac{d^2y(x)}{dx^2} - \tfrac{1}{4}y(x) \simeq 0 \quad \text{for } x \to \infty,$$

so that $y(x)$ behaves as $e^{-x/2}$ at infinity. Separating off the characteristic power behaviour at the origin as well as the exponential at infinity we are left with

$$y(x) = x^{s_+ + 1}e^{-x/2}w(x). \tag{45}$$

From this we calculate

$$\frac{d^2y}{dx^2} = \big[s_+(s_+ + 1)w + \tfrac{1}{4}x^2w - (s_+ + 1)xw$$
$$+ (2s_+ + 2 - x)xw' + x^2w''\big]x^{s_+ - 1}e^{-x/2}.$$

Inserting this into eq. (42) and using the relation (43) the function $w(x)$ is found to satisfy the differential equation

$$xw''(x) + (2s_+ + 2 - x)w'(x) - (s_+ + 1 - \eta)w(x) = 0. \tag{46}$$

This is Kummer's differential equation for the confluent hypergeometric function. The solution regular at the origin is

$${}_1F_1(a;b;x) = 1 + \frac{a}{b}x + \frac{a(a+1)}{b(b+1)}\frac{x^2}{2!} + \frac{a(a+1)(a+2)}{b(b+1)(b+2)}\frac{x^3}{3!} + \cdots, \tag{47}$$

with

$$a = s_+ + 1 - \eta, \qquad b = 2s_+ + 2.$$

From the well-known asymptotic behaviour of this series,

$${}_1F_1(a;b;x) \sim \frac{\Gamma(b)}{\Gamma(b-a)}(-x)^{-a} + \frac{\Gamma(b)}{\Gamma(a)}e^x x^{a-b}, \tag{48}$$

we see that a must be chosen at a pole of $\Gamma(a)$ so that the second term in eq. (48) vanishes and does not change the exponential decrease at infinity, eq. (45). The Γ-function has poles of first order at the sequence of all non-positive integers. Thus we obtain the condition

$$s_+ + 1 - \eta = -n' \quad (n' = 0,1,2,\ldots).$$

In order to obtain complete analogy to the nonrelativistic case we introduce the main quantum number

$$n \coloneqq n' + l + 1,$$

so that the previous condition reads

$$\eta \stackrel{!}{=} n + s_+ - l.$$

From this result, and inserting the definitions (39) and (41), one deduces the general formula for the energy eigenvalues, viz.

$$E_{nl} = m\left\{1 + \left(\frac{Z\alpha}{n - l - \frac{1}{2} + \sqrt{(l+\frac{1}{2})^2 - (Z\alpha)^2}}\right)^2\right\}^{-1/2}. \tag{49}$$

Expanding this expression in powers of $Z\alpha$, one finds

$$E_{nl} = m\left[1 - \frac{(Z\alpha)^2}{2n^2} - \frac{(Z\alpha)^4}{2n^4}\left(\frac{2n}{2l+1} - \frac{3}{4}\right) + \mathrm{O}\left((Z\alpha)^6\right)\right]. \tag{50}$$

The binding energies of these bound states are given by subtracting the rest mass from eq. (49),

$$B_{nl} = E_{nl} - m,$$

so that the expansion (50) gives

$$-B_{nl} = \frac{(Z\alpha)^2}{2n^2}m - \left(\frac{2n}{2l+1} - \frac{3}{4}\right)\frac{(Z\alpha)^4}{2n^4}m + \mathrm{O}\left((Z\alpha)^6\right). \tag{50$'$}$$

The first term is the familiar expression for the energy levels of the nonrelativistic hydrogen atom and exhibits its well-known dynamical l-degeneracy.

The corresponding eigenfunctions are obtained from eqs. (45) and (47). Their normalization to unity is performed by means of known integrals containing powers, exponentials, and confluent hypergeometric functions.*) We skip this straightforward but somewhat tedious calculation and give here the result,

$$y_{nl}(r) = \left(\frac{k\Gamma(2s_+ + n - l + 1)}{(s_+ + n - l)\Gamma(n - l)}\right)^{1/2}\frac{1}{\Gamma(2s_+ + 2)}$$

$$\times (2kr)^{s_+ + 1} e^{-kr}\,{}_1F_1(l + 1 - n; 2s_+ + 2; 2kr). \tag{51}$$

We conclude this section by verifying that these solutions go over into the well-known wave functions of the nonrelativistic hydrogen atom if $(Z\alpha)^2 \ll 1$. Indeed, in this case,

$$k = \sqrt{m^2 - E^2} \simeq \sqrt{2m(m - E)} \simeq \sqrt{2m\frac{(Z\alpha)^2}{2n^2}m} = \frac{Z\alpha}{n}m,$$

$$s_+ \simeq l,$$

*) See e.g. Ref. R5.

so that

$$x = 2kr \simeq 2r/na_{\mathrm{B}},$$

with $a_{\mathrm{B}} = 1/Z\alpha m$ the Bohr radius. When inserted into eq. (51) this gives

$$y_{nl}^{\mathrm{n.r.}}(r) = \left(\frac{(l+n)!}{a_{\mathrm{B}}(n-l-1)!}\right)^{1/2} \frac{1}{n(2l+1)!} \times \left(\frac{2r}{na_{\mathrm{B}}}\right)^{l+1} \mathrm{e}^{-r/na_{\mathrm{B}}} {}_1F_1(-n+l+1; 2l+2; 2r/na_{\mathrm{B}}). \tag{52}$$

Finally, we remind the reader that in an atom, with a nucleus of finite mass M, the particle mass m must be replaced with the reduced mass (38).

3. Pionic and kaonic atoms

The physics of mesonic atoms is determined by three major features: the static Coulomb interaction with the nuclear partner, the multiple scattering inside the nucleus due to the meson–nucleon strong interaction, and all possible inelastic reactions between the meson and the nucleus which eventually lead to disappearance of the particle from its bound state. We studied the case of Coulomb interaction with a point charge in the previous section. The multiple scattering due to strong interactions can be described by means of optical potentials which are constructed according to the theory developed in sec. 3 of Chap. I. The elementary pion–nucleon and kaon–nucleon scattering amplitudes, finally, which enter the optical potential can be taken from sec. 1.5 above.

In the case of kaons this description of the atom is practically complete as the elementary system kaon plus nucleon has inelastic channels open right at threshold, such as for instance

$$\mathrm{K}^-\mathrm{N} \to \pi\Lambda, \qquad \mathrm{K}^-\mathrm{N} \to \pi\Sigma, \qquad \mathrm{K}^-\mathrm{N} \to \pi\pi\Lambda, \tag{53}$$

etc. As a consequence the scattering amplitudes around threshold are complex and the optical potential has a strong absorptive part even below the kaon–nucleon elastic threshold.

For pions this is not so: Free pion–nucleon scattering at low energies is dominantly elastic so that the scattering lengths and volumes (34) are real. The optical potential built on the basis of these quantities does not contain any absorption as yet. In principle, the pion which has strangeness and baryon number zero may be absorbed on a single nucleon,

$$\pi^- + \mathrm{N}_{\mathrm{bound}} \to \mathrm{N}', \tag{54}$$

provided this nucleon is bound in a nucleus (so that the momentum mismatch is compensated for by the residual nucleus). It can be shown, however, that the absorption mechanism (54) on a single nucleon is strongly suppressed in a real nucleus as compared to other absorption reactions. This is so because it liberates a momentum transfer which is large as compared to typical nucleon momenta in nucleon bound states. Loosely speaking such high momentum transfer finds only a very small Fourier component in the nucleon's bound state wave function and the transition rate comes out very small. Instead it is more favourable to share the momentum that the pion deposits in the nucleus between at least two bound nucleons, via the reaction

$$\pi^- + (N_1N_2)_{\text{bound}} \to N_1' + N_2'. \tag{55}$$

Indeed, this absorption mechanism on *two* nucleons is the dominant one in most practical situations. Therefore, for pionic atoms we must generalize the optical potential by adding absorptive terms to it which incorporate the mechanism (55) in at least a phenomenological manner.

In this section we start by defining the observable quantities which are typical for the influence of strong interactions on the bound states of mesonic atoms. By collecting the results of the previous sections and of Chap. I we then formulate the optical potential for pions and kaons in a quantitative manner and indicate how one proceeds in computing the complex energy eigenvalues of mesonic atoms. In the course of discussion we illustrate these matters through a set of selected results for pionic and kaonic atoms.

3.1. *Observables and their extraction from the data*

Suppose the strong interaction of the bound meson with the nucleus is described by the optical potential (I.82), or by the expressions (I.91) and (I.95) if effects of second order are to be included. For simplicity we assume the proton and neutron density to be the same so that the optical potential depends on a common spherical matter density $\rho(r)$ of the size of the nuclear charge distribution. This interaction is short-ranged and becomes effective only if the mesonic bound state has some overlap with the nuclear density. This overlap may be estimated by taking the meson bound states to be nonrelativistic, hydrogenlike wave functions (52) for circular orbits $(n = l + 1)$. For s-wave interaction we have

$$\begin{aligned} I_s &\equiv \int R^2_{n\,l=n-1}(r)\rho(r)r^2\,\mathrm{d}r \\ &\simeq \frac{2^{2n}}{n^{2n+2}(2n-1)!\,a_B^{2n+1}4\pi}\langle r^{2(n-1)}\mathrm{e}^{-2r/na_B}\rangle. \end{aligned} \tag{56}$$

For the p-wave term of the optical potential one finds after some calculation

$$\begin{aligned} I_p &\equiv \int \left[F_1^2(r) + F_2^2(r)\right]\rho(r)r^2\,\mathrm{d}r \\ &\simeq \frac{(n-1)\cdot 2^{2n}}{n^{2n+2}(2n-2)!\,a_B^{2n+1}4\pi}\langle r^{2(n-2)}\mathrm{e}^{-4r/(2n-1)a_B}\rangle, \end{aligned} \tag{57}$$

where the functions $F_i(r)$ arise from the gradient operators of the potential (I.82) as applied to the mesonic wave function and are given by*) (Scheck 1972)

$$F_1(r) \simeq -\sqrt{\frac{n}{2n-1}}\frac{1}{na_{\mathrm{B}}}R_{n,n-1}(r), \tag{58a}$$

$$F_2(r) \simeq \sqrt{\frac{n-1}{2n-1}}\left(\frac{2n-1}{r}-\frac{1}{na_{\mathrm{B}}}\right)R_{n,n-1}(r), \tag{58b}$$

and where $R_{nl}(r) = (1/r)y_{nl}^{\mathrm{n.r.}}(r)$, cf. eq. (52).

The quantities

$$\langle r^{\kappa}\mathrm{e}^{-\alpha r}\rangle := 4\pi\int_0^{\infty} r^{\kappa}\rho(r)\mathrm{e}^{-\alpha r}r^2\,\mathrm{d}r \tag{59}$$

are generalized moments of the nuclear mass density. Clearly, the important feature in the state dependence of I_{s}, I_{p} is provided by the centrifugal tail r^{2l} of the wave function: the overlap increases strongly with decreasing l. Furthermore, it can be shown that the cascade runs primarily through circular orbits. As a consequence, the influence of the strong interaction will be felt only in the few lowest (circular) bound states where the meson penetrates the nuclear interior. In the upper part of the spectrum, the states with higher angular momentum l will be hydrogenlike to a very good approximation and will be unaffected by the strong interaction.

Depending on the strength of the absorptive part in the optical potential and on the nuclear charge number, the meson's cascade may not reach the 1s or even the 2p states because the meson is absorbed through a reaction of the type (53) or (55). The absorptive interaction causes a broadening of the Coulomb energy levels which again depends on the amount of overlap between the mesonic wave function and the nuclear density. In Fig. 1 we sketch in a qualitative manner the level diagram of a typical (light) kaonic or pionic atom: The basic pattern of the spectrum is given by the Coulomb energies (49). Superimposed to this we see the effects of the strong interaction which shifts the energy levels, upwards or downwards, and causes an absorptive level broadening which comes in addition to the radiative width.

Our discussion and this qualitative figure indicate clearly what the relevant observables are: the shift ε_{nl} and the strong interaction width Γ_{nl}. The shift, in particular, is defined as the difference between the relativistic energy (49) in the field of a point charge Ze and the actual energy eigenvalue in the Coulomb field of the extended nuclear charge distribution $\rho_{\mathrm{c}}(r)$ and of the optical potential which is caused by the strong interaction:

$$\varepsilon_{nl} := E_{nl}(\text{point charge } Ze) - E_{nl}(\rho_{\mathrm{c}}(r) \text{ and strong interaction}). \tag{60}$$

In an actual experiment one measures the atomic X-rays (which stem predominantly from E1 transitions). If it is large enough (of the order of, say, 1 to 10 keV), the strong interaction width Γ_{nl} is read off directly from the broadening of the transition X-ray. For higher orbits it must be deduced from the measured relative intensities of these X-rays. Indeed, the absorption width $\Gamma_{n_i, l_i = n_i - 1}$ of a given circular orbit

*) In particular, formula (57) is worked out in Ebersold et al. (1978).

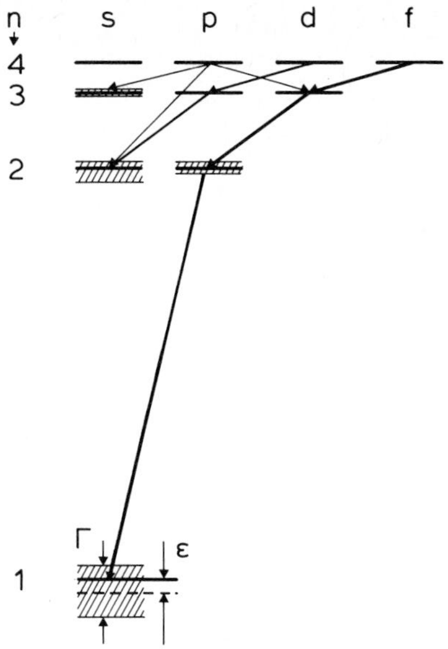

Fig. 1. Level spectrum of a pionic or kaonic atom indicating qualitatively the level broadenings and shifts caused by strong interactions. The arrows show typical E1-X-ray transitions, the most intense lines being indicated by full arrows.

reduces the yield

$$Y = \frac{I_\gamma(n_i \to n_i - 1)}{\sum\limits_{n' > n_i} \{ I_\gamma(n' \to n_i) + I_{\text{Auger}}(n' \to n_i) \}} \tag{61}$$

of the radiative transition $n_i \to n_i - 1$ because of a finite probability for the meson to be absorbed, instead of continuing its radiative cascade to the next lower state $(n_i - 1, l = n_i - 2)$.

Let $(n_c, l_c = n_c - 1)$ be the lowest circular orbit which is observable through a study of the atomic cascade. We call this the *critical level*. Because of the rapid variation of shifts and widths one usually observes no more than the quantities

$$\varepsilon_{n_c n_c - 1}, \quad \Gamma_{n_c n_c - 1} \quad \text{and} \quad \Gamma_{n_c + 1 n_c} \tag{62}$$

in a given nucleus. Only in a few cases has it been possible to also determine

$\varepsilon_{n_c + 1 n_c}$ and, possibly, $\Gamma_{n_c + 2 n_c + 1}$.

Examples are shown below, in secs. 3.3 and 3.5.*)

*)An indirect access to more deeply bound states is possible if state mixing occurs. This is worked out by Leon (1976), while examples are given by Leon et al. (1979).

The calculation of the intensities of the atomic Auger and γ-transitions proceeds along standard lines, provided the initial distribution over n and l, right after the meson has been trapped in the nuclear Coulomb field, is known. Some care must be taken, however, in calculating the transition amplitudes: In contrast to the electron, the meson is heavy so that recoil effects on the nucleus may be important. As an example, we consider an electric dipole (E1) X-ray transition in an atom with a meson of mass m and a nucleus of mass M.

Let O be the center-of-mass of the nucleus, S the center-of-mass of the system nucleus plus meson, and let m_{M} and m_A be the masses of the meson and the nucleus, respectively. The electric dipole operator must be expressed in coordinates of the overall center-of-mass system. As the meson carries charge $-e$ and the nucleus carries charge $+Ze$, the dipole operator is

$$\boldsymbol{D} = -e\boldsymbol{s}_{\mathrm{M}} + Ze\boldsymbol{s}_{\mathrm{O}}. \tag{63}$$

The wave functions of the bound meson, on the other hand, are expressed in terms of coordinates $\boldsymbol{r}_{\mathrm{M}}$ which refer to the *nuclear* center-of-mass system in O because the nuclear charge and density distributions are given with respect to that system. Hence we must rewrite the operator $\boldsymbol{D}$, eq. (63), in terms of the coordinates $\boldsymbol{r}_{\mathrm{M}}$. From Fig. 2 we have

$$m_{\mathrm{M}}\boldsymbol{s}_{\mathrm{M}} + m_A\boldsymbol{s}_{\mathrm{O}} = 0,$$

$$m_{\mathrm{M}}\boldsymbol{r}_{\mathrm{M}} = -(m_{\mathrm{M}} + m_A)\boldsymbol{s}_{\mathrm{O}},$$

so that

$$\boldsymbol{s}_{\mathrm{O}} = -\frac{m_{\mathrm{M}}}{m_A + m_{\mathrm{M}}}\boldsymbol{r}_{\mathrm{M}}, \qquad \boldsymbol{s}_{\mathrm{M}} = \frac{m_A}{m_A + m_{\mathrm{M}}}\boldsymbol{r}_{\mathrm{M}}. \tag{64}$$

From this one determines $\boldsymbol{D}$, viz.

$$\boldsymbol{D} = -e\left(1 + \frac{(Z-1)m_{\mathrm{M}}}{m_A + m_{\mathrm{M}}}\right)\boldsymbol{r}_{\mathrm{M}}. \tag{65}$$

Thus, the transition matrix elements

$$\langle n'l'|\boldsymbol{D}|nl\rangle$$

calculated on the basis of atomic bound state wave functions in the frame centered

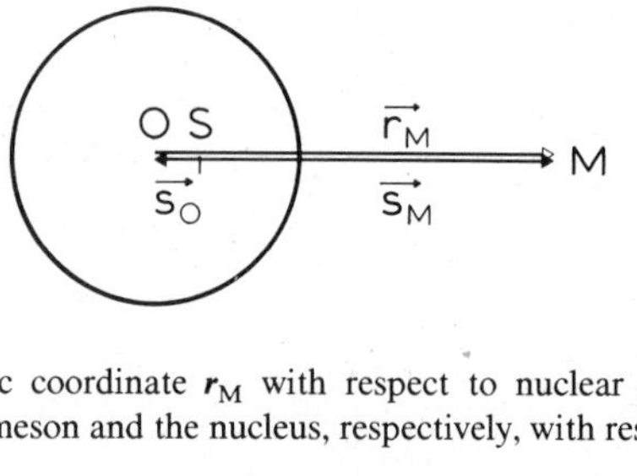

Fig. 2. Definition of mesonic coordinate $\boldsymbol{r}_{\mathrm{M}}$ with respect to nuclear center of mass, as well as of coordinates $\boldsymbol{s}_{\mathrm{M}}$ and $\boldsymbol{s}_{\mathrm{O}}$, of the meson and the nucleus, respectively, with respect to their common center of mass S.

in O, must be multiplied by the enhancement factor

$$1+\frac{(Z-1)m_{\mathrm{M}}}{m_A+m_{\mathrm{M}}}. \tag{66}$$

Physically, this result expresses the fact that it is not the meson alone which jumps from the initial to the final orbit in presence of an inert nucleus but, rather, that the system as a whole makes the transition. The nucleus, being charged, contributes to the transition dipole moment with a weight factor given by its charge Ze. [A similar effect is well-known in the nuclear shell model by the name "effective charges".] In pionic atoms this recoil effect is small. In kaonic atoms, and even more so in antiprotonic atoms, however, it is important. As a matter of example, consider kaonic helium, kaonic oxygen, and antiprotonic lead, for which cases the enhancement factor (66) is, respectively,

$$1.12\ (\mathrm{K}^-, Z=2), \qquad 1.22\ (\mathrm{K}^-, Z=8), \qquad 1.38\ (\bar{\mathrm{p}}, Z=82).$$

3.2. Optical potential for pionic atoms

From the expressions (I.82) or, equivalently, (I.91) and (I.96) we know that the optical potential based on the elementary s- and p-wave scattering amplitudes has the general structure

$$U(\boldsymbol{r})=-\frac{4\pi A}{2\bar{m}}\left\{q(\boldsymbol{r})-\nabla\frac{\alpha(\boldsymbol{r})}{1+\frac{4}{3}\pi\xi\alpha(\boldsymbol{r})}\nabla\right\}. \tag{67}$$

[Here we have chosen the iterated form of the Lorentz–Lorenz correction, cf. eq. (I.96) and discussion. This parametrization is customary in the analysis of atomic data.] This potential is to be inserted into the Klein–Gordon equation (eqs. (I.81) and (36))

$$\left\{\Delta+(E-V(\boldsymbol{r}))^2-\bar{m}^2\right\}\psi(\boldsymbol{r})=2\bar{m}U(\boldsymbol{r})\psi(\boldsymbol{r}), \tag{68}$$

where $\bar{m}$ is the reduced mass and where V now denotes the Coulomb potential of the nuclear extended charge distribution,

$$V(\boldsymbol{r})=-Z\alpha\int \mathrm{d}^3r'\frac{1}{|\boldsymbol{r}-\boldsymbol{r}'|}\rho_{\mathrm{p}}(\boldsymbol{r}'). \tag{69}$$

Here $\rho_{\mathrm{p}}(\boldsymbol{r})$ denotes the proton density, normalized to 1,

$$\int \mathrm{d}^3r\,\rho_{\mathrm{p}}(\boldsymbol{r})=1.$$

In case of a spherically symmetric distribution these equations become

$$V(r)=-4\pi Z\alpha\left\{\frac{1}{r}\int_0^r\rho_{\mathrm{p}}(r')r'^2\,\mathrm{d}r'+\int_r^\infty\rho_{\mathrm{p}}(r')r'\,\mathrm{d}r'\right\}, \tag{69'}$$

$$4\pi\int_0^\infty\rho_{\mathrm{p}}(r')r'^2\,\mathrm{d}r'=1.$$

Denoting the neutron distribution by $\rho_{\mathrm{n}}(\boldsymbol{r})$ and normalizing it to 1, the nuclear mass distribution $\rho(\boldsymbol{r})$ is

$$\rho(\boldsymbol{r}) = \frac{1}{A}\{Z\rho_p(\boldsymbol{r}) + (A-Z)\rho_n(\boldsymbol{r})\}. \tag{70}$$

In terms of these densities, the functions $q(\boldsymbol{r})$ and $\alpha(\boldsymbol{r})$ have the explicit form

$$q(\boldsymbol{r}) = \left(1 + \frac{m_\pi}{m_N}\right)\left\{b_0\rho(\boldsymbol{r}) + \frac{1}{A}b_1\left[Z\rho_p(\boldsymbol{r}) - (A-Z)\rho_n(\boldsymbol{r})\right]\right\} + \left(1 + \frac{m_\pi}{2m_N}\right)B_0 A\rho^2(\boldsymbol{r}), \tag{71}$$

$$\alpha(\boldsymbol{r}) = \left(1 + \frac{m_\pi}{m_N}\right)^{-1}\left\{c_0\rho(\boldsymbol{r}) + \frac{1}{A}c_1\left[Z\rho_p(\boldsymbol{r}) - (A-Z)\rho_n(\boldsymbol{r})\right]\right\} + \left(1 + \frac{m_\pi}{2m_N}\right)^{-1} C_0 A\rho^2(\boldsymbol{r}). \tag{72}$$

These formulae need some further commenting:

(i) The first term on the r.h.s. of eq. (71) originates from the first two terms in the expressions for the amplitude (32), when these are inserted into the expressions for the optical potential as derived in Chap. I. b_0 and b_1 are given in terms of the elementary pion–nucleon s-wave scattering lengths by eqs. (33), except for medium corrections of the kind derived in eq. (I.91) which, if important, would replace b_0 (and similarly b_1) by effective, renormalized values. The extra factor $(1 + m_\pi/m_N)$ takes account, in a rough approximation, of the kinematic transformation from the pion–*nucleon* c.m. system to the pion–*nucleus* c.m. system. This may be seen as follows: In the collision of the pion on a single nucleon the kinematic recoil is transmitted to the nucleus as a whole. For the construction of the optical potential we need the pion scattering amplitude expressed in the pion nucleus c.m. system. As the nucleus is much heavier than the pion this latter system practically coincides with the laboratory system (up to terms of order $1/A$), in which the nucleus and (to some approximation also) the individual nucleon on which the pion scatters, are at rest. The amplitude (32), on the other hand, refers to the pion–*nucleon* c.m. system. For purely forward scattering the amplitudes in the laboratory system and in the c.m. system are related by

$$f_{\text{lab}} = \frac{q_{\text{lab}}}{q} f_{\text{c.m.}}, \tag{73}$$

where q_{lab}, q denote the absolute values of the pion three-momentum in the laboratory and c.m. systems, respectively. This relation follows from the optical theorem (17b) which has the same form in the two systems. Some calculation shows that

$$\frac{q_{\text{lab}}}{q} = \sqrt{\frac{s}{m_N^2}} = \frac{1}{m_N}\sqrt{m_N^2 + m_\pi^2 + 2m_N E_\pi} \simeq 1 + \frac{m_\pi}{m_N}, \tag{74}$$

as $E_\pi \simeq m_\pi$.

(ii) The third term on the r.h.s. of eq. (71) takes account of the absorption on two nucleons, by the processes

$$\pi^- + (\mathrm{pp}) \to \mathrm{p} + \mathrm{n} \quad \text{and} \quad \pi^- + (\mathrm{np}) \to \mathrm{n} + \mathrm{n}, \tag{75}$$

the constant B_0 being an effective, complex number. This term is a purely phenomenological description of the absorption process, the physical picture being that the pion is absorbed on a correlated pair of two nucleons (or two-nucleon cluster) of mass $2m_{\mathrm{N}}$ in the nucleus.

If one wishes to discriminate between the two main absorption channels (75), one must make the replacement

$$B_0 A \rho^2(\boldsymbol{r}) \to A \left\{ \left(\frac{Z}{A} \right)^2 B_{\mathrm{pp}} \rho_{\mathrm{p}}^2(\boldsymbol{r}) + \frac{Z(A-Z)}{A^2} B_{\mathrm{pn}} \rho_{\mathrm{p}}(\boldsymbol{r}) \rho_{\mathrm{n}}(\boldsymbol{r}) \right\} \tag{76}$$

in eq. (71). For example, if the ratio γ of the capture rates on proton–neutron pairs and on proton–proton pairs is known, then $B_{\mathrm{pn}} = \gamma B_{\mathrm{pp}}$. If, furthermore, proton and neutron densities are the same, then the effective parameter B_0 is related to B_{pp} and B_{pn} by

$$B_0 = \frac{1}{A^2} Z(Z + \gamma(A-Z)) B_{\mathrm{pp}} \quad (B_{\mathrm{pn}} = \gamma B_{\mathrm{pp}}). \tag{77}$$

In practice, this procedure, as yet, is not too useful because the ratio γ is not well known and because there are other competing absorption mechanisms, for instance, on clusters heavier than the two-nucleon system.

(iii) The p-wave terms (72) have the same structure as the s-wave terms (71). The kinematic factor is the inverse of the term (74) because the p-wave amplitude in the c.m. system is proportional to $\boldsymbol{q} \cdot \boldsymbol{q}'$. The gradient operators in eq. (67) act on the pion wave function in the laboratory system and, therefore, are equivalent to $q_{\mathrm{lab}} \cdot q'_{\mathrm{lab}}$. This must be corrected by the factor (74). The effective absorption term, parametrized by the constant C_0, may be decomposed into its components describing p-wave absorption on proton–proton and proton–neutron pairs, respectively, in complete analogy to eq. (76).

(iv) In principle, the expression (72) should also contain spin dependent terms, due to the last term in the amplitude (32). These terms contain the nucleon spin operator $\boldsymbol{\sigma}$. In a real nucleus they give rise to typical nuclear *one*-body matrix elements, in contrast to all other terms which involve coherent sums over all nucleons. As a consequence, such terms in the optical potential are expected to be of order $1/A$ as compared to the others. Except for the lightest atoms, these spin dependent terms are unimportant, in practice.

(v) It is a matter of choice whether or not the absorptive p-wave term (third term on r.h. side of eq. (72)) is included in the iterated form of the Lorentz–Lorenz effect as indicated in eq. (67). In any event C_0, or for that matter C_{pp} and C_{pn}, and, very similarly, B_0 are no more than effective parameters which are to be determined from a fit to experiment.

The optical potential (67), with $q(\boldsymbol{r})$ and $\alpha(\boldsymbol{r})$ as given by eqs. (71) and (72), contains all important physical features of pionic atoms. The parameters of the real part of the potential (more precisely: the first two terms on the right-hand of eqs. (71) and (72), respectively) are given by the free pion–nucleon scattering amplitudes up to, moderately small, medium corrections which are calculable.

The absorptive part, at this stage, is rather more phenomenological in nature. It describes what is believed to be the dominant absorption process but the parameters must be fitted to experiment. Microscopic calculations of pion absorption in nuclei are necessarily complicated ones. Nevertheless some attempts have been made, yielding fair agreement with the empirical parameters of the optical potential.

On the whole, the optical potential (67) provides an excellent description of the wealth of data from pionic atoms, all over the periodic table. However, there are also some failures in individual cases, in particular when finer details are being studied. At the time of this writing, the subject of pionic atoms is far from being closed and much more theoretical and experimental work remains to be done.

3.3. Pionic atoms: parameters and typical results

The measured pion–nucleon s-wave scattering lengths are, typically, in units of the pion's Compton wave length

$$\lambda_\pi = m_\pi^{-1} \triangleq \hbar/m_\pi c = 1.41 \text{ fm}$$

(Nagels et al. 1976),

$$a_3 - a_1 \simeq -0.26 m_\pi^{-1}, \qquad a_1 + 2a_3 \simeq -0.02 m_\pi^{-1}. \tag{78a}$$

The p-wave scattering volumes are

$$\begin{aligned} a_{31} - a_{11} &\simeq 0.042 m_\pi^{-3}, & a_{33} - a_{13} &\simeq 0.30 m_\pi^{-3}, \\ a_{11} + 2a_{31} &\simeq -0.16 m_\pi^{-3}, & a_{13} + 2a_{33} &\simeq 0.36 m_\pi^{-3}. \end{aligned} \tag{78b}$$

When inserted into eqs. (33) and (34) this gives

$$b_0 \simeq 0, \qquad b_1 \simeq -0.09 m_\pi^{-1}, \tag{79}$$

$$c_0 \simeq 0.19 m_\pi^{-3}, \qquad c_1 \simeq 0.21 m_\pi^{-3}. \tag{80}$$

These numbers reveal several remarkable properties. The symmetric s-wave scattering length b_0 is practically zero. The p-wave scattering volumes, on the other hand, are large: the Δ-resonance is felt already at threshold.

b_0 being so small, and c_0 being large, means that the p-wave term of the optical potential (67) is important. In fact, except for the atomic s-states, the energy shift ε_{nl}, for $l > 0$, is determined primarily by the p-wave potential. Only the atomic states with $l = 0$ do not feel the p-wave potential. Furthermore, as b_0 vanishes, the isospin dependent term, proportional to b_1, is important whenever the shift is dominated by the s-wave potential. Indeed, the comparison of ε_{1s} for light nuclei with isospin $\frac{1}{2}$ and for neighbouring nuclei with isospin zero shows clear evidence for this term in the optical potential (Ericson et al. 1969).

We summarize the main characteristics of the experimental findings: The pionic 1s-level is observable from hydrogen ($Z = 1$) to about ^{23}Na ($Z = 11$). The *shift* ε_{1s} *is*

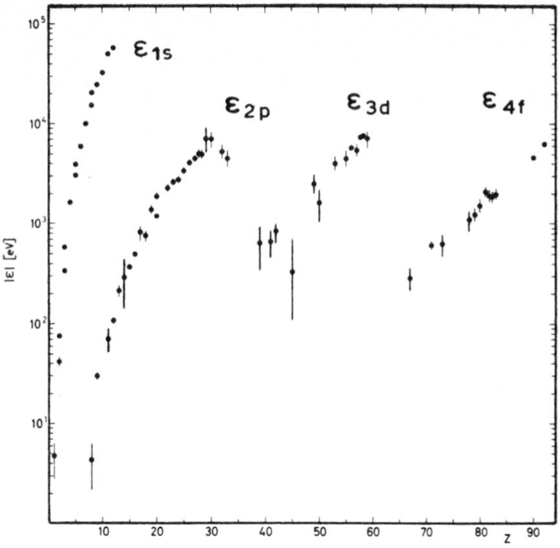

Fig. 3. Energy shifts $|\varepsilon|$ of pionic levels. ε_{1s} is negative (i.e. repulsive); ε_{2p}, ε_{3d}, ε_{4f} are positive (i.e. attractive).

always negative which means, on account of the definition (60), that the effective interaction is *repulsive*. The p-wave, gradient potential has practically no influence on s-level shifts. Further, the absorptive potential which always gives rise to an effective repulsion, is relatively weak. Therefore, the renormalized constant b_0 must be non-zero and negative. As to the higher atomic states, the 2p level is observable from about $Z = 13$ to about $Z = 33$; the 3d level from $Z = 40$ to about $Z = 82$; the 4f level from about $Z = 67$ on up. In all these cases the shift is found to be *positive*. This is in agreement with the expectation based on the value of c_0, eq. (80), which corresponds to a strongly *attractive p-wave potential*. In Figures 3 and 4 we show some results for 1s- and 2p-states, respectively. Assuming $\xi = 1$ the data is well fitted by values of c_0, b_1 and c_1, very close to the ones predicted from free πN scattering, cf. eqs. (79) and (80). The effective value of b_0 is found to be of the order of *⁾

$$(b_0)_{\text{effective}} \simeq -0.03 m_\pi^{-1},$$

while the absorptive part is characterized by values of typically

$$B_0 \simeq \mathrm{i}\,0.04 m_\pi^{-4}, \qquad C_0 \simeq \mathrm{i}\,0.08 m_\pi^{-6}.$$

3.4. Practical aspects in analyzing pionic atoms

3.4.1. Computation of bound state energies

Here we wish to comment on the practical problem of solving the Klein–Gordon equation (68) and of finding the complex energy eigenvalues of the pionic bound

*⁾Recent data and fits to the data are found in Konijn et al. (1978), Batty et al. (1979), Tauscher (1977).

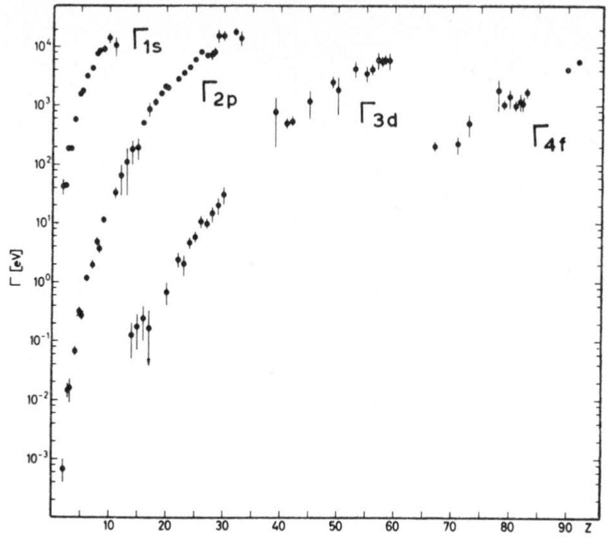

Fig. 4. Absorption widths of pionic levels.

states. The basic assumption is that shift and width of a given orbit (n, l) can be calculated, in a quasi-stationary approximation, by solving the static equation (68) for its eigenvalues E_{nl}, for which the radial wave function is regular both at $r=0$ and $r=\infty$. In this static approximation ε_{nl} and Γ_{nl} are obtained by equating

$$E_{nl} = \mathrm{Re}(E_{nl}) + \mathrm{i}\,\mathrm{Im}(E_{nl}) = \mathrm{Re}(E_{nl}) - \frac{i}{2}\Gamma_{nl}, \tag{81}$$

$$E_{nl} = E_{nl}(\text{point charge } Ze) - \mathrm{Re}(E_{nl}), \tag{82a}$$

$$\Gamma_{nl} = -2\,\mathrm{Im}(E_{nl}). \tag{82b}$$

The domain of validity of this procedure is not obvious: The optical potential being complex, the equation (68) is not hermitean. Presumably, the static approximation is a good one as long as

$$\Gamma_{nl} \ll |\Delta E_{\text{atomic}}|, \tag{83}$$

where ΔE_{atomic} is some typical excitation energy of the atomic system; for instance $\Delta E = E_{nl} - E_{n-1,l-1}$, or some other energy separation to a level that could be admixed to the state with quantum numbers (n, l). The atom being a complicated dynamical system, it is not easy to state the conditions by which the condition (83) holds. Clearly, if the absorption is increased strongly, the static approximation is bound to fail eventually.

How does one proceed about solving eq. (68) practically? For simplicity consider a spherical nucleus with spin $J=0$. In this case the nucleon densities and, accordingly, the optical potential as well as the Coulomb potential are spherically symmetric; the pionic wave function separates into a radial and an angular part,

$$\psi(\boldsymbol{r}) = R_{nl}(r)Y_{lm}(\hat{r}) \equiv \frac{1}{r} y_{nl}(r) Y_{lm}(\hat{r}).$$

The radial function $y(r)$ obeys the differential equation

$$\frac{\mathrm{d}^2 y}{\mathrm{d}r^2} - \left[\frac{l(l+1)}{r^2} + \bar{m}^2 - (E - V(r))^2\right] y = 2\bar{m} U(r) y, \tag{84}$$

with $V(r)$ given by eq. (69′) and $U(r)$ by eq. (67). The p-wave part of the optical potential, eq. (67), contains gradient operators which act on the wave function $y(r)$. It is not difficult to show that one can transform eq. (84) to an equivalent differential equation which no longer contains derivatives, by means of the substitution

$$y(r) = \frac{1}{\sqrt{1 + \bar{\alpha}(r)}} z(r), \tag{85}$$

where

$$\bar{\alpha}(r) = \frac{\alpha(r)}{1 + \frac{4}{3}\pi\xi\alpha(r)}.$$

If the substitution (85) is introduced into eq. (84) one obtains

$$\frac{\mathrm{d}^2 z}{\mathrm{d}r^2} - \left[\frac{l(l+1)}{r^2} + \frac{\bar{m}^2 - (E - V(r))^2}{1 + \bar{\alpha}(r)}\right] z$$

$$= \frac{-4\pi A}{1 + \bar{\alpha}(r)} \left\{ q(r) - \frac{1}{4} \frac{1}{1 + \bar{\alpha}} \left(\frac{\mathrm{d}}{\mathrm{d}r}\bar{\alpha}\right)^2 + \frac{1}{2}\left(\frac{\mathrm{d}^2}{\mathrm{d}r^2}\bar{\alpha} + \frac{2}{r}\frac{\mathrm{d}}{\mathrm{d}r}\bar{\alpha}\right)\right\} z, \tag{86}$$

where the derivatives ∇ and Δ now only act on $\bar{\alpha}(r)$ but not on the wave function.

The optical potential, being proportional to the nuclear densities, rapidly goes to zero outside the nuclear radius R_0. At the same time, the Coulomb potential for $r > R_0$ goes over into the potential (37) of a point charge. There is a clear separation of the r-scale into two domains: An *inner domain*, $r \leq R_{\mathrm{M}}$, where the complex optical potential is different from zero and the Coulomb potential $V(r)$ is the potential of an *extended* charge distribution. An *outer domain*, $r > R_{\mathrm{M}}$, where the optical potential vanishes and where $V(r)$ is the potential of a *point* charge. R_{M} is a matching radius, somewhat larger than the nuclear radius, e.g. $R_{\mathrm{M}} \simeq 3$–$4\ R_0$. The problem to be solved is then the following:

a) *For* $r \leq R_M$: Find a complex wave function $y_i(r)$ which is regular at the origin and has the correct centrifugal behavior*)

$$y_i(r) \underset{r \to 0}{=} a_0 r^{l+1}.$$

This is usually done by numerical integration.

*) The centrifugal tail is not the one of eq. (45) because $V(r)$, for an extended charge, is regular at the origin.

b) *For* $r > R_M$: Find a solution $y_0(r)$ of eq. (36′) for arbitrary complex values of the energy E, which is regular at infinity, i.e. which decreases exponentially for large r like

$$y_0(r) \underset{r\to\infty}{\sim} e^{-r\,\mathrm{Re}\,k}, \quad k=\sqrt{\overline{m}^2-E^2}\,.$$

c) Vary the complex energy E until the inner and the outer solution match at $r = R_M$, i.e. until the logarithmic derivatives coincide,

$$\left.\frac{y_i'}{y_i}\right|_{r=R_M} = \left.\frac{y_0'}{y_0}\right|_{r=R_M} \quad \text{at } E=E_{nl}. \tag{87}$$

This is then the energy eigenvalue.

Exponentially decreasing solutions for the potential of a point charge do indeed exist, even if the energy is chosen to be complex. This can be seen as follows: As we saw in sec. 2 above, one type of solutions of eq. (36′) (regular at the origin), has the structure (45)

$$y_{\mathrm{I}}(r) = x^{s+1}e^{-x/2}{}_1F_1(a;\,b;\,x), \tag{88a}$$

where

$$x=2kr, \qquad a=s+1-\eta, \qquad b=2s+2,$$

with s and η as defined by eqs. (44a) and (41), respectively. Note that $k=(\overline{m}^2-E^2)^{1/2}$ is now complex. However, if ${}_1F_1(a;\,b;\,x)$ is a solution of Kummer's equation (46), then so is the function $x^{1-b}{}_1F_1(1+a-b;\,2-b;\,x)$. Hence

$$y_{\mathrm{II}}(r) = x^{-s}e^{-x/2}{}_1F_1(-s-\eta;\,-2s;\,x) \tag{88b}$$

is another solution, for the same angular momentum and energy, of eq. (36′) which is linearly independent of the solution (88a). For arbitrary E, both solutions increase like $e^{x/2}$ at $|x|\to\infty$ because of the second term in the asymptotic expansion of ${}_1F_1(a;\,b;\,x)$, cf. eq. (48). Thus, for large r, we have

$$y_{\mathrm{I}}(r) \sim \frac{\Gamma(2s+2)}{\Gamma(s+1-\eta)}x^{-\eta}e^{x/2} + O(e^{-x/2}),$$

$$y_{\mathrm{II}}(r) \sim \frac{\Gamma(-2s)}{\Gamma(-s-\eta)}x^{-\eta}e^{x/2} + O(e^{-x/2}). \tag{89}$$

Therefore, the specific combination

$$y_\infty(r) = \frac{\Gamma(-2s-1)}{\Gamma(-s-\eta)}y_{\mathrm{I}} + \frac{\Gamma(2s+1)}{\Gamma(s+1-\eta)}y_{\mathrm{II}} \tag{90}$$

will have the required asymptotic behaviour proportional to $e^{-x/2}$ because upon using $\Gamma(z+1)=z\Gamma(z)$ one sees that the exponentially increasing parts (89) cancel out.

Remark: The linear combination (90) is a Whittaker function $W(\eta, s+\frac{1}{2})$ with one complex parameter and a complex argument. There exist useful recursion and differentiation formulae for these functions which are needed in the matching condition (87). The same method, applied to the Dirac equation with real eigenvalues, is discussed in Chap. IV, sec. 6.3.

3.4.2. Distortion effects and interplay of s- and p-wave potentials

In the previous section we described practical procedures for obtaining the complex energy eigenvalues of the Klein–Gordon equation (68). In table 2 we list a number of case studies (A) to (D), and one experiment (E), for the specific example of pionic lutetium, $^{175}\mathrm{Lu}_\pi$: In (A) to (D), ε_{nl} and Γ_{nl} are calculated by two methods:

(i) from the expectation value of the optical potential in the hydrogenlike state (n, l) described by the relativistic functions (51) in the field of a point-charge (first-order perturbation theory);

(ii) by solving the Klein–Gordon equation (68) numerically as described above, ("exact").

The theoretical numbers, obtained under various assumptions about the optical potential, are instructive with regard to a number of important questions which we shall now discuss.

a) *Damping by absorptive potential.* Intuitively one expects the imaginary, absorptive part of the optical potential to damp the pionic wave function inside the nucleus and, as a consequence thereof, to reduce the net attraction as compared to the situation without absorption. That this is indeed so may be seen by comparing the lines "exact" of cases (C) and (D): When the absorption is switched on, the attractive shift is reduced, by ~ 6% for the 4f-state, by ~ 23% by the 3d-state. The deeper lying state 3d is affected much more than the 4f-level which is more peripheral.

Table 2
Comparison of shifts and widths calculated by integration of the complex Klein–Gordon equation (68) ("exact"), and by perturbation theory ("pert.th.") on the basis of relativistic, hydrogenlike wave functions. The calculated values are taken from Koch et al. (1980), the data is taken from Ebersold et al. (1978). All energies are in keV.

$^{175}\mathrm{Lu}_\pi$ $(Z=71)$ Optical Potential		4f-state		3d-state	
		ε_{43}	Γ_{43}	ε_{32}	Γ_{32}
(A) $c_0 = c_1 = 0$	pert.th.	−0.152	0.165	−33.4	32.1
	exact	−0.140	0.117	−24.9	13.7
(B) $b_0 = b_1 = 0$	pert.th.	0.599	0.157	75.5	31.4
	exact	0.696	0.623	44.8	154.8
(C) $B_0 = C_0 = 0$	pert.th.	0.446	0	39.1	0
(no absorption)	exact	0.427	0	24.5	0
(D) full	pert.th.	0.446	0.157	39.1	31.4
potential	exact	0.403	0.243	18.9	35.5
(E) experiment		0.597(70)	0.200(70)	—	—

b) *Distortion*. Comparison of the exact calculation of ε_{nl} and Γ_{nl} with the predictions of first-order perturbation theory in cases (A), (B), and (D) shows that the distortion of the pionic wave function can have important effects on shift and width of the state. Perturbation theory may be an acceptable, though rough, approximation for the peripheral 4f-state (with the noticeable exception of Γ_{4f}, in case (B)), but it definitely fails for the deeply bound state 3d.

c) *Interplay of s- and p-wave potential.* A specific feature of pionic atoms is the simultaneous presence of a moderately repulsive scalar potential (from the pion-nucleon s-wave amplitude) and a strongly attractive, derivative potential (from the pion–nucleon p-wave amplitude). In states like the 4f and 3d, for a heavy nucleus such as lutetium, both potentials are important, as may be seen by comparing cases (A), (B) and (D). If the p-wave potential is turned off, case (A), the shifts ε_{4f} and ε_{3d} are negative, i.e. repulsive, the width are less than half their final values, case (D). If the s-wave potential is turned off, case (B), ε_{4f} and ε_{3d} are found large and positive, reflecting the strong attraction in the p-wave potential. At the same time, the widths become larger, by a factor of almost three in case of the 4f, by a factor of more than four in case of the 3d state, as compared to the combined effect of both potentials, case (D).

The final values, case (D), are the results of a delicate balance between the two parts of the potential. It is easy to understand why the pion feels both the s-wave and the p-wave optical potential: even though the pion moves in a state with $l = 2$ or $l = 3$, with respect to the center-of-mass of the nucleus, it moves predominantly in relative s- and p-waves with respect to individual nucleons. As may be seen from the results of table 2, the relative importance of s-wave potential or p-wave potential is greater in the 3d-state than in the 4f state.

3.5. *Kaonic atoms: optical potential and typical results*

In some respect the optical potential for negative kaons in nuclei is very much simpler than the case of pions. In contrast to the case of the pion–nucleon interaction, the $\overline{\mathrm{K}}\mathrm{N}$ s-wave scattering lengths are large, while the p-wave scattering volumes are small. The optical potential is dominated by the s-wave term, the p-wave gradient interaction may be neglected to a good approximation. Furthermore, as emphasized in the introduction to sec. 3, there are important absorption channels on *single* nucleons, like those indicated in (53), which are open right at threshold. Absorption on two nucleons or larger clusters is relatively unimportant and may be neglected as compared to the absorption on single nucleons.

On the other hand, there is a new feature in the $\overline{\mathrm{K}}\mathrm{N}$ system at low energies: The baryonic resonance $\Lambda(1405)$ with the following characteristics (I: isospin, J^π: spin and parity, S: strangeness),

$$\Lambda(1405)\begin{cases} I = 0, J^\pi = \tfrac{1}{2}^-, S = -1, \\ m_\Lambda = (1405 \pm 5)\ \mathrm{MeV}, \Gamma_\Lambda = (40 \pm 10)\ \mathrm{MeV} \end{cases} \tag{91}$$

couples strongly to the isospin zero channel of the $\mathrm{K}^-\mathrm{p}$ system. Actually, to a good approximation, the Λ behaves like an ordinary, nonrelativistic bound state of a kaon and a proton.*) It is, as yet, not well understood how to incorporate such a bound state-like resonance of the elementary amplitude into the optical potential.**) We ignore the problem here, but will be prepared to possibly find an effective s-wave scattering length from kaonic atom data, rather different from the value as obtained from the free kaon–nucleon scattering.

The optical potential for negative kaons then is (Ericson et al. 1970)

$$U(\boldsymbol{r}) = -\frac{4\pi A}{2\bar{m}} q(\boldsymbol{r}), \tag{92}$$

where $\bar{m}$ is the reduced mass of the kaon–nucleus system and where $q(\boldsymbol{r})$ is given by

$$q(\boldsymbol{r}) = \left(1 + \frac{m_{\mathrm{K}}}{m_{\mathrm{N}}}\right)\left\{a(\mathrm{K^-p})\frac{Z}{A}\rho_{\mathrm{p}}(\boldsymbol{r}) + a(\mathrm{K^-n})\frac{A-Z}{A}\rho_{\mathrm{n}}(\boldsymbol{r})\right\}. \tag{93}$$

From relations (5a) and (5b) the $\mathrm{K^-p}$ and $\mathrm{K^-n}$ scattering lengths can be expressed in terms of the isospin 0 and 1 channels as follows:

$$a(\mathrm{K^-p}) = \tfrac{1}{2}(a_0 + a_1),$$
$$a(\mathrm{K^-n}) = a_1. \tag{94}$$

In particular, if neutron and proton densities are assumed to be equal, the function $q(\boldsymbol{r})$ is proportional to an average scattering length $\bar{a}$, viz.

$$q(\boldsymbol{r}) = \left(1 + \frac{m_{\mathrm{K}}}{m_{\mathrm{N}}}\right)\bar{a}\rho(\boldsymbol{r}), \tag{93'}$$

with

$$\bar{a} = \frac{Z}{2A}(a_0 + a_1) + \frac{A-Z}{A}a_1. \tag{95}$$

The analysis of free $\bar{\mathrm{K}}$–nucleon scattering data gives slightly different results depending on whether it is based on a constant scattering length formalism or on the K-matrix formalism. Typical results are: $a_0 = (-1.7 + \mathrm{i}0.7)\,\mathrm{fm}$, $a_1 = (0 + \mathrm{i}0.6)\,\mathrm{fm}$. For $Z \simeq \frac{1}{2}A$ we would then obtain a number of the order of

$$\tfrac{1}{4}(a_0 + a_1) + \tfrac{1}{2}a_1 \simeq (-0.4 + \mathrm{i}0.7)\,\mathrm{fm}. \tag{96}$$

Fits to the shifts and widths of kaonic atoms give, typically (Koch et al. 1972, Batty et al. 1979)

$$\bar{a} = (+0.34 + \mathrm{i}0.81)\,\mathrm{fm}. \tag{97}$$

Whilst the imaginary part of $\bar{a}$ agrees roughly with the value expected from free scattering, eq. (96), the real part has the opposite sign of the free value: The effective

*) exercise 11.

**) For an interesting attempt which leads to an effective potential of similar structure as the one for pions, eq. (67), see Brockmann et al. (1978).

interaction with the nucleus is attractive rather than repulsive. The discrepancy is due to the presence of the hyperon resonance $\Lambda(1405)$ in the isospin zero channel. The effects of this resonance are lumped into the effective scattering length $\bar{a}$.

We show some typical results below. An interesting new feature of kaonic atoms as compared to pionic atoms is this: Even though the effective $\bar{a}$, eq. (97), gives rise to a strongly *attractive* optical potential (92), the shifts are always *negative*, i.e. the atomic bound state are repelled by the strong interaction. This peculiar behaviour is a consequence of the large imaginary part of $\bar{a}$. The absorptive potential is very strong (see exercise 13) and, therefore, effectively leads to a negative, repulsive energy shift.

As the mass of the kaon is more than three times larger than the mass of the pion, the kaon in a state of given n and l, moves considerably closer to the nucleus than the pion. As a consequence, in a given element, the kaonic critical level is usually one (sometimes two) circular orbits above the pionic one.

In Figs. 5 and 6 we show examples of measured and calculated energy shifts and widths of kaonic atoms, for various nuclear partners. The agreement of the data with the theoretical values is generally good. Here again, very much as for pions, detailed and more accurate data, and better microscopic analysis of the optical potential should be achieved in the future. So far, the optical potential is still a semi-phenomenological one and needs to be better understood so that its functional form and its parameters become predictable to a better accuracy.

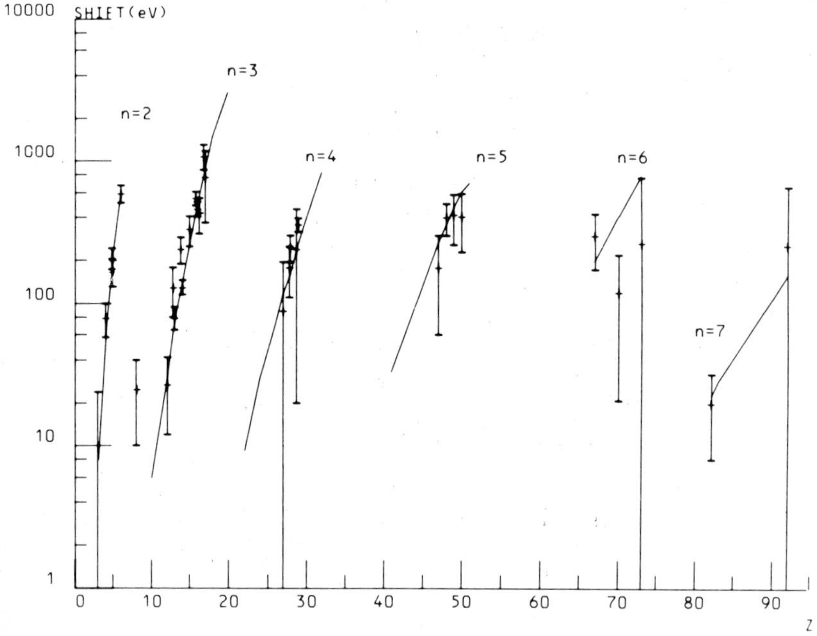

Fig. 5. Shifts of kaonic levels. Compilation of data from Batty, Exotic Atoms, a Review (Rutherford RL-80-094).

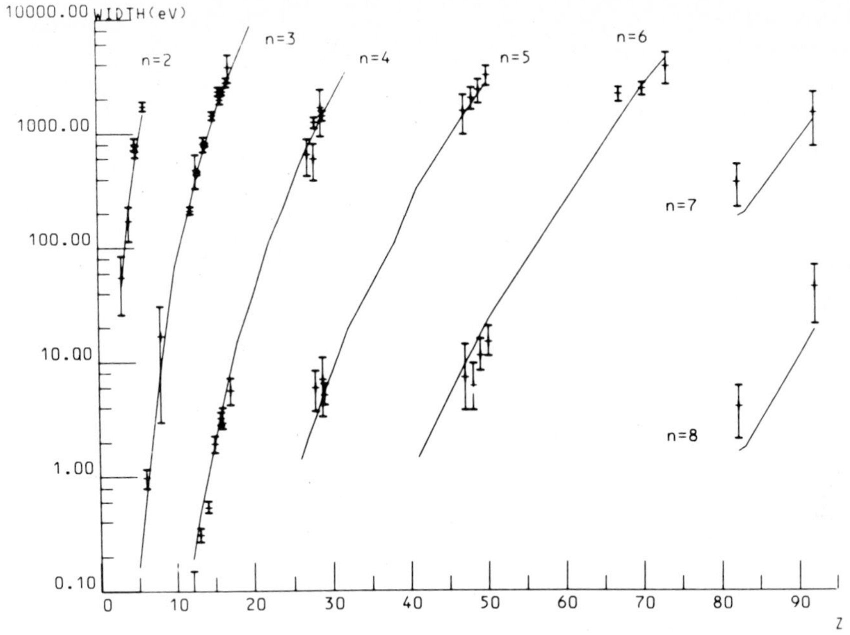

Fig. 6. Widths of kaonic levels.

4. Pionic and kaonic atoms with deformed nuclei

In this section we consider an interesting application of mesonic atoms to a problem concerning the ground state structure of strongly deformed nuclei: electrostatic and strong hyperfine interactions. These effects give good access to nonspherical matter densities in the nuclear ground state. They also illustrate the use of exotic atoms as a tool in nuclear physics.

Suppose we study a nucleus such as ${}^{176}_{71}$Lu which has spin $J = 7$ and whose ground state has a strong deformation of quadrupole shape. By this we mean that the electric quadrupole moment

$$Q_s = \sqrt{\frac{16\pi}{5}} \int d^{3A}r\, \phi^2_{J,M=J}(\boldsymbol{r}_1 \cdots \boldsymbol{r}_A) \sum_{i=1}^{Z} r_i^2 Y_{20}(\hat{r}_i) \tag{98}$$

is the largest static multipole moment in the nuclear ground state $|JM\rangle$. In the case of ^{176}Lu we have $Q_s = 5.71\,\mathrm{b}$ [$1\,\mathrm{b} = 10^{-24}\ \mathrm{cm}^2$]. As a first consequence of the fact that $J \geq 1$ and that the ground state is no more spherically symmetric, the Coulomb potential (69) aquires a quadrupole interaction term

$$V_{E2}(\boldsymbol{r}) = -Z\alpha \frac{4\pi}{5} \int d^3r' \frac{r_<^2}{r_>^3} \sum_m Y^*_{2m}(\hat{r}) Y_{2m}(\hat{r}') \rho_p(\boldsymbol{r}'), \tag{99}$$

which is obtained by expanding $|\boldsymbol{r} - \boldsymbol{r}'|^{-1}$ in spherical harmonics and where $r_<, r_>$

denote, respectively, the smaller and the larger of $|\boldsymbol{r}|$ and $|\boldsymbol{r}'|$. The proton density, when computed in the magnetic substate with $M=J$, is

$$\rho_{\mathrm{p}}(\boldsymbol{r})=\frac{1}{Z}\int \mathrm{d}^{3A}r\phi_{JJ}^{2}(\boldsymbol{r}_1\cdots\boldsymbol{r}_A)\sum_{i=1}^{Z}\delta(\boldsymbol{r}-\boldsymbol{r}_i). \tag{100}$$

When expanding this function in terms of spherical harmonics $Y_{\kappa\mu}$, only those with even κ can contribute because of parity, and only $\mu=0$ can occur since initial and final state have the same projection $M=J$. Thus we may write

$$\rho_{\mathrm{p}}(\boldsymbol{r})=\rho_0^{(\mathrm{p})}(r)+\sum_{\kappa=1}^{[J]}\sqrt{\frac{4\kappa+1}{16\pi}}\,\rho_{2\kappa}^{(\mathrm{p})}(r)Y_{2\kappa,0}(\hat{r}). \tag{101}$$

The factor in the sum on the r.h.s. of eq. (101) is chosen in such a way that the second moment of $Z\rho_2^{(\mathrm{p})}$ is the spectroscopic quadrupole moment (98) without any further factor, viz.

$$Q_{\mathrm{s}}=Z\int_0^{\infty}\mathrm{d}r\,r^2\left[r^2\rho_2^{(\mathrm{p})}(r)\right]. \tag{98$'$}$$

The presence of the quadrupole interaction (99) has two important effects:

(i) Neither the nuclear spin J nor the meson's orbital angular momentum l are good quantum numbers. Instead, the atomic states must be classified by the total angular momentum $F=l+J$, which takes on the values

$$F=l+J,\, l+J-1,\ldots,|l-J|.$$

(ii) The unperturbed state l (defined for the monopole part of the Coulomb potential) is split in a series of hyperfine states which are characterized by the eigenvalues of F.

In first order perturbation theory the additional quadrupole energy of a state $(J,l)F$ is given by

$$\Delta E_{\mathrm{c}}(F)=-Z\alpha\frac{4\pi}{5}\sqrt{\frac{5}{16\pi}}\int_0^{\infty}\mathrm{d}r\,r^2R_{nl}^2(r)$$
$$\times\int_0^{\infty}\mathrm{d}r'\,r'^2\rho_2^{(\mathrm{p})}(r')\frac{r_<^2}{r_>^3}\,\langle (J,l)F|\sum_m Y_{2m}^{*}(\hat{r})Y_{2m}(\hat{r}')|(J,l)F\rangle. \tag{102}$$

The angular integrals can be done by means of standard angular momentum algebra [R6, R7] and can be written in terms of the following explicit function:

$$C(J,l,F)=\frac{3X(X+1)-4J(J+1)l(l+1)}{2J(2J-1)l(2l-1)}, \tag{103}$$

where

$$X\coloneqq F(F+1)-J(J+1)-l(l+1). \tag{104}$$

In accord with tradition in atomic physics the function $C(J, l, F)$ is defined in such a way that for $F = l + J$ (the "stretched case")

$$C(J, l, F = J + l) = 1.$$

In terms of this function one has

$$\Delta E_c(F) = A_2 C(J, l, F), \tag{105}$$

where A_2, the electric quadrupole hyperfine constant, is given by

$$A_2 = \tfrac{1}{2} Z\alpha \frac{l}{2l+3} \times \left\{ \int_0^\infty \mathrm{d}r\, r^2 R_{nl}^2(r) \left[\frac{1}{r^3} \int_0^r \mathrm{d}r'\, r'^4 \rho_2^{(\mathrm{p})}(r') + r^2 \int_r^\infty \mathrm{d}r' \frac{1}{r'} \rho_2^{(\mathrm{p})}(r') \right] \right\}, \tag{106}$$

and where R_{nl} is the mesonic bound state wave function, normalized to 1:

$$\int_0^\infty \mathrm{d}r\, r^2 R_{nl}^2(r) = 1. \tag{107}$$

Before we proceed we note two important properties of A_2, eq. (106): The observable pionic and kaonic orbits penetrate the nucleus very little. This means, in first approximation, that in eq. (99) one may take $r_< = r'$, $r_> = r_{\mathrm{meson}}$, or, equivalently, that the first integral in the square brackets on the r.h.s. of eq. (106) may be extended to infinity, while the second may be neglected. In this approximation A_2 becomes proportional to the spectroscopic quadrupole moment (98′).

$$A_2 \simeq \tfrac{1}{2} \alpha Q_s \frac{l}{2l+3} \int_0^\infty \mathrm{d}r\, r^2 R_{nl}^2(r) \frac{1}{r^3} \tag{108}$$

(see exercise 14). Secondly, as R_{nl} is complex, A_2 acquires a small imaginary part which must be interpreted as a contribution to the width. In summary, the quadrupole hyperfine constant is determined predominantly by the spectroscopic quadrupole moment, the corrections due to the finite size of the charge distribution (cf. eq. (106)) being small. At the same time, there is a moderate distortion effect on A_2, due to the deviation of the actual, complex wave function R_{nl} from what it would be if there were no optical potential.

4.1. *Quadrupole hyperfine structure in pionic atoms*

Let us now turn to the non-spherical contributions in the optical potential (67) which arise in the case of a permanently deformed nucleus (Scheck 1972, Koch et al. 1980). For simplicity we assume that neutron and proton densities are the same and that these densities are given by eq. (101). If the quadrupole term dominates we have

$$\rho(\boldsymbol{r}) = \rho_{\mathrm{p}}(\boldsymbol{r}) = \rho_{\mathrm{n}}(\boldsymbol{r}) \simeq \rho_0(r) + \sqrt{\frac{5}{16\pi}}\, \rho_2(r) Y_{20}. \tag{109}$$

Inserting this decomposition into the expression (67) for the optical potential and expanding consistently to second order in $\rho_2(r)$ leads to an analogous decomposition of the optical potential

$$U(\boldsymbol{r}) = U_{(0)}(r) + U_{(2)}(\boldsymbol{r}). \tag{110}$$

with

$$U_{(0)}(r) = -\frac{4\pi A}{2\bar{m}}\left\{q_{00}(r) - \nabla\frac{\alpha_{00}(r)}{1+\frac{4}{3}\pi\xi\alpha_{00}(r)}\nabla\right\}. \tag{111}$$

Here, the functions $q_{00}(r)$ and $\alpha_{00}(r)$ are given by the expressions (71) and (72), respectively, with all densities replaced with the common monopole density $\rho_0(r)$. Inserting this potential (111) into the Klein–Gordon equation (68) and solving this equation as described above yields the unperturbed shift and width ε_{00}, Γ_{00} of an atomic state with given quantum numbers (n, l).

The quadrupole potential $U_{(2)}(\boldsymbol{r})$ contains all terms proportional to $\rho_2(r)$, up to and including terms of second order in $\rho_2(r)$. This expansion contains the product of two spherical harmonics of like arguments

$$Y_{20}(\hat{r})Y_{20}(\hat{r}) = \frac{1}{\sqrt{4\pi}}\left(Y_{00} + \frac{2\sqrt{5}}{7}Y_{20} + \frac{6}{7}Y_{40}\right).$$

Thus, $U_{(2)}$ contains both monopole and quadrupole parts,

$$U_{(2)}(r) = -\frac{4\pi A}{2\bar{m}}\left\{q_{02}(r) - \nabla\alpha_{02}(r)\nabla + \sqrt{\frac{5}{16\pi}}\,Y_{20}(\hat{r})\left[q_2(r) - \nabla\alpha_2(r)\nabla\right]\right\}, \tag{112}$$

where

$$q_{02}(r) = \left(1+\frac{m_\pi}{2m_N}\right)B_0\frac{5}{64\pi^2}\rho_2^2(r), \tag{113a}$$

$$\alpha_{02}(r) = \frac{5}{64\pi^2}\frac{1}{D^2(r)}\left[\left(1+\frac{m_\pi}{2m_N}\right)^{-1}C_0\rho_2^2(r) - \frac{4}{3}\pi\xi\frac{A^2(r)}{D(r)}\right]. \tag{113b}$$

The auxiliary functions $D(r)$ and $A(r)$ are defined by

$$D(r) = 1 + \tfrac{4}{3}\pi\xi\alpha_{00}(r),$$

$$A(r) = \left(1+\frac{m_\pi}{m_N}\right)^{-1}\left\{c_0\rho_2(r) + c_1\frac{2Z-A}{A}\rho_2(r)\right\} + 2\left(1+\frac{m_\pi}{2m_N}\right)^{-1}C_0\rho_0(r)\rho_2(r).$$

The monopole terms in $U_{(2)}$ gives rise to additional monopole shift and width $\delta\varepsilon_0$

and $\delta\Gamma_0$, respectively. In first order perturbation theory these are

$$\delta\varepsilon_0 + \tfrac{1}{2}\mathrm{i}\delta\Gamma_0 = \frac{4\pi A}{2\bar{m}}\left\{\int_0^\infty \mathrm{d}r\, r^2 R_{nl}^2(r) q_{02}(r) + \int_0^\infty \mathrm{d}r\, r^2\left[F_1^2(r) + F_2^2(r)\right]\alpha_{02}(r)\right\}, \tag{114}$$

where $F_1(r)$ and $F_2(r)$ contain derivatives of $R_{nl}(r)$ and arise from the gradient potential,

$$F_1(r) = \sqrt{\frac{l+1}{2l+1}}\left(\frac{\mathrm{d}}{\mathrm{d}r} - \frac{l}{r}\right)R_{nl}(r), \tag{115a}$$

$$F_2(r) = \sqrt{\frac{l}{2l+1}}\left(\frac{\mathrm{d}}{\mathrm{d}r} + \frac{l+1}{r}\right)R_{nl}(r). \tag{115b}$$

[A special case ($l = n-1$ and nonrelativistic approximation) of these formulae was used in eq. (58) above.]

In the quadrupole term of $U_{(2)}(\boldsymbol{r})$ the functions $q_2(r)$ and $\alpha_2(r)$ are given by

$$q_2(r) = \left(1 + \frac{m_\pi}{m_\mathrm{N}}\right)\left\{b_0\rho_2(r) + b_1\frac{2Z-A}{A}\rho_2(r)\right\} + \left(1 + \frac{m_\pi}{2m_\mathrm{N}}\right)B_0\left[2\rho_0(r)\rho_2(r) + \frac{5}{28\pi}\rho_2^2(r)\right], \tag{116}$$

$$\alpha_2(r) = \frac{1}{D^2(r)}\left\{\left[A(r) + \frac{5}{28\pi}\left(1 + \frac{m_\pi}{2m_\mathrm{N}}\right)^{-1} C_0\rho_2^2(r)\right] - \frac{1}{D(r)}\tfrac{4}{3}\pi\xi\frac{5}{28\pi}A^2(r)\right\}, \tag{117}$$

with $A(r)$ and $D(r)$ as defined above.

The quadrupole term of $U_{(2)}(\boldsymbol{r})$, eq. (112), has the same angular structure as the electric quadrupole interaction. To see this, remember that the optical potential was based on the assumption of zero-range interactions between the pion and the nucleons. Thus, the interaction of the pion with a given nucleon i is proportional to

$$\delta(\boldsymbol{r}_\pi - \boldsymbol{r}_i) = \sum_{\lambda\mu}\frac{\delta(r_\pi - r_i)}{r_\pi r_i}Y_{\lambda\mu}^*(\hat{r}_\pi)Y_{\lambda\mu}(\hat{r}_i). \tag{118}$$

When taking the expectation value in the nuclear ground state with $M = J$, only even multipoles of eq. (118) survive. In the situation that we consider here, only the monopole and quadrupole terms are important. The angular structure of the term with $\lambda = 2$ is indeed the same as in $V_{\mathrm{E}2}(\boldsymbol{r})$, eq. (99). As a consequence, the nonspherically symmetric part of the optical potential causes an additional *quadrupole hyperfine structure* which has the same pattern as the electric hyperfine splitting, eq. (105). More precisely,

$$E_{nl}(\text{point charge}) - E_{n(l,J)F} = \left(\varepsilon_0 + \tfrac{1}{2}\mathrm{i}\Gamma_0\right) + \left(\varepsilon_2 + \tfrac{1}{2}\mathrm{i}\Gamma_2 - A_2\right)C(J,l,F), \quad (119)$$

with $\varepsilon_0 = \varepsilon_{00} + \delta_{\varepsilon 0}$ and $\Gamma_0 = \Gamma_{00} + \delta\Gamma_0$. In first order (non-relativistic) perturbation theory one finds

$$\varepsilon_2 + \tfrac{1}{2}\mathrm{i}\Gamma_2 = -\frac{5A}{4\bar{m}}\frac{l}{2l+3}\int_0^\infty \mathrm{d}r\, r^2 \times \left\{ R_{nl}^2(r) q_2(r) + \frac{\alpha_2(r)}{2l+1}\left[\frac{(l+2)(2l-1)}{l+1}F_1^2(r) + \frac{(l-1)(2l+3)}{l}F_2^2(r) + \frac{6}{\sqrt{l(l+1)}}F_1(r)F_2(r)\right]\right\}. \quad (120)$$

Clearly, in order to evaluate these expressions one needs a model for the deformed mass densities (109) from which $\delta_0(r)$ and $\delta_2(r)$ can be obtained. For example, one may describe the nucleus in the framework of the rotator model and one may assume the matter density to be, in the body fixed frame of reference,

$$\bar{\rho}(\boldsymbol{r}) = N\{1 + \exp(4\ln 3 \cdot x(\boldsymbol{r}))\}^{-1}, \quad (121a)$$

with

$$x(\boldsymbol{r}) = \frac{1}{t}\left[r - c(1 + \beta Y_{20}(\hat{\boldsymbol{r}}))\right]. \quad (121b)$$

Here c and t are parameters which characterize the radius of half-density and the skin thickness, respectively, whilst β is the deformation parameter. The details are worked out in the literature quoted above.

In order to get a feeling for the spatial structure of the mesonic atom with a deformed nucleus, we have plotted the functions $\rho_0(r)$ and $\rho_2(r)$ in fig. 7 (as calculated from the model density (121) and for $Z = 71$). Also shown is the square of the pionic 4f-wave function $y_{4f}^2 = r^2 R_{4f}^2(r)$, on the same scale. The figure demonstrates that the strong interaction, quadrupole hyperfine structure, eq. (120), probes the nuclear quadrupole mass density in the extreme nuclear surface. Fig. 8 finally, shows the details of the hyperfine structure for the case of the pionic 3d state in ^{176}Lu.

The effects were measured in a number of elements. In most cases, both the sign and the magnitude of the effects are found to be in good agreement with the predictions (Ebersold et al. 1978, Konijn et al. 1978, Batty et al. 1981, Konijn et al. 1981).

4.2. *The case of kaonic atoms*

A very similar analysis applies also to kaonic atoms. In fact, in this case, the optical potential is much simpler, for the reasons discussed in sec. 3.5. For the atoms

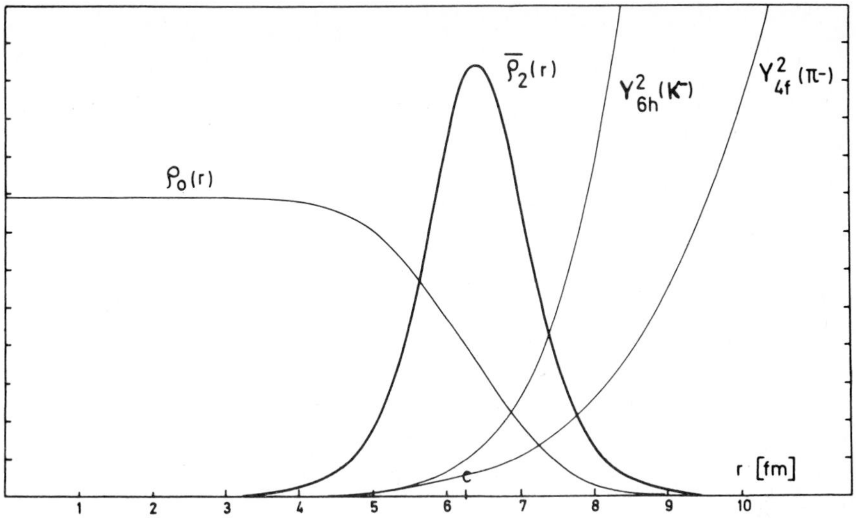

Fig. 7. Spherical and quadrupole density in the body-fixed system as calculated from the model density (121). Also shown are pionic and kaonic densities in 4f and 6h states, respectively.

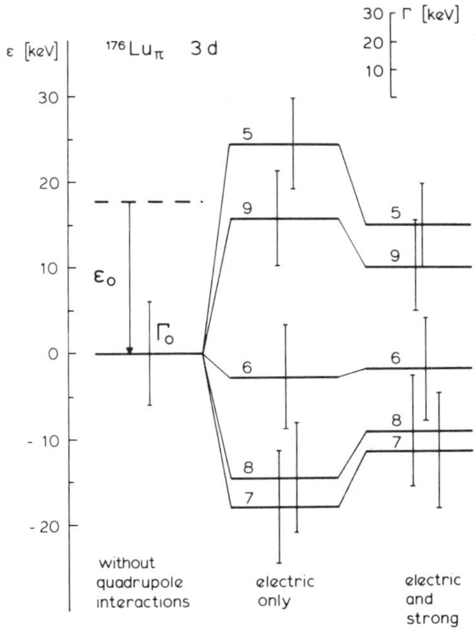

Fig. 8. Electric and strong interaction quadrupole hyperfine structure in pionic lutetium 176. The scale for the widths is shown in the insert in the upper right corner. (Taken from Koch et al. 1980.)

with deformed nuclei it reads

$$U(\boldsymbol{r}) = U_{(0)}(r) + U_{(2)}(\boldsymbol{r}), \tag{122}$$

with

$$U_{(0)}(r) = -\frac{4\pi A}{2\bar{m}}\left(1+\frac{m_{\mathrm{K}}}{m_{\mathrm{N}}}\right)\bar{a}\rho_0(r), \tag{123a}$$

$$U_{(2)}(\boldsymbol{r}) = -\frac{4\pi A}{2\bar{m}}\left(1+\frac{m_{\mathrm{K}}}{m_{\mathrm{N}}}\right)\bar{a}\sqrt{\frac{5}{16\pi}}\,\rho_2(r)Y_{20}(\hat{r}). \tag{123b}$$

In first order perturbation theory we have

$$\varepsilon_2 + \tfrac{1}{2}\mathrm{i}\Gamma_2 = -\frac{5A}{4\bar{m}}\left(1+\frac{m_{\mathrm{K}}}{m_{\mathrm{N}}}\right)\frac{l}{2l+3}\int_0^{\infty}\mathrm{d}r\, r^2R_{nl}^2(r)\rho_2(r). \tag{124}$$

Like in the previous case, eq. (124) shows that these effects probe the quadrupole matter density in the extreme nuclear surface region. This is illustrated by fig. 7 which also shows the squared kaonic 6h wave function $y_{6\mathrm{h}}^2 = r^2R_{6\mathrm{h}}^2(r)$ for $_{71}$Lu, in relation to the quadrupole density $\rho_2(r)$.

Generally, the effects are larger in kaonic atoms than in pionic atoms because the kaonic potential does not have the compensating tendencies of s-wave and p-wave potentials. Measured values in selected nuclei can be found in the literature.

References

Batty, C.J., S.F. Biagi, E. Friedman, S.D. Hoath, J.D. Davies, G.J. Pyle, G.T.A. Squier, D.M. Asbury, and A. Gubermann, 1979, *Nucl. Phys.* A322, 445.

Batty, C.J., S.F. Biagi, M. Blecher, S.D. Hoath, R.A.J. Riddle, B.L. Roberts, J.D. Davies, G.J. Pyle, G.T.A. Squier, D.M. Asbury, and A.S. Clough, 1979, *Nucl. Phys.* A329, 407.

Batty, C.J., S.F. Biagi, R.A.J. Riddle, B.L. Roberts, G.J. Pyle, G.T.A. Squier, D.M. Asbury, and A.S. Clough, 1981, *Nucl. Phys.* A355, 383.

Brockmann, R., W. Weise, and L. Tauscher, 1978, *Nucl. Phys.* A308, 365.

Ebersold, P., B. Aas, W. Dey, R. Eichler, H.J. Leisi, W.W. Sapp, and F. Scheck, 1978, *Nucl. Phys.* A296, 493.

Ericson, T.E.O. and M. Krell, 1969, *Nucl. Phys.* B11, 521.

Ericson, T.E.O. and F. Scheck, 1970, *Nucl. Phys.* B19, 450.

Koch, J.H. and M.M. Sternheim, 1972, *Phys. Rev. Lett.* 28, 1061.

Koch, J.H. and F. Scheck, 1980, *Nucl. Phys.* A340, 221.

Konijn, J., J.K. Panman, J.H. Koch, W. van Doesburg, G.T. Ewan, T. Johansson, G. Tibell, K. Fransson, and L. Tauscher, 1978, *Nucl. Phys.* A326, 401.

Konijn, J., W. van Doesburg, G.T. Ewan, T. Johansson, and G. Tibell, 1981, *Nucl. Phys.* A360, 187.

Leon, M., 1976, *Nucl. Phys.* A260, 461.

Leon, M., J.N. Bradbury, P.A.M. Gram, R.L. Hutson, M.E. Schillaci, C.K. Hargrove, and J.J. Reidy, 1979, *Nucl. Phys.* A322, 397.

Nagels, M.M., J.J. de Swart, H. Nielsen, G.C. Oades, H.L. Petersen, B. Tromborg, G. Gustafson, A.C. Irving, C. Jarlskog, W. Pfeil, H. Pilkuhn, F. Steiner, and L. Tauscher, 1976, *Nucl. Phys.* B109, 1.

Scheck, F., 1972, *Nucl. Phys.* B42, 573.

Tauscher, L., 1977, *Proc. International School on the Physics of Exotic Atoms*, Erice.

Exercises

1. Consider all elastic and charge exchange scattering amplitudes of two hadrons with isospins $I^{(1)}$ and $I^{(2)}$, respectively. How many of such amplitudes are there and how many relations between them are there if the T-matrix is a scalar under isospin?

2. Consider an exotic atom with an Ω^- bound in the field of a proton. What are the typical energy scales of its atomic cascade? What could be the principal decay channels of the system?

3. Consider a nucleon–antinucleon bound state $|\mathrm{N\bar{N}}; I, I_3, l, S\rangle$ with relative angular momentum l, total spin S, and coupled to good isospin I, I_3. Express the states $|\mathrm{p\bar{p}}; l, S\rangle$ and $|\mathrm{n\bar{n}}; l, S\rangle$ in terms of the states with good I. Show that the former are eigenstates of G-parity, $G = C\exp(\mathrm{i}\pi I_2)$, if $I_3 = 0$, the eigenvalues being $(-)^{l+S+I}$.

4. Study the selection rules in the decay of the coupled states $|\mathrm{N\bar{N}}; I, I_3 = 0, l, S\rangle$ as well as of the states $|\mathrm{p\bar{p}}; l, S\rangle$ and $|\mathrm{n\bar{n}}; l, S\rangle$ into n pions, which follow from the results of exercise 3. (Pions have negative G-parity.)

5. a) Knowing that the Δ-resonance carries isospin $\frac{3}{2}$ and occurs in the charge states $\Delta^{++}, \Delta^{+}, \Delta^{0}, \Delta^{-}$, calculate the ratio of the decay widths $\Gamma(\Delta^{++} \to \pi + \mathrm{p})$ and $\Gamma(\Delta^{-} \to \pi^{-}\mathrm{n})$.

b) The deuteron being an isosinglet, calculate the ratio of the cross sections $\sigma(\mathrm{pp} \to \pi + \mathrm{d})$ and $\sigma(\mathrm{pn} \to \pi^0\mathrm{d})$ (the kinematic conditions being the same in the two reactions).

6. Assume that pion–nucleon scattering be completely dominated by the $l = 1$, $I = J = \frac{3}{2}$ partial wave. Calculate the angular distribution in the c.m. system, for unpolarized nucleon target. Repeat the calculation for $l = 2$, $I = J = \frac{3}{2}$ and compare to the previous results.

7. Making use of eq. (IV.150), estimate the cascade time for an atom with antiprotons.

8. In first Born approximation the scattering amplitude is given by

$$f(\theta) = -\frac{2\overline{m}}{4\pi}\int \mathrm{d}^3 r \mathrm{e}^{-\mathrm{i}\boldsymbol{k}\boldsymbol{r}} V(r) \mathrm{e}^{\mathrm{i}\boldsymbol{k}'\boldsymbol{r}} \equiv \sum_{l=0}^{\infty} f_l(\theta).$$

Expanding the two exponentials separately in terms of spherical harmonics and Bessel functions, prove the expression

$$f_l(\theta) = -\frac{2\overline{m}}{4\pi}(2l+1)\int j_l^2(kr) V(r) r^2 \mathrm{d}r \, P_l(\hat{k}\cdot\hat{k}').$$

9. Let A_0 be the π–nucleus scattering length,

$$A_0 \simeq A b_0 + (N - Z) b_1,$$

and let $V(r)$ be the equivalent potential which yields the π^-–nucleus scattering amplitude in first Born approximation. Show that

$$A_0 = -\frac{\overline{m}}{2\pi}\int V(r) r^2 \mathrm{d}r.$$

Calculate the energy shift ΔE_{n0} of the s-states and compare $\Delta E_{n0}/E_{n0}$ with actual values of the shifts in s-states.

10. Following the method of exercise 9 show that for p-states

$$\Delta E_{n1}/E_{n1} \simeq -\left(1-\frac{1}{n^2}\right)\frac{4}{n}\frac{A_1}{a_\pi^3},$$

where a_π is the Bohr radius. Can you explain why this naive estimate of p-state shifts fails quantitatively? What about d- and f-states?

11. At low energies the scattering phase $\delta(\kappa)$ is related to the scattering length a and the effective range r_0 by ctg $\delta = 1/a\kappa + \frac{1}{2}r_0\kappa$. Use this formula in expressing the partial wave amplitude $f=(1/\kappa)e^{i\delta}\sin\delta$ in terms of a, r_0 and κ. Apply the formula that you obtain to $\overline{\mathrm{K}}$N s-wave scattering in the isospin zero channel. If $a=(-1.7+i0.7)$ fm and if r_0 is neglected, show that the formula predicts a bound state. Compare to the parameters of the resonance (91).

12. Calculate the densities $\rho_0(r)$ and $\rho_2(r)$, eqs. (101) and (109), for the example

$$\rho(\boldsymbol{r}) = N\{1+\exp(x(\boldsymbol{r})\,4\ln 3)\}^{-1},$$

where

$$x(\boldsymbol{r}) = \frac{1}{t}\left[r - c(1+\beta Y_{20})\right].$$

Calculate the normalization factor N up to second order in β.

13. Assume the nuclear densities to be homogeneous ones. Estimate the pionic s-wave and p-wave potentials for the parameters (79) and (80). Do the same for the kaonic potential (92), inserting the parameters (97).

14. Calculate the expectation value $\langle 1/r^3\rangle$ for circular orbits, $n=l+1$, in pointlike Coulomb field. Compare the results for relativistic and nonrelativistic wave functions.

15. Show that the operators defined in eqs. (21) are indeed projection operators and project onto states with $j=l\pm\frac{1}{2}$, respectively. *Hint*: Make use of the relations for $\boldsymbol{t}\cdot\boldsymbol{\tau}$ and $\boldsymbol{\sigma}\cdot\boldsymbol{l}$ on p. 59.

Chapter III

FERMION FIELDS AND THEIR PROPERTIES

In this chapter we study the description of particles with spin 1/2 in a Lorentz invariant theory. The Dirac equation is derived in its natural framework: the spinor representations of SL(2, $\mathbb{C}$). The properties of this equation and of its solutions are discussed in detail. The quantization of Dirac fields is developed in the light of covariance and causality, and all important consequences of quantization are worked out.

Sections 1 to 7 contain the fundamentals while the remaining sections 8 and 9 deal with questions of practical importance in embedding fermions in a more comprehensive theory of elementary particles: the description of masses in the Lagrange density; the description of spin in the language of (covariant) density matrices; and the coupling of a charged Dirac field to the electromagnetic field. The case of the coupling to a non-Abelian local gauge field is treated in Chap. V, in the context of weak interactions.

1. Lorentz group and SL(2, $\mathbb{C}$)

The natural basis for the construction of a Lorentz invariant theory of free fermions (i.e. particles with spin 1/2) is provided by the isomorphism that links the group of proper, orthochronous Lorentz transformations to SL(2, $\mathbb{C}$), the linear group in two complex dimensions. This connection is completely analogous to the relation between the rotation group SO(3) and the special, unimodular group SU(2) in two dimensions. It is not difficult to construct spinor representations of SL(2, $\mathbb{C}$). One then discovers, in fact, that SL(2, $\mathbb{C}$)—in contrast to SU(2)—admits *two* inequivalent spinor representations both of which are relevant for the Lorentz covariant description of fermions. This connection is essential for the understanding of the properties of Dirac theory, such as the space–time structure of Dirac fields, the particle–antiparticle symmetry, and the relation between spin and statistics. These matters are no more difficult to understand than the traditional, historical approach to Dirac theory but from a logical point of view this approach is much more satisfactory. The reader should not be frightened by the few group theoretical terms that we will use and which are necessary for the understanding of these matters. The text is sufficiently self-contained so that it should be understandable even on a rudimentary basis of knowledge in group theory.

We begin by working out the precise relationship between $L_+^\uparrow$, the proper orthochronous Lorentz group, and SL(2, $\mathbb{C}$). This we do by establishing the following correspondence between the four-vectors x^μ in the space–time continuum and the hermitean matrices in two complex dimensions.

To any four-vector $x = \{x^0, \boldsymbol{x}\} \equiv \{x^\mu\}$ let there correspond a two-dimensional hermitean matrix X which is constructed as follows:

$$X_{ik} := x^0 \delta_{ik} + \boldsymbol{x}(\boldsymbol{\sigma})_{ik} = \begin{pmatrix} x^0 + x^3 & x^1 - \mathrm{i}x^2 \\ x^1 + \mathrm{i}x^2 & x^0 - x^3 \end{pmatrix}. \tag{1}$$

$\boldsymbol{\sigma}$ denotes the three Pauli matrices*⁾

$$\sigma^{(1)} = \begin{pmatrix} 0 & 1 \\ 1 & 0 \end{pmatrix}, \quad \sigma^{(2)} = \begin{pmatrix} 0 & -\mathrm{i} \\ \mathrm{i}, & 0 \end{pmatrix}, \quad \sigma^{(3)} = \begin{pmatrix} 1 & 0 \\ 0 & -1 \end{pmatrix}. \tag{2}$$

Eq. (1) can be written in a somewhat more compact notation if we define

$$\sigma^0 := \mathbb{1} = \begin{pmatrix} 1 & 0 \\ 0 & 1 \end{pmatrix} \tag{3}$$

and introduce the contravariant or covariant set of four matrices, respectively,

$$\sigma^\mu := \{\sigma^0, -\boldsymbol{\sigma}\}. \tag{4a}$$

$$\sigma_\mu := g_{\mu\nu}\sigma^\nu = \{\sigma^0, \boldsymbol{\sigma}\}, \tag{4b}$$

so that eq. (1) reads

$$X = x^0 \mathbb{1} + \boldsymbol{x} \cdot \boldsymbol{\sigma} = x_\mu \sigma^\mu = x^\mu \sigma_\mu. \tag{5}$$

A first observation is that the invariant norm of x^μ, $x^2 = (x^0)^2 - (\boldsymbol{x})^2$, is identical with the determinant of X,

$$\det X = (x^0)^2 - (\boldsymbol{x})^2. \tag{6}$$

The prescription (1) or (5) establishes an *isomorphism* between the space of the vectors x^μ (Minkowski space) and the space of the two-dimensional hermitean matrices $\mathscr{H}(2)$.

Let us consider a Lorentz transformation $\Lambda \in L_+^\uparrow$, (i.e., a proper, orthochronous Lorentz transformation) that takes vector x into vector y,

$$\Lambda : x \mapsto y, \quad y = \Lambda x.$$

The connection between the corresponding matrices $Y = y_\mu \sigma^\mu$ and X must be of the form (recall that X, Y are hermitean)

$$Y = AXA^\dagger, \tag{7}$$

where A is a nonsingular complex 2×2 matrix. $A^\dagger$ denotes the hermitean conjugate matrix. In order to ensure invariance of the norm $y^2 = x^2$, we impose the condition**⁾

$$\det A = 1. \tag{8}$$

The 2×2 complex matrices with determinant 1 form a group: the *special linear group* (or: *unimodular group*) in two complex dimensions, denoted by SL(2, $\mathbb{C}$). The

*⁾We write the "number" of any Pauli matrix (2) in parentheses to avoid confusion with upper and lower indices as in eqs. (4).

**⁾Eq. (7) only requires $|\det A| = 1$. However, as $A(\Lambda)$ must be deformable continuously into $\pm$ the unit matrix, we must require condition (8).

matrix A, eq. (7), is a function of the Lorentz transformation Λ. Actually, if $A(\Lambda)$ is a representative of Λ in $\mathscr{H}(2)$ then so is $-A(\Lambda)$, since the relation (7) is invariant under this change of sign and since $-A(\Lambda)$ belongs to SL(2, $\mathbb{C}$), too. The precise correspondence between the group of proper orthochronous Lorentz transformations $L_+^\uparrow$ and SL(2, $\mathbb{C}$) is established by the following two theorems.

Theorem (I): For any $\Lambda \in L_+^\uparrow$ there exists a matrix $A(\Lambda) \in \mathrm{SL}(2,\mathbb{C})$ such that if

$$y = \Lambda x \quad \text{then} \quad Y = A(\Lambda) X A^\dagger(\Lambda),$$

A is determined up to a sign.

Conversely,

Theorem (II): To any $A \in \mathrm{SL}(2,\mathbb{C})$ corresponds a unique $\Lambda \in L_+^\uparrow$. This Λ is the image of both A and $-A$.

These theorems express the fact that the mapping of SL(2, $\mathbb{C}$) onto $L_+^\uparrow$ is a homomorphism. Furthermore, as the set of elements $Z = \{\mathbb{1}, -\mathbb{1}\}$ forms an invariant subgroup of SL(2, $\mathbb{C}$), i.e.

$$AZA^{-1} = Z \quad \forall A \in \mathrm{SL}(2,\mathbb{C}), \tag{9}$$

it is useful to introduce the factor group denoted $\mathrm{SL}(2,\mathbb{C})/Z$. This is the group of the cosets $\{AZ\}$ of the invariant subgroup Z. Thus, theorems (I) and (II) establish the isomorphism

$$L_+^\uparrow \cong \mathrm{SL}(2,\mathbb{C})/\{\mathbb{1}, -\mathbb{1}\} \tag{10}$$

between the proper orthochronous Lorentz group and the factor group $\mathrm{SL}(2,\mathbb{C})/Z$.

Theorem (II) is obvious from eqs. (7) and (8). Indeed, any $A \in \mathrm{SL}(2,\mathbb{C})$ induces a transformation $X \to Y$ such that $y^2 = x^2$. This means that x and y are related by a Lorentz transformation $\Lambda(A)$. As A can be deformed continuously into unity, Λ must belong to $L_+^\uparrow$.

The explicit calculation of Λ from a given A is somewhat technical and shall not be worked out here.*)

The proof of theorem (I) proceeds by explicit construction of $A(\Lambda)$ for given $\Lambda \in L_+^\uparrow$. According to theorem (A1)**) Λ can be decomposed, in a unique way, into a boost $L(\boldsymbol{v})$, followed by a rotation R, $\Lambda = R \cdot L(\boldsymbol{v})$. Thus it is sufficient to consider these two kinds of transformations separately. In the following we denote the image of R in SL(2, $\mathbb{C}$) by $U(R)$, the image of $L(\boldsymbol{v})$ by $H(\boldsymbol{v})$, in order to underline the fact that $U(R)$ will turn out to be unitary, while $H(\boldsymbol{v})$ will be found to be hermitean.

(i) *Rotations.* It is useful to parametrize the rotation R by a set of Euler angles ϕ, θ, ψ which we define as indicated in Fig. 1. One then verifies, through explicit

*) This construction can be found for example in Ref. R8.

**) See App. A, p. 375.

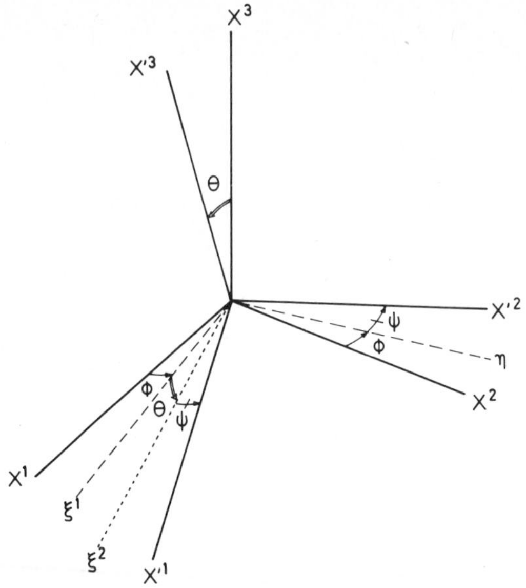

Fig. 1. Definition of Euler's angles. First a rotation by the angle ϕ about the x^3-axis, then a rotation by the angle θ about the η-axis, and finally a rotation by the angle ψ about the x'^3-axis.

calculation, that

$$R \leftrightarrow \pm U(R), \tag{11}$$

with

$$U(R) = e^{(i/2)\psi\sigma^{(3)}} e^{(i/2)\theta\sigma^{(2)}} e^{(i/2)\phi\sigma^{(3)}} \tag{12}$$

do indeed represent the rotation R. For instance, for a rotation about the 3-axis one has

$$Y = AXA^\dagger = \begin{pmatrix} e^{i\phi/2} & 0 \\ 0 & e^{-i\phi/2} \end{pmatrix} \begin{pmatrix} x^0 + x^3 & x^1 - ix^2 \\ x^1 + ix^2 & x^0 - x^3 \end{pmatrix} \begin{pmatrix} e^{-i\phi/2} & 0 \\ 0 & e^{i\phi/2} \end{pmatrix}$$

$$= \begin{pmatrix} x^0 + x^3 & e^{i\phi}(x^1 - ix^2) \\ e^{-i\phi}(x^1 + ix^2) & x^0 - x^3 \end{pmatrix},$$

which can be rewritten in the more familiar form

$$y^0 = x^0, \qquad y^1 = x^1 \cos\phi + x^2 \sin\phi,$$

$$y^3 = x^3, \qquad y^2 = -x^1 \sin\phi + x^2 \cos\phi.$$

Similarly, for a rotation about the 2-axis we use

$$e^{(i/2)\theta\sigma^{(2)}} = \mathbb{1}\cos\theta/2 + i\sigma^{(2)}\sin\theta/2 = \begin{pmatrix} \cos\theta/2 & \sin\theta/2 \\ -\sin\theta/2 & \cos\theta/2 \end{pmatrix}$$

and verify that $\exp\{(i/2)\theta\sigma^{(2)}\} X \exp\{-(i/2)\theta\sigma^{(2)}\}$, when written out in terms of

Cartesian coordinates, does indeed yield the expected result:

$$y^0 = x^0, \qquad y^1 = x^1\cos\theta - x^3\sin\theta,$$

$$y^2 = x^2, \qquad y^3 = x^3\cos\theta + x^1\sin\theta.$$

(The reader should verify this.)

As the composition of three Euler rotations $R_3(\psi)R_2(\theta)R_3(\phi)$ is the most general case, we have thus shown that for any rotation $\Lambda = R$ there exists a matrix $U(R) \in \mathrm{SL}(2,\mathbb{C})$. This matrix is given by expression (1).

In fact, $U(R)$ is *unitary* and, therefore, belongs to SU(2), the special group of unitary transformations U in two dimensions. SU(2) is a subgroup of SL(2, ℂ). This is no surprise when one recalls that the rotation group SO(3) is a subgroup of $\mathrm{L}_+^\uparrow$. Thus we recover the well-known isomorphism

$$\mathrm{SO}(3) \cong \mathrm{SU}(2)/\{\mathbb{1}, -\mathbb{1}\} \tag{13}$$

that is the basis for constructing the representations of the rotation group.*)

(ii) *Special Lorentz transformations*. Let $\boldsymbol{v}$ be the velocity vector characterizing the "boost" $L(\boldsymbol{v})$. Then $\boldsymbol{w} := \boldsymbol{v}/|\boldsymbol{v}|$ is the unit vector in the direction of the boost, and

$$\boldsymbol{v} = \boldsymbol{w}\,\mathrm{tgh}\,\lambda, \tag{14}$$

where λ is the rapidity. The image of $L(\boldsymbol{v})$ is

$$L(\boldsymbol{v}) \leftrightarrow \pm H(\boldsymbol{v}) \tag{15}$$

with

$$H(\boldsymbol{v}) = e^{(1/2)\lambda\boldsymbol{\sigma}\cdot\boldsymbol{w}}$$

This can be seen as follows: By an appropriate rotation of the space coordinate, choose the 3-direction to coincide with $\boldsymbol{w}$. Then

$$H(\boldsymbol{v}) = \begin{pmatrix} e^{\lambda/2} & 0 \\ 0 & e^{-\lambda/2} \end{pmatrix}$$

and

$$Y = HXH^\dagger = \begin{pmatrix} e^{\lambda/2} & 0 \\ 0 & e^{-\lambda/2} \end{pmatrix}\begin{pmatrix} x^0 + x^3 & x^1 - ix^2 \\ x^1 + ix^2 & x^0 - x^3 \end{pmatrix}\begin{pmatrix} e^{\lambda/2} & 0 \\ 0 & e^{-\lambda/2} \end{pmatrix}$$

$$= \begin{pmatrix} e^{\lambda}(x^0 + x^3) & x^1 - ix^2 \\ x^1 + ix^2 & e^{-\lambda}(x^0 - x^3) \end{pmatrix},$$

so that

$$y^1 = x^1, \qquad y^0 = x^0\cosh\lambda + x^3\sinh\lambda,$$

$$y^2 = x^2, \qquad y^3 = x^0\sinh\lambda + x^3\cosh\lambda,$$

as expected. So $\pm H(\boldsymbol{v})$, eq. (15), is indeed the image of the given boost $L(\boldsymbol{v})$.

*) See e.g. Ref. R9.

Using the known properties of the Pauli matrices we can also write

$$H(\boldsymbol{v}) = \mathbb{1}\cosh(\lambda/2) + \boldsymbol{\sigma}\cdot\boldsymbol{w}\sinh(\lambda/2). \tag{16}$$

Note, in particular, that $H(\boldsymbol{v})$ is a linear combination of the hermitean matrices $\{\sigma^\mu\}$, with real coefficients, and, therefore, is *hermitean*.

In summary: Any Lorentz transformation $\Lambda \in L_+^\uparrow$ can be decomposed uniquely into a boost followed by a rotation,

$$\Lambda = R\cdot L(\boldsymbol{v}),$$

with R and $\boldsymbol{v}$ obtained as described in Appendix A. Its image under the isomorphism (10) is $\pm A(\Lambda) \in \mathrm{SL}(2,\mathbb{C})$, where

$$A(\Lambda) = U(R)H(\boldsymbol{v}), \tag{17}$$

and where $U(R)$ and $H(\boldsymbol{v})$ are given by eqs. (12) and (16), respectively. Note that we have recovered the fact that any matrix of $\mathrm{SL}(2,\mathbb{C})$ can be written as a product of a unitary and a hermitean matrix with determinant 1. This is the generalization of the well-known decomposition of a complex number into its modulus (which is a real number) and a phase factor.

2. Spinor representations and spinor fields, "dotted" and "undotted" spinors

We can now proceed to the construction of spinor representations of $\mathrm{SL}(2,\mathbb{C})$ and, thereby, spinor representations of the Lorentz group. By definition, a spinor is a two component object which transforms under rotations of the space coordinates with the unitary matrix $D^{(1/2)}(\psi,\theta,\phi)$. More precisely, let χ_m ($m = +1/2$ or $-1/2$) denote the basis of an irreducible representation of the rotation group with $j = 1/2$. A state vector ψ in this space can be expanded in terms of this basis,

$$\psi = \sum_m c_m \chi_m. \tag{18}$$

In accordance with our conventions, the transformation law for the expansion coefficients is

$$c'_\mu = \sum_m D^{(1/2)}_{\mu m}(\psi,\theta,\phi)c_m, \tag{19}$$

while the base transforms with the transformation matrix $\tilde{D}^{-1}$, contragredient to D. As $\tilde{D}^{-1} = D^*$, we have

$$\chi'_\mu = \sum_m D^{(1/2)*}_{\mu m}\chi_m. \tag{20}$$

With respect to $\mathrm{SL}(2,\mathbb{C})$ we already noted that there are two inequivalent spinor representations of $\mathrm{SL}(2,\mathbb{C})$. It is not difficult to construct these representations on the basis of the results of the preceding paragraph. Let $A(\Lambda)$ be the image of an arbitrary Lorentz transformation $\Lambda \in L_+^\uparrow$. One sees at once that there are four types of spinors, whose behaviour under a given Lorentz transformation is indicated in the following table*):

Spinors	Transformation matrix under L.T.	
$\{c_a\}=\{c_1,c_2\}$	$A(\Lambda)$	(21a)
$\{c^a\}=\{c^1,c^2\}$	$\tilde{A}^{-1}(\Lambda)$	(21b)
$\{c_A\}=\{c_{\mathrm{I}},c_{\mathrm{II}}\}$	$A^*(\Lambda)$	(21c)
$\{c^A\}=\{c^{\mathrm{I}},c^{\mathrm{II}}\}$	$\hat{A}:=(\tilde{A}^*)^{-1}=(A^\dagger(\Lambda))^{-1}$	(21d)

As is evident, spinors $\{c^a\}$ with the transformation law (21b) are contragredient to spinors $\{c_a\}$ with the transformation law (21a). Similarly, spinors $\{c^A\}$ with transformation law (21d) and spinors $\{c_A\}$ with transformation law (21c) are contragredient to each other.

If we take Λ to be a *rotation* R, i.e. if Λ belongs to the subgroup SO(3) of $L_+^\uparrow$, then $A(\Lambda=R)\equiv U(R)$ is a unitary matrix. In this case we have $\hat{A}=A$, and $\tilde{A}^{-1}=A^*$. Therefore, we find that with respect to rotations

$$\{c_a\}\sim\{c^A\},\qquad \{c^a\}\sim\{c_A\}, \tag{22}$$

where the symbol $\sim$ means "transforms as". Obviously, this is not true for those Lorentz transformations Λ which include boosts and for which $\hat{A}$ is not equal to A.

From the theory of the rotation group**) it is known that any irreducible spherical tensor of arbitrary rank κ can be transformed to a contragredient one by applying to it a rotation of an angle π about the 2-axis. With standard phase conventions for the representation of angular momentum operators, the general transformation matrix to contragredience is

$$(\mathrm{e}^{\mathrm{i}\pi J_2})_{\mu m}=D^{(\kappa)}_{\mu m}(0,\pi,0)=(-)^{\kappa-\mu}\delta_{\mu,-m}. \tag{23}$$

For integer κ this matrix is symmetric, for half-integer it is antisymmetric.

Specifically, for spinors ($\kappa=\frac{1}{2}$) the relation is***)

$$c^a=\left(\mathrm{e}^{\mathrm{i}\pi\sigma^{(2)}/2}\right)^{ab}c_b=(\mathrm{i}\sigma^{(2)})^{ab}c_b. \tag{24}$$

Let us introduce the notation

$$\varepsilon:=\mathrm{e}^{\mathrm{i}\pi\sigma^{(2)}/2}=\mathrm{i}\sigma^{(2)}=\begin{pmatrix}0&1\\-1&0\end{pmatrix}. \tag{25}$$

and let $\varepsilon^{ab}\equiv\varepsilon$, so that $c^a=\varepsilon^{ab}c_b$. Then evidently $\varepsilon_{ab}=\varepsilon^{-1}$ with

$$\varepsilon^{-1}=\tilde{\varepsilon}=-\varepsilon. \tag{26}$$

So far, the relationship (24) is proven only with respect to rotations. It is not difficult, however, to show that it is true also for boosts, and therefore for any $\Lambda\in L_+^\uparrow$ (see exercise 12). In the case of SL(2, $\mathbb{C}$) spinors there is another and perhaps

*) We recall: $\tilde{A}$ is the transposed, A^* the complex conjugate of A, and $A^\dagger=\tilde{A}^*$ is the hermitean conjugate.

**) See e.g. Ref. R4, where this is called the "U-transformation".

***) Summation over repeated, contragredient indices is implied.

more direct way of seeing this, by noting that both

$$\left(b_1c^1 + b_2c^2\right) \quad \text{and} \quad \left(b_1c_2 - b_2c_1\right) = \det\begin{pmatrix} b_1 & b_2 \\ c_1 & c_2 \end{pmatrix}$$

are SL(2, ℂ) invariants.*) From this observation we conclude that

$$c^1 \sim c_2 \quad \text{and} \quad c^2 \sim -c_1$$

under any $\Lambda \in L_+$. Thus transition to contragredience is indeed effected by the transformation (24).

It is clear that there must be analogous relationships for spinors of the second kind, viz.

$$c_A = \varepsilon_{AB} c^B. \tag{27}$$

As c^a and c_A have the same behaviour under rotations, it is convenient to fix relative signs by taking

$$\varepsilon_{AB} = \varepsilon^{ab} \equiv \varepsilon, \tag{28}$$

so that the inverse transformation is $\varepsilon^{AB} = \varepsilon_{ab} = \varepsilon^{-1} = \tilde{\varepsilon} = -\varepsilon$. Using the fact that the transposed of ε is the same as its inverse, we note the relationships

$$\tilde{A}^{-1} = \varepsilon A \tilde{\varepsilon}, \tag{29a}$$

$$\hat{A} = \varepsilon A^* \tilde{\varepsilon}. \tag{29b}$$

From these relations one derives, in particular,

$$\varepsilon = A \varepsilon \tilde{A}, \qquad \varepsilon = A^* \varepsilon A^\dagger. \tag{30}$$

These relations show explicitly that ε is *invariant* under all SL(2, ℂ) transformations. In fact, one verifies that ε is the only tensor which is invariant under SL(2, ℂ). We shall make use of this important fact below, in connection with the quantization of the Dirac field.

This spinor calculus was introduced by van der Waerden (van der Waerden, 1929). Following van der Waerden, spinors of the second kind are often denoted thus: $\{c_{\dot{a}} = c_{\dot{1}}, c_{\dot{2}}\}$, i.e. by giving them dotted indices, in order to distinguish them from spinors of the first kind which carry undotted indices. We find it less confusing to label them with capital letters and roman numerals instead. Thus our notation is: *small* indices and *arabic* numerals refer to spinors of the first kind ("undotted spinors"); *capital* indices and *roman* numerals refer to spinors of the second kind ("dotted spinors").

The following rather obvious remarks are useful for the sequel.

(i) Indices of a spinor or, more generally, of a tensor carrying spinor indices, can be "raised" or "lowered" in the standard manner, by multiplying the spinor (or tensor) with ε and $\tilde{\varepsilon}$, respectively.

*) SL(2, ℂ) is characterized by the invariant skew-symmetric scalar product $b_a \varepsilon^{ab} c_b$ and hence is isomorphic to Sp(2, ℂ), the symplectic group in two complex dimensions.

(ii) Any equation containing spinors that is to be Lorentz invariant (more precisely: invariant with respect to $L_+^\uparrow$) must contain the same number of small indices and the same number of capital indices on either side.

(iii) Complex conjugation means exchanging small and capital indices (compare eqs. (21a) with (21c), (21b) with (21d)).

(iv) Only the summation over *contragredient* indices, either two small or two capital (Verjüngung) is an invariant operation. Summation over two unlike indices cannot be invariant. As a consequence tensors of the type $t^{ab\ldots m;\,AB\ldots M}$ which are separately symmetric in their small as well as in their capital indices, are irreducible. Indeed all tensors of the type $\varepsilon_{ab}t^{ab\ldots m;\,AB\ldots M}$ and $\varepsilon_{AB}t^{ab\ldots m;\,AB\ldots M}$ vanish if t has the indicated symmetry.

For a tensor it is irrelevant in which order the group of small indices and the group of capital indices are written, relative to each other. For example $t^{aB}=t^{Ba}$, $t^{ab,C}=t^{C,ab}=t^{aCb}$.

(v) Returning to the definition (5) we started from, it is clear that $X=\sigma_\mu x^\mu$ is such a tensor with one small and one capital index. Indeed, when

$$x\mapsto x'=\Lambda x \quad \text{then} \quad X\mapsto X'=A(\Lambda)XA^\dagger(\Lambda),$$

which means that X, when written out explicitly, carries indices as indicated here:

$$X\equiv X_{aB}=(\sigma_\mu)_{aB}x^\mu=(\sigma^\mu)_{aB}x_\mu. \tag{31}$$

Equipped with this knowledge we can now define two classes of *spinor fields* by means of the transformation behaviour under SL(2, $\mathbb{C}$).

Spinor fields of the *first* kind are denoted by

$$\phi(x)=\begin{pmatrix}\phi_1(x)\\ \phi_2(x)\end{pmatrix}. \tag{32a}$$

Under $\Lambda\in L_+^\uparrow$ they transform according to

$$\phi(x)\mapsto\phi'_a(x'=\Lambda x)=A(\Lambda)_a{}^b\phi_b(x). \tag{32b}$$

Spinor fields of the *second* kind are denoted by

$$\chi(x)=\begin{pmatrix}\chi^{\mathrm{I}}(x)\\ \chi^{\mathrm{II}}(x)\end{pmatrix} \tag{33a}$$

and transform according to the law

$$\chi(x)\mapsto\chi'^A(x'=\Lambda x)=(A^*(\Lambda))^A{}_B\chi^B(x). \tag{33b}$$

(Note that $(A^*)^A{}_B=\varepsilon^{AM}A^{*\,N}_M\varepsilon_{NB}=\tilde{A}^{*\,-1}\equiv\hat{A}$.)

The spinor fields have the same transformation behaviour under rotations R. Indeed, for $\Lambda=R$ they transform with the matrix $D^{(1/2)}(\psi,\theta,\phi)$, as indicated in eq. (19). They behave differently, however, under more general transformations $\Lambda\in L_+^\uparrow$, for which $\tilde{A}^{-1}$ and A^* are not the same.

For later purpose and in order to get some practice in working with SL(2, $\mathbb{C}$) spinor fields of first and second kind, we recapitulate the remarks (i) to (v) above and write out some examples in applying them to spinor fields. According to remark (iii) complex conjugation transforms small into capital, and capital into small indices, viz.

$$(\phi_a(x))^* = \phi^*{}_A(x),$$

$$(\chi^A(x))^* = \chi^{*a}(x). \tag{34a}$$

This applies whenever the index has its normal position, i.e. is a *lower*, *small* index (i.e., a spinor of first kind), or an *upper*, *capital* index (i.e. a spinor of second kind). If that index was raised, or lowered, by means of the ε-tensor (28), then there is an extra minus sign which follows from the rules but should not be forgotten:

$$(\phi^a)^* = (\varepsilon^{ab}\phi_b)^* = \varepsilon^{ab}\phi^*{}_B = -\varepsilon^{AB}\phi^*{}_B = -\phi^{*A},$$

$$(\chi_A)^* = (\varepsilon_{AB}\chi^B)^* = \varepsilon_{AB}\chi^{*b} = -\varepsilon_{ab}\chi^{*b} = -\chi^*{}_a. \tag{34b}$$

This minus sign is a consequence of the convention (28), according to which $\varepsilon^{AB} = -\varepsilon^{ab}$. [Had we chosen the convention $\varepsilon^{AB} = \varepsilon^{ab}$ instead, there would be no minus sign here but it would appear in other places. Cf. discussion of parity, charge conjugation and time reversal in sec. 5 below.]

Products of the kind $\phi_a\varepsilon^{ab}\phi_b$, $\chi^A\varepsilon_{AB}\chi^B$ etc. are Lorentz invariant. One shows easily that they can be written in several, equivalent ways, for instance,

$$\phi_a\varepsilon^{ab}\phi_b = \phi_a\phi^a = -\phi^b\phi_b, \qquad \chi^A\varepsilon_{AB}\chi^B = \chi^A\chi_A = -\chi_B\chi^B, \tag{35a}$$

and similarly

$$\phi_a\chi^{*a} = -\phi^a\chi^*{}_a, \qquad \phi^*{}_A\chi^A = -\phi^{*A}\chi_A. \tag{35b}$$

Before closing this section, we shall show that the two kinds of spinors can be related by means of space reflexion P, or equivalently, by time reversal T. By definition we have

$$P\{x^0, \boldsymbol{x}\} = \{x^0, -\boldsymbol{x}\}, \tag{36a}$$

$$T\{x^0, \boldsymbol{x}\} = \{-x^0, \boldsymbol{x}\}. \tag{36b}$$

Let $X = \sigma^\mu x_\mu = \sigma_\mu x^\mu$ and $X^P = \sigma_\mu (Px)^\mu$, then it follows that

$$XX^P = (x^0\mathbb{1} + \boldsymbol{x}\cdot\boldsymbol{\sigma})(x^0\mathbb{1} - \boldsymbol{x}\cdot\boldsymbol{\sigma}) = (x^0)^2 - \boldsymbol{x}^2 = x^2. \tag{37}$$

One verifies easily that, if we define the set $\hat{\sigma}^\mu$ in analogy to $\hat{A}$, eq. (29)*), then

$$\hat{\sigma}^\mu := \varepsilon\sigma^{\mu *}\varepsilon^{-1} = \sigma_\mu = (\mathbb{1}, \boldsymbol{\sigma}) \tag{38}$$

and, therefore, that

$$\varepsilon X^*\varepsilon^{-1} = X^P.$$

Under a Lorentz transformation Λ_A, we have, $Y = \sigma_\mu(\Lambda_A x)^\mu = AXA^\dagger$, and, since $\hat{A} = \varepsilon A^*\varepsilon^{-1}$, $Y^P = \hat{A}X^P\hat{A}^\dagger$. This last relation, which can also be written as follows

*) Note, however, that whilst $\sigma^0 = \mathbb{1}$ belongs to SL(2, ℂ) the Pauli matrices $\boldsymbol{\sigma}$ do not.

(using $P^2 = 1$),

$$\sigma_\mu((P\Lambda_A P)Px)^\mu = \hat{A}\sigma_\mu(Px)^\mu \hat{A}^\dagger,$$

says that the Lorentz transformation $\Lambda_{\hat{A}}$ pertaining to $\hat{A}$, and the original Lorentz transformation Λ_A are related by the equation

$$\Lambda_{\hat{A}} = P\Lambda_A P. \tag{39}$$

Finally, by defining $X^T := \sigma_\mu(TX)^\mu$ and observing that $X^T = -X^P$, we find in a similar manner

$$\Lambda_{\hat{A}} = T\Lambda_A T. \tag{40}$$

These relations show that space reflexion and time reversal relate spinors of the first kind to spinors of the second kind and vice versa. On the other hand, we know that we can reach any homogeneous Lorentz transformation by multiplying $L_+^\uparrow$ by P, T and PT. Therefore, in order to construct a spinor equation that is to be covariant under the *full* Lorentz group, we shall need spinor fields of *both* kinds.

The exact relationship between $\phi_a(x)$ and $\chi^B(x)$ under space reflexion is not fixed a priori, within a phase factor $e^{i(\pi/2)n}$. This is so because we have always the freedom to combine the operation of space reflexion with a complete rotation by 2π. Whilst this makes no difference for integer spin, it yields an extra minus sign in the case of half-integral spin. Therefore, space reflexion applied twice to a spinor field can be chosen to yield plus or minus that same field. The only restriction is that the same phase convention must be chosen for all spinors, in order to ascertain the correct transformation behaviour of bilinears in the spinor fields.

These remarks illustrate the fact that it is not meaningful to assign an *absolute* intrinsic parity to a fermion. Only relative parities can be physically relevant. We come back to this below.

3. Dirac equation for free particles

As a more technical preparation to what follows, we recall the definitions (3), (4) and (38):

$$\sigma^\mu := (\sigma^0, -\boldsymbol{\sigma}), \qquad \hat{\sigma}^\mu := \varepsilon\sigma^{\mu *}\varepsilon^{-1} = (\sigma^0, \boldsymbol{\sigma}),$$

where σ^0 is the two dimensional unit matrix and $\boldsymbol{\sigma} = \{\sigma^{(i)}\}$ is a short-hand notation for the Pauli matrices. With the aid of the well-known relation

$$\sigma^{(i)}\sigma^{(j)} = \delta^{ij} + i\sum_k \varepsilon^{ijk}\sigma^{(k)} \tag{41}$$

we verify the important relationship

$$(\sigma^\mu)_{aB}(\hat{\sigma}^\nu)^{Bc} + (\sigma^\nu)_{aB}(\hat{\sigma}^\mu)^{Bc} = 2g^{\mu\nu}\delta_a{}^c. \tag{42}$$

Furthermore, we can define derivatives of spinors, in analogy to eq. (31), by introducing*)

$$\partial_{aB} := \left(\sigma_\mu \partial^\mu\right)_{aB} = \left(\sigma^\mu \partial_\mu\right)_{aB} = \mathbb{1}\partial_0 - \boldsymbol{\sigma}\cdot\nabla. \tag{43}$$

The nature and position of indices are the same as in eq. (31) since $\partial^\mu = \partial/\partial x_\mu$ behaves like x^μ under Lorentz transformations and, therefore, with $x' = \Lambda x$, we have

$$\left(\sigma^\mu \partial'_\mu\right) = A(\Lambda)\left(\sigma^\mu \partial_\mu\right) A^\dagger(\Lambda),$$

We can also construct the matrix $\hat{\sigma}^\mu \partial_\mu$ by multiplying the complex conjugate of eq. (43) from the left with ε and from the right with ε^{-1}. Complex conjugation converts small into capital indices and vice versa, while ε raises indices. As a result, it follows that $\hat{\sigma}^\mu \partial_\mu$ carries indices as indicated here:

$$\partial^{Ab} := \left(\hat{\sigma}^\mu \partial_\mu\right)^{Ab} = \mathbb{1}\partial^0 + \boldsymbol{\sigma}\cdot\nabla. \tag{44}$$

Making use of the relations (42) we note that

$$\partial^{Ab}\partial_{bC} = \delta^A{}_C \partial_\mu \partial^\mu = \delta^A{}_C \Box \tag{45}$$

and similarly

$$\partial_{aB}\partial^{Bc} = \delta_a^c \Box, \tag{45'}$$

where $\Box = \partial_\mu \partial^\mu = (\partial^0)^2 - \Delta$ is the (four-dimensional) d'Alembert operator.

Equipped with these tools we can now proceed to derive the Dirac equation.

A relativistic wave equation (more precisely: a system of equations) that is to provide a quantum-mechanical description of free spin-1/2 particles has to meet the following conditions:

(i) The equation must be linear, homogeneous and of first order in the time derivative. Linearity is imposed by the superposition principle; first-order time derivatives are necessary if a probability interpretation is to be possible.

(ii) The equation must be covariant under the full Lorentz group. As we have noted above this implies that it must contain both spinor fields of the first kind $\phi_a(x)$ as well as spinor fields of the second kind $\chi^A(x)$.

(iii) In order to ensure the correct relativistic energy–momentum relation, both kinds of fields must obey the Klein–Gordon equation for the mass m of the particles

$$(\Box + m^2)\phi_a(x) = 0, \tag{46a}$$

$$(\Box + m^2)\chi^B(x) = 0. \tag{46b}$$

The simplest system of linear homogeneous differential equations of first order which meets these conditions reads

$$\mathrm{i}\left(\sigma^\mu \partial_\mu\right)_{aB}\chi^B(x) = m\phi_a(x), \tag{47a}$$

$$\mathrm{i}\left(\hat{\sigma}^\mu \partial_\mu\right)^{Bc}\phi_c(x) = m\chi^B(x). \tag{47b}$$

This set of four coupled differential equations constitutes what is called the *Dirac equation(s)* for the force-free case. It describes the free motion of spin-1/2 particles

*) Note the minus sign in front of $\boldsymbol{\sigma}\cdot\nabla$ which is due to $\partial_\mu = (\partial^0, \nabla)$ being the covariant derivative.

of arbitrary mass m, including the limit of mass zero, in which case eqs. (47) reduce to the Weyl equations.

Let us then verify that the Dirac equation does indeed obey the conditions listed above. First, it is evident that eqs. (47) are covariant under $L_+^\uparrow$, by construction. They are also invariant under parity P and time reversal T, as these operations transform χ into ϕ, $(\sigma^\mu \partial_\mu)$ into $(\hat{\sigma}^\mu \partial_\mu)$ and vice versa. (The details are worked out below, Sec. 5.) Thus they are covariant with respect to the full Lorentz group.

Second, it is easily verified that each component $\phi_a(x)$ satisfies the Klein–Gordon equation. Indeed, applying the operator $\mathrm{i}(\sigma^\mu \partial_\mu)$ to eq. (47b) we obtain for the left-hand side, using eq. (45),

$$-(\sigma^\mu \partial_\mu)_{aB}(\hat{\sigma}^\nu \partial_\nu)^{Bc}\phi_c(x) = -\delta_a^c \Box \phi_c(x) = -\Box \phi_a(x).$$

For the right-hand side we make use of eq. (47a),

$$m\mathrm{i}(\sigma^\mu \partial_\mu)_{aB}\chi^B(x) = m^2 \phi_a(x).$$

Thus we obtain $\Box \phi_a(x) + m^2 \phi_a(x) = 0$.

Similarly, by applying the operator $\mathrm{i}(\hat{\sigma}_\mu \partial^\mu)$ to eq. (47a) we verify that the components $\chi^B(x)$ also satisfy the Klein–Gordon equation.

It is customary and useful to group the two spinors $\phi_a(x)$ and $\chi^B(x)$ together into one four-component Dirac spinor

$$\Psi(x) := \begin{pmatrix} \phi_a(x) \\ \chi^B(x) \end{pmatrix} = \begin{pmatrix} \phi_1(x) \\ \phi_2(x) \\ \chi^{\mathrm{I}}(x) \\ \chi^{\mathrm{II}}(x) \end{pmatrix}. \tag{48}$$

Eqs. (47) can then be written in a more compact form, upon introduction of a set of four 4×4 matrices,

$$\gamma^\mu := \begin{pmatrix} 0 & \sigma^\mu \\ \hat{\sigma}^\mu & 0 \end{pmatrix}, \tag{49}$$

$$(\mathrm{i}\gamma^\mu \partial_\mu - m\mathbb{1}_4)\Psi(x) = 0, \tag{50}$$

where $\mathbb{1}_4$ is now the four-dimensional unit matrix. For later purposes it will be convenient to define one further matrix

$$\gamma_5 := \mathrm{i}\gamma^0\gamma^1\gamma^2\gamma^3 \tag{51}$$

(which is also written γ^5, without distinction as to the position of the index 5).

With σ^μ and $\hat{\sigma}^\mu$ as defined before, we have

$$\gamma^0 = \begin{pmatrix} 0 & \mathbb{1} \\ \mathbb{1} & 0 \end{pmatrix}, \quad \gamma^i = \begin{pmatrix} 0 & -\sigma^{(i)} \\ \sigma^{(i)} & 0 \end{pmatrix}, \quad \gamma^5 = \begin{pmatrix} \mathbb{1} & 0 \\ 0 & -\mathbb{1} \end{pmatrix}. \tag{52}$$

From eq. (42) we obtain the important anticommutators for the Dirac matrices

$$\gamma^\mu\gamma^\nu + \gamma^\nu\gamma^\mu = 2g^{\mu\nu}\mathbb{1}_4, \tag{53}$$

which imply, in particular, that

$$(\gamma^0)^2 = \mathbb{1}_4, \qquad (\gamma^i)^2 = -\mathbb{1}_4 \quad (i=1,2,3).$$

Writing out explicitly the spinor indices, the matrices γ^μ of eq. (49) have the structure

$$\gamma^\mu = \begin{pmatrix} 0 & (\sigma^\mu)_{aA} \\ (\hat{\sigma}^\mu)^{Aa} & 0 \end{pmatrix}. \tag{49'}$$

In particular, if we multiply the four γ-matrices to form γ_5, as indicated in the definition (51), we find

$$\gamma_5 = \mathrm{i}\begin{pmatrix} (\sigma^0\hat{\sigma}^1\sigma^2\hat{\sigma}^3)_a{}^b & 0 \\ 0 & (\hat{\sigma}^0\sigma^1\hat{\sigma}^2\sigma^3)^A{}_B \end{pmatrix} = \begin{pmatrix} \delta_a{}^b & 0 \\ 0 & -\delta^A{}_B \end{pmatrix}. \tag{51'}$$

This explicit representation (51′) of γ_5 leads us to two important remarks:

(i) According to (51′) γ_5 maps the spinor $(\phi_a, \chi^A)^T$ onto the spinor $(\phi_a, -\chi^A)^T$. If we compare this to the action of the unit matrix

$$\mathbb{1} = \begin{pmatrix} \delta_a{}^b & 0 \\ 0 & \delta^A{}_B \end{pmatrix}$$

we see that the combinations

$$\tfrac{1}{2}(\mathbb{1} + \gamma_5) = \begin{pmatrix} \delta_a{}^b & 0 \\ 0 & 0 \end{pmatrix}, \qquad \tfrac{1}{2}(\mathbb{1} - \gamma_5) = \begin{pmatrix} 0 & 0 \\ 0 & \delta^A{}_B \end{pmatrix} \tag{54}$$

project onto spinors of the first kind and spinors of the second kind, respectively.

(ii) From what we said above it is clear that it would not be meaningful to distinguish between γ_5 with upper or lower index.

From eqs. (49′) and (51′) we see that the structure of a general matrix M that maps the spinor $\Psi = (\phi_b, \chi^B)^T$ onto another spinor $\Psi' = (\phi'_a, \chi'^A)^T$ must be the following:

$$M = \begin{pmatrix} M_a{}^b & M_{aB} \\ M^{Ab} & M^A{}_B \end{pmatrix}.$$

This is important to know when we wish to form invariants, or Lorentz *co*variants (i.e. vectors, tensors etc.), in terms of Ψ and its hermitean conjugate $\Psi^\dagger$. Actually, $\Psi^\dagger$ is not the appropriate, conjugate spinor because of its spinorial structure which is

$$\Psi^\dagger(x) = (\phi^*{}_A(x) \quad \chi^{*a}(x))$$

and which cannot be contracted with $M\Psi(x)$. On the other hand, we know that, for instance, $\chi^{*a}\phi_a$ and $\phi^*{}_A\chi^A$ are invariants, cf. eqs. (35). The correct position of the indices is achieved if instead of $\Psi^\dagger$ we introduce

$$\overline{\Psi(x)} := (\phi^*{}_A \quad \chi^{*a})\begin{pmatrix} 0 & \delta^A{}_B \\ \delta_a{}^b & 0 \end{pmatrix}$$

$$= (\chi^{*b}(x) \quad \phi^*{}_B(x)). \tag{55}$$

This can also be written as follows:

$$\overline{\Psi(x)} = \left(\phi^*{}_A(\hat{\sigma}^0)^{Ac} \quad \chi^{*a}(\sigma^0)_{aC}\right)\begin{pmatrix} 0 & (\sigma^0)_{cB} \\ (\hat{\sigma}^0)^{Cb} & 0 \end{pmatrix}$$

$$\equiv \left(\phi^{*c} \quad \chi^*{}_C\right)\gamma^0.$$

We then see that

$$\overline{\Psi(x)}\,\Psi(x) = \chi^{*a}\phi_a + \phi^*{}_A\chi^A,$$

$$\overline{\Psi(x)}\,\gamma_5\Psi(x) = \chi^{*a}\phi_a - \phi^*{}_A\chi^A$$

are invariant under proper, orthochronous Lorentz transformations. We note that $\overline{\Psi(x)}\,\Psi(x)$ is even, $\overline{\Psi(x)}\gamma_5\Psi(x)$ is odd under space reflection (see the discussion at the end of the previous section), so that the first is a Lorentz *scalar*, the second a Lorentz *pseudoscalar*. Similarly,

$$\overline{\Psi(x)}\,\gamma^\alpha\Psi(x) = \chi^{*a}(\sigma^\alpha)_{aB}\chi^B + \phi^*{}_A(\hat{\sigma}^\alpha)^{Ab}\phi_b$$

is a Lorentz vector, etc. We shall elaborate on this in somewhat more detail in Chap. V, secs. 2 and 6, below.

The form (49), or (52), of γ_μ is just one possible representation of the Dirac matrices—the so-called *high-energy representation*. This is a *natural* choice because in this representation the reduction of the full Dirac field into irreducible (two-component) spinors is complete, see eq. (48). There are, of course, other, equivalent, representations of the γ-matrices two of which we now discuss.

Suppose we subject $\Psi(x)$ to a linear, nonsingular substitution S:

$$\Psi(x) \to \Psi'(x) = S\Psi(x),$$

so that the Dirac equation (50) becomes

$$\left(\mathrm{i}\gamma'^\mu\partial_\mu - m\mathbb{1}_4\right)\Psi'(x) = 0,$$

with $\gamma'^\mu = S\gamma^\mu S^{-1}$. Obviously, the anticommutators (53) are invariant under any such substitution. For instance, the choice

$$S = \frac{1}{\sqrt{2}}\begin{pmatrix} \mathbb{1} & \mathbb{1} \\ \mathbb{1} & -\mathbb{1} \end{pmatrix} = S^{-1}$$

leads to the representation

$$\gamma'^0 = \begin{pmatrix} \mathbb{1} & 0 \\ 0 & -\mathbb{1} \end{pmatrix}, \qquad \gamma'^i = \begin{pmatrix} 0 & \sigma^{(i)} \\ -\sigma^{(i)} & 0 \end{pmatrix}, \qquad \gamma'_5 = \begin{pmatrix} 0 & \mathbb{1} \\ \mathbb{1} & 0 \end{pmatrix}. \tag{56}$$

This representation which is called the *standard representation* is particularly useful for weakly relativistic situations in electron and muon physics.

Both representations (52) and (56) belong to the class of representations in which the Dirac matrices have the following hermiticity properties,

$$(\gamma^0)^\dagger = \gamma^0, \qquad (\gamma^i)^\dagger = -\gamma^i, \tag{57a}$$

$$\gamma^0(\gamma^\mu)^\dagger\gamma^0 = \gamma^\mu. \tag{57b}$$

In this book we use only this class of representations; there are, however, other representations in which these properties do not hold.

The following relation is always true (and follows from eq. (53)):

$$\gamma^\mu \gamma_5 + \gamma_5 \gamma^\mu = 0. \tag{58}$$

Thus γ_5 *anticommutes* with any product of an *odd* number of Dirac matrices, but *commutes* with any product of an *even* number of Dirac matrices.

One other representation is useful in the discussion of relativistic single particle problems such as bound states in an external potential, scattering of a particle in external fields etc. We start from the standard representation (55) (omitting the primes) and set

$$\beta \equiv \gamma^0 = \begin{pmatrix} \mathbb{1} & 0 \\ 0 & -\mathbb{1} \end{pmatrix},$$

$$\alpha^{(i)} := \beta \gamma^i = \begin{pmatrix} 0 & \sigma^{(i)} \\ \sigma^{(i)} & 0 \end{pmatrix}. \tag{59}$$

Then $\gamma^i = \beta\alpha^{(i)}$ and, upon multiplication of the Dirac equation by β, we obtain

$$\mathrm{i}\frac{\partial}{\partial t}\Psi(x) = (-\mathrm{i}\boldsymbol{\alpha}\cdot\nabla + m\beta)\Psi(x). \tag{60a}$$

When written in this form, the Dirac equation shows the closest formal resemblance to the nonrelativistic Schrödinger (or Schrödinger–Pauli) equation, with the matrix

$$\text{“}H_0\text{”} = -\mathrm{i}\boldsymbol{\alpha}\cdot\nabla + m\beta \tag{60b}$$

playing the role of the force-free Hamiltonian. For this reason the form (60) of the equation may be called *Hamiltonian form* of the Dirac equation. This analogy must, however, be understood with great care. Indeed H_0, eq. (60b) does not have the properties required by quantum mechanics of a single particle. Specifically, the spectrum of H_0 is not bounded from below; H_0 has arbitrarily large negative eigenvalues (see below). This is a first hint to the fact that Dirac theory is not a single particle theory and can only be interpreted consistently in its second-quantized form. It is true, however, that the operator (60b) appears in the Hamiltonian density of the Dirac field. When integrated over all space, this density yields the correct Hamiltonian of the Dirac field and does have the properties required by the principles of quantum mechanics.

Before we turn to the derivation of explicit solutions of the Dirac equation, we add a few further comments on these results. The explicit two-component formulation (47) is essential in discussing the basic principles of Dirac theory. In the literature on elementary particle physics one uses mostly the four-component form (50) of the Dirac equation. This form is indeed useful for most practical calculations of processes involving spin-1/2 particles and we make use of it in many chapters of this book.*) It is not so useful, however, for questions about the principles of the

*)We note, however, that quantum electrodynamics and other field theories can equivalently be formulated in the two-component formalism discussed above. This has been worked out, for the case of QED, by L.M. Brown, Proc. of Colorado Theor. Physics Institute, Colorado (1961). In some applications this formalism is simpler than the standard one.

theory. For instance, the covariance of eq. (50), if it is not derived as done here, is not obvious and must be proven explicitly by making use of a theorem of Pauli.*) Questions like: why do we have to describe relativistic spin-1/2 particles by (at least) four-component wave functions?; why do we have to quantize Dirac theory by means of anticommuting field operators? etc., do not have straightforward answers in the four-component theory. In the two-component formulation the covariance of eqs. (47) is evident.

In general, the answers to questions about fundamental properties of the theory are obtained in a more direct and transparent manner than in the four-component formulation.

4. Plane wave solutions of the Dirac equation

In this section we derive plane wave solutions of the Dirac equation. For definiteness, we take eq. (50) in the standard representation (56). We make the ansatz

$$\Psi_\alpha(x) = u_\alpha^{(r)}(\boldsymbol{p})\mathrm{e}^{-\mathrm{i}px} + v_\alpha^{(r)}(\boldsymbol{p})\mathrm{e}^{\mathrm{i}px}, \tag{61}$$

where $px = p^0x^0 - \boldsymbol{p}\boldsymbol{x}$ with $p^0 = (m^2 + \boldsymbol{p}^2)^{1/2}$; u and v are spinor amplitudes characterized by three-momentum $\boldsymbol{p}$ and spin orientation $r = 1$ or 2 with respect to some arbitrary quantization axis. The first term on the r.h.s. of eq. (61) is called *positive frequency part* of $\Psi(x)$; its time dependence is $\mathrm{e}^{-\mathrm{i}Et/\hbar}$. The second term, correspondingly, is called *negative frequency part*. The physical interpretation of the latter will become clear in the context of the quantization of the Dirac field.

It follows from eq. (50) that the spinors u and v obey the equations

$$(\gamma^\mu\partial_\mu - m\mathbb{1})u^{(r)}(\boldsymbol{p}) = 0, \tag{62}$$

$$(\gamma^\mu\partial_\mu + m\mathbb{1})v^{(r)}(\boldsymbol{p}) = 0. \tag{63}$$

It is useful to consider also the equations which follow for the hermitean conjugate spinors. These equations take a particularly simple form if the definition (55) is introduced**)

$$\overline{\Psi(x)} := \Psi^\dagger(x)\gamma^0, \tag{64a}$$

$$\overline{u(\boldsymbol{p})} := u^\dagger(\boldsymbol{p})\gamma^0, \qquad \overline{v(\boldsymbol{p})} := v^\dagger(\boldsymbol{p})\gamma^0. \tag{64b}$$

Note that both $\Psi^\dagger$ and $\overline{\Psi}$, as well as $u^\dagger$, $\bar{u}$, $v^\dagger$, $\bar{v}$ are "row vectors", so that forms like $\bar{u}u$ are simple numbers, whilst $u\bar{u}$ etc. are four-by-four matrices. From eqs. (50), (62), and (63) one derives easily

$$\overline{\Psi(x)}\left(\mathrm{i}\gamma^\mu\overleftarrow{\partial}_\mu + m\mathbb{1}\right) = 0, \tag{50$'$}$$

*) See discussion in sec. V.2.1 below.

**) The theorem says that for any two sets of Dirac matrices which fulfill the anticommutation relations (53) there is a nonsingular matrix S which transforms one set into the other.

where the arrow is meant to indicate that the derivative applies to $\bar{\Psi}$ on the left, i.e. $\bar{\Psi}\gamma^\mu\overleftarrow{\partial}_\mu \equiv \partial_\mu\bar{\Psi}\gamma^\mu$. Further,

$$\overline{u(\boldsymbol{p})}\left(\gamma^\mu p_\mu - m\mathbb{1}\right) = 0, \tag{62'}$$

$$\overline{v(\boldsymbol{p})}\left(\gamma^\mu p_\mu + m\mathbb{1}\right) = 0. \tag{63'}$$

Eqs. (62) and (63), for the case of massive particles $m \neq 0$, can be solved in two steps:

1. Go to the rest system of the particle where $p = (m, \mathbf{0})$. As $p^0 = m$, eq. (62) reduces to

$$\begin{pmatrix} 0 & 0 \\ 0 & \mathbb{1} \end{pmatrix} \begin{pmatrix} u_1(0) \\ u_2(0) \\ u_3(0) \\ u_4(0) \end{pmatrix} = 0,$$

while eq. (63) reduces to

$$\begin{pmatrix} \mathbb{1} & 0 \\ 0 & 0 \end{pmatrix} \begin{pmatrix} v_1(0) \\ v_2(0) \\ v_3(0) \\ v_4(0) \end{pmatrix} = 0,$$

the solutions of which are

$$u^{(r)}(0) = \begin{pmatrix} \chi^{(r)} \\ 0 \\ 0 \end{pmatrix} \quad \text{and} \quad v^{(r)}(0) = \begin{pmatrix} 0 \\ 0 \\ \chi^{(r)} \end{pmatrix},$$

respectively. Here $\chi^{(r)}$ is a two-component object which, in fact, must be a Pauli spinor, well-known from nonrelativistic quantum mechanics.

2. The solutions for arbitrary momentum $\boldsymbol{p}$ can then be expressed in terms of the rest-frame solutions by utilizing the relation

$$\gamma^\mu p_\mu \gamma^\nu p_\nu = \tfrac{1}{2} p_\mu p_\nu \left(\gamma^\mu\gamma^\nu + \gamma^\nu\gamma^\mu\right) = p_\mu g^{\mu\nu} p_\nu = p^2,$$

which follows from (53). It is seen that

$$u^{(r)}(\boldsymbol{p}) = N\left(\gamma^\mu p_\mu + m\mathbb{1}\right) u^{(r)}(0), \tag{65}$$

$$v^{(r)}(\boldsymbol{p}) = -N\left(\gamma^\mu p_\mu - m\mathbb{1}\right) v^{(r)}(0) \tag{66}$$

satisfy eqs. (62) and (63), respectively. The normalization constant N is conveniently chosen so as to obtain

$$u^{(r)\dagger}(\boldsymbol{p}) u^{(s)}(\boldsymbol{p}) = v^{(r)\dagger}(\boldsymbol{p}) v^{(s)}(\boldsymbol{p}) = 2 p_0 \delta_{rs}. \tag{67}$$

This is a covariant normalization of one-particle states.*) It is achieved by taking

*) In some textbooks the normalization of one-fermion states is taken to be p_0/m instead of $2p_0$. Our normalization is the same for fermions and bosons.

$N=(p^0+m)^{-1/2}$. Inserting the explicit representation of γ-matrices into (65) and (66) (here in the standard representation), we have

$$u^{(r)}(\boldsymbol{p})=\sqrt{p^0+m}\begin{pmatrix}\chi^{(r)}\\ \dfrac{\boldsymbol{\sigma}\cdot\boldsymbol{p}}{p_0+m}\chi^{(r)}\end{pmatrix}, \tag{68}$$

$$v^{(r)}(\boldsymbol{p})=\sqrt{p^0+m}\begin{pmatrix}\dfrac{\boldsymbol{\sigma}\cdot\boldsymbol{p}}{p_0+m}\chi^{(r)}\\ \chi^{(r)}\end{pmatrix}. \tag{69}$$

One verifies by explicit calculation that

$$\overline{u^{(r)}(\boldsymbol{p})}u^{(s)}(\boldsymbol{p})=2m\delta_{rs}, \tag{70}$$

$$\overline{v^{(r)}(\boldsymbol{p})}v^{(s)}(\boldsymbol{p})=-2m\delta_{rs}, \tag{71}$$

and also

$$u^{(r)\dagger}(\boldsymbol{p})v^{(s)}(-\boldsymbol{p})=0. \tag{72}$$

Obviously these spinors can be transformed to any other representation of the Dirac matrices by applying the corresponding transformation matrix S to them. (Example: Transformation to high-energy representation by means of S, p. 105.)

Mass zero case: We have derived the spinors in momentum space (68) and (69) by "boosting" the Pauli spinors from the particle's rest system to appropriate momentum $\boldsymbol{p}$, see eqs. (65) and (66). For a massless particle there is no rest system. Nevertheless, we obtain the plane wave solutions describing the force-free motion of a massless spin-1/2 particle by simply taking the limit $m\to 0$ of the solutions (68) and (69).*) We then have, of course, $p^0=|p|$ and the operator

$$\frac{\boldsymbol{\sigma}\cdot\boldsymbol{p}}{p_0}=\frac{\boldsymbol{\sigma}\cdot\boldsymbol{p}}{|\boldsymbol{p}|}=:h \tag{73}$$

becomes the helicity operator.

For the sake of illustration, let us transform the $m=0$ solutions back to the "high-energy" representation (52), by means of the transformation matrix S. We obtain from eq. (68)

$$u^{(r)}(\boldsymbol{p},m=0)=\sqrt{\frac{|\boldsymbol{p}|}{2}}\begin{pmatrix}(\mathbb{1}+h)\chi^{(r)}\\ (\mathbb{1}-h)\chi^{(r)}\end{pmatrix}. \tag{74}$$

Taking for $\chi^{(r)}$ magnetic substates in the direction of $\boldsymbol{p}$, or opposite to it, means taking the eigenvalues of h to be $+1$ or -1, respectively. Thus, either the lower two components or the upper two components vanish.

*)This limiting procedure is only applicable for spin 1/2. For higher spin $J>1/2$ the limit $m\to 0$ is discontinuous: Whilst a massive particle with spin J has $(2J+1)$ magnetic substates, a massless particle can have only two helicity states $h=(\boldsymbol{J}\cdot\boldsymbol{h})/|\boldsymbol{p}|=\pm J$.

Obviously, these solutions can be obtained directly from our eqs. (47): In the limit $m=0$, these equations decouple. With the ansatz $e^{\pm ipx}u(p)$, they are seen to lead to the eigenvalue equation of the helicity operator (73).

5. A few more properties of Dirac spinors

We return to the component form (47) of the Dirac equation and derive a few more properties of its solutions. First we note that if $\Psi(x)=(\phi_a(x),\chi^A(x))^T$ is a solution then so is the spinor

$$\Psi_P(x)=\begin{pmatrix}\chi_a(Px)\\ \phi^A(Px)\end{pmatrix}=\begin{pmatrix}(\sigma^0)_{aB}\chi^B(Px)\\ (\hat{\sigma}^0)^{Ab}\phi_b(Px)\end{pmatrix}=\gamma^0\Psi(Px), \tag{75}$$

Here $Px=(x^0,-\boldsymbol{x})$ is the parity transform of x, cf. eq. (36a); γ^0 is given by eq. (49).

Similarly, one shows that if $\Psi(x)$ is a solution of the Dirac equation (47) then also

$$\Psi_C(x)=\begin{pmatrix}\chi^*{}_a(x)\\ -\phi^{*B}(x)\end{pmatrix}=i\gamma^2\Psi^*(x) \tag{76}$$

is a solution.*) $\Psi_C(x)$ is called the *charge conjugate* spinor of $\Psi(x)$. For the moment, of course, this is only a formal definition. Charge conjugation becomes physically relevant only when the particle interacts e.g. with an external electromagnetic field or with other particles. It can then be shown that the transformation (76) does indeed lead to the wave functions of the corresponding antiparticle.

It is not difficult to prove that Ψ_C, as defined by (76) is a solution of the Dirac equation. Take the complex conjugate of eq. (47a), multiply with ε from the left and insert $\varepsilon\cdot\varepsilon^{-1}$ between the operators $\sigma^\mu\partial_\mu$ and χ (remember rule (iii) on page 99):

$$-i\varepsilon^{AA'}\left(\sigma^{\mu *}\partial_\mu\right)_{A'b'}\varepsilon^{b'b}\varepsilon_{bb''}\chi^{*b''}(x)=m\varepsilon^{AA'}\phi^*{}_{A'}(x).$$

From eq. (38) and with our convention (28) we can write this as

$$i\left(\hat{\sigma}^\mu\partial_\mu\right)^{Ab}\chi^*{}_b(x)=-m\phi^{*A}(x),$$

where χ_B and ϕ^a are defined according to the rules (27) and (24), respectively. The second equation (47b) is treated similarly:

$$-i\varepsilon_{bb'}\left(\hat{\sigma}^{\mu *}\partial_\mu\right)^{b'A'}\varepsilon_{A'A}\varepsilon^{AA''}\phi^*{}_{A''}(x)=m\varepsilon_{bb'}\chi^{*b'}(x),$$

which can be written as

$$i\left(\sigma^\mu\partial_\mu\right)_{bA}\left(-\phi^{*A}(x)\right)=m\chi^*{}_b(x).$$

*) Recall that complex conjugation converts small into capital, capital into small indices, $(\chi^B)^*=\chi^{*b}$, $(\phi_a)^*=\phi^*{}_A$, and that indices are lowered and raised by means of ε-matrices.

Thus, we have shown that

$$\Psi_C(x) = i\gamma^2\Psi^*(x) = \begin{pmatrix} 0 & -\varepsilon \\ \varepsilon & 0 \end{pmatrix}\begin{pmatrix} \phi_a \\ \chi^A \end{pmatrix}^*$$

$$= \begin{pmatrix} 0 & \varepsilon_{ab} \\ -\varepsilon^{BA} & 0 \end{pmatrix}\begin{pmatrix} \phi^*{}_A \\ \chi^{*b} \end{pmatrix} = \begin{pmatrix} \chi^*{}_a \\ -\phi^{*B} \end{pmatrix} = \begin{pmatrix} -(\chi^{II})^* \\ (\chi^{I})^* \\ (\phi_2)^* \\ -(\phi_1)^* \end{pmatrix}$$

is also a solution of eqs. (47).

If one prefers to write the charge conjugate spinor in terms of $\bar{\Psi}$ (cf. eq. (64)), relation (76) reads

$$\Psi_C(x) = i\gamma^2\gamma^0\left(\overline{\Psi(x)}\right)^{\text{transposed}}. \tag{76'}$$

That the matrix

$$C = i\gamma^2\gamma^0 = -C^{-1} = -C^\dagger = \tilde{C} \tag{77}$$

does indeed convert particle wave functions into antiparticle wave functions and vice-versa can be seen by considering the Dirac equation for a fermion with charge e in presence of some *external* electromagnetic fields A_μ,

$$\left(i\gamma^\mu\partial_\mu - e\gamma^\mu A_\mu - m\mathbb{1}\right)\Psi(x) = 0. \tag{78}$$

Take the complex conjugate of this equation, transform the wave function according to eq. (76) (or equivalently eq. (76′)), and observe that

$$(C\gamma^0)(\gamma^\mu)^*(C\gamma^0)^{-1} = -\gamma^\mu. \tag{79}$$

One thereby obtains from (78)

$$\left(i\gamma^\mu\partial_\mu + e\gamma^\mu A_\mu - m\mathbb{1}\right)\Psi_C(x) = 0, \tag{80}$$

which is seen to be the Dirac equation for the antiparticle (charge $-e$) in the same *external* fields.*⁾

The class of spinors for which $\Psi_C(x) = \Psi(x)$ are called *Majorana spinors*. Note that these are not *real* fields (as would be the case for a spin zero field). From what we said above it is clear that such a spinor can only describe a neutral particle, $e = 0$.

*⁾Note that here we apply charge conjugation C only to $\Psi(x)$, but not to the sources of the external fields. Had we done so, we would have found eq. (78) to be invariant under C. The electromagnetic interaction is invariant under C.

Finally, one proves by similar means that with $\Psi(x)$ also the time reversed spinor

$$\Psi_T(x) := \begin{pmatrix} (\sigma^0)_{aA}\varepsilon^{AA'} & \phi^*{}_{A'}(Tx) \\ (\hat{\sigma}^0)^{Bb}\varepsilon_{bb'} & \chi^{*b'}(Tx) \end{pmatrix} \equiv \begin{pmatrix} \phi^*{}_a(Tx) \\ \chi^{*B}(Tx) \end{pmatrix} = \begin{pmatrix} -\phi_2^*(Tx) \\ \phi_1^*(Tx) \\ -\chi^{*II}(Tx) \\ \chi^{*I}(Tx) \end{pmatrix} \tag{81}$$

is a solution of the Dirac equation (47).

The relationship (81) thus reads

$$\Psi_T(x) = \begin{pmatrix} 0 & \sigma^0_{aA}\varepsilon^{AB} \\ \hat{\sigma}^{0Bb}\varepsilon_{bc} & 0 \end{pmatrix} \begin{pmatrix} \chi^{*c}(Tx) \\ \phi^*{}_B(Tx) \end{pmatrix} \equiv T \begin{pmatrix} \chi^{*c}(Tx) \\ \phi^*{}_B(Tx) \end{pmatrix}. \tag{81$'$}$$

The spinor on the r.h.s. is nothing but the transposed of $\overline{\Psi(Tx)}$, see eq. (55), while the matrix T, in the high-energy representation, is equal to

$$T = \gamma_5(\mathrm{i}\gamma^2), \tag{82}$$

One verifies that $T := \mathrm{i}\gamma_5\gamma^2$ has the properties

$$T = \mathrm{i}\gamma_5\gamma^2 = -T^{-1} = -T^\dagger = -\tilde{T}. \tag{83}$$

6. Quantization of Majorana fields

There are many indications for the fact that the Dirac field cannot be a classical field. One indication for this is that a spinor field ϕ or χ changes sign when a complete rotation by 2π is performed on it. Therefore, such a field cannot be a classical observable. Such observables which, of course, must be invariant under complete rotations of the coordinate system, can only depend on bilinear forms in the spinor fields. Another indication is this: when one computes the total energy, i.e. the Hamilton density of the unquantized Dirac field integrated over all space, this energy is found to be zero. (See exercise 13, for the case of Majorana spinors.) Furthermore, if one wants to interpret the Dirac equation in the framework of a single particle theory, in a spirit like in nonrelativistic quantum mechanics of one particle dynamics, one runs into two major difficulties: The energy of a free fermion can assume arbitrarily large negative values. Also, particle and antiparticle appear in an asymmetric way: while free particles are states with positive energy, antiparticle states appear as "holes" in particle levels of negative energy. The interpretation in terms of "particles" and "holes" avoids the negative energies but does not repair the apparent asymmetry in the treatment of particles and antiparticles.*)

All these difficulties disappear if Dirac theory is interpreted in the framework of second quantization. This is what we are going to show next. For the sake of

*) In a system of a *finite* number of fermions the hole theory is a perfectly consistent and useful approach. This is not so in a field theory with *infinitely* many degrees of freedom and in which genuine antiparticles occur.

transparency, we start with the case of Majorana fields, but we will see that the case of more general Dirac fields is no more difficult to treat.

A Majorana field of mass m is defined here by the condition

$$\Psi_C(x) = \Psi(x),$$

which implies

$$\phi_a(x) = \varepsilon_{ab}(\chi^B(x))^* = \varepsilon_{ab}\chi^{*b}(x) = \chi^*{}_a(x). \tag{84}$$

In this case the Dirac equation reduces to

$$\mathrm{i}(\sigma^\mu\partial_\mu)_{aB}\chi^B(x) = m\chi^*{}_a(x), \tag{85a}$$

or, equivalently,

$$\mathrm{i}(\hat{\sigma}^\mu\partial_\mu)^{Bc}\phi_c(x) = -m\phi^{*B}(x). \tag{85b}$$

These equations could have been derived from the following Lagrangian density (c.f. exercise 1):

$$\mathscr{L}_M = \frac{\mathrm{i}}{2}\phi^*{}_B\left(\hat{\sigma}^\mu\overleftrightarrow{\partial}_\mu\right)^{Bb}\phi_b + \frac{m}{2}\left[\phi^*{}_A\varepsilon^{AB}\phi^*{}_B + \phi_a\varepsilon^{ab}\phi_b\right]. \tag{86}$$

From this one derives the "momentum" canonically conjugate to ϕ_a,

$$\pi^a := \frac{\partial\mathscr{L}_M}{\partial(\partial^0\phi_a)} = \frac{\mathrm{i}}{2}\phi^*{}_B(\hat{\sigma}^0)^{Ba}. \tag{87}$$

Therefore, in quantizing the Majorana field we should discuss commutation rules of the operators ϕ as well as ϕ^*. However, as ϕ and ϕ^* are related by eq. (85b), it is sufficient to discuss the commutation of $\phi_a(x)$ and $\phi_b(y)$. The commutation rules for ϕ and π can be derived from these by applying eq. (85b) to them.

Similarly, the commutation rules for the χ field and its conjugate will then follow by means of the relations (84). Let us first consider the *commutator* of $\phi_a(x)$ and $\phi_b(y)$, for which we write

$$[\phi_a(x), \phi_b(y)] = t_{ab}f(x-y). \tag{88}$$

Obviously, the conditions that the right-hand side of eq. (88) must fulfill, are these:

(i) t_{ab} must be an invariant tensor with respect to SL(2, $\mathbb{C}$), i.e. $t_{ab} = A_a{}^m A_b{}^n t_{mn}$.

(ii) $f(x-y)$ must be Lorentz invariant and must satisfy the Klein–Gordon equation for mass m, both in x and y.

(iii) The product $t_{ab}f(x-y)$ must be antisymmetric under the simultaneous interchange of a with b, and x with y.

The only SL(2, $\mathbb{C}$) invariant tensor is $t_{ab} = \varepsilon_{ab}$ which, as we know, is *anti*symmetric in its indices. As to $f(x-y)$ we note that there are precisely two linearly independent, Lorentz invariant solutions of the Klein–Gordon equation for mass m.*) These

*) See Appendix A.

can be chosen as follows:

$$\Delta_0(z;m) = -\frac{\mathrm{i}}{(2\pi)^3}\int\frac{\mathrm{d}^3k}{2\omega_k}(\mathrm{e}^{-\mathrm{i}kz}-\mathrm{e}^{\mathrm{i}kz}), \tag{89}$$

$$\Delta_1(z;m) = \frac{1}{(2\pi)^3}\int\frac{\mathrm{d}^3k}{2\omega_k}(\mathrm{e}^{-\mathrm{i}kz}+\mathrm{e}^{\mathrm{i}kz}). \tag{90}$$

with $\omega_k = (k^2+m^2)^{1/2}$ and $z = x - y$. We note, in particular, that $\Delta_0(z;\ m)$ is *antisymmetric* when z is replaced by $-z$ and vanishes for spacelike separation of x and y. Thus, Δ_0 is a causal distribution.

$\Delta_1(z;\ m)$ on the other hand, is symmetric in z and does not vanish for $z^2 < 0$. Thus, Δ_1 cannot be a causal distribution.

We are now in a difficulty: The quantization rule should be *causal*, i.e. the commutator $[\phi_a(x), \phi_b(y)]$ should vanish whenever $(x-y)$ is spacelike. Therefore we must take the right-hand side of eq. (88) to be

$$\varepsilon_{ab}\Delta_0(x-y;m). \tag{91}$$

This, however, cannot be correct as this last expression is *symmetric* under ($a \leftrightarrow b$, $x \leftrightarrow y$). Therefore, if we insist on quantizing the theory by means of commutators, we have no other choice than to take the commutator (88) to be

$$[\phi_a(x), \phi_b(y)] = c_1\varepsilon_{ab}\Delta_1(x-y;m). \tag{92}$$

Such a theory is not acceptable on physical grounds: it is Lorentz invariant but is in conflict with causality, and therefore, will necessarily lead to physically unacceptable consequences (see also below, Sec. 8).

On the other hand, there is really no compelling reason why one should try to impose commutators on fermion fields. Fermion fields are not observable. Commutators are relevant only for quantum mechanical observables which, as we said before, must be bilinear in fermion fields.

With this remark in mind, it is not difficult to resolve the puzzle. In contrast to eq. (92) the expression (91) is perfectly acceptable: It is Lorentz invariant and vanishes for spacelike $(x-y)$. As it is symmetric under simultaneous interchange of a and b, x and y, it is natural to consider the *anticommutator*

$$\{\phi_a(x), \phi_b(y)\} := \phi_a(x)\phi_b(y) + \phi_b(y)\phi_a(x) \tag{93}$$

of these fields, instead of their commutator, and to require the quantization rule

$$\{\phi_a(x), \phi_b(y)\} = c_0\varepsilon_{ab}\Delta_0(x-y;m), \tag{94}$$

where c_0 is a constant to be determined from eqs. (85). For this purpose, apply the operation $\mathrm{i}(\hat{\sigma}^\mu\partial^x_\mu)^{Aa}$ onto eq. (94), and sum over a. This gives

$$\{\phi^{*A}(x), \phi_b(y)\} = -\frac{\mathrm{i}c_0}{m}(\hat{\sigma}^\mu\partial^x_\mu)^{Aa}\varepsilon_{ab}\Delta_0(x-y;m).$$

If we take $x^0 = y^0$, then

$$\{\phi^{*A}(x), \phi_b(y)\}_{x^0=y^0} = -\frac{\mathrm{i}c_0}{m}\delta^{Aa}\varepsilon_{ab}\partial_0^x\Delta_0(x-y;m)\Big|_{x^0=y^0}$$

$$= \frac{\mathrm{i}c_0}{m}\delta^{Aa}\varepsilon_{ab}\delta(\boldsymbol{x}-\boldsymbol{y}).$$

Multiplying this equation with ε_{BA}, one obtains

$$\{\phi^*{}_B(x), \phi_b(y)\}_{x^0=y^0} = \frac{\mathrm{i}c_0}{m}\delta_{Bb}\delta(\boldsymbol{x}-\boldsymbol{y}).$$

The left-hand side is a positive hermitean operator. Thus, c_0 must be negative, pure imaginary. For dimensional reasons c_0 must have the dimension of an energy. Thus we take

$$c_0 = -\mathrm{i}m. \tag{95}$$

This quantization rule by means of anticommutators leads to a consistent interpretation of the theory. This is shown below for the case of the general Dirac field of which the Majorana field is a special case.

7. Quantization of Dirac fields

The reasoning of the preceding paragraph is readily applied to the more general case of arbitrary Dirac fields. Here, the possibility of quantizing via *commutators* is excluded by the following reason: Suppose we would require the causal commutator

$$[\phi_a(x), \chi^*{}_b(y)] = c_3\varepsilon_{ab}\Delta_0(x-y;m), \tag{96}$$

where c_3 is some number still to be determined. Applying the charge conjugation to this equation*⁾, we would obtain the following sequence of equations,

$$[\chi^*{}_a(x), \phi_b(y)] = c_3\varepsilon_{ab}\Delta_0(x-y;m)$$
$$= -[\phi_b(y), \chi^*{}_a(x)] = -c_3\varepsilon_{ba}\Delta_0(y-x;m)$$
$$= -c_3\varepsilon_{ab}\Delta_0(x-y;m).$$

The last step follows from the symmetry of expression (91). Obviously, $c_3 = 0$, and the ansatz (96) is seen to be inconsistent. Here again, we could repair this inconsistency by replacing the antisymmetric distribution Δ_0 by the symmetric Δ_1. However, this would again lead to an acausal theory. Very much like in the case of Majorana fields the only other possibility is to quantize the theory by means of

*⁾Recall that $C^{-1} = -C$, so that

$$C\phi_a C^{-1} = \chi^*{}_a, \qquad C\chi^B C^{-1} = -\phi^{B*},$$
$$C^{-1}\chi^*{}_a C = \phi_a, \qquad C^{-1}\phi^{B*}C = \chi^B.$$

anticommutators. Thus we postulate

$$\{\phi_a(x), \chi^*{}_b(y)\} = -im\varepsilon_{ab}\Delta_0(x-y;m). \tag{97}$$

This quantization rule leads to a consistent interpretation of Dirac fields in terms of particles and antiparticles with the correct (positive) energy spectrum. These particles obey Fermi–Dirac statistics, so that the Pauli principle is a consequence of the theory.*)

We could prove these statements right away by expanding the two-component fields ϕ and χ in terms of plane wave solutions and in terms of the corresponding creation and annihilation operators. In order to establish the connection with other, more conventional presentations of this subject, we prefer instead to reformulate first the quantization rule (97) in terms of four-component field operators $\Psi(x)$, as defined above in eq. (48). We have

$$\Psi(x) = \begin{pmatrix} \phi_a(x) \\ \chi^B(x) \end{pmatrix}, \qquad \Psi^\dagger(y) = \begin{pmatrix} \phi^*{}_C(y) & \chi^{*d}(y) \end{pmatrix},$$

their anticommutator being

$$\{\Psi(x), \Psi^\dagger(y)\} = \begin{pmatrix} \{\phi_a(x), \phi^*{}_C(y)\} & \{\phi_a(x), \chi^{*d}(y)\} \\ \{\chi^B(x), \phi^*{}_C(y)\} & \{\chi^B(x), \chi^{*d}(y)\} \end{pmatrix}.$$

Each of the anticommutators of two-component fields is a two dimensional matrix. All four of them can be derived from eq. (97) by applying the ε-tensors and/or by making use of the Dirac equation (47). One finds

$$\begin{aligned}\{\phi_a(x), \phi^*{}_C(y)\} &= -\frac{i}{m}\left(\sigma^{\mu*}\partial^y_\mu\right)_{Cb'}\{\phi_a(x), \chi^{*b'}(y)\} \\ &= -\frac{i}{m}\left(\sigma^{\mu*}\partial^y_\mu\right)_{Cb'}\varepsilon^{b'b}\{\phi_a(x), \chi^*{}_b(y)\} = \left(\sigma^{\mu*}\partial^y_\mu\right)_{Ca}\Delta_0(x-y;m) \\ &= \left(\sigma^\mu\partial^y_\mu\right)_{aC}\Delta_0(x-y;m).\end{aligned}$$

Similarly, we have

$$\begin{aligned}\{\phi_a(x), \chi^{*d}(y)\} &= \varepsilon^{db}\{\phi_a(x), \chi^*{}_b(y)\} \\ &= \varepsilon^{db}\varepsilon_{ab}(-im)\Delta_0(x-y;m) = \delta^d_a im\Delta_0(x-y;m), \\ \{\chi^B(x), \phi^*{}_C(y)\} &= \{\phi_c(y), \chi^{*b}(x)\}^* = \delta^B{}_C im\Delta_0(x-y;m), \\ \{\chi^B(x), \chi^{*d}(y)\} &= \frac{i}{m}\left(\hat\sigma^\mu\partial^x_\mu\right)^{Ba}\varepsilon^{db}\{\phi_a(x), \chi^*{}_b(y)\} \\ &= -\left(\hat\sigma^\mu\partial^x_\mu\right)^{Bd}\Delta_0(x-y;m),\end{aligned}$$

so that we find for the anticommutator of Ψ and $\Psi^\dagger$,

$$\{\Psi(x), \Psi^\dagger(y)\} = i\begin{pmatrix} i\sigma^\mu\partial^x_\mu & m\mathbb{1} \\ m\mathbb{1} & i\hat\sigma^\mu\partial^x_\mu \end{pmatrix}\Delta_0(x-y;m),$$

where we have used that $\partial^y_\mu\Delta_0 = -\partial^x_\mu\Delta_0$.

*)This is a special case of the famous spin-statistics theorem (Fierz 1938, Pauli 1940).

The matrix that appears in this last expression can be written as follows (cf. eq. (49)),

$$\left(m\gamma^0 + \mathrm{i}\gamma^\mu\gamma^0\partial_\mu^x\right).$$

By multiplying with γ^0 from the right we obtain, finally,

$$\left\{ \Psi(x), \overline{\Psi(y)} \right\} = \mathrm{i}\left(m\mathbb{1} + \mathrm{i}\gamma^\mu\partial_\mu^x\right)\Delta_0(x-y;\, m). \tag{98}$$

Thus we have arrived at a compact notation of the quantization rule (97), formulated in terms of the field operator $\Psi(x)$ and its adjoint.

Let us now derive a few consequences of the quantization of the Dirac field by means of anticommutators. Firstly, let us look at eq. (98) for the special case of equal times, $x^0 = y^0$. With the aid of the formulae (A.17) for the covariant causal distribution Δ_0, we find

$$\left\{ \Psi_\alpha(x), \Psi_\beta^\dagger(y)\right\}_{x^0=y^0} = \delta_{\alpha\beta}\delta(\boldsymbol{x}-\boldsymbol{y}). \tag{99}$$

Secondly, we expand $\Psi(x)$ and $\overline{\Psi(x)}$, as usual, in terms of "normal oscillations", i.e. in terms of plane wave solutions, very much like in eq. (61), viz.

$$\Psi_\alpha(x) = \frac{1}{(2\pi)^{3/2}}$$
$$\times \sum_{r=1}^{2} \int \frac{\mathrm{d}^3 p}{2E_p}\left[a^{(r)}(\boldsymbol{p})u_\alpha^{(r)}(\boldsymbol{p})\mathrm{e}^{-\mathrm{i}px} + b^{(r)\dagger}(\boldsymbol{p})v_\alpha^{(r)}(\boldsymbol{p})\mathrm{e}^{\mathrm{i}px}\right], \tag{100a}$$

$$\overline{\Psi_\alpha(x)} = \frac{1}{(2\pi)^{3/2}}$$
$$\times \sum_{r=1}^{2} \int \frac{\mathrm{d}^3 p}{2E_p}\left[a^{(r)\dagger}(\boldsymbol{p})\overline{u_\alpha^{(r)}(\boldsymbol{p})}\mathrm{e}^{\mathrm{i}px} + b^{(r)}(\boldsymbol{p})\overline{v_\alpha^{(r)}(\boldsymbol{p})}\mathrm{e}^{-\mathrm{i}px}\right]. \tag{100b}$$

Here, $u^{(r)}(\boldsymbol{p})$, $v^{(r)}(\boldsymbol{p})$, are the force-free solutions in momentum space (68) and (69) that we constructed above in sec. 4. r is the spin index, $\boldsymbol{p}$ the momentum and $E_p = (m^2 + \boldsymbol{p}^2)^{1/2}$ the corresponding energy. The inverse formulae expressing the operators $a^{(r)}(\boldsymbol{p})$ and $b^{(r)}(\boldsymbol{p})$ in terms of the field operators are easily derived by using the orthogonality relations (70)–(72). One finds

$$a^{(r)}(\boldsymbol{p}) = \frac{1}{(2\pi)^{3/2}}\int \mathrm{d}^3x\, \mathrm{e}^{\mathrm{i}px} u_\alpha^{(r)\dagger}(\boldsymbol{p})\Psi_\alpha(x), \tag{101a}$$

$$b^{(r)}(\boldsymbol{p}) = \frac{1}{(2\pi)^{3/2}}\int \mathrm{d}^3x\, \mathrm{e}^{\mathrm{i}px}\overline{\Psi_\alpha(x)}\,\gamma^0_{\alpha\beta}v_\beta^{(r)}(\boldsymbol{p}). \tag{101b}$$

In eqs. (100) and (101) we have written out the Dirac spinor indices, for the sake of clarity. Thus, in eqs. (100) the spinors $u(\boldsymbol{p})$ and $v(\boldsymbol{p})$ in momentum space carry the spinor index of the field operators $\Psi(x)$. In eqs. (101) one has to sum over the spinor indices as indicated, the operators $a^{(r)}(\boldsymbol{p})$ and $b^{(r)}(\boldsymbol{p})$ carrying no such index.

In going over from two-component spinors to four-component spinors (48) we lose the clear distinction between co- and contragredient spinor indices of first and second kind. Also the covariance properties of the theory become less transparent than in the two-component formulation. What we gain, however, is a very compact notation that is useful for almost all practical calculations which Dirac spinors. In particular, as long as the order of a product of Dirac spinors and Dirac matrices is respected, we need not write out the spinor indices at all. The sum over first and second kind spinors is automatically contained.

From eqs. (101) and (99), making use of relations (70)–(72) one derives the following anticommutation relations:

$$\{a^{(r)}(\boldsymbol{p}), a^{(s)\dagger}(\boldsymbol{q})\} = 2E_p\delta_{rs}\delta(\boldsymbol{p}-\boldsymbol{q}), \tag{102a}$$

$$\{b^{(r)}(\boldsymbol{p}), b^{(s)\dagger}(\boldsymbol{q})\} = 2E_p\delta_{rs}\delta(\boldsymbol{p}-\boldsymbol{q}), \tag{102b}$$

$$\{a^{(r)}(\boldsymbol{p}), a^{(s)}(\boldsymbol{q})\} = \{b^{(r)}(\boldsymbol{p}), b^{(s)}(\boldsymbol{q})\}$$
$$= \{a^{(r)}(\boldsymbol{p}), b^{(s)}(\boldsymbol{q})\} = \{a^{(r)}(\boldsymbol{p}), b^{(s)\dagger}(\boldsymbol{q})\} = 0. \tag{102c}$$

These anticommutation rules show that we deal here with creation and destruction operators for two kinds of particles that obey Fermi–Dirac statistics.*) In more detail,

$a^{(r)\dagger}(\boldsymbol{p})$ creates a *particle* state with four-momentum $p=(E_p, \boldsymbol{p})$ and spin projection r,

$b^{(r)\dagger}(\boldsymbol{p})$ creates an *antiparticle* state with four-momentum $p=(E_p, \boldsymbol{p})$ and spin projection r,

$a^{(r)}(\boldsymbol{p})$ and $b^{(r)}(\boldsymbol{p})$ are the corresponding annihilation operators.

Thus, for example, applying $a^{(r)\dagger}(p)$ to the vacuum yields a one-particle state

$$a^{(r)\dagger}(\boldsymbol{p})|0\rangle = |\boldsymbol{p}, r\rangle,$$

which is normalized according to the covariant prescription

$$\langle \boldsymbol{p}', r'|\boldsymbol{p}, r\rangle = 2E_p\delta_{rr'}\delta(\boldsymbol{p}-\boldsymbol{p}'). \tag{103}$$

Some of these statements are proven in the following section.

8. Lagrange density of Dirac field, charge, energy, momentum and spin of Dirac particles

It is not difficult to find a Lagrange density whose Euler equations are the Dirac equation for $\Psi(x)$ and its adjoint. In eq. (86) we already found a Lagrange density for the case of Majorana spinors. For the more general case of unrestricted Dirac fields a Lagrangian is

$$\mathscr{L}_{\mathrm{D}} = \frac{\mathrm{i}}{2}\left[\phi^{*}{}_{C}(x)\left(\hat{\sigma}^{\mu}\overleftrightarrow{\partial}_{\mu}\right)^{Ca}\phi_a(x) + \chi^{*d}(x)\left(\sigma^{\mu}\overleftrightarrow{\partial}_{\mu}\right)_{dB}\chi^{B}(x)\right]$$
$$- m_{\mathrm{D}}\left[\phi^{*}{}_{B}(x)\chi^{B}(x) + \chi^{*d}(x)\phi_d(x)\right], \tag{104}$$

*)See e.g. Ref. R3.

where m_{D} is a mass parameter and the quantity in square brackets is called Dirac mass term.*) One verifies easily that $\mathscr{L}_{\mathrm{D}}$ leads to the correct equations (47). When written in terms of four-spinors this Lagrangian takes a very simple form, viz.

$$\mathscr{L}_{\mathrm{D}}=\overline{\Psi(x)}\left[\frac{\mathrm{i}}{2}\gamma^{\mu}\overleftrightarrow{\partial}^{\mu}-m_{\mathrm{D}}\mathbb{1}\right]\Psi(x). \tag{104'}$$

The field variables being $\Psi(x)$ and $\overline{\Psi}(x)$, their conjugate momenta are

$$\Pi(x):=\frac{\partial\mathscr{L}_{\mathrm{D}}}{\partial(\partial_0\Psi)}=\frac{\mathrm{i}}{2}\Psi^{\dagger}(x), \tag{105a}$$

$$\overline{\Pi(x)}:=\frac{\partial\mathscr{L}_{\mathrm{D}}}{\partial(\partial_0\overline{\Psi})}=-\frac{\mathrm{i}}{2}\gamma^{0}\Psi(x). \tag{105b}$$

8.1. Charge of particles and antiparticles

The Lagrangian (104) is invariant under global gauge transformations of the first kind,

$$\Psi(x)\mapsto \mathrm{e}^{\mathrm{i}\alpha}\Psi(x),\qquad \overline{\Psi(x)}\mapsto \mathrm{e}^{-\mathrm{i}\alpha}\overline{\Psi(x)}\,. \tag{106}$$

Taking α to be infinitesimal this means that $\mathscr{L}_{\mathrm{D}}$ is invariant with respect to variations $\delta\Psi=\mathrm{i}\alpha\Psi$, $\delta\mathscr{L}_{\mathrm{D}}=0$. When this is worked out (and making use of the equations of motion) one obtains the conservation condition

$$\partial_{\mu}j^{\mu}(x)=0 \tag{107}$$

for the "current density" operator

$$j^{\mu}(x)=\overline{\Psi(x)}\,\gamma^{\mu}\Psi(x). \tag{108}$$

Note that, at this point, we do not know which physical current density is to be represented by the operator (108) nor do we know whether the divergence condition (107) actually corresponds to a physical conservation law. These questions cannot be answered until we know what the interactions of the fermion field are and how these interactions behave under the same gauge transformations. For a world *with* interactions we will have to consider simultaneous global gauge transformations of all fields that enter the theory, possibly also *local* gauge transformations (i.e. gauge transformations where α becomes dependent on space and time coordinates).

For the moment it may suffice to say that for *leptons*, i.e. for particles which have only electromagnetic and weak (and gravitational) interactions, the operator (108) multiplied with e represents the electromagnetic current density. If the Ψ are taken to be *free* fields (100), then $j^{\mu}(x)$ is the current operator in the sense of perturbation theory, i.e. $ej^{\mu}(x)A_{\mu}(x)$ is the interaction with the Maxwell field, represented by the potentials A_{μ}. In this case we can compute single particle matrix elements of $j^{\mu}(x)$ and of the corresponding charge operator Q. Indeed, if the fields are sufficiently

*) The most general case is treated below in subsection 8.4, at the end of this section.

well-behaved such that $j^\mu(x)$ vanishes at infinity,*) we conclude from the divergence condition (107) that the integral of $j^0(x)$ over all space is a constant of the motion:

$$Q := \int d^3x\, j^0(\boldsymbol{x}; t), \quad \frac{d}{dt} Q = 0. \tag{107'}$$

This follows from

$$\frac{d}{dt} \int d^3x\, j^0(x) = -\int d^3x \sum_{i=1}^{3} \frac{\partial}{\partial x^i} j^i(x) = 0.$$

It is not difficult to compute Q using eqs. (100), as well as the relations (70)–(72). One finds

$$Q = \sum_{r=1}^{2} \int \frac{d^3p}{2E_p} \left[a^{(r)\dagger}(\boldsymbol{p}) a^{(r)}(\boldsymbol{p}) + b^{(r)}(\boldsymbol{p}) b^{(r)\dagger}(\boldsymbol{p}) \right], \tag{109}$$

or, making use of the anticommutators (102b),

$$Q = \sum_{r=1}^{2} \int \frac{d^3p}{2E_p} \left[a^{(r)\dagger}(\boldsymbol{p}) a^{(r)}(\boldsymbol{p}) - b^{(r)\dagger}(\boldsymbol{p}) b^{(r)}(\boldsymbol{p}) \right]. \tag{109'}$$

We now calculate one-particle matrix elements of Q, again making use of the anticommutation rules (102).

For particles created by $a^{(r)\dagger}(\boldsymbol{p})$ we find

$$Q a^{(r)\dagger}(\boldsymbol{p})|0\rangle = +a^{(r)\dagger}(\boldsymbol{p})|0\rangle. \tag{110a}$$

Similarly, for "b"-type particles we have

$$Q b^{(r)\dagger}(\boldsymbol{p})|0\rangle = -b^{(r)\dagger}(\boldsymbol{p})|0\rangle. \tag{110b}$$

These results show that "a"-type particles and "b"-type particles have opposite "charge". This "charge" can be the electric charge, but it can also be any other charge-like quantum number that is respected by the fermion's interactions (lepton number, baryon number etc.). Thus, "a" and "b" particles are charge-conjugate to each other, or, in other terms, they are *antiparticles* of each other. We note, however, the complete symmetry of the theory in the two types of particles. It is only a matter of convention which of them is called particle and which is called antiparticle. This is in contrast to the old "hole theory" where particles are accepted as such but antiparticles are doomed to live in the shadow world of "holes" in particle states with "negative energy" (see also note at the end of this section).

The operator Q can also be applied to a Fock-state of many free fermions, say N particles and M antiparticles. Any such state is eigenstate of Q with eigenvalue ("charge") $(N - M)$. If in every reaction of these particles the "charge" Q is

*) In fact the fields (100) are not well-behaved and, strictly speaking, we should smooth them out with appropriate weight functions.

conserved, we say that Q is "additively conserved". This means that

$$\sum_i Q(i) = \sum_f Q(f),$$

where i counts the particles in the initial state, f counts the particles in the final state.

8.2. *Energy and momentum*

The energy–momentum tensor density is given by the general expression

$$T^{\mu\nu}(x) := \sum_n \frac{\partial \mathscr{L}}{\partial(\partial_\mu \phi_n)} \partial^\nu \phi_n - g^{\mu\nu}\mathscr{L}. \tag{111}$$

It satisfies the continuity equation

$$\partial_\mu T^{\mu\nu}(x) = 0 \tag{112}$$

(use the equations of motion to verify this).

Specifically, the energy density of the Dirac field is given by

$$\mathscr{H}(x) \equiv T^{00}(x) = -\overline{\Psi(x)}\left(\frac{\mathrm{i}}{2}\boldsymbol{\gamma}\cdot\overleftrightarrow{\nabla} - m\mathbb{1}\right)\Psi(x) \tag{113}$$

and the momentum density is given by

$$T^{0m}(x) = \frac{\mathrm{i}}{2}\overline{\Psi(x)}\,\gamma^0 \overleftrightarrow{\partial}^m \Psi(x) = -\frac{\mathrm{i}}{2}\Psi^\dagger \overleftrightarrow{\nabla}_m \Psi. \tag{114}$$

The divergence condition (112) implies that the four-vector

$$(H, \boldsymbol{P}) := \left(\int \mathrm{d}^3x\, T^{00}, \int \mathrm{d}^3x\, T^{0i}\right)$$

is a constant of the motion. The operators H and $\boldsymbol{P}$ are easily calculated by inserting the expansions (100) of the fields in terms of creation and annihilation operators. Consider first H,

$$H = \int \mathrm{d}^3x\, \mathscr{H}(x) = \int \mathrm{d}^3x\, \Psi^\dagger(x)(-\mathrm{i}\boldsymbol{\alpha}\cdot\nabla + m\beta)\Psi(x),$$

where we have integrated the derivative on $\overline{\Psi}$ by parts and have inserted the definitions (59). With (100) and making use again of the orthogonality properties of spinors in momentum space one finds

$$H = \sum_{r=1}^{2} \int \frac{\mathrm{d}^3p}{2E_p} E_p \left[a^{(r)\dagger}(\boldsymbol{p})a^{(r)}(\boldsymbol{p}) - b^{(r)}(\boldsymbol{p})b^{(r)\dagger}(\boldsymbol{p})\right]. \tag{115a}$$

Upon using the anticommutation rule (102b) this can be written, up to an infinite constant,

$$H = \sum_{r=1}^{2} \int \frac{\mathrm{d}^3p}{2E_p} E_p \left[a^{(r)\dagger}(\boldsymbol{p})a^{(r)}(\boldsymbol{p}) + b^{(r)\dagger}(\boldsymbol{p})b^{(r)}(\boldsymbol{p})\right]. \tag{115b}$$

The infinite constant should not worry us as only energy *differences* are physically relevant. Whenever the difference of the energies of any two states is taken, this constant drops out.

Applying the operator H to a single particle state shows that $E_p = (m^2 + \boldsymbol{p}^2)^{1/2}$ is the energy of this state, independently of whether the state contains an "a"-type or "b"-type particle. For calculating the action of H onto a more general Fock state the following commutators are helpful:

$$\left[a^{(r)}(\boldsymbol{p}), H\right] = E_p a^{(r)}(\boldsymbol{p}), \tag{116a}$$

$$\left[b^{(r)}(\boldsymbol{p}), H\right] = E_p b^{(r)}(\boldsymbol{p}), \tag{116b}$$

$$\left[a^{(r)\dagger}(\boldsymbol{p}), H\right] = -E_p a^{(r)\dagger}(\boldsymbol{p}), \tag{116c}$$

$$\left[b^{(r)\dagger}(\boldsymbol{p}), H\right] = -E_p b^{(r)\dagger}(\boldsymbol{p}). \tag{116d}$$

They show that applying $a^\dagger(\boldsymbol{p})$ or $b^\dagger(\boldsymbol{p})$ to any eigenstate Ψ_0 of H increases its energy by the amount E_p: Indeed, let Ψ_α be an eigenstate of H with energy E_α,

$$H|\Psi_\alpha\rangle = E_\alpha|\Psi_\alpha\rangle.$$

Then $a^{(r)\dagger}(\boldsymbol{p})|\Psi_\alpha\rangle$ is also eigenstate of H,

$$\begin{aligned} Ha^{(r)\dagger}(\boldsymbol{p})|\Psi_\alpha\rangle &= \left[a^{(r)\dagger}(\boldsymbol{p})H + E_p a^{(r)\dagger}(\boldsymbol{p})\right]|\Psi_\alpha\rangle \\ &= (E_\alpha + E_p)a^{(r)\dagger}(\boldsymbol{p})|\Psi_\alpha\rangle. \end{aligned}$$

Its energy is seen to be $E_\alpha + E_p$. Exactly the same argument applies to the state $b^{(r)\dagger}(\boldsymbol{p})|\Psi_\alpha\rangle$.

Similarly, application of $a(\boldsymbol{p})$ or $b(\boldsymbol{p})$ leads to a new eigenstate of H whose energy is reduced by the amount E_p.

Quite similarly, the total momentum operator may be calculated from (114) and (100), and is found to be

$$\boldsymbol{P} = \sum_{r=1}^{2} \int \frac{\mathrm{d}^3 p}{2E_p} \boldsymbol{p}\left[a^{(r)\dagger}(\boldsymbol{p})a^{(r)}(\boldsymbol{p}) + b^{(r)\dagger}(\boldsymbol{p})b^{(r)}(\boldsymbol{p})\right]. \tag{117}$$

By using arguments completely analogous to the above, one shows that the operators $a^{(r)\dagger}(\boldsymbol{p})$ or $b^{(r)\dagger}(\boldsymbol{p})$, when applied to a Fock state, add an additional three-momentum $\boldsymbol{p}$, independently of the *type* of particle. Similarly, the corresponding annihilation operators take away three-momentum $\boldsymbol{p}$, independently of the type of particle.

To summarize, we have convinced ourselves that $a^{(r)\dagger}(\boldsymbol{p})$ and $b^{(r)\dagger}(\boldsymbol{p})$ $[a^{(r)}(\boldsymbol{p})$ and $b^{(r)}(\boldsymbol{p})]$ must be interpreted as creation (annihilation) operators for free fermions with four-momentum $p = (E_p = (p^2 + m^2)^{1/2}, \boldsymbol{p})$ and spin projection r. The corresponding "wave functions" in momentum space are given by the spinors $u^{(r)}(\boldsymbol{p})$ and $v_c^{(r)}(\boldsymbol{p})$, respectively, see eqs. (68) and (69). The two kinds of particles are distinguished through their "charge", cf. eqs. (109) and (110). The two types of particles are said to be antiparticles of each other. The formalism is completely symmetric in the two kinds of particles.

Remark. Let us return, for a moment, to the Hamiltonian H of eq. (115a) and note the minus sign in front of the second term. In passing from (115a) to (115b) it was essential that b and $b^{\dagger}$ *anti*commuted, in order to obtain the integrand of eq. (115b) with the two plus signs. Had we taken *commutators* instead of the anti-commutators (100), the second term of eq. (115b) would still be negative. This would mean not only that antiparticle states had negative kinetic energies but also that H, the total energy of the field, had arbitrarily large negative eigenvalues. Both consequences, quite obviously, are not tenable on physical grounds: We know that free electrons and positrons have positive energy, and we know that a physical Hamiltonian must have a spectrum which is bounded from below. In many introductions to Dirac theory these difficulties are quoted as the primary motivation for rejecting commutators of fermion fields and for using anticommutators instead. This procedure, although acceptable, is not satisfactory for it does not reveal the real origin of the difficulty. As we have seen above, enforcing commutators leads unavoidably into a theory which is in conflict with causality. This is the deeper reason why the theory must be quantized by means of anticommutators. The spectrum of H coming out unbounded from below, when commutators are used instead, is a symptom rather than the fundamental cause of the difficulties.

8.3. Spin properties of Dirac particles

As is well-known relativistic motion mixes spin and orbital angular momentum degrees of freedom in a complicated way. In general, when the *total* angular momentum (sum of spin and orbital angular momentum) is conserved in a reaction, neither the total spin nor the total orbital momentum are conserved separately. Therefore, for a *massive* particle, one must go to the rest frame if one wishes to know its spin. In the rest system the orbital angular momentum is zero, so that when we perform a rotation R in that system the particle state will transform with $D^{(S)}(R)$ where S is its spin. From the very construction of force-free solutions of the Dirac equation, we know that massive fermions carry spin $1/2$.

The general case can be treated in several ways. The most transparent is perhaps the method used above which consists in "boosting" the particle state back to its rest system and perform a rotation there.*) Another approach makes use of the Pauli–Lubanski four-vector which is defined by

$$W_{\mu} = -\tfrac{1}{2}\varepsilon_{\alpha\beta\lambda\mu}M^{\alpha\beta}P^{\lambda}$$

and where P^{λ} and $M^{\alpha\beta}$ are the generators of infinitesimal Lorentz transformations. This set of operators generates the little group (i.e. the set of all Lorentz transformations that leave the eigenvalues of the energy–momentum operator P^{μ} invariant) and yields a relativistic description of arbitrary spins. It is also very useful in discussing the spin properties of *massless* particles. In this case one finds that (excluding continuous spin) massless particles are characterized by helicity states rather than

*) See e.g. Ref. R10.

spin states and that the helicity can only have two values $\pm\lambda$. λ, which must be integer or half-integer is said to be the "spin" of the particle.

We do not go into these general matters here and refer the reader to the extensive literature on this subject. Instead, we turn to the more practical question of how to handle the spin and polarization properties of spin 1/2-particles.

8.3.1. The case of massive fermions

Let us first recall the nonrelativistic description of spin and polarization of a massive fermion (in other words we go to the rest system of the particle first). Consider a statistical mixture of spin-1/2 particles polarized parallel or opposite to a given direction

$$\boldsymbol{n} = (\sin\theta\cos\phi, \sin\theta\sin\phi, \cos\theta)$$

in space. The angles θ and ϕ specify the vector $\boldsymbol{n}$ with respect to a given frame of reference K_0.

Had we used, instead, a frame K whose 3-direction coincides with $\boldsymbol{n}$, the density matrix describing this state would be given by

$$\rho|_K = \lambda_+|m_{\boldsymbol{n}} = +\tfrac{1}{2}\rangle\langle m_{\boldsymbol{n}} = +\tfrac{1}{2}| + \lambda_-|m_{\boldsymbol{n}} = -\tfrac{1}{2}\rangle\langle m_{\boldsymbol{n}} = -\tfrac{1}{2}| = \begin{pmatrix} \lambda_+ & 0 \\ 0 & \lambda_- \end{pmatrix}. \tag{118}$$

Here λ_+, λ_- are the statistical weights of the fraction of particles polarized along the positive or negative $\boldsymbol{n}$-direction, respectively. These weights have the properties

$$0 \le \lambda_\pm \le 1, \qquad \lambda_+ + \lambda_- = 1.$$

The states with $\lambda_+ = 1$, $\lambda_- = 0$; $\lambda_+ = 0$, $\lambda_- = 1$ are pure states corresponding to full polarization parallel and opposite to $\boldsymbol{n}$, respectively. The same density matrix, but written out with respect to the original frame, is obtained from (118) through a rotation by the Euler angles (ψ, θ, ϕ)

$$\rho|_{K^0} = D^{(1/2)\dagger}(\psi, \theta, \phi)\, \rho|_K\, D^{(1/2)}(\psi, \theta, \phi), \tag{119}$$

with

$$D^{(1/2)}(\psi, \theta, \phi) = e^{(i/2)\psi\sigma^{(3)}} e^{(i/2)\theta\sigma^{(2)}} e^{(i/2)\phi\sigma^{(3)}}$$
$$= \begin{pmatrix} \cos(\theta/2)e^{(i/2)(\psi+\phi)} & \sin(\theta/2)e^{(i/2)(\psi-\phi)} \\ -\sin(\theta/2)e^{-(i/2)(\psi-\phi)} & \cos(\theta/2)e^{-(i/2)(\psi+\phi)} \end{pmatrix}.$$

The angles θ, ϕ are the same as before, ψ is arbitrary, but drops out in the density matrix. One finds by straightforward calculation,

$$\rho|_{K^0} = \frac{1}{2}\left\{\begin{pmatrix} 1 & 0 \\ 0 & 1 \end{pmatrix} + (\lambda_+ - \lambda_-)\begin{pmatrix} \cos\theta & \sin\theta e^{-i\phi} \\ \sin\theta e^{i\phi} & -\cos\theta \end{pmatrix}\right\},$$

which is, of course, the same as

$$\rho = \tfrac{1}{2}\{\mathbb{1} + (\lambda_+ - \lambda_-)\boldsymbol{\sigma}\cdot\boldsymbol{n}\} \equiv \lambda_+ P_+ + \lambda_- P_-, \tag{119'}$$

where $P_\pm$ are the projectors onto "spin up" and "spin down" states along the direction $\boldsymbol{n}$, respectively,

$$P_\pm := \tfrac{1}{2}(\mathbb{1} \pm \boldsymbol{\sigma}\cdot\boldsymbol{n}). \tag{120}$$

It is convenient to define a polarization vector

$$\boldsymbol{\zeta} := (\lambda_+ - \lambda_-)\boldsymbol{n}, \tag{121}$$

so that

$$\rho = \tfrac{1}{2}(\mathbb{1} + \boldsymbol{\zeta}\cdot\boldsymbol{\sigma}). \tag{122}$$

This vector has the norm

$$\boldsymbol{\zeta}^2 = (\lambda_+ - \lambda_-)^2 = (\mathbb{1} - 2\lambda_-)^2 \le 1.$$

It lies within a sphere of unit radius, the so-called *Poincaré sphere*.*) Its direction gives the direction of predominant polarization, its magnitude the degree of polarization. Thus, if $\boldsymbol{\zeta}$ lies *inside* the sphere, the state is a statistical mixture, if $\boldsymbol{\zeta}$ lies *on* the sphere, the state is pure. Indeed, if we compute the polarization $\boldsymbol{P}$ in the state described by the density matrix (119′), we find

$$\boldsymbol{P} = \langle\boldsymbol{\sigma}\rangle = \mathrm{Sp}(\rho\boldsymbol{\sigma}) = (\lambda_+ - \lambda_-)\boldsymbol{n} \equiv \boldsymbol{\zeta}. \tag{123a}$$

The *degree* of polarization is given by

$$P \equiv |\boldsymbol{P}| = |\boldsymbol{\zeta}| = \frac{\lambda_+ - \lambda_-}{\lambda_+ + \lambda_-} = \lambda_+ - \lambda_-. \tag{123b}$$

Thus far, our description of polarization applies to the rest system of the particle, or, in an approximate way, to weakly relativistic motion. How do these notions generalize when we deal with truly *relativistic* motion of the particle?

For this purpose we must (i) transform the polarization vector $\boldsymbol{\zeta}$ to arbitrary Lorentz frames, (ii) find a covariant four-vector which is the generalization of the spin operator $\boldsymbol{\sigma}$, by constructing the covariant form of the spin projection operators (120), and (iii) find a covariant (Lorentz invariant) expression for the density matrix.

Point (i) is easy to carry out: $\boldsymbol{\zeta}$ being a classical quantity, we simply have to "boost" the vector $(0, \boldsymbol{\zeta})$ to the particle momentum $\boldsymbol{p}$ (cf. eq. (A.10) of Appendix A):

$$s := L(\boldsymbol{p})(0, \boldsymbol{\zeta}) = \left(\frac{1}{m}\boldsymbol{p}\cdot\boldsymbol{\zeta}, \boldsymbol{\zeta} + \frac{\boldsymbol{p}\cdot\boldsymbol{\zeta}}{m(E_p + m)}\boldsymbol{p}\right). \tag{124}$$

This four-vector has the following properties:

$$s^2 = -\boldsymbol{\zeta}^2, \quad \text{i.e.} \quad 0 \le (-s^2) \le 1, \tag{125a}$$

$$s\cdot p = 0. \tag{125b}$$

In particular, the degree of polarization (which obviously is a Lorentz scalar) is

*) This sphere is originally defined in the description of polarized or partially polarized light. The formalism describing polarized electromagnetic waves (or photons) is the same as for spin-1/2 particles. The real quantities $\zeta_1, \zeta_2, \zeta_3$ are called Stokes parameters in electrodynamics.

$P=\sqrt{(-s^2)}$. As ζ is a spin expectation value, cf. eq. (123a), it must be even under parity operation in the particle's rest frame. From this observation we see that s^μ, eq. (124), is an *axial* four-vector.

Ad (ii): The construction of the covariant spin projection operators is somewhat complicated by the fact that the Dirac equation admits solutions of positive and negative frequencies (particles and antiparticles) each of which can have two polarization states. Therefore matrices like $u_\alpha^{(r)}(\boldsymbol{p})\overline{u_\beta^{(r)}(\boldsymbol{p})}$ (or $v_\alpha^{(r)}(\boldsymbol{p})\overline{v_\beta^{(r)}(\boldsymbol{p})}$), which are the analogues of the non-relativistic projector $P_r=|\chi^{(r)}\rangle\langle\chi^{(r)}|$—when taken separately—will project out the spin direction r only for positive (negative) frequency solutions. In other words, both $u\bar{u}$ and $v\bar{v}$ will contain the covariant spin projection operator but multiplied by projectors onto solutions with positive and negative frequency, respectively. With these remarks in mind we proceed as follows:

Let $u(\boldsymbol{p})$ be a particle spinor of momentum $\boldsymbol{p}$, mass m, polarized along an arbitrary direction $\boldsymbol{n}$ in the rest frame of the particle. Let $v(\boldsymbol{p})$ be the corresponding antiparticle spinor with the same three-momentum and let it also be polarized along the same direction $\boldsymbol{n}$. $\boldsymbol{n}$ is a unit vector, $\boldsymbol{n}^2=1$. In the rest system of these particles we have (using the standard representation)

$$u(0)\overline{u(0)}=\frac{1}{2}\begin{pmatrix}\mathbb{1}+\boldsymbol{\sigma}\cdot\boldsymbol{n} & 0\\ 0 & 0\end{pmatrix},$$

$$v(0)\overline{v(0)}=\frac{1}{2}\begin{pmatrix}0 & 0\\ 0 & \mathbb{1}-\boldsymbol{\sigma}\cdot\boldsymbol{n}\end{pmatrix}.$$

Expressing these matrices in terms of γ-matrices one verifies easily that

$$u(0)\overline{u(0)}=\tfrac{1}{2}(\mathbb{1}+\gamma^0)\tfrac{1}{2}(\mathbb{1}-\gamma_5\boldsymbol{n}\cdot\boldsymbol{\gamma}),$$

$$v(0)\overline{v(0)}=-\tfrac{1}{2}(\mathbb{1}-\gamma^0)\tfrac{1}{2}(\mathbb{1}-\gamma_5\boldsymbol{n}\cdot\boldsymbol{\gamma}).$$

We know from eqs. (65) and (66) that $u(\boldsymbol{p})=N(\not{p}+m)u(0)$ and $v(\boldsymbol{p})=N(\not{p}-m)v(0)$ with $N=(E_p+m)^{-1/2}$. Using this we calculate the sum*⁾

$$\begin{aligned}
&u(\boldsymbol{p})\overline{u(\boldsymbol{p})}+v(\boldsymbol{p})\overline{v(\boldsymbol{p})}\\
&=\frac{1}{E_p+m}\left\{(\not{p}+m)u(0)\overline{u(0)}\,(\not{p}+m)+(\not{p}-m)v(0)\overline{v(0)}\,(\not{p}-m)\right\}\\
&=\frac{1}{4(E_p+m)}\{(\not{p}+m)(\mathbb{1}+\gamma_0)(\mathbb{1}-\gamma_5n^i\gamma^i)(\not{p}+m)\\
&\qquad-(\not{p}-m)(\mathbb{1}-\gamma_0)(\mathbb{1}-\gamma_5n^i\gamma^i)(\not{p}-m)\}\\
&=\frac{1}{4(E_p+m)}\{[4m\not{p}+2\not{p}\gamma^0\not{p}+2m^2\gamma^0]\\
&\qquad-\gamma_5n^i[2m(\gamma^i\not{p}-\not{p}\gamma^i)-2m^2\gamma^0\gamma^i+2\not{p}\gamma^0\gamma^i\not{p}]\}.
\end{aligned}$$

By commuting the γ-matrices such as to move $\not{p}$ to the left in each term and using

*⁾We use the "slash" notation, $\not{a}\equiv a^\mu\gamma_\mu$.

$\not{p}\not{p} = m^2$ this can be transformed to

$$= \left\{ \not{p} - \gamma_5 n^i \left[-\not{p}\gamma^i + \not{p}\gamma^0 \frac{p^i}{E_p + m} + \frac{m}{E_p + m} p^i \right] \right\}$$

$$= \not{p} \left\{ \mathbb{1} + \gamma_5 \left[-n^i\gamma^i + n^i p^i \left(\frac{1}{E_p + m}\gamma^0 + \frac{1}{m(E_p + m)} \not{p} \right) \right] \right\}$$

$$= \not{p} \left\{ \mathbb{1} + \gamma_5 \left[\frac{\boldsymbol{n} \cdot \boldsymbol{p}}{m}\gamma^0 - \left(n^i + \frac{\boldsymbol{n} \cdot \boldsymbol{p}}{m(E_p + m)} p^i \right) \gamma^i \right] \right\}$$

$$= \not{p} \{ \mathbb{1} + \gamma_5 \not{n} \},$$

where n is defined in analogy to eq. (124) and is nothing but the four-vector $(0; \boldsymbol{n})$ "boosted" to the system where the particles have momentum $\boldsymbol{p}$. In exactly the same manner one computes the difference

$$u(\boldsymbol{p})\overline{u(\boldsymbol{p})} - v(\boldsymbol{p})\overline{v(\boldsymbol{p})} = m\{\mathbb{1} + \gamma_5 \not{n}\}.$$

From these results follow the important relations

$$u_\alpha(\boldsymbol{p})\overline{u_\beta(\boldsymbol{p})} = \tfrac{1}{2}\{(\not{p} + m\mathbb{1})(\mathbb{1} + \gamma_5\not{n})\}_{\alpha\beta}, \tag{126a}$$

$$v_\alpha(\boldsymbol{p})\overline{v_\beta(\boldsymbol{p})} = \tfrac{1}{2}\{(\not{p} - m\mathbb{1})(\mathbb{1} + \gamma_5\not{n})\}_{\alpha\beta}. \tag{126b}$$

We recall the definition of n^μ:

$$n^\mu = \left(\frac{\boldsymbol{p} \cdot \boldsymbol{n}}{m}, \boldsymbol{n} + \frac{\boldsymbol{n} \cdot \boldsymbol{p}}{m(E_p + m)} \boldsymbol{p} \right), \tag{127}$$

which satisfies $n^2 = -1$, $(np) = 0$.

Let us comment on these results. Obviously, the spin projection operator onto the positive $\boldsymbol{n}$ direction is

$$\pi_{\boldsymbol{n}} = \tfrac{1}{2}(\mathbb{1} + \gamma_5 \gamma^\mu n_\mu) \tag{128}$$

for both particles and antiparticles. In the rest system it reduces to the familiar form

$$\pi_{\boldsymbol{n}}|_{\boldsymbol{p}=0} = \frac{1}{2}\begin{pmatrix} \mathbb{1} + \boldsymbol{\sigma} \cdot \boldsymbol{n} & 0 \\ 0 & \mathbb{1} - \boldsymbol{\sigma} \cdot \boldsymbol{n} \end{pmatrix}. \tag{129}$$

The second term in eq. (128) is the scalar product of n^μ and $\gamma_5\gamma^\mu$. We already know that n^μ is an axial vector; $\gamma_5\gamma^\mu$, when taken between Dirac fields, is also an axial vector, so that $\overline{\Psi}\gamma_5\gamma^\mu\Psi n_\mu$ is a scalar.

As to the other factors in eqs. (126) it should be clear that

$$\Omega_\pm = \pm\frac{1}{2m}(\not{p} \pm m\mathbb{1}) \tag{130}$$

are the projectors onto positive and negative frequency solutions, respectively. The normalization follows from the requirement

$$\Omega_{\pm}^2 = \Omega_{\pm}.$$

From $\not{p}\not{n} = 2(pn) - \not{n}\not{p} = -\not{n}\not{p}$ and from (58) we see that $\not{p}$ commutes with $\gamma_5\gamma^\mu$ and, therefore, that

$$[\pi_n, \Omega_+] = 0 = [\pi_n, \Omega_-],$$

which repeats our statement that π_n is the spin projection operator, independently of whether we deal with a particle or an antiparticle.

It is now easy to answer question (iii): The covariant density matrix that describes an ensemble of *particles* with partial polarization $\boldsymbol{P} = \boldsymbol{\zeta}$ (cf. eq. (123)) is*)

$$\rho = 2m\Omega_+ \tfrac{1}{2}(\mathbb{1} + \gamma_5 \not{s}) = \tfrac{1}{2}(\not{p} + m\mathbb{1})(\mathbb{1} + \gamma_5\not{s}), \tag{131}$$

where s is defined in eq. (124) and has the norm

$$\sqrt{-s^2} = |\boldsymbol{\zeta}| \equiv P = \lambda_+ - \lambda_-.$$

We verify that

$$\rho^2 = 4m^2\Omega_+ \frac{1}{2}\left(\frac{1+\boldsymbol{\zeta}^2}{2}\mathbb{1} + \gamma_5\not{s}\right),$$

which is equal to $2m\rho$ only for $|\boldsymbol{\zeta}| = 1$, and that

$$\mathrm{Sp}\left(\frac{\rho^2}{4m^2}\right) = \tfrac{1}{2}(1+\boldsymbol{\zeta}^2) \le \mathrm{Sp}\left(\frac{\rho}{2m}\right) = 1. \tag{132a}$$

Note that ρ is not hermitean but that instead

$$\gamma^0\rho^\dagger\gamma^0 = \rho. \tag{132b}$$

In eq. (132a) the equality sign holds if $|\boldsymbol{\zeta}| = 1$, i.e. for a pure state. In that case $\rho^2 = 2m\rho$.

Similarly, the density matrix describing antiparticles with polarization $\boldsymbol{\zeta}$ is given by**)

$$\rho = -2m\Omega_- \tfrac{1}{2}(\mathbb{1} + \gamma_5\not{s}) = \tfrac{1}{2}(\not{p} - m\mathbb{1})(\mathbb{1} + \gamma_5\not{s}). \tag{133}$$

It is easy to see that $\rho^\dagger$ describes the parity-mirror state of ρ, i.e. $\boldsymbol{p} \to -\boldsymbol{p}$ but $\boldsymbol{\zeta} \to \boldsymbol{\zeta}$. Since fermions always appear in bilinear forms in any observable, there is no harm in having a non-hermitean density matrix (see exercise 9).

If one insists on having a hermitean density matrix one can take, instead of (131):

$$P := \gamma^0\rho = \tfrac{1}{2}(E_p\mathbb{1} - \boldsymbol{p}\cdot\boldsymbol{\alpha} \pm m\beta)(\mathbb{1} + \gamma_5\not{s}). \tag{134}$$

This matrix has the properties (for particles *and* antiparticles),

$$P^\dagger = P, \qquad \mathrm{Sp}(P) = 2E_p.$$

*)We have normalized the density matrix in accord with the covariant normalization (70) and (67), i.e. $\mathrm{Sp}\,\rho = 2m$.

**)Normalization and sign in agreement with eq. (71).

Thus, its trace is not a Lorentz scalar (the normalization is the covariant one of eqs. (67)).

It is convenient to express the polarization vector $\boldsymbol{\zeta}$ on (or inside) the Poincaré sphere in terms of its component ζ_ℓ along the particle's three-momentum $\boldsymbol{p}$ and its components perpendicular to $\boldsymbol{p}$,

$$\boldsymbol{\zeta} = \zeta_\ell \boldsymbol{p}/|\boldsymbol{p}| + \boldsymbol{\zeta}_\mathrm{t} \quad \text{with} \quad \zeta_\ell = \frac{1}{|\boldsymbol{p}|}\boldsymbol{p}\cdot\boldsymbol{\zeta}. \tag{135a}$$

The four-vector s, eq. (124), is then given by the components

$$s^0 = \frac{1}{m}|\boldsymbol{p}|\zeta_\ell, \qquad \boldsymbol{s}_\ell = \frac{E_p}{m}\zeta_\ell\frac{\boldsymbol{p}}{|\boldsymbol{p}|}, \qquad \boldsymbol{s}_\mathrm{t} = \boldsymbol{\zeta}_\mathrm{t}. \tag{135b}$$

The special cases of longitudinal and transverse polarizations (with respect to the momentum $\boldsymbol{p}$) can be read off these formulae.

A case of special interest is the case of extreme relativistic motion which we discuss separately in the next section.

8.3.2. Extreme relativistic motion and the neutrinos

Suppose first that we deal with a massive particle whose energy is very large as compared to its mass, i.e.

$$E \gg m, \qquad |\boldsymbol{p}| \simeq E.$$

Let us take the 3-direction in the direction of the particle's momentum $\boldsymbol{p}$. Then from eqs. (131) and (133)

$$\rho \simeq \tfrac{1}{2}\{E(\gamma^0-\gamma^3) \pm m\mathbb{1}\}\left\{\mathbb{1} + \gamma_5\left[\frac{E}{m}\zeta_\ell(\gamma^0-\gamma^3) - \zeta_\mathrm{t}^1\gamma^1 - \zeta_\mathrm{t}^2\gamma^2\right]\right\}$$

$$\simeq \tfrac{1}{2}E(\gamma^0-\gamma^3)\{\mathbb{1} + \gamma_5[\mp\zeta_\ell - \zeta_\mathrm{t}^1\gamma^1 - \zeta_\mathrm{t}^2\gamma^2]\}.$$

Here we have used $(\gamma^0-\gamma^3)^2 = (\gamma^0)^2 + (\gamma^3)^2 - \{\gamma^0,\gamma^3\} = 0$ and we have neglected $m\mathbb{1}$ against $E(\gamma^0-\gamma^3)$. The result can also be written as follows:

$$\rho \simeq \tfrac{1}{2}\not{p}\{\mathbb{1} - \gamma_5[\pm\zeta_\ell + \zeta_\mathrm{t}^1\gamma^1 + \zeta_\mathrm{t}^2\gamma^2]\}; \tag{136}$$

the positive sign holds for particles, the negative sign for antiparticles.

The expression (136) which can be used to describe, for example, electrons and positrons at ultra-relativistic energies, shows that such particles can have any partial or full polarization, along their momentum or transverse to it. For instance, a statistical mixture of electrons with positive helicity (statistical weight λ_+) and electrons with negative helicity (statistical weight λ_-) at very high energy is described by

$$\rho \simeq \tfrac{1}{2}\not{p}\{\mathbb{1} - (\lambda_+ - \lambda_-)\gamma_5\}.$$

Expression (136) is also applicable to massless fermions, i.e. to neutrinos. However, there is one essential restriction. The only possible spin states are the ones with positive or negative helicity $h = \pm\lambda$. This can be understood very qualitatively as follows: Massless particles have no rest system. For a *massive* particle we can always go back to its rest system and rotate its spin into any direction we wish by means of the full rotation group. For a *massless* particle, the particle's momentum $\boldsymbol{p}$ singles out a specific spatial direction; the only Lorentz transformations that may remain "good" symmetries are the ones which leave this direction invariant (so-called *little group*): rotations about the spatial direction $\boldsymbol{p}$ and reflections with respect to any plane in the three-dimensional space that contains this direction. Regarding the spin properties of photons, we know that "right"- and "left"-circularly polarized plane waves are described by the polarization vectors (taking the photon momentum $\boldsymbol{q}$ in the 3-direction)

$$\boldsymbol{e}_{\pm 1} = \mp \frac{1}{\sqrt{2}}(e_1 \pm \mathrm{i} e_2).$$

Rotations about the three-axis leave these quantities invariant (except for multiplication by a phase); while reflection with respect to the plane spanned by the 1-axis and the 3-axis transforms one state into the other. As both kinds of transformations are symmetries which leave the Maxwell equations invariant (in homogeneous and isotropic space), also any other transverse photon polarization is possible: any linear superposition of the two helicity states $e_{\pm 1}$ is acceptable (linear polarization, elliptic polarization).

For neutrinos and antineutrinos the possible spin states are the states with helicity $\lambda = \pm \frac{1}{2}$. It appears that *neutrinos* which are produced in weak interactions at moderate energies always have *negative* helicity (they are said to be "left-handed"), whilst *antineutrinos* always have *positive* helicity (they are said to be "right-handed"). Thus, the density matrix for neutrinos (eq. (136) with upper sign and $\zeta_\ell = -1$) as well as for antineutrinos (eq. (136) with lower sign and $\zeta_\ell = +1$) reads

$$\rho^{(\nu)} = \tfrac{1}{2} \not{p}(\mathbb{1} + \gamma_5). \tag{137}$$

It describes a pure state: a neutrino state of negative helicity or an antineutrino state of positive helicity. A priori there is no reason why the massless neutrino (antineutrino) couples to other particles only in left-handed (right-handed) states. In principle massless fermions could have either helicity, or could be in states which are superpositions of the two helicities $\lambda = \pm \frac{1}{2}$. As we have seen above, the density matrix

$$\rho(m = 0) = \tfrac{1}{2} \not{p}(\mathbb{1} - 2\lambda\gamma_5) \tag{138}$$

describes particles with helicity λ, or antiparticles with helicity $-\lambda$. For electrons at ultra-high energies these are indeed possible states.

The fact that for neutrinos and antineutrinos eq. (137) is the only possibility, is of dynamical origin: In connection with the photon's helicity we have said above that the symmetry operations relevant to a massless particle of a given momentum p are rotations about $\boldsymbol{p}$ and reflections with respect to planes containing vector $\boldsymbol{p}$. These reflections, in particular, convert positive into negative helicity and vice versa (exercise 14). Whilst electromagnetic interactions are invariant under both rotations and reflections, this is not so for weak interactions. Weak interactions are invariant under rotations but not under parity or, for our purpose, under reflections with respect to planes that contain $\boldsymbol{p}$. Therefore, the two possible helicity states are not degenerate dynamically. Actually parity violation in the leptonic sector is found to be maximal: one helicity state ($\lambda = +1/2$ for ν, $\lambda = -1/2$ for $\bar{\nu}$) seems to decouple completely from the physical world of particles.

As a consequence the parity transform of $\rho(m=0)$, eq. (138),

$$\rho_P(m=0) = \gamma^0 \rho(m=0;\, \boldsymbol{p} \to -\boldsymbol{p})\gamma^0 = \tfrac{1}{2}\not{p}(\mathbb{1} + 2\lambda\gamma_5) \tag{139}$$

describes states which do not couple in weak interactions. By the same token charge conjugation cannot be a symmetry of weak interactions, for it transforms a left-handed neutrino (right-handed antineutrino) into a left-handed antineutrino (right-handed neutrino) which decouples from other particles.

As an exercise the reader is invited to show that the charge conjugate of (138) is given by

$$\rho_C(m=0) = \tfrac{1}{2}\not{p}(\mathbb{1} + 2\lambda\gamma_5). \tag{140}$$

If, on the other hand, we consider the combined operation of parity and charge conjugation, PC, then we see from eqs. (139) and (140) that $\rho(m=0)$ and, more specifically, $\rho^{(\nu)}$ are invariant under PC. Thus the neutrino and the antineutrino are PC-partners of each other.

Note. For spin-1/2 particles it is customary to denote the helicity states by $h := 2\lambda$, i.e. $h = \pm 1$ instead of $\lambda = \pm 1/2$. We shall adopt this convention in Chapters IV and V below.

8.4. Dirac and Majorana mass terms

In eqs. (86) and (104) we have encountered real mass terms for the case of one single Majorana and one single Dirac field, respectively. These are special cases of the most general mass term that is compatible with Lorentz invariance. As the general case is instructive and provides further insight into the structure of the theory we work it out in some detail.

Let $\phi_a(x)$ and $\chi^A(x)$ be spinors of first and second kind, respectively, without any condition (such as, e.g., eq. (84)) imposed on them. The most general, Lorentz invariant and hermitean Lagrangian (without interaction terms) that can be constructed on the basis of these fields, is the following:

$$\begin{aligned}\mathscr{L} = {} & \frac{\mathrm{i}}{2}\left\{\phi^*{}_A(x)\left(\hat{\sigma}^\mu \overleftrightarrow{\partial}_\mu\right)^{Ab}\phi_b(x) + \chi^{*a}(x)\left(\sigma^\mu \overleftrightarrow{\partial}_\mu\right)_{aB}\chi^B(x)\right\} \\ & + \left\{m_{\mathrm{D}}\chi^*{}_a\varepsilon^{ab}\phi_b - m_{\mathrm{D}}^*\phi^*{}_A\varepsilon^{AB}\chi_B\right\} \\ & + \tfrac{1}{2}\left\{m_1\phi_a\varepsilon^{ab}\phi_b + m_1^*\phi^*{}_A\varepsilon^{AB}\phi^*{}_B\right\} \\ & - \tfrac{1}{2}\left\{m_2\chi^{*a}\varepsilon_{ab}\chi^{*b} + m_2^*\chi^A\varepsilon_{AB}\chi^B\right\}.\end{aligned} \tag{141}$$

Here m_{D}, m_1, m_2 are arbitrary, complex parameters with dimension of masses. Each one of the three mass terms in curly brackets is hermitean by itself, as is easily verified by means of relations (34) and (35). The terms containing m_1, m_2 and their complex conjugates are generalizations of the mass term in the Majorana Lagrangian (86), whilst the terms in m_{D}, m_{D}^* generalize the Dirac case of eq. (104).

In order to clarify the structure of the mass term in eq. (141) we note first that it can be rewritten in a way which eliminates the apparent asymmetry in the spinor fields $\phi(x)$ and $\chi(x)$. Indeed, $\chi^*{}_a = \varepsilon_{ab}\chi^{*b}$ is another spinor of first kind, having the same transformation properties as ϕ_a. Likewise $\phi^{*A} = \varepsilon^{AB}\phi^*{}_B$ is a spinor of second kind, very much like χ^A. It is convenient to introduce the following, more symmetric notation:

$$\phi_a^{(1)}(x) \coloneqq \phi_a(x), \qquad \phi_a^{(2)}(x) \coloneqq \chi^*{}_a(x), \tag{142}$$

as well as to define the following Majorana fields:

$$\Phi^{(1)}(x) \coloneqq \begin{pmatrix} \phi_a(x) \\ -\phi^{*A}(x) \end{pmatrix} \equiv \begin{pmatrix} \phi_a^{(1)}(x) \\ -\phi^{(1)*A}(x) \end{pmatrix}, \tag{143a}$$

$$\Phi^{(2)}(x) \coloneqq \begin{pmatrix} \chi^*{}_a(x) \\ \chi^A(x) \end{pmatrix} \equiv \begin{pmatrix} \phi_a^{(2)}(x) \\ -\phi^{(2)*A}(x) \end{pmatrix}. \tag{143b}$$

[The signs in eqs. (143) follow from the convention for ε_{AB}, from eq. (76) for $\Psi_C(x)$, via the relations (34).] By making use of the defining relation $\hat{\sigma}^\mu = \varepsilon(\sigma^\mu)^*\varepsilon^{-1}$ and of the hermiticity of the matrices σ^μ one shows that

$$\phi^a\left(\sigma^\mu \overleftrightarrow{\partial}_\mu\right)_{aB}\phi^{*B} = \phi^*{}_A\left(\hat{\sigma}^\mu \overleftrightarrow{\delta}_\mu\right)^{Ab}\phi_b$$

and an analogous relation for the kinetic energy of the χ-field. Thus, the kinetic energy in the Lagrangian (141) can be written in the following form:

$$\mathscr{L}_{\mathrm{kin}} = \frac{\mathrm{i}}{4}\sum_{k=1}^{2}\left\{\phi^{(k)*}{}_A(x)\left(\hat{\sigma}^\mu \overleftrightarrow{\partial}_\mu\right)^{Ab}\phi^{(k)}{}_b(x) - \phi^{(k)a}(x)\left(\sigma^\mu \overleftrightarrow{\partial}_\mu\right)_{aB}\phi^{(k)*B}(x)\right\}. \tag{144}$$

The mass terms in m_1 and m_2 have the structure

$$-\tfrac{1}{2}m_{ii}\phi^{(i)a}(x)\phi^{(i)}{}_a(x),$$

with $m_{11} \equiv m_1$, $m_{22} \equiv m_2$. As to the term in m_{D} we note that $\phi^a\chi^*{}_a \equiv \phi^{(1)a}\phi_a^{(2)}$ is the same as $\chi^{*a}\phi_a \equiv \phi^{(2)a}\phi_a^{(1)}$. Therefore all mass terms can be written in the following compact notation:

$$\mathscr{L}_{\mathrm{mass}} = -\frac{1}{2}\left\{\sum_{i,k=1}^{2} m_{ik}\phi^{(i)a}\phi^{(k)}{}_a + \mathrm{h.c.}\right\}, \tag{145}$$

where

$$m_{ik} = \begin{pmatrix} m_1 & m_{\mathrm{D}} \\ m_{\mathrm{D}} & m_2 \end{pmatrix} \eqqcolon M. \tag{146}$$

The mass matrix M, eq. (146), is a symmetric, (in general) complex 2×2 matrix. The physical fermion fields which are described by the free Lagrangian (145) are obtained by a transformation W,

$$\phi_a^{(i)\prime}(x)=\sum_j W_{ij}\phi_a^{(j)}(x) \tag{147}$$

of the fields $\phi^{(i)}$ which leaves invariant $\mathscr{L}_{\text{kin}}$ and transforms $\mathscr{L}_{\text{mass}}$ to diagonal form with real, positive semidefinite eigenvalues. If the kinetic energy $\mathscr{L}_{\text{kin}}$ is to be invariant, W must be unitary,

$$WW^\dagger=\mathbb{1}. \tag{148}$$

In order to diagonalize $\mathscr{L}_{\text{mass}}$ of eq. (145), we must have

$$W^*MW^\dagger=\mathring{M}, \tag{149}$$

with

$$\mathring{M}=\begin{pmatrix}\lambda_1 & 0\\ 0 & \lambda_2\end{pmatrix} \quad \text{and} \quad \lambda_1,\lambda_2\geq 0.$$

It is easy to show that λ_1 and λ_2 can indeed be requested to be positive semidefinite, without loss of generality. To see this, suppose that the eigenvalues $\lambda_j=|\lambda_j|e^{i\alpha_j}$ were indeed complex. Let

$$P=\begin{pmatrix}e^{i\alpha_1/2} & 0\\ 0 & e^{i\alpha_2/2}\end{pmatrix}$$

so that

$$\mathring{M}=P\mathring{M}'P \quad \text{with} \quad \mathring{M}'=\begin{pmatrix}|\lambda_1| & 0\\ 0 & |\lambda_2|\end{pmatrix},$$

Multiplying eq. (149) with P^{-1}, both from the right and from the left, and observing that $P^{-1}=P^*=P^\dagger$, we obtain

$$(PW)^*M(PW)^\dagger=\mathring{M}'.$$

Now, if W is unitary, so is $W':=PW$, and our assertion is proved. Therefore, from now on we shall assume $\lambda_1,\lambda_2\geq 0$. As it stands, eq. (149) is not a standard diagonalization prescription because M is not hermitean (the equation is multiplied by W^*, not W, from the left). However, by taking its hermitean conjugate which is

$$WM^\dagger\tilde{W}=\mathring{M}, \tag{149'}$$

by multiplying this equation with eq. (149) from the right and using the unitarity of W, we may transform this eigenvalue problem to

$$W(M^\dagger M)W^\dagger=\mathring{M}^2, \tag{150}$$

i.e. to the standard problem of diagonalizing the hermitean matrix $(M^\dagger M)$. This is what we now proceed to do. We set

$$m_1=\mu_1 e^{i\varphi_1},\qquad m_2=\mu_2 e^{i\varphi_2},\qquad m_D=\mu_D e^{i\varphi_D}$$

with μ_1, μ_2, μ_D real and positive. By transforming, in a first step, the fields $\phi_a^{(i)} \to \phi_a^{(i)} e^{i\varphi_i/2}$, the matrix M becomes

$$M \to M' = \begin{pmatrix} \mu_1 & \mu_D e^{i\phi} \\ \mu_D e^{i\phi} & \mu_2 \end{pmatrix} \tag{146'}$$

with

$$\phi := \varphi_D - \tfrac{1}{2}(\varphi_1 + \varphi_2). \tag{151}$$

This gives

$$M'^{\dagger} M' = \begin{pmatrix} \mu_1^2 + \mu_D^2 & \mu_D(\mu_1 e^{i\phi} + \mu_2 e^{-i\phi}) \\ \mu_D(\mu_1 e^{-i\phi} + \mu_2 e^{i\phi}) & \mu_2^2 + \mu_D^2 \end{pmatrix},$$

whose eigenvalues are easily calculated,

$$\begin{aligned} (\lambda_{1/2})^2 &= \tfrac{1}{2}\left\{ \mu_1^2 + \mu_2^2 + 2\mu_D^2 \pm \sqrt{(\mu_1^2 - \mu_2^2)^2 + 4\mu_D^2(\mu_1^2 + \mu_2^2 + 2\mu_1\mu_2\cos(2\phi))} \right\} \\ &= \tfrac{1}{4}\left(\sqrt{(\mu_1 + \mu_2)^2 + 4\mu_D^2 \sin^2\phi} \pm \sqrt{(\mu_1 - \mu_2)^2 + 4\mu_D^2\cos^2\phi} \right)^2. \end{aligned} \tag{152}$$

Thus we obtain

$$\lambda_1 = \tfrac{1}{2}\left\{ \sqrt{(\mu_1 + \mu_2)^2 + 4\mu_D^2\sin^2\phi} + \sqrt{(\mu_1 - \mu_2)^2 + 4\mu_D^2\cos^2\phi} \right\}, \tag{153a}$$

$$\lambda_2 = \tfrac{1}{2}\left| \sqrt{(\mu_1 + \mu_2)^2 + 4\mu_D^2\sin^2\phi} - \sqrt{(\mu_1 - \mu_2)^2 + 4\mu_D^2\cos^2\phi} \right|, \tag{153b}$$

with ϕ defined in eq. (151).

The matrix W can be determined from eqs. (149) and (150). Noting that it must have the general form

$$W = e^{i\delta} \begin{pmatrix} \cos\alpha e^{i\beta} & \sin\alpha e^{i\gamma} \\ -\sin\alpha e^{-i\gamma} & \cos\alpha e^{-i\beta} \end{pmatrix} \tag{154a}$$

[i.e. a product of a SU(2) matrix and a phase], one finds after some calculation

$$\operatorname{tg} 2\alpha = \frac{2\mu_D \Delta}{\mu_1^2 - \mu_2^2} \tag{154b}$$

with

$$\Delta := \sqrt{\mu_1^2 + \mu_2^2 + 2\mu_1\mu_2\cos(2\phi)},$$

$$\operatorname{tg}(\beta - \gamma) = \frac{\mu_2 - \mu_1}{\mu_1 + \mu_2} \operatorname{tg}\phi \tag{154c}$$

$$\operatorname{tg}(2\beta + 2\gamma) = \frac{\mu_1\mu_2\sqrt{(\mu_1^2 - \mu_2^2)^2 + 4\mu_D^2\Delta^2}\,\sin(2\phi)}{2\mu_1^2\mu_2^2 - \mu_D^2\Delta^2 + \mu_1\mu_2(\mu_1^2 + \mu_2^2)\cos(2\phi)}, \tag{154d}$$

$$\operatorname{tg} 4\delta = \frac{-\mu_D^2\Delta^2 \sin 2\phi}{\mu_1\mu_2(\mu_1^2 + \mu_2^2) + (2\mu_1^2\mu_2^2 - \mu_D^2\Delta^2)\cos(2\phi)}. \tag{154e}$$

Let us now interpret these results. After diagonalization, as described above, the mass term in the Lagrangian (145) becomes

$$\mathscr{L}_{\text{mass}} = -\frac{1}{2}\sum_{k=1}^{2}\lambda_k\{\phi^{(k)\prime a}\phi_a^{(k)\prime} + \phi^{(k)\prime *}{}_A\phi^{(k)\prime * A}\}. \tag{155}$$

Here, $\phi_a^{(i)\prime}$ is given by eq. (147). When expressed in terms of four-component spinors (143), the mass eigenstates are

$$\Phi'^{(k)}(x) = \begin{pmatrix} \sum_l W_{kl}\phi_a^{(l)}(x) \\ -\sum_l W_{kl}^*\phi^{(l)*A}(x) \end{pmatrix}. \tag{156}$$

When written in terms of these Majorana fields the mass term becomes simply

$$-\frac{1}{2}\sum_{k=1}^{2}\lambda_k\overline{\Phi'^{(k)}(x)}\Phi'^{(k)}(x)$$

and the transformed Lagrangian reads

$$\mathscr{L} = \frac{1}{2}\sum_{k=1}^{2}\left\{\frac{\mathrm{i}}{2}\overline{\Phi'^{(k)}(x)}\gamma^\alpha\overleftrightarrow{\partial}_\alpha\Phi'^{(k)}(x) - \lambda_k\overline{\Phi'^{(k)}(x)}\Phi'^{(k)}(x)\right\}, \tag{157}$$

i.e. it has the familiar form (104′).

Cases of special interest are the ones where λ_1 and λ_2 are degenerate. From eqs. (153) we see that this happens *either*

(i) if $\mu_1 = \mu_2$, $\mu_D = 0$; *or*

(ii) if $\mu_1 = \mu_2 =: \mu$ and

$$\phi = \varphi_D - \tfrac{1}{2}(\varphi_1 + \varphi_2) = (2n+1)\pi/2.$$

The first case is trivial and gives $\lambda_1 = \lambda_2 = \mu_1 = \mu_2$. The second is particularly interesting and has $\lambda_1 = \lambda_2 = (\mu^2 + \mu_D^2)^{1/2}$. In either case we can introduce a Dirac field by combining $\Phi^{(1)}$ and $\Phi^{(2)}$:

$$\Psi(x) := \frac{1}{\sqrt{2}}(\Phi^{(1)}(x) + \mathrm{i}\Phi^{(2)}(x)). \tag{158}$$

The orthogonal combination

$$\frac{1}{\sqrt{2}}(\Phi^{(1)}(x) - \mathrm{i}\Phi^{(2)}(x)) = \Psi_C(x)$$

is then the charge conjugate of Ψ. The Lagrangian (157) reduces to the Lagrangian of a single Dirac field.

This discussion shows that, in some sense, Majorana fields are more fundamental than Dirac fields. The general Lagrangian (141) describes two Majorana fields, eq. (156), with mass eigenvalues λ_1, λ_2, eqs. (153), respectively. Only if these eigenvalues are equal can the two fields be combined to a Dirac field and its charge conjugate.*)

*)It is interesting to remark that these considerations can be generalized to an arbitrary number of fields.

The essential difference between these two cases is the following:

(i) If $\lambda_1 \neq \lambda_2$, the basic fields are Majorana fields and, therefore, cannot carry any additively conserved charge quantum number. In other words, if the theory with interactions admits an absolutely and additively conserved charge Q, then the two Majorana fields must belong to the eigenvalue zero of that charge. If this is not so, then the mass terms conserve that charge Q only modulo 2 because they cannot connect states which differ by $\Delta Q = \pm 2$.

(ii) If $\lambda_1 = \lambda_2$ the basic fields are one Dirac field and its charge conjugate Ψ_C. These fields being different, Ψ can carry any nonvanishing charge Q, Ψ_C then carrying charge $-Q$.

9. Charged fermion fields in interaction with electromagnetic fields

9.1. External field case

In many situations electromagnetic interactions of charged leptons with some other charged system can be treated in the *external field approximation*. If the system with which the lepton interacts is very heavy, it will be able to absorb or to provide three-momentum in reactions with the lepton without altering its own state. If, in addition, it is dynamically inert its internal structure will not intervene in such reactions. In those cases the effect of the system on a charged lepton can be represented, to a certain approximation, by classical external vector potentials $A_\mu^{\text{ext}}(x)$. A case of special relevance for atomic, nuclear and particle physics is the electromagnetic interaction of electrons and muons with nuclei. As far as the *kinematics* is concerned, the nucleus is so much heavier than the electron or the muon that in a reaction with these leptons its recoil is usually not important. The nucleons act like an external macroscopic source of electric and magnetic fields which can absorb or produce any mismatch of three-momentum that there may be in a given reaction. Well-known examples for such situations are: Bremsstrahlung in matter; atomic bound states of electrons or muons; elastic and inelastic scattering of electrons or muons on nuclei at intermediate energies.

Regarding the *dynamics*, the internal structure of the nucleus (other than its initial and final state involved in the reaction) is usually unimportant. This is so because the virtual intermediate excitation of higher states of the nucleus necessitates (at least) a two-step process and, therefore, is of higher order in the fine structure constant. Such effects which are called *nuclear polarizability shift* in atoms, and *dispersion corrections* in electron scattering, are generally small and may be added as a correction to the results of the external field approximation from which one started.

Let $A_\mu^{\text{ext}}(t, \boldsymbol{x})$ be an external, classical four-vector potential describing a given set of classical electric and magnetic fields. The coupling of a lepton of charge Q to these fields is found by the substitution

$$\partial_\mu \rightarrow \partial_\mu + \mathrm{i}QA_\mu^{\text{ext}} \tag{159}$$

in the particle's equations of motion. This is the so-called "minimal coupling" rule on which we shall comment below. If we apply this prescription to the free Dirac equation (50), we obtain

$$\left(\mathrm{i}\gamma^{\mu}\partial_{\mu} - Q\gamma^{\mu}A_{\mu}^{\mathrm{ext}} - m\mathbb{1}\right)\Psi(x) = 0. \tag{160}$$

For example, let us take a stationary external electric field and let us consider stationary solutions of this equation (positive frequency), viz.

$$A_{\mu}^{\mathrm{ext}}(x) = (\phi(\boldsymbol{x}); \mathbf{0}),$$

$$\Psi(x) = \mathrm{e}^{-\mathrm{i}Et}\psi(\boldsymbol{x}),$$

$$\left[(E - V(\boldsymbol{x}))\gamma^{0} + \mathrm{i}\boldsymbol{\gamma}\cdot\nabla - m\mathbb{1}\right]\psi(\boldsymbol{x}) = 0, \tag{160$'$}$$

where we have set $V(\boldsymbol{x}) = Q\phi(\boldsymbol{x})$.

For an electron or muon $Q = -e$, whilst for a point-like nucleus of charge number Z, $\phi(\boldsymbol{x}) = Ze/|\boldsymbol{x}|$, so that

$$\left[(E + Ze^{2}/|\boldsymbol{x}|)\gamma^{0} + \mathrm{i}\boldsymbol{\gamma}\cdot\nabla - m\mathbb{1}\right]\psi(\boldsymbol{x}) = 0.$$

Equivalently, by multiplying eq. (160′) with $\gamma^{0} \equiv \beta$ (in standard representation) from the left, one has

$$E\psi(\boldsymbol{x}) = \left[-\mathrm{i}\boldsymbol{\alpha}\cdot\nabla + V(\boldsymbol{x}) + m\beta\right]\psi(\boldsymbol{x}), \tag{161}$$

where, for a point-like nucleus,

$$V(\boldsymbol{x}) = -Ze^{2}/|\boldsymbol{x}|. \tag{162}$$

If the spatial extension of the nuclear charge density $\rho_{\mathrm{c}}(\boldsymbol{x})$ cannot be neglected as compared to the typical size of the electron or muon state, we have instead

$$V(\boldsymbol{x}) = -Ze^{2}\int \mathrm{d}^{3}x' \frac{\rho_{\mathrm{c}}(\boldsymbol{x}')}{|\boldsymbol{x} - \boldsymbol{x}'|}. \tag{163}$$

We take the nuclear charge density $\rho_{\mathrm{c}}(x)$ normalized to unity,

$$\int \mathrm{d}^{3}x\, \rho_{\mathrm{c}}(\boldsymbol{x}) = 1. \tag{164}$$

In particular, if this density is spherically symmetric, the potential is also spherically symmetric and is given by*)

$$V(r) = -4\pi Ze^{2}\left[\frac{1}{r}\int_{0}^{r}\rho_{\mathrm{c}}(r')r'^{2}\,\mathrm{d}r' + \int_{r}^{\infty}\rho_{\mathrm{c}}(r')r'\,\mathrm{d}r'\right]. \tag{165}$$

It is eq. (161) that is the most convenient one in treating atomic bound states of charged leptons as well as scattering off nuclei at low energies. The representation (59) is particularly well adapted in this case as it divides the Dirac spinors naturally into "large" (i.e. nonrelativistic) and "small" (i.e. relativistic) components. Also, in

*)For a point-like nucleus, placed at the origin,

$$\rho_{\mathrm{c}}(\boldsymbol{x}) = \delta(\boldsymbol{x}) = (1/r^{2})\delta(r)\delta(\cos\theta - 1)\delta(\phi),$$

in which case both eq. (163) and eq. (165) go over into eq. (162).

this form eq. (161) comes closest to the nonrelativistic Schrödinger–Pauli equation.*) This is so because β which multiplies the dominant mass term, is an "even" matrix, that is, has the structure $\begin{pmatrix} x & 0 \\ 0 & x \end{pmatrix}$ whilst the α_i which multiply the components of the three-momentum are "odd", that is, have the structure $\begin{pmatrix} 0 & x \\ x & 0 \end{pmatrix}$. Thus, β connects upper with upper, as well as lower with lower two-spinors, whilst the α_i connect upper with lower, and lower with upper two-spinors. For particle solutions the lower two-spinor is of order $p/m \sim v/c$ relative to the upper. In the limit $p \to 0$ the upper two-spinor goes over into the nonrelativistic Schrödinger wave function multiplied by a Pauli spinor.

The other extreme situation is the one where the particle's motion is highly relativistic, that is where $E \gg m$. In that case the mass term in eq. (160′) is negligibly small. It is then more convenient to write eq. (160′) in a representation where both γ^0 and the γ^i are "odd"**), as in this case the equations of motion of upper and lower two-spinors decouple from each other. The "high-energy representation" (52) has the required property. Writing

$$\psi(\boldsymbol{x}) = \begin{pmatrix} \phi(\boldsymbol{x}) \\ \chi(\boldsymbol{x}) \end{pmatrix}$$

in this representation, eq. (160′) becomes

$$[\mathrm{i}\boldsymbol{\sigma}\cdot\nabla + (E - V(\boldsymbol{x}))]\,\phi(\boldsymbol{x}) = m\chi(\boldsymbol{x}), \tag{166a}$$

$$[-\mathrm{i}\boldsymbol{\sigma}\cdot\nabla + (E - V(\boldsymbol{x}))]\,\chi(\boldsymbol{x}) = m\phi(\boldsymbol{x}). \tag{166b}$$

When the mass term is neglected, these equations decouple completely. Quite obviously, they follow directly from the Dirac equation (47) if harmonic time dependence $\mathrm{e}^{-\mathrm{i}Et}$ is introduced.

We shall encounter both situations below and shall make use of either representation in our treatment of bound state problems and of relativistic scattering.

9.2. *Interaction with the quantized Maxwell field*

We start from the combined Lagrange densities of the *free* Dirac field and the *free* Maxwell field

$$\mathscr{L}_0(x) = \mathscr{L}_\mathrm{D}(x) + \mathscr{L}_\gamma(x), \tag{167}$$

where $\mathscr{L}_\mathrm{D}(x)$ is given by eq. (104), whilst $\mathscr{L}_\gamma(x)$ is

$$\mathscr{L}_\gamma(x) = -\tfrac{1}{4}F_{\mu\nu}(x)F^{\mu\nu}(x), \tag{168}$$

where $F^{\mu\nu} = f^{\mu\nu}$ and

$$f^{\mu\nu}(x) := \partial^\mu A^\nu(x) - \partial^\nu A^\mu(x) \tag{169}$$

*) See e.g. Ref. R11.

**) Obviously, γ^0 and γ^i cannot be even simultaneously.

is the covariant electromagnetic field tensor,

$$f^{\mu\nu} = \begin{pmatrix} 0 & -E^1 & -E^2 & -E^3 \\ E^1 & 0 & -B^3 & B^2 \\ E^2 & B^3 & 0 & -B^1 \\ E^3 & -B^2 & B^1 & 0 \end{pmatrix}, \tag{169'}$$

so that $E^i = -f^{0i}$, $B^i = -\varepsilon^{ijk} f^{jk}$,

$$f^{lm} = -\varepsilon^{lmn} B^n,$$

$A_\mu(x)$ is the quantized Maxwell field.

The minimal substitution rule,

$$\partial_\alpha \Psi \to \partial_\alpha \Psi + \mathrm{i} Q A_\alpha \Psi, \tag{170a}$$

$$\partial_\alpha \overline{\Psi} \to \partial_\alpha \overline{\Psi} - \mathrm{i} Q A_\alpha \overline{\Psi}, \tag{170b}$$

when introduced into $\mathscr{L}_0(x)$, eq. (167), leads to the interaction Lagrangian

$$\mathscr{L}_0 \to \mathscr{L}(x) = \mathscr{L}_\mathrm{D}(x) + \mathscr{L}_\gamma(x) - Q\overline{\Psi(x)}\, \gamma^\alpha \Psi(x) A_\alpha(x). \tag{171}$$

The interaction term is seen to be the scalar product of the four-vector potential and the particle's electromagnetic current

$$j^\alpha_{\mathrm{e.m.}}(x) = Q\overline{\Psi(x)}\, \gamma^\alpha \Psi(x), \tag{172}$$

a result which is well-known from classical electrodynamics. As it stands the Lagrangian density (171) implies that the free particle described by the field $\Psi(x)$ has a "normal" magnetic moment

$$\mu = g \frac{Q}{2m} \frac{1}{2} = \frac{Q}{2m}, \tag{173}$$

so that its g-factor is equal to 2. Any deviation from this so-called Dirac value comes about through radiative corrections and, therefore, is of order $\mathrm{O}(\alpha)$.

This is not difficult to prove. Consider the interaction of the particle with a stationary magnetic field $\boldsymbol{B}$, which we describe by a three-vector potential $\boldsymbol{A}$. For simplicity we consider scattering of the particle from an initial momentum $\boldsymbol{p}$ to a final momentum $\boldsymbol{p}'$, through the interaction

$$\langle \boldsymbol{p}' | \boldsymbol{j}_{\mathrm{e.m.}}(0) | \boldsymbol{p} \rangle \cdot \boldsymbol{A}(\boldsymbol{q}). \tag{174}$$

We consider the matrix element in the Breit system which is defined by the requirement

$$\boldsymbol{p} + \boldsymbol{p}' = 0.$$

This system of reference is particularly convenient since the limit of taking the squared four-momentum transfer $q^2 = (p' - p)^2 \to 0$ to zero leads us automatically to the rest system of the particle. At the same time eq. (174) gives the corresponding nonrelativistic expression. It is not difficult to verify (in the standard representation) the following:

$$\langle -\boldsymbol{p}|j^i_{\mathrm{e.m.}}(0)|\boldsymbol{p}\rangle$$
$$=\frac{1}{(2\pi)^3}Q\overline{u(-\boldsymbol{p})}\,\gamma^i u(\boldsymbol{p})$$
$$=\frac{Q}{(2\pi)^3}u^\dagger(-\boldsymbol{p})\begin{pmatrix}0 & \sigma^{(i)}\\ \sigma^{(i)} & 0\end{pmatrix}u(\boldsymbol{p})$$
$$=\frac{2Q}{(2\pi)^3}\mathrm{i}\varepsilon^{ijk}p^j(\chi^\dagger\sigma^k\chi).$$

Therefore, the scattering amplitude which describes magnetic back scattering is

$$T=\frac{\mathrm{i}}{(2\pi)^3}2Q\varepsilon^{ijk}p^j(\chi^\dagger\sigma^k\chi)A^i(\boldsymbol{q}=2\boldsymbol{p}).$$

Let us compare this to the nonrelativistic scattering amplitude which we would obtain from the well-known interaction of a magnetic moment $\boldsymbol{\mu}$ in a constant external field $\boldsymbol{B}$,

$$T_{\mathrm{n.r.}}=(\psi_{-\boldsymbol{p}},H_{\mathrm{int}}\psi_{\boldsymbol{p}})$$

with

$$H_{\mathrm{int}}=-\boldsymbol{\mu}\cdot\boldsymbol{B}=-g\frac{Q}{2m}\boldsymbol{S}\cdot\boldsymbol{B},\quad B^i=\varepsilon^{ijk}\partial_j A^k(\boldsymbol{x})$$

and

$$\psi_{\boldsymbol{p}}=\sqrt{\frac{2m}{(2\pi)^3}}\,\mathrm{e}^{\mathrm{i}\boldsymbol{p}\cdot\boldsymbol{x}}\chi.$$

(The factor $\sqrt{2m}$ is included in order to have the correct normalization when the nonrelativistic limit $\langle \boldsymbol{p}'|\boldsymbol{p}\rangle$ is taken.) One finds by partial integration

$$T_{\mathrm{n.r.}}=gQ\tfrac{1}{2}(\chi^\dagger\sigma^i\chi)\varepsilon^{ijk}2\mathrm{i}\,p_j A^k\frac{1}{(2\pi)^3}$$
$$=\frac{\mathrm{i}gQ}{(2\pi)^3}\varepsilon^{kji}p^j(\chi^\dagger\sigma^i\chi)A^k(2\boldsymbol{p}).$$

Comparing $T_{\mathrm{n.r.}}$ and T, we se that indeed $g=2$. In very much the same way we can show that the corresponding antiparticle has the opposite magnetic moment

$$\mu_{\text{antiparticle}}=-Q/2m.$$

As it stands the Lagrangian $\mathscr{L}(x)$, eq. (171), describes quantum electrodynamics of a fermion as well as its antiparticle of given physical mass m, charge Q and normal g factor, $g=2$. Of course, we can generalize it immediately to an arbitrary number of fermions (of the same charge and g-factor but different mass),

$$\mathscr{L}_{\mathrm{QED}}=\sum_{\mathrm{f}}\mathscr{L}_{\mathrm{D}}(x,m_{\mathrm{f}})+\mathscr{L}_{\gamma}(x)-\sum_{\mathrm{f}}Q(\mathrm{f})\overline{\Psi}_{\mathrm{f}}\gamma^{\alpha}\Psi_{\mathrm{f}}A_{\alpha}, \tag{175}$$

so as to be able to describe electrons, muons, τ-leptons etc. ($Q_{\mathrm{e}}=Q_{\mu}=Q_{\tau}=-e$). Appendix C summarizes the Feynman rules for quantum electrodynamics of spin-1/2 fermions, as defined by the Lagrange density (171).

We note here in passing that predictions of this theory for all electromagnetic properties of electrons and muons are in perfect agreement with experiment. In particular, tests of low-energy properties of electrons and muons have been pushed to the *natural limit* of quantum electrodynamics, i.e. up to the point where weak and strong interactions start to interfere with the pure electromagnetic interactions. We shall come back to this statement and quote some examples at various occasions in the course of this book.

9.3. *Some remarks on these results*

(i) The result (173) says that the g-factor of electrons and muons is $g=2$ (except for radiative corrections). This has always been considered a major success of Dirac's theory of electrons and muons. While this is certainly true, it is sometimes said also that $g=2$ is a consequence of the minimal substitution rule (170). Unfortunately this latter statement is not correct. It is true that if we postulate the specific form (175) for the Lagrangian of leptonic QED then $g=2$ follows from it, as we have seen above. However, there is nothing that forbids us to add an arbitrary four-divergence $\partial_{\mu}M^{\mu}$ to the free Lagrangian, as we know that this will not alter the equations of motion (supposing M^{μ} sufficiently well-behaved). For instance, we may wish to take $\mathscr{L}'_{\mathrm{D}}=\mathscr{L}_{\mathrm{D}}+\partial_{\mu}M^{\mu}$ instead of $\mathscr{L}_{\mathrm{D}}$, with the choice

$$M^{\mu}=-\mathrm{i}\frac{a}{8m}\overline{\Psi(x)}\,\sigma^{\mu\nu}\overleftrightarrow{\partial}_{\nu}\Psi(x).$$

For a free particle this does not make any difference. $\mathscr{L}'_{\mathrm{D}}$ yields the same equations of motion as $\mathscr{L}_{\mathrm{D}}$. However, if we introduce the coupling to the Maxwell field through minimal substitution, we find $\mathscr{L}_{\mathrm{QED}}$ of eq. (175) but supplemented by the term*)

$$\frac{a}{4m}Q\overline{\Psi(x)}\,\sigma^{\mu\nu}\Psi(x)f_{\mu\nu}. \tag{176}$$

This new term describes the interaction of an anomalous magnetic moment of magnitude $2a$. Thus, the new "minimal" theory describes fermions whose g-factor is

$$g'=2(1+a). \tag{177}$$

In the light of this remark there does not seem to be anything special about the value $g=2$. The Dirac value $g=2$ refers to the specific Lagrangian (175). The minimal coupling prescription must be applied to the specific Lagrangian (104).

*) This is worked out in Ref. R12.

(ii) If a strongly interacting fermion (that is a baryon) couples to a photon the vertex

$$\langle p'|j^{\alpha}_{\text{e.m.}}(0)|p\rangle$$

is renormalized through strong interactions and must be analyzed in terms of electric and magnetic form factors $F_i(q^2=(p-p')^2)$ (see below). However, if the momentum transfer is sufficiently small, we may treat the baryon like a point-like Dirac particle that carries a given anomalous magnetic moment a. In this case we may use the theory sketched above by simply adding to it the Pauli term (176).

Such a situation is encountered, for instance, in the atomic bound states of exotic atoms with antiprotons or Σ^- particles. In these baryonic atoms the fine structure is determined by the term

$$V_{ls}=\frac{1}{2m^2}(1+2a)\boldsymbol{l}\cdot\boldsymbol{s}\frac{1}{r}\frac{\mathrm{d}V_{\mathrm{c}}}{\mathrm{d}r}, \tag{178}$$

where V_{c} is the Coulomb potential created by the nucleus. The fine structure may in fact be used to measure the anomaly a. This has been done for the case of the Σ^- with the result (Roberts et al. 1974, Dugan et al. 1975),

$$\mu(\Sigma^-)=-1.48\pm0.37.$$

(iii) One might wonder whether the choice of the plus sign in the minimal substitution rule

$$\partial_\alpha\to\partial_\alpha+\mathrm{i}Q(\mathrm{f})A_\alpha \tag{179}$$

is unique or not, $Q(\mathrm{f})$ being the charge of the particle,

$$Q(\mathrm{p})=+|e|,\qquad Q(\mathrm{e})=Q(\mu)=Q(\tau)=-|e|.$$

Quite obviously this sign determines for example the sign of the potential term in eq. (161) and, therefore, is related to the fact that like charges repell, unlike charges attract each other. In fact, the sign is fixed by two standard conventions in electrodynamics: (i) it is customary to let the electric field $\boldsymbol{E}$ of a *positive* point charge point *outward*; (ii) the sign of the four-potential is fixed such that $\boldsymbol{E}=-\nabla A^0$.

(iv) Minimal coupling and the requirement of gauge invariance of QED are intimately connected. Suppose we subject a charged spinor field $\Psi(x)$ to a local gauge transformation,

$$\Psi(x)\to\mathrm{e}^{\mathrm{i}\Lambda(x)}\Psi(x),\qquad \overline{\Psi(x)}\to\mathrm{e}^{-\mathrm{i}\Lambda(x)}\overline{\Psi(x)}. \tag{180}$$

Whilst the mass term in the free Lagrangian $\mathscr{L}_{\mathrm{D}}$ is obviously invariant under this transformation, the kinetic energy term

$$\bar{\Psi}\frac{\mathrm{i}}{2}\gamma^\alpha\overleftrightarrow{\partial}_\alpha\Psi \tag{181}$$

is not. It goes over into

$$\bar{\Psi}\frac{\mathrm{i}}{2}\gamma^\alpha\overleftrightarrow{\partial}_\alpha\Psi+\bar{\Psi}\gamma^\alpha\Psi\,\partial_\alpha\Lambda(x). \tag{181'}$$

If on the other hand we consider the "minimally substituted" kinetic energy instead,

$$\frac{\mathrm{i}}{2}\{\bar{\Psi}\gamma^\alpha(\partial_\alpha + \mathrm{i}QA_\alpha)\Psi - ((\partial_\alpha - \mathrm{i}QA_\alpha)\bar{\Psi})\gamma^\alpha\Psi\}, \qquad (182)$$

then we see that the extra term in (181′) can be absorbed into the four-vector potential by the substitution

$$A_\alpha - \tfrac{1}{Q}\partial_\alpha\Lambda(x) =: A'_\alpha. \qquad (183)$$

This, however, is a gauge transformation of the four-vector potential which leaves invariant $\mathscr{L}_{\mathrm{M}}$, the Lagrangian of the Maxwell fields. Thus, transformations (180) and (183), taken together, leave invariant the coupled system of Maxwell and Dirac fields. For this it is essential that the modified kinetic energy (182) contain what is called the "covariant derivative"

$$\vec{D}_\alpha := \vec{\partial}_\alpha + \mathrm{i}QA_\alpha,$$
$$\overleftarrow{D}_\alpha := \overleftarrow{\partial}_\alpha - \mathrm{i}QA_\alpha, \qquad (184)$$

We have touched here upon a general feature of gauge fields (A_μ) in interaction with matter fields (Ψ) to which we shall return below in connection with gauge theories of more general nature.

References

Dugan, G., Y. Asano, M.Y. Chen, S.C. Cheng, E. Hu, L. Lidofsky, W. Patton, C.S. Wu, V. Hughes, and D. Lu, 1975, *Nucl. Phys.* A254, 396.

Fierz, M., 1938, *Helv. Phys. Acta* 12, 3.

Pauli, W., 1940, *Phys. Rev.* 58, 716.

Roberts, B.L., C.R. Cox, M. Eckhause, J.R. Kane, R. Welsh, D.A. Jenkins, W.C. Lam, P.D. Barnes, R.A. Eisenstein, J. Miller, R.B. Sutton, A.R. Kunselman, and R.J. Powers, 1974, *Phys. Rev. Lett.* 32, 1265.

Van der Waerden, B.L., 1929, *Göttinger Nachrichten*, p. 100. See also O. Laporte and G. E. Uehlenbeck, *Phys. Rev.* 37 (1931) 1380.

Exercises

1. Derive the Euler–Lagrange equations of L_{M}, eq. (86).

2. Verify that the matrices σ^μ are both unitary and hermitean.

3. Prove eq. (23) by explicit calculation of the exponential series for the case $\kappa = \frac{1}{2}$. Show that the transformation $U \equiv D(0, \pi, 0)$ does indeed effect the transition from cogredience (i.e. transformation behaviour $D(\theta_i)$) to contragredience (i.e. transformation behaviour $\tilde{D}^{-1} = D^*$), and vice versa. *Hint*: Prove first the relations

$$UJ_i + J_i^* U = 0 \quad \text{and} \quad U\mathrm{e}^{\mathrm{i}\boldsymbol{\alpha}\boldsymbol{J}}U^{-1} = \mathrm{e}^{-\mathrm{i}\boldsymbol{\alpha}\boldsymbol{J}^*}.$$

(Note the dependence on the phase convention for the angular momentum matrices.)

4. Study plane wave solutions of eqs. (47) for the case $m = 0$ and show that they are eigenstates of helicity.

5. Carry out the calculation that leads to eq. (109). Do the same for eq. (117).

6. Derive the Euler–Lagrange equations for the Lagrangian (141) and compare to eqs. (47).

7. Study the behaviour of the Lagrangian (141) under parity P, charge conjugation C, and time reversal T. Show that it is invariant under the combined transformation $\Theta = PCT$

8. Prove the relation

$$(\sigma^\alpha)_{aB}(\sigma_\alpha)_{cD} = -2\varepsilon_{ac}\varepsilon_{BD}.$$

9. Physical amplitudes are always bilinear functions of spinor fields. Show: The fact that the density matrix ρ is not hermitean does not conflict with observables being real.

10. For which orientation of ζ and in which basis is the density matrix (131) diagonal? Interpret the answer. Write down the explicit form of ρ for polarization along the momentum.

11. Choose a basis $(n^{(0)}, n^{(1)}, n^{(2)}, n^{(3)})$ in four-dimensional momentum space such that (i) $n^{(0)\mu} = (1/m)p^\mu$; (ii) the vectors $n^{(i)\alpha}$ are spacelike; (iii) $(n^{(\alpha)} \cdot n^{(\beta)}) = g^{\alpha\beta}$ (orthogonality); (iv) $n_\mu^{(\alpha)} g_{\alpha\beta} n_\nu^{(\beta)} = g_{\mu\nu}$ (completeness). For example:

$$n^{(1)} = (0, \boldsymbol{n}^1), \qquad n^{(2)} = (0, \boldsymbol{n}^2), \qquad n^{(3)} = \left(\frac{|\boldsymbol{p}|}{m}, \frac{E_p \boldsymbol{p}}{m|\boldsymbol{p}|}\right)$$

with $\boldsymbol{n}^1 \perp \boldsymbol{n}^2 \perp \boldsymbol{p}$. Calculate the polarization along the direction $\boldsymbol{n}^i$ for a state that is characterized by a point ζ on or inside the Poincaré sphere.

12. Show that the relationship (24) holds true for boosts.

13. Construct the Hamilton density for a free Majorana field. Show: When integrated over all space this density gives zero.

14. Show: Reflections with respect to planes perpendicular to the momentum $\boldsymbol{p}$ exchange positive and negative helicity states.

Chapter IV

ELECTROMAGNETIC PROCESSES AND INTERACTIONS

The electron, the muon, and their neutrinos are important tools in testing the structure of the fundamental electromagnetic and weak interactions. On the other hand, if these interactions are known, they serve as ideal probes for the internal structure of complex hadronic targets such as nucleons and nuclei. Although electroweak interactions should in fact be discussed as a whole and on the same footing, purely electromagnetic interactions play a distinctive role, for obvious experimental reasons: At low and intermediate energies the effective electromagnetic coupling is larger by many orders of magnitude than the weak couplings, so that electromagnetic processes are measurable to much higher accuracy than purely weak processes.

The *fundamental* aspects of unified electroweak interactions are discussed below, in Chap. V. The present chapter deals primarily with *applications* of charged leptons to problems of nucleon and nuclear structure, and to selected precision tests of quantum electrodynamics (QED) at low momentum transfers. In most of these applications the electromagnetic interactions effectively appear in the form of external fields in the leptonic particle's Dirac equation. This is the domain where the physics of (electromagnetically) interacting leptons can still be described in the framework of an effective, though relativistic, single particle theory. In contrast to this, the topics discussed in the last chapter V, will make use of the full intrinsic many-body nature of Dirac theory.

1. Electron scattering off a composite target: Qualitative considerations

Electron scattering at high and very high energies is an important tool for the investigation of the structure of various strongly interacting particles (hadrons). Among these only proton, neutron and nuclei can be prepared as targets in scattering experiments. Hence most of what we know about internal hadronic structure concerns protons and neutrons. Nevertheless there is also some information on long-lived hadrons such as pions from electron-positron colliding beam experiments in which pairs of such hadrons are created.*)

If electron scattering off a nucleon or a nucleus is to give more information on the target than just its electric charge, the electron's de Broglie wavelength $\lambda\!\!\!^{-} = 1/k$ must have a magnitude comparable to the *spatial size* of the nucleon or nucleus, respectively. The radius of the proton is about $r_p \simeq 0.86$ fm; the charge radius of nuclei is

*) The essential difference between these two types of experiments is that in electron scattering the invariant momentum transfer is *spacelike* whilst in electron–positron annihilation it is *timelike*.

approximately

$$r_c \simeq 1.1 \text{ fm } A^{1/3} \tag{1}$$

(A being the nuclear mass number). Thus $\lambda\!\!\!^{-}$ should be of the order of, or smaller than, about 1 fm. Hence its momentum must be of the order of or greater than

$$k = \hbar c/\lambda\!\!\!^{-} \simeq 200 \text{ MeV}.$$

Obviously, at these energies the electron is highly relativistic, its energy is very large as compared to its rest mass

$$E = \sqrt{\boldsymbol{k}^2 + m^2} \gg m.$$

In fact, in most cases we will neglect the mass altogether. We then deal with a massless charged fermion which behaves in a way somewhat similar to neutrinos, with the exception that the electron spin can assume any direction. The following simple estimates may serve to illustrate qualitatively what one learns from the study of *elastic scattering* and of *inelastic scattering to discrete excited states* in a hadronic target (nucleon or nucleus). A more detailed and quantitative analysis of these processes follows in the next sections.

Consider first elastic scattering of an electron off an extended object with spherically symmetric charge density $\rho(r)$ and total charge Ze. $\rho(r)$ shall be normalized to unity,

$$\int \rho(r)\,\mathrm{d}^3r = 4\pi \int_0^\infty \rho(r) r^2\,\mathrm{d}r = 1. \tag{2}$$

The corresponding electrostatic potential $\phi(r)$ is related to the charge density through Poisson's equation,

$$\Delta\phi(r) = -Ze\rho(r).$$

We calculate the differential cross section for elastic scattering in Born approximation and, for the moment, neglect the spin of the electron. In fact, in electromagnetic scattering of electrons at very high energies, the spin of the electron is not essential.*) We shall show below that for $E \gg m$, a spherically symmetric, parity even, potential does not lead to polarization of an initially unpolarized electron beam. This is in contrast to low energies of the order of the mass, $E \simeq m$, where there is polarization through spin–orbit coupling (Mott scattering). With this approximation the electron is then described by a Klein–Gordon equation with external electrostatic potential

$$V(r) = -e\phi(r).$$

To order $(Z\alpha)$ the scattering amplitude is given by

$$f(\boldsymbol{p}', \boldsymbol{p}) \simeq -\frac{k}{2\pi} \int \mathrm{e}^{-\mathrm{i}\boldsymbol{p}'\boldsymbol{r}}(-e\phi(r))\mathrm{e}^{\mathrm{i}\boldsymbol{p}\boldsymbol{r}}\,\mathrm{d}^3r \tag{3}$$

($\boldsymbol{p}$, $\boldsymbol{p}'$ being the initial and final momenta of the electron, respectively, $k = |\boldsymbol{p}| = |\boldsymbol{p}'|$).

*) This is true because electromagnetic interactions are invariant under parity.

Introducing the momentum transfer $\boldsymbol{q} := \boldsymbol{p} - \boldsymbol{p}'$, for which

$$q \equiv |\boldsymbol{q}| = 2k\sin(\theta/2) \tag{4}$$

(θ being the scattering angle), the amplitude $f(\boldsymbol{q})$ can be expressed in terms of the charge density $\rho(r)$ by integrating by parts twice and making use of Poisson's equation.*) One obtains

$$f(\boldsymbol{q}) \simeq \frac{Ze^2k}{2\pi q^2}\int e^{i\boldsymbol{q}\boldsymbol{r}}\rho(r)\,d^3r = \frac{2Z\alpha k}{q^2}\int e^{i\boldsymbol{q}\boldsymbol{r}}\rho(r)\,d^3r, \tag{5}$$

where we have replaced $e^2/4\pi$ by α (natural units).

Thus, the differential cross section is

$$\left(\frac{d\sigma}{d\Omega}\right)_{\text{no spin}} = |f(\boldsymbol{q})|^2 \simeq \left(\frac{Z\alpha}{2k}\right)^2 \frac{1}{\sin^4(\theta/2)}|F(\boldsymbol{q})|^2,$$

with the *charge form factor* $F(\boldsymbol{q})$ defined as follows:

$$F(\boldsymbol{q}) := \int \rho(r) e^{i\boldsymbol{q}\boldsymbol{r}}\,d^3r. \tag{6}$$

When the electron spin is included, the scattering cross section for a spin zero target just gets another factor $\cos^2(\theta/2)$, so that we obtain

$$\frac{d\sigma}{d\Omega} = \left(\frac{d\sigma}{d\Omega}\right)_{\text{Mott}} |F(\boldsymbol{q})|^2, \tag{7}$$

with

$$\left(\frac{d\sigma}{d\Omega}\right)_{\text{Mott}} = \left(\frac{Z\alpha}{2k}\right)^2 \frac{\cos^2(\theta/2)}{\sin^4(\theta/2)}. \tag{8}$$

The Mott cross section (8) is derived below, including the necessary kinematics if recoil of the struck target becomes important. However, already at this point, we can read off a few qualitative physical features from these formulae.

(i) In the forward direction, $\boldsymbol{q} = 0$, the form factor $F(\boldsymbol{q})$ is equal to one, by virtue of the normalization condition (2): $F(0) = 1$. The same result obtains for *all* momenta $\boldsymbol{q}$ if the charge distribution is concentrated in a point,

$$\rho_{\text{point}}(r) = \frac{1}{4\pi}\frac{1}{r^2}\delta(r) \to F(\boldsymbol{q}) \equiv 1 \quad \forall \boldsymbol{q}. \tag{9}$$

Thus, the Mott cross section (8) describes the scattering off a point charge Ze placed at the origin.

(ii) If the charge is not pointlike the cross section (7) is modulated by the form factor $F(\boldsymbol{q})$. It is this form factor which contains information about the target beyond its charge Ze. Hence, elastic electron scattering measures, in essence, the *spatial distribution* of the charge density $\rho(r)$. In particular, if Born approximation is

*) In order to do this in a mathematically correct manner the $1/r$ potential must be multiplied by a convergence factor, say $e^{-\alpha r}$, and the limit $\alpha \to 0$ must be taken at the end.

applicable, then eqs. (6) and (7) show that the cross section is just the square of the Fourier transform of $\rho(r)$.

(iii) If the region over which $\rho(r)$ is appreciably different from zero, is characterized by a radius R, the momentum transfer must be chosen such that

$$qR \gtrsim 1. \tag{10}$$

Indeed, if q is chosen too small, i.e. $qR \ll 1$ the form factor does not yet deviate much from unity and little information is obtained. If q is too large, i.e. $qR \gg 1$, the exponential in eq. (6) oscillates rapidly and $F(\boldsymbol{q})$ becomes unmeasurably small. Thus q must be tuned to the size of the extended object that one wants to map. The quantitative details and the nature of the information obtained by means of elastic scattering are worked out in some of the following sections (secs. 4, 5).

(iv) A similar situation is encountered when we consider inelastic scattering to discrete excited states. In that case the ground state charge density (which we assumed spherically symmetric, for simplicity) is replaced by a transition density $\rho_{\mathrm{fi}}(\boldsymbol{r})$ or more complicated functionals of charge and current densities of the target particle. Equivalently, the elastic charge form factor is replaced by inelastic form factors. Like in the case of elastic scattering the cross section depends on a leptonic part which is known and some kinematics, whilst the hadronic structure is contained in the form factors.

In the simplest case (electric charge scattering on a nucleus to a discrete excited state) the form factor for the transition from state i to state f, and with multipolarity λ, will be proportional to

$$F_{\mathrm{fi}}(q) \propto \int_0^\infty \rho_{\mathrm{fi}}(r)\, j_\lambda(qr)\, r^2\, \mathrm{d}r. \tag{11}$$

This follows from expanding the exponential $\mathrm{e}^{\mathrm{i}\boldsymbol{q}\boldsymbol{r}}$ in terms of spherical harmonics and from the selection rules for angular momentum imposed by the spins and parities of initial and final target state. The expression (11) is reminiscent of the transition amplitude for the corresponding photoexcitation of state f. In that case q is replaced by the photon energy $k_\gamma = E_\mathrm{i} - E_\mathrm{f}$. For given energy this is a fixed number. Furthermore, in many cases $k_\gamma r$ is small compared to one over the domain where $\rho_{\mathrm{fi}}(r)$ is appreciably different from zero. In this case the Bessel function in eq. (11) can be replaced by its limiting form for small argument,

$$j_\lambda(k_\gamma r) \simeq (k_\gamma r)^\lambda/(2\lambda+1)!!. \tag{12}$$

Thus, the γ-transition depends essentially only on one specific *moment* of the transition charge density. The power behaviour $(k_\gamma r)^\lambda$ with $k_\gamma r \ll 1$ limits the γ-transition to the lowest possible multipolarity.

In *electron* scattering, on the contrary, q is a variable momentum transfer which can be chosen as large as one wishes. This means that a complete mapping of $F_{\mathrm{fi}}(q)$ and, thereby, of $\rho_{\mathrm{fi}}(r)$ can be obtained, at least in principle. Also, all multipolarities compatible with the angular momentum and parity selection rules can contribute on equal footing if q is chosen appropriately. For instance, an E5 electro-excitation can have as large a cross section as an E2 transition. Finally, if both the momentum transfer and the energy transfer to the target are chosen sufficiently large one reaches a new domain where the electron starts to probe, in a rather general sense, the *constituent structure of the target*. In the case of scattering on nuclei this is called *quasi-free scattering*, expressing the fact that the energy transfer is much larger than the binding energy of the nucleons in the nucleus so that the nucleus behaves like a cloud of almost free nucleons.

In the case of *nucleons* the analogous domain is called the *deep inelastic* region. Although there are essential differences to the case of nuclei the leading idea is the same as in quasi-free scattering on nuclei: Deep inelastic scattering where both momentum and energy transfers are large, must be sensitive to interactions within the target at very small distances.

2. Elastic scattering on a spin zero target, Born approximation

The elastic scattering of an electron on a spin zero target is the simplest case. It can be dealt with by means of standard Green function techniques of potential scattering without having to invoke covariant perturbation theory and Feynman rules. The result is the correct covariant cross section in Born approximation and contains many of the essential features of more complicated situations such as scattering on spin-1/2 targets or the like. Because of this simplicity we consider this case first.

The "spin-zero target" may be a pion or any nucleus whose spin is zero.

We start with the kinematics of fig. 1. Let

$$k = (E; \boldsymbol{k}), \qquad k' = (E'; \boldsymbol{k}') \tag{13}$$

be the four-momenta of the electron in the *laboratory system* before and after the scattering. Similarly,

$$p = (M; 0), \qquad p' = \left(\sqrt{M^2 + \boldsymbol{p}'^2}\,; \boldsymbol{p}'\right) \tag{14}$$

denote the four-momenta of the target before and after the scattering process, respectively, with M the target mass. As the electron energy is chosen large as compared to its rest mass, $E \simeq |\boldsymbol{k}|$ and $E' \simeq |\boldsymbol{k}'|$ and the square of the four-momentum transfer is

$$q^2 \coloneqq -(k - k')^2 = -(E - E')^2 + (\boldsymbol{k} - \boldsymbol{k}')^2 \simeq 4EE'\sin^2(\theta/2) \tag{15}$$

(q^2 is the same as $-t$, see below, eq. (36)).

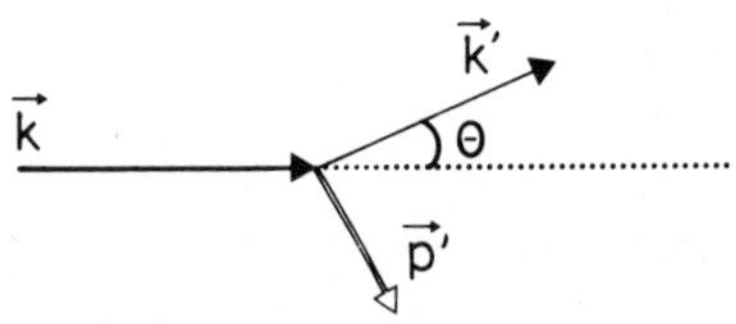

Fig. 1. Kinematics of electron scattering in the laboratory system.

Setting $m_e \simeq 0$, energy–momentum conservation gives the relationship

$$E' = E\frac{1}{1+2(E/M)\sin^2(\theta/2)}. \tag{16}$$

We note that the denominator of eq. (16) is a typical recoil term. Whether or not this recoil term is important depends on the mass of the target. For a nucleus, for instance, $M \simeq A \cdot 940$ MeV will in general be large as compared to E if the electron energy is chosen to be a few hundred MeV.

Suppose we can characterize the ground state of the target by a spherically symmetric charge distribution. In the case of a nucleus of charge number Z this is

$$Ze\rho(r) = \langle \Psi_0 | e \sum_{i=1}^{Z} \delta(\boldsymbol{r} - \boldsymbol{r}_i) | \Psi_0 \rangle$$

with Ψ_0 the (spherical) ground state wave function.*) The corresponding electrostatic potential $V(r)$ is given by eq. (III.165) and must be inserted into the Dirac equation (III.166) in which we neglect the mass term on the righthand side.

We wish to construct solutions of eqs. (III.166) which describe the scattering of electrons whose incoming momentum is directed along the positive 3-axis. In order to avoid the complications due to the infinite range of the Coulomb potential we assume, as usual, that $\lim_{r\to\infty} rV(r) = 0$. This means that we multiply the Coulomb potential by a screening factor and that we take the correct limit at the end of our calculation. As is well-known, using this mathematically incorrect procedure we miss the typical constant and logarithmic phases of the Coulomb scattering problem. However, very much like in the nonrelativistic case, these complications are irrelevant for our discussion; they may be inserted in the final results.

For $r \to \infty$ the two-component spinors $\phi(\boldsymbol{x})$ and $\chi(\boldsymbol{x})$ must be eigenstates of helicity. Thus, in the *center-of-mass system* they have the form

$$\phi(\boldsymbol{x}) \sim u_+(0,0)\mathrm{e}^{\mathrm{i}k^*z} + f(\theta,\varphi)\frac{\mathrm{e}^{\mathrm{i}k^*r}}{r}u_+(\theta,\varphi), \tag{17a}$$

$$\chi(\boldsymbol{x}) \sim u_-(0,0)\mathrm{e}^{\mathrm{i}k^*z} + g(\theta,\varphi)\frac{\mathrm{e}^{\mathrm{i}k^*r}}{r}u_-(\theta,\varphi), \tag{17b}$$

where $k^* := |\boldsymbol{k}_{\text{c.m.}}| = |\boldsymbol{k}'_{\text{c.m.}}| \simeq E^*$ and where

$$u_\pm(\theta,\varphi) = D^{(1/2)*}(0,\theta,\varphi)u_\pm(0,0),$$

$$u_+(0,0) = \begin{pmatrix}1\\0\end{pmatrix}, \qquad u_-(0,0) = \begin{pmatrix}0\\1\end{pmatrix} \tag{18}$$

(cf. eq. (III.119)). f and g denote the scattering amplitudes whose squares give the cross section in the c.m. system. $u_\pm(0,0)$ are positive and negative helicity eigenstates, respectively, of the incoming state whose momentum points in the positive

*)In the case of an elementary particle $\rho(r)$ is the Fourier transform of the form factor, see below. This leaves open the question whether or not $\rho(r)$ may be computable from a constituent picture.

3-direction. Similarly

$$u_+(\theta,\varphi)=\begin{pmatrix} e^{-i\varphi/2}\cos(\theta/2) \\ e^{i\varphi/2}\sin(\theta/2) \end{pmatrix}, \qquad u_-(\theta,\varphi)=\begin{pmatrix} -e^{-i\varphi/2}\sin(\theta/2) \\ e^{i\varphi/2}\cos(\theta/2) \end{pmatrix} \tag{19}$$

are the helicity eigenstates along the outgoing momentum.

Two important properties follow from the equations (III.166) with m set equal to zero:

(i) Eqs. (III.166a) and (III.166b) are completely decoupled. The potential $V(r)$, which is spherically symmetric, cannot change the helicity.

(ii) From invariance under *parity* and under *rotations* one shows that $f(\theta,\varphi)$ and $g(\theta,\varphi)$ must be equal (see exercise 1). Thus, *a fast electron cannot be polarized by a spherically symmetric potential.* [Note that this is not so for energies comparable to the rest mass. At such energies the spin orbit force can indeed flip the spin of the electron (Mott scattering).] Therefore, it is sufficient to solve one of the eqs. (III.166), for example the first one, which reads

$$(i\boldsymbol{\sigma}\cdot\nabla + k^*)\phi(\boldsymbol{x}) = V(r)\phi(\boldsymbol{x}). \tag{20}$$

We solve this equation with the aid of the appropriate Green function which satisfies the equation

$$(i\boldsymbol{\sigma}\cdot\nabla_x + k^*)G(\boldsymbol{x}-\boldsymbol{x}') = \delta(\boldsymbol{x}-\boldsymbol{x}'), \tag{21}$$

whereby

$$\phi(\boldsymbol{x}) = u_+(0,0)e^{ik^*z} + \int G(\boldsymbol{x}-\boldsymbol{x}')V(r')\phi(\boldsymbol{x}')\,d^3x'. \tag{22}$$

The Green function with the correct asymptotic behaviour is given by

$$G(\boldsymbol{x}-\boldsymbol{x}') = (i\boldsymbol{\sigma}\cdot\nabla_x - k^*)\frac{e^{ik^*|\boldsymbol{x}-\boldsymbol{x}'|}}{4\pi|\boldsymbol{x}-\boldsymbol{x}'|}. \tag{23}$$

This follows from the equations

$$(i\boldsymbol{\sigma}\cdot\nabla + k^*)(i\boldsymbol{\sigma}\cdot\nabla - k^*) = -(\Delta + k^{*2}),$$

$$(\Delta + k^{*2})\frac{e^{ik^*|\boldsymbol{x}-\boldsymbol{x}'|}}{4\pi|\boldsymbol{x}-\boldsymbol{x}'|} = -\delta(\boldsymbol{x}-\boldsymbol{x}').$$

First Born approximation means replacing $\phi(\boldsymbol{x}')$ on the r.h.s. of eq. (22) by the incoming plane wave. Thus

$$\phi(\boldsymbol{x}) \simeq u_+(0,0)e^{ik^*z} + \frac{1}{4\pi}\int d^3x'(i\boldsymbol{\sigma}\cdot\nabla_x - k^*)\frac{e^{ik^*|\boldsymbol{x}-\boldsymbol{x}'|}}{|\boldsymbol{x}-\boldsymbol{x}'|}V(r')u_+(0,0)e^{ik^*z'},$$

from which we must now extract the scattering amplitude $f(\theta,\varphi)$. For finite r', but taking the limit $r \to \infty$, we have

$$\left(\mathrm{i}\boldsymbol{\sigma}\cdot\nabla_x - k^*\right)\frac{\mathrm{e}^{\mathrm{i}k^*|\boldsymbol{x}-\boldsymbol{x}'|}}{|\boldsymbol{x}-\boldsymbol{x}'|} = -\left(\mathrm{i}\boldsymbol{\sigma}\cdot\nabla_{x'} + k^*\right)\frac{\mathrm{e}^{\mathrm{i}k^*|\boldsymbol{x}-\boldsymbol{x}'|}}{|\boldsymbol{x}-\boldsymbol{x}'|}$$

$$\underset{r\to\infty}{\sim} -\left(\mathrm{i}\boldsymbol{\sigma}\cdot\nabla_{x'} + k^*\right)\frac{\mathrm{e}^{\mathrm{i}k^*r}}{r}\mathrm{e}^{-\mathrm{i}k^*\hat{\boldsymbol{x}}\cdot\boldsymbol{x}'}.$$

If we set $\boldsymbol{k}' := k^*\cdot\hat{\boldsymbol{x}} = k^*\boldsymbol{x}/r$, this gives for $r \to \infty$

$$\phi(\boldsymbol{x}) \sim u_+(0,0)\mathrm{e}^{\mathrm{i}k^*z} - \frac{\mathrm{e}^{\mathrm{i}k^*r}}{r}\frac{1}{4\pi}\int \mathrm{d}^3x'\mathrm{e}^{\mathrm{i}\boldsymbol{q}\boldsymbol{x}'}V(r')(\boldsymbol{\sigma}\cdot\boldsymbol{k}' + k^*)u_+(0,0) \tag{24}$$

with $\boldsymbol{q} := \boldsymbol{k} - \boldsymbol{k}'$. The r.h.s. of eq. (24) is to be identified with the general form (17a), giving

$$f(\theta,\varphi) = -\frac{1}{4\pi}\int \mathrm{d}^3x'\mathrm{e}^{\mathrm{i}\boldsymbol{q}\boldsymbol{x}'}V(r')u_+^\dagger(\theta,\varphi)(\boldsymbol{\sigma}\cdot\boldsymbol{k}' + k^*)u_+(0,0).$$

The scalar product under the integral sign is easily worked out by making use of the equation satisfied by the spinor $u_+(\theta,\varphi)$:

$$u_+^\dagger(\theta,\varphi)(\boldsymbol{\sigma}\cdot\boldsymbol{k}' - k^*) = 0.$$

One finds

$$u_+^\dagger(\theta,\varphi)(\boldsymbol{\sigma}\cdot\boldsymbol{k}' + k^*)u_+(0,0) = 2k^*u_+^\dagger(\theta,\varphi)u_+(0,0) = 2k^*\mathrm{e}^{\mathrm{i}\varphi/2}\cos(\theta/2),$$

which then gives

$$f(\theta,\varphi) = -\frac{k^*}{2\pi}\mathrm{e}^{\mathrm{i}\varphi/2}\cos(\theta/2)\int \mathrm{d}^3r\,\mathrm{e}^{\mathrm{i}\boldsymbol{q}\boldsymbol{r}}V(r).$$

Alternatively, we may integrate by parts and make use of Poisson's equation $\Delta V(r) = Ze^2\rho(r)$ to obtain

$$f(\theta,\varphi) = \frac{2Z\alpha k^*}{q^2}\mathrm{e}^{\mathrm{i}\varphi/2}\cos(\theta/2)\int \mathrm{d}^3r\,\mathrm{e}^{\mathrm{i}\boldsymbol{q}\cdot\boldsymbol{r}}\rho(r) \tag{25}$$

($\alpha = \mathrm{e}^2/4\pi \simeq \frac{1}{137}$). The differential cross section in the center-of-mass system is then given by

$$\left(\frac{\mathrm{d}\sigma}{\mathrm{d}\Omega}\right)_{\mathrm{c.m.}} = |f(\theta,\varphi)|^2 = \left.\frac{\mathrm{d}\sigma}{\mathrm{d}\Omega}\right|_{\mathrm{Mott}} F^2(\boldsymbol{q}), \tag{26}$$

where

$$\left.\frac{\mathrm{d}\sigma}{\mathrm{d}\Omega}\right|_{\mathrm{Mott}} = \frac{(Z\alpha)^2\cos^2(\theta/2)}{4k^{*2}\sin^4(\theta/2)} \tag{27}$$

is the Mott cross section, eq. (8), θ is the scattering angle, in the c.m. system and $F(\boldsymbol{q})$ is the (real) charge form factor of the charge distribution $\rho(r)$ (Mott, 1929).

3. A few properties of form factor and cross section

Before we proceed to elastic scattering on targets with nonvanishing spin we wish to discuss a few properties of the results obtained in the previous section.

The order of magnitude of the Mott cross section is easily estimated. Take, for instance, $k^* = 200$ MeV $\triangleq (200/\hbar c)$ fm$^{-1} = 1.01$ fm^{-1}, $\theta = 90°$ and $Z = 82$. This gives

$$\left.\frac{\mathrm{d}\sigma}{\mathrm{d}\Omega}\right|_{\mathrm{Mott}} = (82)^2 \times 2.6 \times 10^{-31}\ \mathrm{cm}^2 = 1.7 \times 10^{-27}\ \mathrm{cm}^2.$$

A few properties of the form factor have already been mentioned above, in sec. 1. If the target is a nucleus the charge distribution can be calculated from the ground state wave function Ψ_0,

$$\rho(\boldsymbol{r}) = \frac{1}{Z}\int \mathrm{d}^3r_1 \int \mathrm{d}^3r_2 \ldots \int \mathrm{d}^3r_A \sum_{\nu=1}^{Z} \delta(\boldsymbol{r} - \boldsymbol{r}_\nu)|\Psi_0(\boldsymbol{r}_1 \ldots \boldsymbol{r}_A)|^2,$$

so that

$$F(q^2) = \frac{1}{Z}\int \mathrm{d}^3r_1 \int \mathrm{d}^3r_2 \ldots \int \mathrm{d}^3r_A \sum_{\nu=1}^{Z} \mathrm{e}^{\mathrm{i}\boldsymbol{q}\cdot\boldsymbol{r}_\nu}|\Psi_0(\boldsymbol{r}_1 \ldots \boldsymbol{r}_A)|^2. \tag{28}$$

We verify the property $F(0) = 1$ which expresses the fact that forward scattering depends only on the total charge of the target. If $\rho(\boldsymbol{r})$ is spherically symmetric, the form factor (6) can be written as follows. We make use of the expansion

$$\mathrm{e}^{\mathrm{i}\boldsymbol{q}\cdot\boldsymbol{r}} = 4\pi \sum_{l=0}^{\infty} \mathrm{i}^l j_l(qr) \sum_{m=-l}^{+l} Y_{lm}^*(\hat{\boldsymbol{q}}) Y_{lm}(\hat{\boldsymbol{r}}) \tag{29}$$

and of the orthogonality property of the spherical harmonics to obtain*)

$$\begin{aligned} F(q^2) &= 4\pi \int_0^\infty \rho(r)\, j_0(qr) r^2\, \mathrm{d}r \\ &= \frac{4\pi}{q}\int_0^\infty \rho(r) \sin(qr) r\, \mathrm{d}r. \end{aligned} \tag{30}$$

If (qr) is small over the domain where $\rho(r)$ is appreciably different from zero, one may expand the form factor in powers of q,

$$\begin{aligned} F(q^2) &= 4\pi \int_0^\infty \rho(r) r^2\, \mathrm{d}r - \tfrac{1}{6}q^2 4\pi \int_0^\infty \rho(r) r^4\, \mathrm{d}r + \mathcal{O}(q^4) \\ &= 1 - \tfrac{1}{6}q^2 \langle r^2 \rangle_{\mathrm{r.m.s.}} + \mathcal{O}(q^4 \langle r^4 \rangle). \end{aligned} \tag{31}$$

Here $\langle r^2 \rangle_{\mathrm{r.m.s.}}$ denotes the root-mean-square radius

$$\langle r^2 \rangle_{\mathrm{r.m.s.}} = -6\frac{\partial F(q^2)}{\partial q^2} = 4\pi \int_0^\infty [\rho(r) r^2] r^2\, \mathrm{d}r. \tag{32}$$

*)If $\rho(\boldsymbol{r})$ is not spherically symmetric the formalism developed for inelastic scattering below may be consulted.

As the momentum transfer increases, more and more moments $\langle r^{2n} \rangle$ come into play. Eventually, if the form factor is known for all momenta q^2, all even moments are determined. This is equivalent to saying that the charge distribution $\rho(r)$ has been mapped completely and is obtained from the form factor by

$$\rho(r) = \frac{1}{(2\pi)^3} \int \mathrm{d}^3 q \, \mathrm{e}^{-\mathrm{i}\boldsymbol{q}\cdot\boldsymbol{r}} F(q^2), \tag{33}$$

In the case of *nuclei* $\rho(r)$ is given by the wave functions of the protons in the nuclear ground state. The r.m.s. radius is then the average r.m.s. radius of the protons (so-called *charge radius*),

$$\langle r^2 \rangle_{\text{r.m.s.}} = \frac{1}{Z} \sum_{i=1}^{Z} \langle r_i^2 \rangle. \tag{34}$$

In the case of the *nucleon* it is not clear a priori what causes its finite charge distribution. The finite extension of charge within the proton may be due to the virtual meson cloud surrounding the proton, and/or to a bound state substructure in which case the charge density reflects some properties of the ground state of the proton's constituents. In any event, the primary physical quantity is the form factor, not the charge density. It is the form factor which describes the particle's coupling to the photon (Coulomb field) and which enters in the expressions for scattering amplitudes and cross sections. Once the form factor is given, we may *define* the charge density through eq. (33) and, in particular, the r.m.s. radius through the derivative of $F(q^2)$, cf. eq. (32).

In sec. 2 we have calculated the cross section in the c.m. system. It is not difficult to transform it to the laboratory system or any other system of reference. For that purpose it is useful to write first the cross section in a Lorentz invariant form. Let us introduce Lorentz-scalar variables (Mandelstam variables), s and t, and let us write these, in the c.m. frame, in terms of k^* and θ (neglecting the electron mass).

$$s = (k+p)^2 = M^2 + 2k^*\sqrt{M^2 + k^{*2}} + 2k^{*2},$$

$$t = (k-k')^2 = -2k^{*2}(1-\cos\theta) \equiv -q^2.$$

Inverting these equations one obtains

$$k^* = (s - M^2)/2\sqrt{s}\,,$$

$$\cos^2(\theta/2) = \left[(s-M^2)^2 + st\right]/(s-M^2)^2.$$

The invariant cross section $\mathrm{d}\sigma/\mathrm{d}t$ is calculated by means of

$$\frac{\mathrm{d}\sigma}{\mathrm{d}t} = \left(\frac{\mathrm{d}\sigma}{\mathrm{d}(\cos\theta)}\right)_{\text{c.m.}} \frac{\mathrm{d}\cos\theta}{\mathrm{d}t} = \int_0^{2\pi} \mathrm{d}\varphi \left(\frac{\mathrm{d}\sigma}{\mathrm{d}\Omega}\right)_{\text{c.m.}} \frac{\mathrm{d}\cos\theta}{\mathrm{d}t},$$

obtaining

$$\frac{\mathrm{d}\sigma}{\mathrm{d}t} = \frac{4\pi(Z\alpha)^2}{t^2} F^2(t) \frac{(s-M^2)^2 + st}{(s-M^2)^2}. \tag{35}$$

It is now easy to calculate the cross section in the laboratory system, where

$$s = M^2 + 2ME,$$
$$t = -2EE'(1-\cos\theta) = -2E^2 \frac{1-\cos\theta}{1+(E/M)(1-\cos\theta)}. \qquad (36)$$

Thus

$$\frac{\mathrm{d}t}{\mathrm{d}(\cos\theta)} = \frac{2E^2}{[1+(E/M)(1-\cos\theta)]^2},$$
$$\frac{(s-M^2)^2+st}{(s-M^2)^2} = \frac{\cos^2(\theta/2)}{1+(E/M)(1-\cos\theta)}.$$

Finally, knowing that

$$\int \frac{\mathrm{d}\sigma}{\mathrm{d}\Omega}\,\mathrm{d}\varphi = \frac{\mathrm{d}\sigma}{\mathrm{d}t}\,\frac{\mathrm{d}t}{\mathrm{d}\cos\theta}$$

we find

$$\left(\frac{\mathrm{d}\sigma}{\mathrm{d}\Omega}\right)_{\mathrm{lab}} = \left(\frac{Z\alpha}{2E}\right)^2 \frac{\cos^2(\theta/2)}{\sin^4(\theta/2)} F^2(q^2) \frac{1}{1+2(E/M)\sin^2(\theta/2)}. \qquad (35')$$

Thus, the cross section in the laboratory system contains the recoil factor $\{1+2(E/M)\sin^2(\theta/2)\}^{-1}$ which we encountered earlier in eq. (16).

Clearly, the method of transforming the cross section from one frame of reference to another that we have developed here, is quite general. It consists in writing first the cross section in Lorentz invariant form (viz. eq. (35)), and then in specializing to any desired frame of reference. In this context, we refer also to the general formulae collected in App. B.

4. Elastic scattering on nucleons

4.1. Current matrix elements and form factors

The scattering cross section (26) and (35) that we have derived in the last two sections holds for any spin zero target, a nucleus, a pion or any other elementary or composite particle with spin zero. The result is fully covariant and thus may also be derived in the framework of covariant perturbation theory. In other words the same cross section must be obtained from the formal Feynman rules for quantum electrodynamics which are summarized in App. C (see exercise 2).

If a particle has no internal structure caused by interactions other than electromagnetic, we say it is *pointlike*. For example, to the best of our knowledge, electron and muon seem to be such particles. In this case the Feynman rules apply as they are given in App. C. In particular, at any photon–fermion vertex we have to write a factor γ_α, the Lorentz index having to be contracted with the photon polarization vector $\varepsilon^{\alpha}(k)$ for an external photon, or with one of the indices of the photon propagator $D^{\alpha\beta}(k)$ for an internal photon line. This is a reflection of the fact that the photon couples to the electromagnetic current j_{α}^{em}, and that for a point-like fermion (on its mass shell) the matrix element of j_{α}^{em} leading from momentum state p to momentum state p' is given by

$$\langle p'|j_{\alpha}^{\text{e.m.}}(0)|p\rangle = \frac{1}{(2\pi)^3} Q\overline{u(p')}\gamma_{\alpha}u(p). \tag{37}$$

If, to the contrary, the particle does have internal structure due to other interactions not described by QED,*) then the Feynman rules are incomplete. They cannot tell us the explicit form of the particle's coupling to an (external or internal) photon line. Nevertheless, these couplings can be reduced considerably and can be expressed in terms of a few real Lorentz scalar functions (form factors) which allow to parametrize the internal structure of the particle in a very condensed form.**) This is achieved by making use of some general properties of the electromagnetic current, such as its behaviour under Lorentz transformations, time reversal, current conservation, hermiticity, isospin content etc.

We exemplify these matters for the case of proton and neutron, i.e. strongly interacting particles of spin 1/2. The spin zero case is similar though somewhat simpler. It is left as an exercise for the reader. The relevant matrix element of the electromagnetic current is

$$\langle p', s'|j_{\alpha}^{\text{em}}(x)|p, s\rangle, \tag{38}$$

where initial and final states are on-mass-shell states of given momentum and spin, and $p^2 = p'^2 = M^2$. Let us work out the restrictions on this matrix element that follow from current conservation, from the space–time structure of the electromagnetic current and hermiticity.

(i) *Current conservation.* Let $x'_{\mu} = x_{\mu} + a$ be an arbitrary translation, that transforms a given field operator according to

$$F(x) \to F'(x') = U(a)F(x)U^{-1}(a) = F(x+a). \tag{39}$$

Using the generalized Heisenberg equations of motion

$$-i\partial^{\mu}F(x) = [P^{\mu}, F(x)], \tag{40}$$

where P^{μ} are the four energy–momentum operators, one shows that (see exercise 3)

$$U(a) = \exp\{ia_{\mu}P^{\mu}\}. \tag{41}$$

If we consider a matrix element of $F(x)$ between specific eigenstates of energy and momentum, we can make use of eq. (39) to transform $F(x)$ to any other point

*) For example, hadrons are composite objects and, to some extent, they are also "dressed" by pion clouds.

**) Note that also a pointlike particle of QED builds up form factors i.e. internal structure by interaction with the Maxwell field. These effects are calculable from QED in higher order perturbation theory.

$x' = x + a$ of Minkowski space, viz.

$$\begin{aligned}\langle q_f|F(x)|qi\rangle &= \langle q_f|U^{-1}(a)F(x+a)U(a)|q_i\rangle \\ &= \langle U(a)q_f|F(x+a)|U(a)q_i\rangle \\ &= e^{ia\cdot(q_i-q_f)}\langle q_f|F(x+a)|q_i\rangle.\end{aligned} \tag{42}$$

In particular, we may take $a_\mu = -x_\mu$ to obtain

$$\langle q_f|F(x)|q_i\rangle = e^{-ix\cdot(q_i-q_f)}\langle q_f|F(0)|q_i\rangle. \tag{42'}$$

Thus, the x-dependence of any such matrix element between eigenstates of four-momentum is a simple exponential; the remaining factor $\langle q_f|F(0)|q_i\rangle$ does not depend on x anymore.

Suppose $F(x)$ is a current operator $J_\alpha(x)$ and suppose that the divergence of J_α is known, $\partial_\alpha J^\alpha(x) = \phi(x)$. Then from eq. (42′),

$$\begin{aligned}\langle q_f|\partial^\alpha J_\alpha(x)|q_i\rangle &= (\partial^\alpha e^{-i(q_i-q_f)\cdot x})\langle q_f|J_\alpha(0)|q_i\rangle \\ &= \langle q_f|\phi(x)|q_i\rangle = e^{-i(q_i-q_f)x}\langle q_f|\phi(0)|q_i\rangle.\end{aligned}$$

Thus, one finds the relation

$$(q_i - q_f)^\alpha\langle q_f|J_\alpha(0)|q_i\rangle = i\langle q_f|\phi(0)|q_i\rangle. \tag{43}$$

If the current is conserved—this is the case for the electromagnetic current—we obtain the condition

$$(q_i - q_f)^\alpha\langle q_f|j_\alpha(0)|q_i\rangle = 0 \tag{44}$$

(we drop the superscript "em" for simplicity).

(ii) *Covariance*. As j_α is a Lorentz-vector, its matrix elements between nucleon states must also transform as Lorentz vectors. For the construction of such vectors we have at our disposal the vectors p'_α and p_α as well as the γ-matrices and combinations thereof. The only vectors that can be formed are*)

$$\overline{u(p')}\gamma_\alpha u(p), \qquad (p'-p)^\beta\overline{u(p')}\sigma_{\alpha\beta}u(p),$$

$$(p+p')_\alpha\overline{u(p')}u(p), \qquad (p-p')_\alpha\overline{u(p')}u(p),$$

where we have defined

$$\sigma_{\alpha\beta} := \tfrac{1}{2}i(\gamma_\alpha\gamma_\beta - \gamma_\beta\gamma_\alpha). \tag{45}$$

That the first two of these are indeed *vector* operators is not difficult to show. Indeed $\overline{\Psi(x)}\gamma_\alpha\Psi(x)$ is a vector operator as should be clear from Chap. III. Similarly, $\overline{\Psi(x)}\sigma_{\alpha\beta}\Psi(x)$ is a tensor operator. Knowing that p_α and p'_α are vectors and making use of the expansion (III.100), the assertion is proved. Among these covariants only three are independent. The external particles are on their mass shell and obey the free Dirac equation. In this case one has the Gordon identity relating the first three covariants (exercise 14).

*) The same forms apply for antiparticle states, with $u(p)$ replaced by $v(p)$.

Thus, the most general covariant decomposition must have the form (dropping the spin indices s, s', for the sake of clarity),

$$\langle p'|j_\alpha(0)|p\rangle = \frac{1}{(2\pi)^3}\overline{u(p')}\left\{\gamma_\alpha F_1(q^2)+\frac{\mathrm{i}}{2M}\sigma_{\alpha\beta}q^\beta F_2(q^2)+\frac{1}{2M}q_\alpha F_3(q^2)\right\}u(p), \quad (46)$$

where q is the momentum transfer $q := p'-p$. (We have set $Q=1$, taking out a factor $|e|$ to be inserted at each photon vertex.) The functions F_i must be Lorentz scalars and, thus, can only depend on Lorentz scalar quantities such as p^2, p'^2 and q^2. Since $p^2=p'^2=m^2$, these are in fact constants, so the only true variation must be in the variable q^2. The divergence condition (44) implies that $F_3(q^2)$ must vanish identically for $q^2 \neq 0$. The other two terms on the r.h.s. of eq. (46) fulfill this condition separately, as one easily verifies by means of the Dirac equations (III.62, 62′).

(iii) *Hermiticity of electromagnetic current*. By definition we have

$$\langle p|j_\alpha^\dagger(0)|p'\rangle = \langle p'|j_\alpha(0)|p\rangle^*. \quad (48a)$$

As j_α is a hermitean operator this is also equal to

$$\langle p|j_\alpha(0)|p'\rangle. \quad (48b)$$

The expression (48a), more explicitly, gives

$$\left(u(p')^\dagger\gamma_0\left[\gamma_\alpha F_1+\frac{i}{2M}\sigma_{\alpha\beta}(p'-p)^\beta F^2\right]u(p)\right)^*$$
$$= u^\dagger(p)\left[(\gamma_0\gamma_\alpha)^\dagger F_1^* - \frac{i}{2M}(\gamma_0\sigma_{\alpha\beta})^\dagger(p'-p)^\beta F_2^*\right]u(p')$$
$$= \overline{u(p)}\left\{\gamma_0\gamma_\alpha^\dagger\gamma_0 F_1^* + \frac{\mathrm{i}}{2M}\gamma_0\sigma_{\alpha\beta}^\dagger\gamma_0(p-p')^\beta F_2^*\right\}u(p'),$$

where we have inserted $(\gamma_0)^2=1$ between $u^\dagger(p)$ and the square brackets and have interchanged p and p' in the second term. We know from eq. (III.57b) that $\gamma_0\gamma_\alpha^\dagger\gamma_0=\gamma_\alpha$. Using the definition (45) one sees that also $\gamma_0\sigma_{\alpha\beta}^\dagger\gamma_0=\sigma_{\alpha\beta}$. As (48a) must equal (48b) we conclude that the form factors F_1 and F_2 are real:

$$F_1^*(q^2)=F_1(q^2), \qquad F_2^*(q^2)=F_2(q^2). \quad (49)$$

We add a few comments on these results: The form factor F_2 has actually been defined such as to make it real by the choice of the factor i in front of the second term of eq. (46). Without this factor F_2 would have come out pure imaginary. Similarly, the factor $1/2M$ is a matter of convention, chosen so as to give $F_2(q^2)$ the same dimension as $F_1(q^2)$. In applying the divergence condition (44) we have used the fact that the two spinors belong to the same mass. If the external fermions have different masses the relations following from current conservation look different. Similarly, if the electromagnetic current is taken between two *different* particles, hermiticity does not suffice to derive reality properties of form factors. However, if the interactions are invariant under time reversal, the combination of hermiticity and time reversal invariance implies again reality conditions of the form factors. Examples of this will be met below in the context of weak interactions.

4.2. *Derivation of cross section*

The physical interpretation of the form factors F_1 and F_2 is discussed below. For the moment, we note that $F_1(q^2=0)=1$ is the charge of the particle in units of $|e|$, see below. We first turn to the computation of the differential cross section,

$$\mathrm{d}\sigma = \frac{(2\pi)^4\delta^{(4)}(p+k-p'-k')}{\left[2E_k/(2\pi)^3\right]\left[2E_p/(2\pi)^3\right]|\boldsymbol{v}_{12}|}\frac{1}{4}\sum_{\text{spins}}|T_{\text{fi}}|^2\frac{\mathrm{d}^3k'}{2E_{k'}}\frac{\mathrm{d}^3p'}{2E_{p'}}. \tag{50}$$

T_{fi} is the T-matrix element, to be obtained from Feynman rules for lowest order perturbation theory. $\boldsymbol{v}_{12}$ is the relative velocity of electron and nucleon in the initial state. It is useful to calculate $\mathrm{d}\sigma$ first in the center-of-mass system and then, in a second step, to write it in a manifestly invariant form from which the cross section in any system of reference can be obtained. In the c.m. system,

$$k=\left(\sqrt{m^2+q^{*2}}\,;\boldsymbol{q}\right),\qquad p=\left(\sqrt{M^2+q^{*2}}\,;-\boldsymbol{q}\right).$$

Introducing the invariant variables s, t, u one has

$$s=(p+k)^2=(p'+k')^2=m^2+M^2+2(pk)=m^2+M^2+2(p'k'), \tag{51a}$$

$$t=(k-k')^2=(p'-p)^2=2m^2-2(kk')=2M^2-2(pp'), \tag{51b}$$

$$u=(k-p')^2=(p-k')^2=m^2+M^2-2(p'k)$$
$$=m^2+M^2-2(pk')=2m^2+2M^2-t-s. \tag{51c}$$

In particular, in the c.m. system

$$q^* := |\boldsymbol{q}| = \frac{1}{2\sqrt{s}}\sqrt{(s-M^2-m^2)^2-4M^2m^2}\,, \tag{52}$$

$$t=-2q^{*2}(1-\cos\theta). \tag{53}$$

The Møller factor in the denominator of eq. (50) is an invariant and can be written as

$$E_kE_p|\boldsymbol{v}_{12}|=\sqrt{(pk)^2-p^2k^2}$$
$$=\frac{1}{2}\sqrt{(s-M^2-m^2)^2-4M^2m^2}=q^*\sqrt{s}\,.$$

As we are interested in $\mathrm{d}\sigma/\mathrm{d}\Omega$ or, equivalently, $\mathrm{d}\sigma/\mathrm{d}t$, all other variables in the final state must be integrated over. The integration over $\boldsymbol{p}'$ may be done first, giving $\boldsymbol{p}'=-\boldsymbol{k}'\equiv\boldsymbol{q}'$:

$$\mathrm{d}\sigma=\frac{(2\pi)^{10}}{16q^*\sqrt{s}\,E_k'E_p'}\frac{1}{4}\sum|T_{\text{fi}}|^2\delta^{(1)}(W-E_k-E_p)q^{*\prime 2}\mathrm{d}q^{*\prime}\mathrm{d}\Omega,$$

where

$$W = \sqrt{M^2 + q^{*\prime 2}} + \sqrt{m^2 + q^{*\prime 2}}.$$

$q^{*\prime}$ is the modulus of the three-momentum in the final state. By the remaining δ-function it becomes equal to q^*. This leaves us with the integration over $q^{*\prime}$ or, equivalently, over W, provided we replace

$$\mathrm{d}q^{*\prime} = \frac{\mathrm{d}q^{*\prime}}{\mathrm{d}W}\mathrm{d}W = \frac{E_{p'}E_{k'}}{q^{*\prime}W}\mathrm{d}W.$$

This gives

$$\frac{\mathrm{d}\sigma}{\mathrm{d}\Omega} = \frac{1}{16s}(2\pi)^{10}\frac{1}{4}\sum |T_{\mathrm{fi}}|^2. \tag{54}$$

The invariant quantity $\mathrm{d}\sigma/\mathrm{d}t$ is obtained from this by integrating over the azimuth φ and by replacing

$$\frac{\mathrm{d}\sigma}{\mathrm{d}(\cos\theta)} = \frac{\mathrm{d}t}{\mathrm{d}(\cos\theta)}\frac{\mathrm{d}\sigma}{\mathrm{d}t} = 2q^{*2}\frac{\mathrm{d}\sigma}{\mathrm{d}t}.$$

With the expression (52) for q^{*2}, this gives

$$\frac{\mathrm{d}\sigma}{\mathrm{d}t} = \frac{(2\pi)^{11}}{8\left[(s - M^2 - m^2)^2 - 4M^2m^2\right]}\frac{1}{4}\sum |T_{\mathrm{fi}}|^2. \tag{55}$$

The next step is the construction of the T-matrix element and the calculation of the spin summation. The Feynman rules give the R-matrix R_{fi}, which is related to T_{fi} by eq. (B3) of App.B. So we have

$$R_{\mathrm{fi}} = \frac{-\mathrm{i}e^2}{(2\pi)^2}\int \mathrm{d}^4\kappa\, \overline{u(k')}\gamma_\alpha u(k)(\kappa^2 + \mathrm{i}\varepsilon)^{-1} g^{\alpha\beta}\langle p'|j_\beta(0)|p\rangle$$
$$\times (2\pi)^3\delta^{(4)}(k - \kappa - k')\delta^{(4)}(p + \kappa - p').$$

The integration over κ, the momentum of the virtual photon, yields $\kappa = k - k' = p' - p$ and leaves us with one δ-function for overall energy–momentum conservation. Thus

$$T_{\mathrm{fi}} = \frac{-e^2}{(2\pi)^3}\overline{u(k')}\gamma^\alpha u(k)\frac{1}{t}\langle p'|j_\alpha(0)|p\rangle.$$

For the actual calculation of the cross section it is useful to rewrite the nucleonic matrix element by means of the Gordon identity

$$\langle p'|j_\alpha(0)|p\rangle = \frac{1}{(2\pi)^3}\overline{u(p')}\left\{(F_1 + F_2)\gamma_\alpha - \frac{1}{2M}(p + p')_\alpha F_2\right\}u(p). \tag{46$'$}$$

The spin summations are best carried out by means of the trace techniques of App.

D. Here we have to calculate the expression

$$\mathcal{M} := \frac{1}{4}\sum_{\text{spins}}\left|\left(\overline{u(k')}\gamma^\alpha u(k)\right)\left(\overline{u(p')}\left[\gamma_\alpha(F_1+F_2)-\frac{1}{2M}P_\alpha F_2\right]u(p)\right)\right|^2$$

$$= \tfrac{1}{4}\mathrm{Sp}\{\gamma^\alpha(\not k+m)\gamma^\beta(\not k'+m)\}$$

$$\times\,\mathrm{Sp}\left\{\left[(F_1+F_2)\gamma_\alpha-\frac{P_\alpha}{2M}F_2\right](\not p+M)\left[(F_1+F_2)\gamma_\beta-\frac{P_\beta}{2M}F_2\right](\not p'+M)\right\},$$

where we have set $p+p'=P$. There are basically four expressions to be calculated, viz.

(a) $\tfrac{1}{4}\mathrm{Sp}\{\gamma^\alpha(\not k+m)\gamma^\beta(\not k'+m)\}\mathrm{Sp}\{\gamma_\alpha(\not p+M)\gamma_\beta(\not p'+M)\}$

$$=8\{2m^2M^2-M^2(kk')-m^2(pp')+(pk)(p'k')+(pk')(p'k)\},$$

which from eqs. (51), is equal to

$$=2\{2(s-M^2-m^2)^2+2st+t^2\}.$$

(b) $\tfrac{1}{4}\mathrm{Sp}\{\gamma^\alpha(\not k+m)\not P(\not k'+m)\}\mathrm{Sp}\{\gamma_\alpha(\not p+M)(\not p'+M)\}$

$$=4M\{m^2P^2+2(Pk)(Pk')-P^2(kk')\}$$

$$=8M\{(s-M^2-m^2)^2+t(s-m^2)\}.$$

(c) $\tfrac{1}{4}\mathrm{Sp}\{\not P(\not k+m)\not P(\not k'+m)\}$

$$=m^2P^2+2(Pk)(Pk')-P^2(kk')$$

$$=2\{(s-M^2-m^2)^2+t(s-m^2)\}.$$

(d) $\tfrac{1}{4}\mathrm{Sp}\{(\not p+M)(\not p'+M)\}=M^2+(pp')=2M^2-\tfrac{1}{2}t.$

All traces in $\mathcal{M}$ are reducible to these prototypes. We find

$$\mathcal{M}=4\left[(s-M^2-m^2)^2+t(s-m^2)\right]$$

$$\times\left[(F_1+F_2)^2-2F_1F_2-F_2^2-\frac{t}{4M^2}F_2^2\right]$$

$$+2(t^2+2m^2t)(F_1+F_2)^2.$$

Inserting this into eq. (55) and replacing $e^2/4\pi=\alpha$, we obtain finally

$$\frac{\mathrm{d}\sigma}{\mathrm{d}t}=\frac{4\pi\alpha^2}{t^2}\frac{(s-M^2-m^2)^2+t(s-m^2)}{(s-M^2-m^2)^2-4M^2m^2}$$

$$\times\left\{F_1^2(t)-\frac{t}{4M^2}F_2^2(t)+\frac{t(t+2m^2)}{2\left[(s-M^2-m^2)^2+t(s-m^2)\right]}\left(F_1(t)+F_2(t)\right)^2\right\}. \tag{56}$$

This formula was first derived by Rosenbluth (1950). It applies to the scattering of any charged lepton of mass m on a complex target with spin 1/2. The target structure is contained entirely in the Lorentz scalar functions $F_1(t)$ and $F_2(t)$, the significance of which will become clear below.

As an exercise let us use eq. (56) to derive the differential cross section in the laboratory system for electron scattering on the nucleon. The electron energy shall be chosen so large that the electron mass can be neglected. Setting $m=0$, the variables s and t in the laboratory system are

$$s \simeq M^2 + 2ME,$$

$$t \simeq -2EE'(1-\cos\theta) = -2E^2 \frac{1-\cos\theta}{1+(E/M)(1-\cos\theta)}.$$

From this one obtains

$$\frac{t^2}{2\left[(s-M^2)^2+ts\right]} \simeq \frac{-tE^2(1-\cos\theta)}{2M^2E^2(1+\cos\theta)} = -\frac{t}{2M^2}\mathrm{tg}^2(\theta/2),$$

$$\frac{\mathrm{d}t}{\mathrm{d}(\cos\theta)} = \frac{2E^2}{(1+(E/M)(1-\cos\theta))^2},$$

so that

$$\left(\frac{\mathrm{d}\sigma}{\mathrm{d}\Omega}\right)_{\mathrm{lab}} = \left(\frac{\alpha}{2E}\right)^2 \frac{\cos^2(\theta/2)}{\sin^4(\theta/2)} \frac{1}{1+2(E/M)\sin^2(\theta/2)}$$

$$\times\left\{F_1^2(t) - \frac{t}{4M^2}F_2^2(t) - \frac{t}{2M^2}\left[F_1(t)+F_2(t)\right]^2\mathrm{tg}^2(\theta/2)\right\} \tag{57}$$

This is the generalization of formula (35′) to a target with spin 1/2.

4.3. *Properties of form factors*

The form factors $F_1(t)$ and $F_2(t)$ are defined through the covariant decomposition (46) of the nucleonic one-particle matrix element of the electromagnetic current operator $j_\alpha(x)$. In this section we work out the physical interpretation of these form factors. For this purpose it is convenient to consider the matrix element (46) in the specific frame of reference where the sum of spatial three-momenta of initial and final nucleon states vanishes, i.e. $\boldsymbol{p}+\boldsymbol{p}'=0$. In this frame the limit of vanishing four-momentum transfer, $q^2 \to 0$, leads us automatically into the rest frame of the particle. It is in the particle's rest frame that we are able to relate F_1 and F_2 to static properties of nucleons.

(i) *Electric form factor*. From eq. (46) and (46′) the charge density (fourth component of $j_\alpha(x)$) is given by

$$\langle \boldsymbol{p}|j_0(0)|-\boldsymbol{p}\rangle = \frac{1}{(2\pi)^3}\overline{u(\boldsymbol{p})}\left\{(F_1+F_2)\gamma_0 - \frac{E_p}{M}F_2\right\}u(-\boldsymbol{p}).$$

Inserting the explicit form of the spinors (III.68) in the standard representation, one finds

$$\overline{u(\boldsymbol{p})}\,\gamma_0 u(-\boldsymbol{p}) = u^\dagger(\boldsymbol{p})u(-\boldsymbol{p}) = 2M,$$

$$\overline{u(\boldsymbol{p})}\,u(-\boldsymbol{p}) = 2E_p,$$

independently of the spin direction. Furthermore, $\boldsymbol{q} = 2\boldsymbol{p}$ and $t = (p'-p)^2 = -4\boldsymbol{p}^2 = -4E_p^2 + 4M^2$. Thus

$$(2\pi)^3\langle \boldsymbol{p}|j_0(0)|-\boldsymbol{p}\rangle = 2M(F_1+F_2) - \frac{2E_p^2}{M}F_2 = 2M\left[F_1 + \frac{t}{4M^2}F_2\right].$$

On the basis of this result it is natural to define

$$G_E(t) := F_1(t) + \frac{t}{4M^2}F_2(t) \tag{58}$$

as the *electric form factor* of the nucleon. It is easy to verify that for the proton

$$F_1^{(p)}(0) = G_E^{(p)}(0) = 1, \tag{59a}$$

which expresses the fact that the proton carries one unit of the positive elementary charge. Indeed, we know that

$$\langle p'|\int d^3x\, j_0(x)|p\rangle = 1\langle p'|p\rangle = 2E_p\delta(\boldsymbol{p}-\boldsymbol{p}').$$

On the other hand, from the decomposition (46), and using translation invariance as in eq. (42′),

$$\int d^3x\,\langle p'|j_0(x)|p\rangle = (2\pi)^3\delta(\boldsymbol{p}-\boldsymbol{p}')\langle p'|j_0(0)|p\rangle$$
$$= \delta(\boldsymbol{p}-\boldsymbol{p}')F_1^{(p)}(0)\,u^\dagger(\boldsymbol{p})u(\boldsymbol{p}) = 2E_pF_1^{(p)}(0)\delta(\boldsymbol{p}-\boldsymbol{p}').$$

Thus $F_1^{(p)}(0) = 1$.

Similarly for the neutron

$$F_1^{(n)}(0) = G_E^{(n)}(0) = 0. \tag{59b}$$

The r.m.s. radii of F_1 and G_E, which are defined by eq. (32) are not the same, however. One finds the relationship

$$\langle r^2\rangle_{G_E} = \langle r^2\rangle_{F_1} + \frac{1}{4M^2}F_2(0). \tag{60}$$

Here, $F_2(0)$ is found to be the anomalous magnetic moment below.

(ii) *Magnetic form factor*. We know from sec. III.9.2 that the magnetic properties are obtained from matrix elements of the spatial current density, i.e. from

$$\langle \boldsymbol{p}|j^i(0)|-\boldsymbol{p}\rangle = \frac{1}{(2\pi)^3}(F_1+F_2)\overline{u(\boldsymbol{p})}\,\gamma^i u(-\boldsymbol{p})$$

(where we have used eq. (46′)). With the explicit solutions (III.68) we have

$$\overline{u(\boldsymbol{p})}\,\gamma^i u(-\boldsymbol{p}) = u^\dagger(\boldsymbol{p})\begin{pmatrix} 0 & \sigma^{(i)} \\ \sigma^{(i)} & 0 \end{pmatrix} u(-\boldsymbol{p})$$

$$= \chi^\dagger\left[-\sigma^{(i)}(\boldsymbol{\sigma}\cdot\boldsymbol{p}) + (\boldsymbol{\sigma}\cdot\boldsymbol{p})\sigma^{(i)}\right]\chi$$

$$= -2\varepsilon_{ikl}p_k\chi^\dagger\sigma^{(l)}\chi = -\varepsilon_{ikl}q_k\chi^\dagger\sigma^{(l)}\chi,$$

$$\langle \boldsymbol{p}|j^i(0)|-\boldsymbol{p}\rangle = -\frac{1}{(2\pi)^3}(F_1 + F_2)\varepsilon_{ikl}q_k\chi^\dagger\sigma^{(l)}\chi.$$

Comparing this to the formulae in sec. III.9.2 we see that the combination

$$G_{\mathrm{M}}(t) \coloneqq F_1(t) + F_2(t) \tag{61}$$

may be interpreted as the *magnetic form factor*.*) In particular, $G_{\mathrm{M}}(0)$ is equal to the total magnetic moment of the particle; $F_1(0)$ gives the "normal" magnetic moment; $F_2(0)$ the "anomalous" magnetic moment. In the case of proton and neutron,

$$F_1^{(\mathrm{p})}(0) = 1, \qquad F_2^{(\mathrm{p})}(0) \equiv \mu_{\mathrm{an}}^{\mathrm{p}} = 1.7928456(11), \tag{62a}$$

$$F_1^{(\mathrm{n})}(0) = 0, \qquad F_2^{(\mathrm{n})}(0) \equiv \mu_{\mathrm{an}}^{\mathrm{n}} = -1.91304184(88). \tag{62b}$$

When rewritten in terms of the Sachs form factors (58) and (61) the differential cross section (57) reads

$$\left(\frac{\mathrm{d}\sigma}{\mathrm{d}\Omega}\right)_{\mathrm{lab}} = \left(\frac{\alpha}{2E}\right)^2 \frac{\cos^2(\theta/2)}{\sin^4(\theta/2)}\,\frac{1}{1+2(E/M)\sin^2(\theta/2)}$$

$$\times\left\{\frac{1}{1-t^2/4M^2}G_{\mathrm{E}}^2(t) - \frac{t}{4M^2}\left[\frac{1}{1-t^2/4M^2} + 2\mathrm{tg}^2(\theta/2)\right]G_{\mathrm{M}}^2(t)\right\}. \tag{57′}$$

Therefore, if one plots the quantity in curly brackets at fixed t and multiplied by $\mathrm{ctg}^2(\theta/2)$, as a function of $\mathrm{ctg}^2(\theta/2)$, the data must fall onto a straight line with slope and intercept, respectively, as follows

$$\frac{G_{\mathrm{E}}^2 + \tau G_{\mathrm{M}}^2}{1+\tau} \quad \text{and} \quad 2\tau G_{\mathrm{M}}^2$$

where $\tau = -t/4M^2$.

4.4. Isospin analysis of nucleon form factors

Isospin invariance is an approximate symmetry of strong interactions. It is a spectrum symmetry in the sense that strongly interacting particles can be classified in mass degenerate multiplets of the isospin group SU(2). While the strong interactions are invariant under isospin transformations, the electromagnetic interactions are not. That is, the strong interactions transform like scalars under isospin transformations, the electromagnetic interaction $j^\alpha A_\alpha$ does not. Nevertheless, it may be

*) $G_{\mathrm{E}}(t)$ and $G_{\mathrm{M}}(t)$ are also called Sachs form factors.

Table 1

	Isospin I	I_3	Spin Parity J^π	Charge conjugation C	G-Parity G	Analogue vector meson states
$j_\alpha^{(0)}(x)$	0	0	$\begin{Bmatrix}0^+\\1^-\end{Bmatrix}$	$-$	$-$	$\omega(783)$ $\phi(1020)$
$j_\alpha^{(1)}(x)$	1	0	$\begin{Bmatrix}0^+\\1^-\end{Bmatrix}$	$-$	$+$	$\rho(770)$

expanded in terms of multipole operators in isospin space, viz.

$$j_\alpha(x) = j_\alpha^{(0)}(x) + j_\alpha^{(1)}(x) + \cdots, \tag{63}$$

where $j_\alpha^{(0)}$ denotes an isoscalar operator, $j_\alpha^{(1)}$ denotes the third component of an isovector operator. There are good indications that the electromagnetic current operator contains only isoscalar and isovector operators, i.e. that the expansion (63) ends with the second term, but a priori this is not known. The two terms on the right hand side carry the quantum numbers of the vector mesons ω, ϕ and ρ, respectively, as summarized in Table 1.

In the case of nucleons it is easy to isolate the isoscalar and isovector parts of the nucleon form factors. Proton and neutron form a doublet of isospin, the proton is assigned $I_3 = +1/2$, the neutron $I_3 = -1/2$. Let $O_{\mu=0}^{(\kappa)}$ be any tensor operator with isospin κ and three component $\mu = 0$ in isospin, having nonvanishing matrix elements between one nucleon states. Then from the Wigner–Eckart theorem, we have

$$\langle \tfrac{1}{2} I_3 | O_\mu^{(\kappa)} | \tfrac{1}{2} I_3 \rangle = (-)^{1/2 - I_3} \begin{pmatrix} \frac{1}{2} & \kappa & \frac{1}{2} \\ -I_3 & \mu & I_3 \end{pmatrix} (\tfrac{1}{2} \| O^{(\kappa)} \| \tfrac{1}{2}), \tag{64}$$

where $(\frac{1}{2}\|O^{(\kappa)}\|\frac{1}{2})$ denotes the reduced matrix element. The first two $3j$-symbols with $\mu = 0$ are given by [R6, R7]

$$\begin{pmatrix} \frac{1}{2} & 0 & \frac{1}{2} \\ -I_3 & 0 & I_3 \end{pmatrix} = \frac{(-)^{1/2 - I_3}}{\sqrt{2}} \begin{pmatrix} \frac{1}{2} & 1 & \frac{1}{2} \\ -I_3 & 0 & I_3 \end{pmatrix} = \sqrt{1/6}\,.$$

All other such symbols for $\kappa \geq 2$ vanish because the triangle rule is not fulfilled. Thus

$$\langle \tfrac{1}{2} I_3 | \sum_\kappa O_0^{(\kappa)} | \tfrac{1}{2} I_3 \rangle = \frac{1}{\sqrt{2}} (\tfrac{1}{2} \| O^{(0)} \| \tfrac{1}{2}) + (-)^{1/2 - I_3} \sqrt{\frac{1}{6}}\, (\tfrac{1}{2} \| O^{(1)} \| \tfrac{1}{2}). \tag{65}$$

As a consequence, we see that the isoscalar may be isolated by taking the sum of matrix elements (65) over I_3, while the isovector is isolated by taking the difference. This leads to the following definitions of isoscalar and isovector nucleon form factors

$$F_i^{(s)} := \tfrac{1}{2}\left(F_i^{(p)} + F_i^{(n)}\right) \qquad (i = 1, 2). \tag{66a}$$

$$F_i^{(v)} := \tfrac{1}{2}\left(F_i^{(p)} - F_i^{(n)}\right) \tag{66b}$$

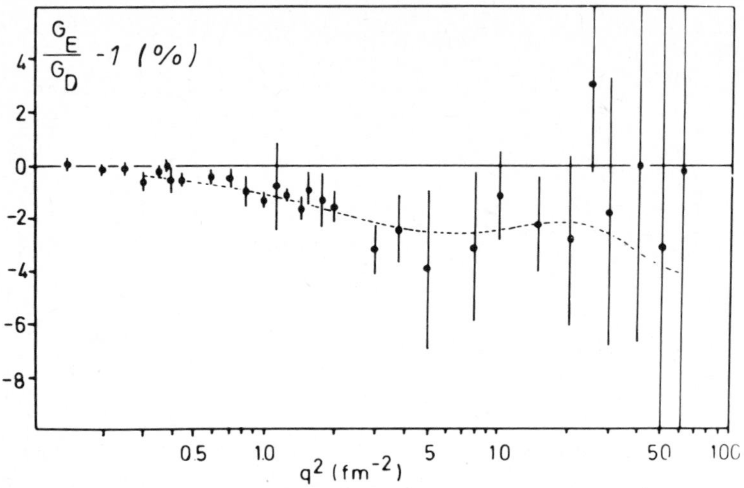

Fig. 2. The ratio G_E^P/G_D in percent deviation from 1, versus q^2. Figure taken from Simon et al. (1980).

Analogous definitions may be introduced for the electric and magnetic form factors (58) and (61).

It is customary to show the measured electric form factor G_E^P of the proton for low momentum transfers, divided by the so-called *dipole* fit:

$$G_D(q^2) = \frac{1}{\left(1 + q^2/q_0^2\right)^2} \quad \text{with} \quad q_0^2 = 18.23\ \text{fm}^{-2}. \tag{67}$$

This dipole dependence of the form factors $G_E^P(q^2)$ and $G_M^P(q^2)/\mu^P$ on q^2 is only of historical interest in so far as early data seemed to be in good agreement with this

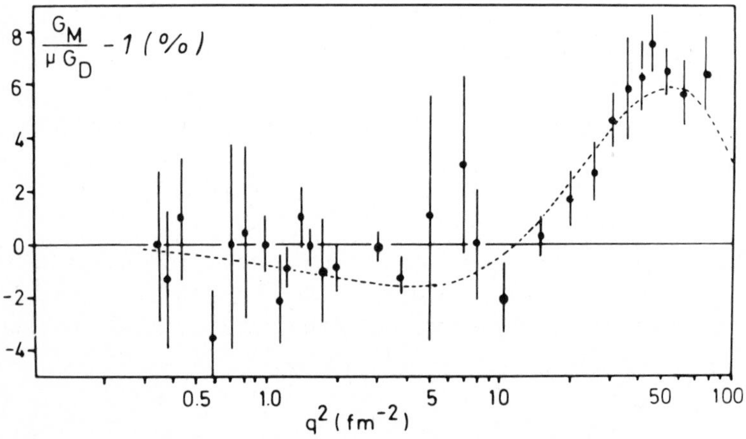

Fig. 3. The ratio $G_M^P/\mu^P G_D$ versus q^2. Figure taken from Simon et al. (1980).

simple ansatz. Better, empirical fits, useful for theoretical analyses, are found in the literature (Borkowski et al. 1976). Nevertheless, it is customary to plot the data in terms of this formula. From the data one deduces the following value for the r.m.s radius of the proton (Simon et al. 1980):

$$\langle r_{\mathrm{E}}^2 \rangle_{\mathrm{p}}^{1/2} = (0.862 \pm 0.012)\ \mathrm{fm}.$$

Figs. 2 and 3 show the electric and magnetic form factors $G_{\mathrm{E}}^{\mathrm{P}}$ and $G_{\mathrm{M}}^{\mathrm{P}}/\mu^{\mathrm{P}}$ of the proton, in units of G_{D}, eq. (67), and in the form of percent deviations from that

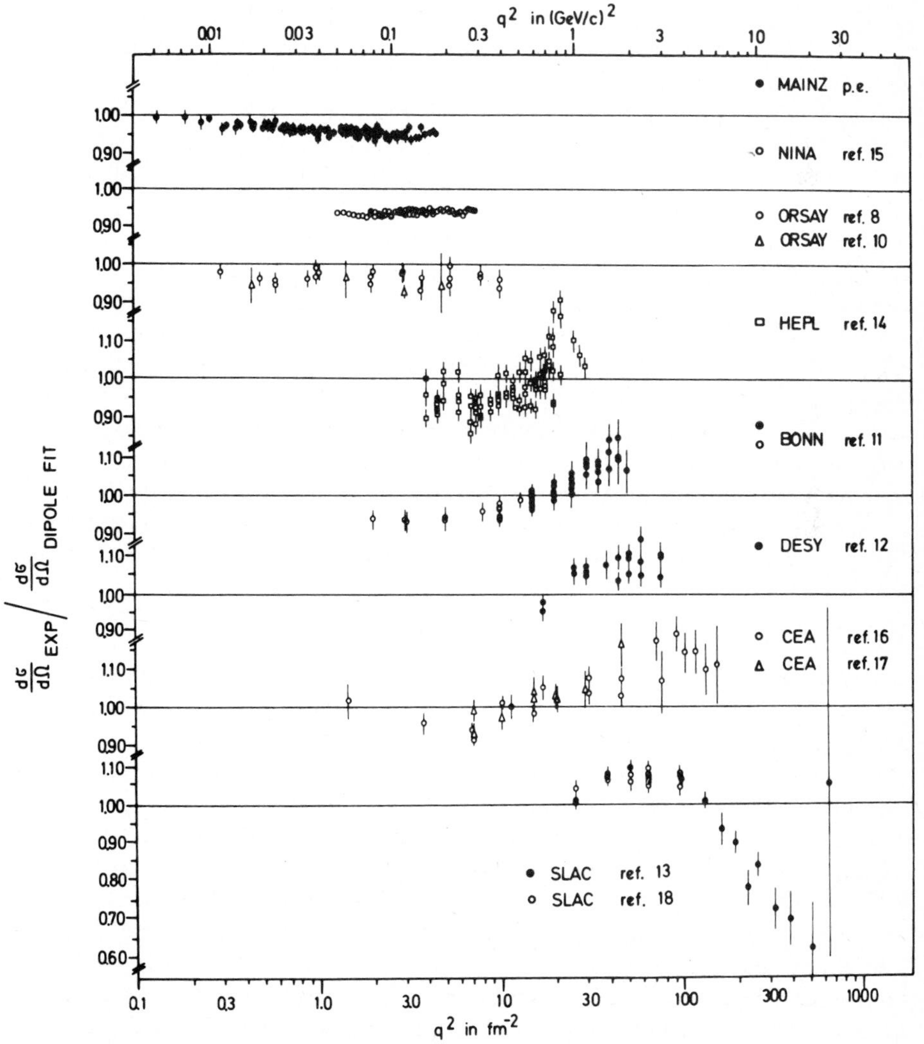

Fig. 4. Ratio of electron–proton scattering cross sections to the dipole formula, as measured in various laboratories, versus q^2. Compilation taken from Borkowski et al. (1975).

ansatz. Fig. 4, finally, shows all existing data for the electron–proton elastic cross section, again in units of the dipole fit, i.e. in units of the cross section (57′) with

$$G_E^p = G_M^p/\mu^p \equiv G_D(q^2).$$

5. Elastic and inelastic electron scattering on nuclei

In secs. 2 and 4 we have derived the cross section for elastic scattering of electrons on spin zero and spin 1/2 targets. These formulae hold for any kind of target, composite or "elementary". The internal structure of the target is hidden in the Lorentz invariant form factors whose definition is based only on Lorentz covariance and current conservation.

We now extend these results to elastic scattering on nuclei of arbitrary spin, as well as to inelastic scattering to discrete nuclear excited states. As we have seen, *elastic* scattering on a target with spin zero depends on one single form factor, the electric form factor. In case of a target with spin 1/2 there are two form factors: the electric and the magnetic dipole form factor. For a target with spin $J=1$ or higher there are, in addition, electric quadrupole form factors or more generally, form factors of multipolarity λ up to $\lambda_{max} = 2J$. This is a consequence of conservation of angular momentum which requires that nuclear initial and final state spins form a triangle with λ, the multipolarity of the form factor, viz. $\boldsymbol{J}_i + \boldsymbol{\lambda} + \boldsymbol{J}_f = 0$. In addition, conservation of parity selects the kind of multipoles that can contribute to elastic scattering, viz. electric monopole, magnetic dipole, electric quadrupole etc.

In studying inelastic scattering to discrete excited states of the nucleus we encounter a very similar situation. The initial state with spin and parity $J_i^{\pi_i}$ (this is generally the nuclear ground state), goes over into a final state with spin and parity $J_f^{\pi_f}$ through excitation by means of multipole fields with multipolarity λ and π_λ such that

$$\boldsymbol{J}_i + \boldsymbol{\lambda} + \boldsymbol{J}_f = 0, \qquad \pi_i \pi_\lambda = \pi_f.$$

Because of the close similarity of these two situations we treat them within the same formalism. Everything that follows for inelastic scattering below can always be specialized to elastic scattering by taking $J_f = J_i$ and $\pi_f = \pi_i$.

In a first step we calculate the cross sections as before, using Born approximation in describing initial and final states of the electron. This approximation is very good for the proton, $Z=1$. Depending on the accuracy of experiments one wishes to analyze, Born approximation may still be acceptable for light nuclei, up to about $Z \triangleq 10$. With increasing charge number of the nucleus it becomes less reliable; for heavy nuclei such as lead ($Z=82$) it fails badly, as it neglects the strong distortion of initial and final electron waves due to the nuclear Coulomb field. In this case a complete partial wave analysis on the basis of exact eigenstates in the nuclear electric field must be worked out. Nevertheless, it can be seen that the cross sections calculated in Born approximation contain all relevant qualitative physical features of the scattering process. The exact cross sections, obtained from partial wave analysis, differ *quantitatively* from the Born cross sections but their structure carries the same (qualitative) physical information. Thus, Coulomb distortion is a problem of purely technical nature, and we leave its discussion to a later section.

5.1. Multipole fields

The theory of electron scattering on nuclei leading to final states with definite spin and parity is based on an expansion of the virtual photon interaction in terms of multipole fields. In this subsection we collect a few results and formulae which are needed in the sequel [R14, R15]:

$$(\nabla^2 + k^2)\boldsymbol{B}(\boldsymbol{r}) = 0. \tag{68}$$

This equation follows from the wave equation for any field $\boldsymbol{B}(\boldsymbol{r}, t)$ with harmonic time dependence,

$$\boldsymbol{B}(\boldsymbol{r}, t) = \mathrm{e}^{\mathrm{i}\omega t}\boldsymbol{B}(\boldsymbol{r}).$$

$\hbar\omega$ is the energy, $\hbar k$ the momentum, k the wave number of this field, and $\hbar\omega = \hbar ck$ or, in natural units, $\omega = k$. $\boldsymbol{B}(\boldsymbol{r})$ can be expanded in terms of a complete set of solutions of the Helmholtz equation with definite angular momentum and definite parity. These basic solutions are called *multipole fields*. They form a complete and orthogonal set of vector functions in the radial variable r and on the unit sphere.

Let us begin with the definition of *vector spherical harmonics*. Let ζ_m ($m = +1, 0, -1$) be spherical unit vectors, defined in terms of Cartesian unit vectors by

$$\begin{aligned} \zeta_0 &= \boldsymbol{e}_3, \\ \zeta_1 &= -\frac{1}{\sqrt{2}}(\boldsymbol{e}_1 + \mathrm{i}\boldsymbol{e}_2), \\ \zeta_{-1} &= \frac{1}{\sqrt{2}}(\boldsymbol{e}_1 - \mathrm{i}\boldsymbol{e}_2). \end{aligned} \tag{69}$$

These vectors obey the symmetry relation

$$\zeta_m^* = (-)^m \zeta_{-m} \tag{70a}$$

and the orthogonality relation

$$\zeta_m^* \zeta_{m'} = \delta_{mm'}. \tag{70b}$$

As defined in eq. (69) the ζ_m transform as a spherical tensor of rank one under rotations. The vector spherical harmonics are then defined as follows:

$$\boldsymbol{T}_{JlM} := \sum_{m_l m_s} (lm_l, 1m_s|JM) Y_{lm_l} \zeta_{m_s}, \tag{71}$$

where $(lm_l, 1m_s|JM)$ denotes the Clebsch–Gordan coefficients that couple the angular momenta l and 1 to $J = l+1, l, l-1$. By construction, $\boldsymbol{T}_{JlM}$ transform under rotations with the unitary rotation matrices $D_{MM'}^{(J)*}(\phi, \theta, \Psi)$, i.e. they are spherical tensors of rank J. In addition they have vector character due to the fact that they contain the spherical unit vectors ζ. The m-th spherical component is given by

$$(\boldsymbol{T}_{JlM})_m \equiv (\boldsymbol{T}_{JlM}\cdot \zeta_m)$$
$$= (-)^m (l\,M+m, 1-m|JM)\,Y_{l\,M+m}.$$

The index l indicates the behaviour of the $\boldsymbol{T}_{JlM}$ under the parity operation. From their definition (71) one sees that under space reflection

$$\boldsymbol{T}_{JlM}(\pi-\theta, \varphi+\pi) = (-)^l \boldsymbol{T}_{JlM}(\theta, \varphi). \tag{72}$$

It is easy to verify the orthogonality property

$$\int \mathrm{d}\Omega \left(\boldsymbol{T}^*_{J'l'M'}\boldsymbol{T}_{JlM}\right) = \delta_{JJ'}, \delta_{ll'}, \delta_{MM'}, \tag{73}$$

which follows from the orthogonality of the ordinary spherical harmonics and of the vectors ζ, as well as from some known properties of Clebsch–Gordan coefficients. Finally, we note that the vector harmonics also form a *complete* set of vector-like functions on the unit sphere. (The completeness follows from completeness of spherical harmonics.) Some special cases are

$$\boldsymbol{T}_{10M} = \frac{1}{\sqrt{4\pi}}\zeta_M,$$

$$\boldsymbol{T}_{01M} = -\frac{1}{\sqrt{4\pi}}\frac{\boldsymbol{r}}{|\boldsymbol{r}|}.$$

Returning to the Helmholtz equation (68), we now construct solutions with definite angular momentum and definite parity. These have the form

$$\boldsymbol{B}_{JlM}(\tilde{r}) = f_l(r)\boldsymbol{T}_{JlM}(\theta, \varphi).$$

Inserting this ansatz into eq. (68) we are led to a differential equation for the function $f_l(r)$ alone which reads

$$\left\{\frac{1}{r^2}\frac{\mathrm{d}}{\mathrm{d}r}\left(r^2\frac{\mathrm{d}}{\mathrm{d}r}\right) + k^2 - \frac{l(l+1)}{r^2}\right\} f_l(r) = 0,$$

or with $z = kr$,

$$\left\{\frac{\mathrm{d}^2}{\mathrm{d}z^2} + \frac{2}{z}\frac{\mathrm{d}}{\mathrm{d}z} + 1 - \frac{l(l+1)}{z^2}\right\} f_l(z) = 0. \tag{74}$$

This equation is well-known in the theory of Bessel functions. We choose the following fundamental system of solutions,

$$f_l^{\mathrm{I}}(r) = j_l(kr) \quad \text{spherical Bessel function,}$$

$$f_l^{\mathrm{II}}(r) = h_l^{(1)}(kr) \quad \text{spherical Hankel function of first kind}$$
$$= j_l(kr) + \mathrm{i}n_l(kr) \tag{75}$$

[$n_l(kr)$ is a spherical Neumann function]. This specific set is chosen in view of the physical boundary conditions that we require for the multipole fields: f_l^{I} is the solution regular at the origin $r=0$, whilst f_l^{II} describes asymptotically outgoing spherical waves,*)

$$h_l^{(1)}(x) \underset{x\to\infty}{\sim} \frac{1}{x}\exp\left\{i(x-(l+1))\frac{\pi}{2}\right\}. \tag{76}$$

Finally, the functions f_l satisfy the completeness relation

$$\int_0^\infty f_l^*(k'r)f_l(kr)r^2\,\mathrm{d}r = \frac{\pi}{2k^2}\delta(k-k'). \tag{77}$$

Equipped with this knowledge we can now define a set of *multipole fields*

(i) *magnetic multipole fields*

$$\boldsymbol{A}_{lm}(\mathrm{M}) := f_l(kr)\boldsymbol{T}_{J=llm}; \tag{78}$$

(ii) *electric multipole fields*

$$\boldsymbol{A}_{lm}(\mathrm{E}) := -\sqrt{\frac{l}{2l+1}}\,f_{l+1}(kr)\boldsymbol{T}_{ll+1m} + \sqrt{\frac{l+1}{2l+1}}\,f_{l-1}(kr)\boldsymbol{T}_{ll-1m}; \tag{79}$$

(iii) *longitudinal multipole fields*

$$\boldsymbol{A}_{lm}(\mathrm{L}) := \sqrt{\frac{l+1}{2l+1}}\,f_{l+1}(kr)\boldsymbol{T}_{ll+1m} + \sqrt{\frac{l}{2l+1}}\,f_{l-1}(kr)\boldsymbol{T}_{ll-1m}. \tag{80}$$

The properties of these fields, as well as the reason for the nomenclature, are discussed extensively in the literature on multipole fields [R14, R15]. In particular, differential properties of them can best be discussed by means of the techniques of angular momentum algebra for which we refer to standard monographs [R4, R6, R16]. Especially useful is the "gradient formula" which reads

$$\begin{aligned}\nabla f(r)Y_{lm} = {} & -\sqrt{\frac{l+1}{2l+1}}\left(\frac{\mathrm{d}f}{\mathrm{d}r} - l\frac{f}{r}\right)\boldsymbol{T}_{ll+1m} \\ & + \sqrt{\frac{l}{2l+1}}\left(\frac{\mathrm{d}f}{\mathrm{d}r} + \frac{l+1}{r}f\right)\boldsymbol{T}_{ll-1m}.\end{aligned} \tag{81}$$

By means of these techniques one shows that *magnetic* and *electric* multipole fields are divergenceless

$$\nabla\cdot\boldsymbol{A}_{lm}(\tau) = 0, \quad \tau = \mathrm{E},\mathrm{M}. \tag{82}$$

Both potentials vanish for $l = 0$. This is evident in the case of eq. (79); in the case of eq. (78) it follows from eq. (71) since $(00,1\mathrm{m}|00)$ vanishes. Thus, electric and magnetic multipole fields are transverse, i.e. fulfill eq. (82), and may be used to describe photon states of definite angular momentum and parity. According to eq. (72) their parities are $(-)^l$ and $(-)^{l+1}$, respectively. However, as the interaction with matter involves always the product of such vector fields with the current operator which is itself odd under parity, the rules for parity change in electromag-

*)These and further properties of spherical Bessel and Hankel functions can be found in Ref. R1 and in other monographs on special functions.

netic transitions are

$$(-)^{l+1} \text{ in magnetic } (\mathrm{M}l) \text{ transitions,}$$

$$(-)^{l} \text{ in electric } (\mathrm{E}l) \text{ transitions.}$$

The longitudinal fields (80) are not divergenceless (hence their name), and they exist also for $l=0$. Thus, while there are no transverse monopole fields, a longitudinal field can carry total angular momentum zero.

On the basis of the orthogonality and completeness relations (73) and (77) one shows easily that the multipole fields fulfill the orthogonality relations

$$\int_0^{\infty} r^2 \mathrm{d}r \int \mathrm{d}\Omega\, \boldsymbol{A}^*_{l'm'}(\boldsymbol{r};\tau')\boldsymbol{A}_{lm}(\boldsymbol{r};\tau) = \frac{\pi}{2k^2}\delta(k-k')\delta_{\tau\tau'}\delta_{ll'}\delta_{mm'}. \tag{83}$$

The importance of these results lies in the fact that a given vector field $\boldsymbol{F}(\boldsymbol{r})$ which is sufficiently regular, can be expanded in terms of orthogonal multipole fields whose parity and angular momentum properties are simple. Whenever matrix elements of $\boldsymbol{F}$ between states of definite angular momentum and parity are to be calculated, only one or a few terms of the multipole expansion give nonvanishing contributions.

Finally, using the techniques of angular momentum algebra, one can show that the multipole fields can also be written in terms of the orbital angular momentum operator applied to spherical harmonics, in terms of curl and gradient of such functions. This provides equivalent representations that are sometimes useful.

$$\boldsymbol{A}_{lm}(\mathrm{M}) = f_l(kr)\frac{1}{\sqrt{l(l+1)}}\boldsymbol{l}Y_{lm}, \tag{78'}$$

$$\begin{aligned}\boldsymbol{A}_{lm}(\mathrm{E}) &= -\frac{i}{k}\nabla\wedge\left(f_l(kr)\boldsymbol{T}_{llm}\right)\\ &= -\frac{i}{k\sqrt{l(l+1)}}\nabla\wedge\boldsymbol{l}\left(f_l(kr)Y_{lm}\right),\end{aligned} \tag{79'}$$

$$\boldsymbol{A}_{lm}(\mathrm{L}) = \frac{1}{k}\nabla\left(f_l(kr)Y_{lm}\right). \tag{80'}$$

The notation (78)–(80) is more useful in calculating matrix elements between states of good angular momentum and parity because one can then make use of the Wigner–Eckart theorem and all the tricks of angular momentum algebra. The representation (78′)–(80′), on the other hand, is very useful if one wants to use identities of vector calculus in order to transform interaction terms to a more convenient form.

5.2. Theory of electron scattering

There are several ways of deriving the Hamiltonian that describes the interaction of (arbitrarily relativistic) electrons with a static target.

(i) One may analyze the scattering process in a semi-classical treatment, starting from a retarded interaction between two given charge and current densities.

$$H_{\text{ret}} = \int d^3 r_n \int d^3 r_e \frac{e^{ik|\boldsymbol{r}_n - \boldsymbol{r}_e|}}{|\boldsymbol{r}_n - \boldsymbol{r}_e|} \{ \rho_n(\boldsymbol{r}_n)\rho_e(\boldsymbol{r}_e) - \boldsymbol{j}_n(\boldsymbol{r}_n)\boldsymbol{j}_e(\boldsymbol{r}_e)\}. \tag{84}$$

One then expands the retarded Green function that appears in eq. (84) in terms of transverse and longitudinal multipole fields. The procedure is conceptually simple but technically somewhat involved and we refer to the literature for details [R20] (Scheck 1966).

(ii) Alternatively, one can calculate the interaction in the framework of quantum electrodynamics, formulated in the Coulomb gauge. The interaction is then given by the *instantaneous* electrostatic Coulomb interaction plus the terms arising from the exchange of virtual but still transverse photons between the electron and the target [R17]. In this case, instead of using plane waves it is appropriate to expand and quantize the transverse photon field in terms of the magnetic and electric multipole fields (78), (79).

In either case the scattering matrix element is found to be

$$\begin{aligned}
&\langle f; k' | H_{\text{int}} | i; k \rangle \\
&= \langle f, k' | \sum_{l=1}^{\infty} \sum_{m=-l}^{+l} \frac{4\pi i}{l(l+1)} \int d^3 r_n \int d^3 r_e \\
&\times \left\{ \frac{1}{k} \left[\boldsymbol{j}_n \cdot \nabla \wedge \boldsymbol{l} \begin{pmatrix} j_l(kr_<) \\ h_l^{(1)}(kr_>) \end{pmatrix} Y_{lm}^*(\hat{\boldsymbol{r}}_n) \right] \left[\boldsymbol{j}_e \cdot \nabla \wedge \boldsymbol{l} \begin{pmatrix} h_l^{(1)}(kr_>) \\ j_l(kr_<) \end{pmatrix} Y_{lm}(\hat{\boldsymbol{r}}_e) \right] \right. \\
&\left. + k \left[\boldsymbol{j}_n \cdot \boldsymbol{l} \begin{pmatrix} j_l(kr_<) \\ h_l^{(1)}(kr_>) \end{pmatrix} Y_{lm}^*(\hat{\boldsymbol{r}}_n) \right] \left[\boldsymbol{j}_e \cdot \boldsymbol{l} \begin{pmatrix} h_l^{(1)}(kr_>) \\ j_l(kr_<) \end{pmatrix} Y_{lm}(\hat{\boldsymbol{r}}_e) \right] \right\} \\
&+ \int d^3 r_n \int d^3 r_e \frac{1}{r_>} \rho_n(\boldsymbol{r}_n) \rho_e(\boldsymbol{r}_e) | i, k \rangle.
\end{aligned} \tag{85}$$

In this expression i and f denote initial and final state of the target, respectively; for example, i stands for initial total momentum p of the target and for all other target quantum numbers such as angular momentum, parity and any other internal quantum numbers as there may be. The notation $r_<$ and $r_>$ serves as a short-hand for the requirement that

if $r_e > r_n$, the combination $j_l(kr_n)h_l^{(1)}(kr_e)$, and

if $r_e < r_n$, the combination $h_l^{(1)}(kr_n)j_l(kr_e)$

must be taken in the first two terms on the r.h.s. of eq. (85). This reflects the correct boundary conditions which are built into eq. (85): It is always the smaller of the two radial variables that is to be inserted into j_l, the function regular at the origin, whilst the larger of the two is to be taken in $h_l^{(1)}$, the function that describes outgoing spherical waves.

Evidently, if we wish to explore the internal structure of the target, we have to choose the momentum transfer and, therefore, the electron energy high enough so that the electron penetrates sizeably into the target. In this case the integrations in eq. (85) are entangled in a nontrivial way. The matrix element does not factor in a target structure function and a leptonic factor. This is unlike Coulomb excitation where penetration is unimportant (and, in fact, often unwanted) and where the lowest order cross section does factor in target and projectile properties.

The expression (85) is fairly general. It applies equally well to elastic and inelastic scattering. It holds independently of what basis we choose for the initial and final electron states. If we take plane waves we shall obtain the transition matrix element in Born approximation. In this case k is the magnitude of the three-momentum in the center-of-mass system. If we wish to include the Coulomb distortion in the electronic states, we have to evaluate eq. (85) with eigenstates of the electron in the static Coulomb field created by the target. In this case k is the asymptotic wave number determined by the electron energy.

The charge and current densities which appear in eq. (85) are still very general. For the electron, we clearly have to set

$$\rho_e(\boldsymbol{r}) = -e\gamma^0\delta(\boldsymbol{r}-\boldsymbol{r}_e) \equiv -e\beta\delta(\boldsymbol{r}-\boldsymbol{r}_e), \tag{86a}$$

$$\boldsymbol{j}_e(\boldsymbol{r}) = -e\gamma^0\boldsymbol{\gamma}\delta(\boldsymbol{r}-\boldsymbol{r}_e) \equiv -e\boldsymbol{\alpha}\delta(\boldsymbol{r}-\boldsymbol{r}_e). \tag{86b}$$

Concerning the target, however, there are many options. For example, let the target be a nucleus treated in a nonrelativistic scheme. If we know the nuclear wave function in terms of states of individual nucleons then

$$\rho_n(\boldsymbol{r}) = \sum_{i=1}^{A} e_i\delta(\boldsymbol{r}-\boldsymbol{r}_i), \tag{87a}$$

$$\boldsymbol{j}_n(\boldsymbol{r}) = \sum_{i=1}^{A}\left[e_i\boldsymbol{v}_i\delta(\boldsymbol{r}-\boldsymbol{r}_i) + \frac{e}{2m}g_s^i(\nabla\wedge\boldsymbol{s}^i)\delta(\boldsymbol{r}-\boldsymbol{r}_i)\right]. \tag{87b}$$

These operators are then to be taken between nonrelativistic nuclear states (e.g. shell model states). If the nucleus is to be described by some set of effective collective coordinates (such as vibrator coordinates in case of collective matter oscillations, or the set of Euler angles in case of rigid rotator motion etc.) then the densities are semi-classical functions containing the collective coordinates in a way determined by the underlying model.

In the following two sections we present the essential steps of the derivation of cross sections in Born approximation (sec. 5.3) and including Coulomb distortion (sec. 5.4).

5.3. *Born approximation*

Before proceeding to the calculation of the matrix element (85) in first Born approximation it is useful to transform the interaction to a more convenient form. The electric part of the interaction (first term on the right-hand side of eq. (85)) is transformed by means of a well-known relation for spherical Bessel and Hankel functions [R18],

$$j_l(kr_<)h_l^{(1)}(kr_>)=\frac{2k}{\mathrm{i}\pi}\int_0^\infty \mathrm{d}q\,\frac{j_l(qr)\,j_l(qr')}{q^2-k^2}+\frac{1}{\mathrm{i}k}\,\frac{r_<^l r_>^{-l-1}}{2l+1}. \tag{88}$$

Likewise, the magnetic term (second term of eq. (85)) is transformed by means of the relation [R18]

$$j_l(kr_<)h_l^{(1)}(kr_>)=\frac{2}{\mathrm{i}\pi k}\int_0^\infty \frac{j_l(qr)\,j_l(qr')}{q^2-k^2}\,q^2\,\mathrm{d}q. \tag{89}$$

Finally, the electric term that comes from the second term on the right-hand side of eq. (88), i.e.

$$\langle \mathrm{f};k'|\int \mathrm{d}^3r_\mathrm{n}\int \mathrm{d}^3r_\mathrm{e}\,\boldsymbol{j}_\mathrm{n}\cdot\nabla\wedge\boldsymbol{l}\begin{pmatrix} r_<^l & \\ & Y_{lm}^* \\ r_>^{-l-1} & \end{pmatrix}\boldsymbol{j}_\mathrm{e}\cdot\nabla\wedge\boldsymbol{l}\begin{pmatrix} r_>^{-l-1} & \\ & Y_{lm} \\ r_<^l & \end{pmatrix}|\mathrm{i};k\rangle$$

can be further transformed by means of the relation [R16]

$$\nabla\wedge\boldsymbol{l}(r^\alpha Y_{lm})=\mathrm{i}(l+1)\nabla(r^\alpha Y_{lm}),\quad \alpha=l,-l-1. \tag{90}$$

Partial integration allows to shift the nabla operators onto the current densities so that, eventually, the continuity equations

$$\langle \mathrm{f}|\nabla\cdot\boldsymbol{j}_\mathrm{n}|\mathrm{i}\rangle=-\mathrm{i}k\langle \mathrm{f}|\rho_\mathrm{n}|\mathrm{i}\rangle,$$

$$\langle k'|\nabla\cdot\boldsymbol{j}_\mathrm{e}|k\rangle=+\mathrm{i}k\langle k'|\rho_\mathrm{e}|k\rangle$$

may be used. In this manner we obtain the equivalent expression for the scattering matrix element (85):

$$\begin{aligned}
&\langle \mathrm{f};k'|H_\mathrm{int}|\mathrm{i};k\rangle\\
&\quad=\langle \mathrm{f};k'|\sum_{l=1}^{\infty}\sum_{m=-l}^{+l}\frac{8}{l(l+1)}\int \mathrm{d}^3r_\mathrm{n}\int \mathrm{d}^3r_\mathrm{e}\int_0^\infty \frac{\mathrm{d}q'}{q'^2-k^2}\\
&\qquad\times\{\boldsymbol{j}_\mathrm{n}\cdot\nabla\wedge\boldsymbol{l}(j_l(q'r_\mathrm{n})Y_{lm}^*(\hat{\boldsymbol{r}}_\mathrm{n}))\boldsymbol{j}_\mathrm{e}\cdot\nabla\wedge\boldsymbol{l}(j_l(q'r_\mathrm{e})Y_{lm}(\hat{\boldsymbol{r}}_\mathrm{e}))\\
&\qquad\quad+q'^2\boldsymbol{j}_\mathrm{n}\cdot\boldsymbol{l}(j_l(q'r_\mathrm{n})Y_{lm}^*(\hat{\boldsymbol{r}}_\mathrm{n}))\boldsymbol{j}_\mathrm{e}\cdot\boldsymbol{l}(j_l(q'r_\mathrm{e})Y_{lm}(\hat{\boldsymbol{r}}_\mathrm{e}))\}\\
&\qquad+\sum_{l=0}^{\infty}\sum_{m=-l}^{+l}\frac{4\pi}{2l+1}\int \mathrm{d}^3r_\mathrm{n}\int \mathrm{d}^3r_\mathrm{e}\frac{r_<^l}{r_>^{l+1}}Y_{lm}^*(\hat{\boldsymbol{r}}_\mathrm{n})Y_{lm}(\hat{\boldsymbol{r}}_\mathrm{e})\rho_\mathrm{n}(\boldsymbol{r}_\mathrm{n})\rho_\mathrm{e}(\boldsymbol{r}_\mathrm{e})|\mathrm{i};k\rangle.
\end{aligned} \tag{91}$$

In the last term of eq. (91) the sum over l runs from zero to infinity; the terms from one to infinity stem from the transformed electric term (see above), whilst the term with $l=0$ is the monopole term of eq. (85). (Note that $Y_{00}=1/\sqrt{4\pi}$.) This last term, summed over *all l*, is nothing but the instantaneous Coulomb interaction

$$\int d^3r_n \int d^3r_e \frac{1}{|\boldsymbol{r}_n - \boldsymbol{r}_e|} \rho_n(\boldsymbol{r}_n)\rho_e(\boldsymbol{r}_e) \tag{92}$$

expanded in a multipole series. Therefore, the first two terms on the right-hand side of eq. (91) must represent the contributions stemming from the exchange of transverse virtual photons with momentum q'.

To calculate (91) in Born approximation means to take

$$|k\rangle = u(k)e^{i\boldsymbol{k}\cdot\boldsymbol{r}_e} \quad \text{and} \quad |k'\rangle = u(k')e^{i\boldsymbol{k}'\cdot\boldsymbol{r}_e}.$$

Inserting this into eq. (91) one can then perform the integration over the electron coordinates $\boldsymbol{r}_e$ and over the (virtual photon) momentum q. These calculations are somewhat lengthy and tedious. We therefore give only one example that shows the technique but do not work out all the details. Take for example the instantaneous interaction term (92) and use the relation (cf. exercise 4)

$$\frac{1}{|\boldsymbol{r}_n - \boldsymbol{r}_e|} = \frac{4\pi}{(2\pi)^3} \int d^3q' \frac{e^{i\boldsymbol{q}'\cdot(\boldsymbol{r}_n - \boldsymbol{r}_e)}}{\boldsymbol{q}'^2}. \tag{93}$$

Integration over $\boldsymbol{r}_e$ gives, together with (86a),

$$\begin{aligned}
&\langle f; k'| \int d^3r_n \int d^3r_e \frac{1}{|\boldsymbol{r}_n - \boldsymbol{r}_e|} \rho_n(\boldsymbol{r}_n)\rho_e(\boldsymbol{r}_e)|i; k\rangle \\
&\quad = -4\pi e u^\dagger(k')u(k) \sum_{l,m} (-)^m i^l \frac{4\pi}{q^2} Y_{lm}(\hat{\boldsymbol{q}}) \\
&\qquad \times \int d^3r_n \langle f|\rho_n(\boldsymbol{r}_n)|i\rangle j_l(qr_n) Y_{l-m}(\hat{\boldsymbol{r}}_n),
\end{aligned}$$

where $\boldsymbol{q}$ is the three-momentum transfer, $\boldsymbol{q} = \boldsymbol{k} - \boldsymbol{k}'$ and $q = |\boldsymbol{q}|$. This last equation suggests to define a multipole form factor of the target

$$M(\mathrm{C}l, m; q) := \frac{(2l+1)!!}{q^l} \int d^3r \rho_n(\boldsymbol{r}) j_l(qr) Y_{lm}(\hat{\boldsymbol{r}}). \tag{94}$$

The factor in front of the integral has been chosen such that in the limit $q \to 0$, $M(\mathrm{C}l, m; q)$ goes over into the static l-pole moment of the target charge density.

The transverse electric and magnetic terms in eq. (91) are worked out in a similar way. In analogy to eq. (94) we are led to define

electric multipole form factors

$$M(\mathrm{E}l, m; q) := \frac{(2l+1)!!}{q^{l+1}(l+1)} \int d^3r \boldsymbol{j}_n(\boldsymbol{r}) \cdot \nabla \wedge \boldsymbol{l}(j_l(qr)Y_{lm}(\hat{\boldsymbol{r}})), \tag{95}$$

and *magnetic multipole form factors*

$$M(\mathrm{M}l, m; q) := -i\frac{(2l+1)!!}{q^l(l+1)} \int d^3r \boldsymbol{j}_n(\boldsymbol{r}) \cdot \boldsymbol{l}(j_l(qr)Y_{lm}(\hat{\boldsymbol{r}})). \tag{96}$$

Here again, the factors have been chosen so that for $qr \ll 1$ the form factors go over into the corresponding transverse electric and magnetic multipole terms which describe the corresponding transition induced by photons [R7] (in the approximation of long wave lengths).

We are concerned here only with the scattering cross section for unpolarized electrons and we do not discriminate the spin orientation of the electron in the final state. Therefore, we have to calculate the incoherent sum of squared matrix elements over all initial and final spin projections and have to divide by $2(2J+1)$, J being the nuclear spin, in order to account for the average over spin orientations in the initial state. The spin summation in the electronic part is done by means of the trace techniques (see Appendix). This is a straightforward but somewhat lengthy calculation that we do not wish to develop here as the details can be found in the literature. The result is, in the unpolarized case [R19] (de Forest et al. 1966),

$$\frac{d\sigma}{d\Omega} = \sum_{l=0}^{\infty} \frac{d\sigma_{El}}{d\Omega} + \sum_{l=1}^{\infty} \frac{d\sigma_{Ml}}{d\Omega}, \tag{97a}$$

where

$$\frac{d\sigma_{El}}{d\Omega} = \alpha^2 \frac{4\pi(l+1)}{l[(2l+1)!!]^2} \frac{q^{2l}}{E^2} \left\{ \frac{l}{l+1} B(Cl;q) V_L(\theta) + B(El;q) V_T(\theta) \right\}, \tag{97b}$$

$$\frac{d\sigma_{Ml}}{d\Omega} = \alpha^2 \frac{4\pi(l+1)}{l[(2l+1)!!]^2} \frac{q^{2l}}{E^2} B(Ml;q) V_T(\theta). \tag{97c}$$

Here the functions $V_L(\theta)$ and $V_T(\theta)$ stem from the summation over electron spins. In the high-energy limit ($E \gg m_e$) they are

$$V_L(\theta) \simeq \frac{\cos^2(\theta/2)}{4\sin^4(\theta/2)}, \qquad V_T(\theta) \simeq \frac{1+\sin^2(\theta/2)}{8\sin^4(\theta/2)}. \tag{98}$$

The functions $B(\tau l; q)$ are the spin-averaged, squared nuclear matrix elements of the operators (94)–(96), viz.

$$B(\tau l; q) = \sum_{M_f, m} \left| \langle J_f M_f | M(\tau l, m; q) | J_i M_i \rangle \right|^2, \tag{99}$$

with $\tau \equiv$ C (longitudinal Coulomb multipoles), $\tau \equiv$ E (electric multipoles), or $\tau \equiv$ M (magnetic multipoles). As before, q is the magnitude of the three-momentum transfer, whilst E is the initial electron energy. Strictly speaking, the expressions (97) hold in the center-of-mass frame. However, they may equally well be applied in the laboratory system provided the typical recoil terms of order of E/M, with M the target mass, can be neglected.

We note that there are no interference terms in the cross sections (97). The magnetic multipoles do not interfere with the electric and longitudinal multipoles because they have different parity selection rules. The interference terms between electric and longitudinal multipoles disappear when the spin average is taken. Eqs. (97b) and (97c) which hold for *elastic* as well as for *inelastic* scattering, demonstrate quite clearly the important new feature of electron scattering on an extended target: The scattering cross section depends on the momentum transfer q which can be chosen arbitrarily large. Thus, looking back at eqs. (94) to (96), the electron probes the *spatial structure* of charge and current densities within the target. In the case of elastic scattering, these are the charge, the electric and magnetic current densities of the ground state. In the case of inelastic scattering, we are probing off-diagonal matrix elements of these operators between the (initial) ground state and some (final) excited states of the target. Because of the close analogy to the elastic case one often calls these matrix elements *transition charge* and *transition current densities*.

The *selection rules* that apply to elastic and inelastic electron scattering derive from angular momentum conservation and from the behaviour of the transition operators and of the nuclear states under parity and time reversal.

Let

$$(J_{\mathrm{i}}, \pi_{\mathrm{i}}) \quad \text{and} \quad (J_{\mathrm{f}}, \pi_{\mathrm{f}})$$

be the spins and parities of initial and final nuclear states, respectively, and let us consider a given multipole operator $M(\tau l)$, eqs. (94)–(96), taken between these states. Angular momentum conservation implies that $(J_{\mathrm{i}}, J_{\mathrm{f}}, l)$ form a triangle $|J_{\mathrm{i}} - J_{\mathrm{f}}| \leq l \leq J_{\mathrm{i}} + J_{\mathrm{f}}$ with $l \geq 1$ for transverse electric and magnetic multipoles. Parity conservation implies that $\pi_{\mathrm{i}} \cdot \pi_{\mathrm{f}} = (-)^l$ for longitudinal and transverse electric multipoles, $\pi_{\mathrm{i}} \cdot \pi_{\mathrm{f}} = (-)^{l+1}$ for transverse magnetic multipoles. Finally, hermiticity of the electromagnetic current and invariance under time reversal give the additional relation for the nuclear reduced matrix element

$$(J_{\mathrm{i}} \| M(\tau l) \| J_{\mathrm{f}}) = (-)^{J_{\mathrm{i}} - J_{\mathrm{f}} + l + \eta} (J_{\mathrm{f}} \| M(\tau l) \| J_{\mathrm{i}}), \tag{100}$$

with $\eta = 0$ for $\tau = \mathrm{C}$; $\eta = 1$ for $\tau = \mathrm{E}$ or M (Donnelly et al. 1975). If the nuclear states are also eigenstates of isospin with eigenvalues I_{i}, I_{f}, respectively, there is an additional phase factor $(-)^{I_{\mathrm{i}} - I_{\mathrm{f}}}$ in eq. (100). In this case the reduced matrix element implies reduction with respect to both angular momentum and isospin (Donnelly et al. 1975).

For elastic scattering, in particular, J_{i} and J_{f} are identical. From eq. (100) we then must have $(-)^{l+\eta} = +1$, $l + \eta$ must be even. Thus, only *even longitudinal* and only *odd transverse magnetic multipoles* contribute to elastic scattering.

As an illustration let us consider some examples:

(i) Elastic scattering on $^3\mathrm{He}(J^\pi = \frac{1}{2}^+)$ and $^{209}\mathrm{Bi}(J^\pi = \frac{9}{2}^-)$. Only the following multipoles give nonvanishing contributions

^{3}He: C0, M1,

^{209}Bi: C0, C2, C4, C6, C8; M1, M3, M5, M7, M9.

(ii) Inelastic scattering from the ground state to the electric dipole giant resonance in $^{16}\mathrm{O}(J_{\mathrm{i}}^{\pi_{\mathrm{i}}} = 0^+ \rightarrow J_{\mathrm{f}}^{\pi_{\mathrm{f}}} = 1^-)$. In this case only the multipoles E1 and C1 can contribute.

These selection rules for El and Ml transitions are the same as for the corresponding photonic electric and magnetic multipole transitions. (There are no Cl transitions in the case of real photons.) There is, however, one essential difference between photo- and electroexcitation: In the case of photonic processes, the momentum transfer q is replaced by

$$k = E_f - E_i \triangleq (E_f - E_i)/\hbar c,$$

the photon momentum (or energy). In most practical cases $k \cdot r$ is smaller than 1, for r of the order of, or smaller than, the typical size of the nucleus. Thus, as a consequence of the behaviour of the spherical Bessel function for small argument (see eq. (12)), a transition of high multipolarity l_H is suppressed relative to a transition of low multiplicity l_L by a typical factor

$$\frac{(kr)^{l_H}}{(2l_H+1)!!} \Big/ \frac{(kr)^{l_L}}{(2l_L+1)!!}. \tag{101}$$

For example, an E3 γ-transition amplitude is suppressed relative to an E1 transition by a factor of the order of $\frac{1}{3}(kr)^2$. Thus, the lowest multipolarity which is compatible with the selection rules will also be the dominant one.

No such ordering of successive multipoles occurs in electron scattering. Indeed, the modulus of the momentum transfer q can become arbitrarily large and the quantity $q \cdot r$ can assume any value, greater or smaller than one. So, in general, high multipoles can be equally important as low multipoles. In the limit of small momentum transfer only, $q \to 0$, we recover the ordering of photonic multipole transitions. Actually, apart from the electronic kinematic factors, El and Ml transition probabilities in electron scattering must go over, in the limit $q \to 0$, into the corresponding γ-transition probabilities. This is called the "photon point."

Examples for the use of Born approximation in the description of electron scattering as well as an analysis of the information carried by elastic and inelastic form factors are postponed to secs. 5.5 and 5.6 till we have completed the discussion of Coulomb distortion, so that we can compare the two methods of analysis, at the same time. For the moment, it may suffice to stress that, whilst a γ-transition gives us just one *moment* of the transition charge density or current density, the matrix elements of the multipole operators (94)–(96) which are relevant for the cross sections for electron scattering (97) yield continuous information on these quantities. As the momentum transfer is varied, these matrix elements probe the spatial structure of nuclear charge and current density. If it were possible to measure the cross sections up to very large momentum transfers we would eventually obtain a complete mapping of $\rho_n(\boldsymbol{r})$ and $\boldsymbol{j}_n(\boldsymbol{r})$.

In practice, this is not possible, however. The elastic cross section to any specific excited state falls off faster than q^{-4} with increasing momentum transfer and, at some point, becomes unmeasurably small. Furthermore, as the energy transfer and the momentum transfer increase (into what is called the "deep inelastic" region), the number of possible final states increases so much that one may not be able to follow up one particular excitation. One then rather measures fully *inclusive scattering* (i.e. summing over all final states), or *semi-inclusive scattering* (where some property of the final state is recorded). The cross section for inclusive scattering quickly starts to dominate over all exclusive reaction channels.

In summary, only limited information on the spatial structure of the target densities is obtained in practice. At low momentum transfer, one starts probing the nuclear periphery, i.e. the charge density in the neighbourhood of the nuclear radius (distance at which the density has dropped to about half its central value). As the momentum transfer increases, more and more information on the nuclear interior appears in the elastic scattering cross sections. The densities at the origin remain always the least well known. Examples of practical analysis are given below.

5.4. The problem of Coulomb distortion

Born approximation in the calculation of electron scattering has the great advantage of being simple and transparent. The form factors are the Fourier transforms of the nuclear matrix elements of charge and current densities and thus provide us with a direct mapping of these important quantities. On the other hand, Born approximation also has serious deficiencies: It does not take into account the distortion of the electron waves in the static Coulomb field of the nucleus. Depending on the accuracy of the available data, the neglect of Coulomb distortion may be tolerable for light nuclei, $Z = 1$ to ~ 10. However, for larger values of the nuclear charge these effects quickly become large and must be taken into account. What are the most prominent effects of Coulomb distortion?

(i) It is not difficult to see that a form factor can have zeroes at physical values of the momentum transfer q. These diffraction zeroes which reflect specific properties of the charge and current densities in coordinate space will also appear in the cross section.*) The cross section will then exhibit a typical diffraction pattern. The diffraction zeroes, strictly speaking, are not realistic and do not appear in the exact expression for the cross section. This is easy to understand in a qualitative manner. Suppose we describe the scattering amplitude in terms of partial waves. Partial waves carry definite angular momenta. Classically, the angular momentum, with respect to the center of the nucleus, is proportional to the impact parameter b times the momentum transfer q. Thus low partial waves penetrate into the nucleus, high partial waves pass by far outside, whilst some intermediate partial waves graze the nuclear edge. In Born approximation the diffraction zeroes come about as a result of destructive interference of partial wave amplitudes. On the other hand, if the effect of the static Coulomb field is taken into account, low partial waves are more distorted than intermediate partial waves, while very high partial waves will be affected only very little. As a consequence, the interference of the partial waves of Born approximation is perturbed. The diffraction zeroes disappear and are replaced with *diffraction minima* of the scattering amplitude of nonvanishing value. Furthermore, the position of these minima will be displaced from the positions of the Born zeroes.

*)This is true if the cross section depends only on one form factor and, in the case of inelastic scattering, if retardation effects are neglected. If it contains several form factors which have their zeroes at different q^2 then cross sections do not go to zero. Obviously, this does not invalidate our discussion.

(ii) A second important effect of Coulomb distortion may also be understood qualitatively. The static field is attractive. Therefore, the exact partial waves are attracted towards the nuclear interior. In terms of Born approximation, this means, effectively, that at least the low and intermediate partial waves are scattered at a higher effective energy k. As the previous zero occurs at a fixed value of the product

$$q \cdot R \simeq 2kR \sin(\theta/2) = \text{const.}$$

(R being the nuclear size parameter), the diffraction minimum is expected at a somewhat lower value of the scattering angle. As a simple example, consider elastic scattering on a spinless nucleus whose charge density is taken to be

$$\rho(r) = \frac{3}{4\pi R_0^3} \Theta(R_0 - r) \quad \text{(homogeneous density)}. \tag{102}$$

It is not difficult to calculate the charge form factor from eqs. (6) and (30), for this density (see exercise 5). One finds

$$F(q^2) = \frac{3}{z} j_1(z) = \frac{3}{z^3} (\sin z - z \cos z), \tag{103}$$

where $z = q \cdot R_0$. The cross section in Born approximation is given by eq. (7), and depends only on the variable

$$qR_0 \simeq 2kR_0 \sin(\theta/2).$$

Thus if we choose the energy such that $k_1 R_0^1 = k_2 R_0^2$ for two different nuclei with charges Z_1, Z_2 and radii R_0^1, R_0^2, the quantity $(k/Z)^2\, \mathrm{d}\sigma/\mathrm{d}\Omega$ will be the same for the two cases. This is illustrated by Fig. 5 which shows the scaled Born cross sections for $Z = 20$ (calcium) and $Z = 82$ (lead), with $kR_0 = 6.48$ (curve marked 3). The

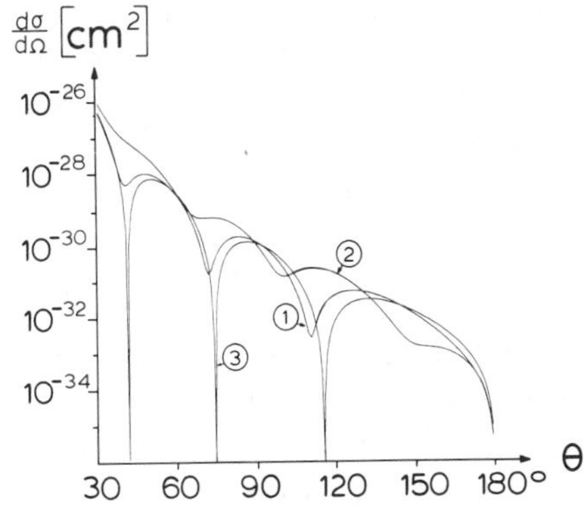

Fig. 5. Differential cross sections for elastic electron scattering on calcium ($Z = 20$, $A = 40$) and lead ($Z = 82$, $A = 208$). Curve 1: exact result for Ca; curve 2: exact result for Pb; curve 3: cross section in first Born approximation for Ca and Pb. The two cross sections for Ca are multiplied by the factor $(82\, R_0(\mathrm{Pb})/20 R_0(\mathrm{Ca}))^2$, so that the cross sections in Born approximation of Ca and Pb coincide. In all three cases the nuclear charge density is assumed to be homogeneous, cf. e.g. (102). The radius for lead is $R_0(\mathrm{Pb}) = 4.26$ fm, the energy is $E = 300$ MeV, so that $kR_0 = 6.48$.

spherical Bessel function of order 1 has zeroes at $z_1 = 4.493$, $z_2 = 7.725$, $z_3 = 10.904$, etc.,*) that is, the form factor and the Born cross section vanish at $\theta_1 = 40.6°$, $\theta_2 = 73.2°$, $\theta_3 = 114.7°, \ldots$. For comparison, the figure also shows the cross sections calculated by means of a full partial wave analysis. (The cross section for calcium is multiplied by the same scale factor as in Born approximation.) We note that, even with the cross section on calcium being rescaled, the exact cross sections do not coincide.

The zeroes are replaced with minima whose position is shifted towards lower scattering angles. The shift is larger for $Z = 82$ than for $Z = 20$.

One may ask whether these strong distortion effects render useless the method of Born approximation in electron scattering. Fortunately, this is not so. As will be seen below the physics information contained in specific features of the cross section is the same, independently of whether Born approximation or the more exact partial wave analysis is used. Coulomb distortion is a technical complication which does not obscure the connection between properties of the nuclear densities and the cross sections. The technical and conceptual simplicity of Born approximation can be made use of in many systematic investigations. Only when comparison with the data is made must the distortion be taken into account.**)

5.5. *Partial wave analysis for elastic scattering*

On our way to construct the partial wave decomposition of electron scattering amplitudes we need central field solutions of the Dirac equation carrying definite angular momentum. According to the formulae of App. E these have the form

$$\psi_{\kappa m}(r, \theta, \varphi) = \begin{pmatrix} g_\kappa(r)\, \varphi_{\kappa m} \\ \mathrm{i} f_\kappa(r)\, \varphi_{-\kappa m} \end{pmatrix}, \tag{104}$$

where κ is Dirac's quantum number. The radial wave functions g_κ and f_κ obey the system of differential equations

$$\frac{\mathrm{d}f_\kappa}{\mathrm{d}r} = \frac{\kappa - 1}{r} f_\kappa - (E - V(r) - m) g_\kappa,$$

$$\frac{\mathrm{d}g_\kappa}{\mathrm{d}r} = -\frac{\kappa + 1}{r} g_\kappa + (E - V(r) + m) f_\kappa. \tag{105}$$

In the case of electron scattering at high energies we can neglect the mass term in eq. (105). In this case we have the following symmetry relations

$$g_{-\kappa}(r) \simeq f_\kappa(r), \qquad f_{-\kappa}(r) \simeq -g_\kappa(r), \tag{106}$$

which can be read off from (105). Furthermore, it is appropriate to use the high-energy representation (III.166) of the Dirac equation. Using the transformation

*) See Ref. R1, Table 10.6.

**) See for example Sick (1974).

matrix S, p. 105, the central field solutions (104) now appear in the form

$$\psi_{\kappa m} = \begin{pmatrix} \phi_{\kappa m} \\ \chi_{\kappa m} \end{pmatrix} = \frac{1}{\sqrt{2}} \begin{pmatrix} g_\kappa \varphi_{\kappa m} + \mathrm{i} f_\kappa \varphi_{-\kappa m} \\ g_\kappa \varphi_{\kappa m} - \mathrm{i} f_\kappa \varphi_{-\kappa m} \end{pmatrix}. \tag{107}$$

For vanishing mass the symmetry relations (106) apply and thus

$$\psi_{-\kappa m} = -\frac{\mathrm{i}}{\sqrt{2}} \begin{pmatrix} g_\kappa \varphi_{\kappa m} + \mathrm{i} f_\kappa \varphi_{-\kappa m} \\ -g_\kappa \varphi_{\kappa m} + \mathrm{i} f_\kappa \varphi_{-\kappa m} \end{pmatrix}.$$

Furthermore, we know from sec. 2, eqs. (17) and (20), that the two asymptotic helicity states have the same scattering amplitudes. Thus, it is sufficient to study two-component spinors $\phi_{\kappa m}$ (e.g. upper two components of (107)) for *positive* κ. Replacing $\kappa > 0$ by $j = \kappa - 1/2$ and introducing $F_j(r) = r f_\kappa(r), G_j(r) = r g_\kappa(r)$, eqs. (105) (in the limit $m = 0$) go over into

$$\frac{\mathrm{d}F_j}{\mathrm{d}r} = \frac{j+1/2}{r} F_j - (E - V(r)) G_j,$$

$$\frac{\mathrm{d}G_j}{\mathrm{d}r} = -\frac{j+1/2}{r} G_j + (E - V(r)) F_j. \tag{105'}$$

As an example consider the case $V(r) \equiv 0$. From (105′) one derives

$$\frac{\mathrm{d}^2 G_j}{\mathrm{d}r^2} + \left(E^2 - \frac{\kappa(\kappa+1)}{r^2} \right) G_j = 0 \quad (\kappa = j + 1/2),$$

whose solutions can be expressed in terms of the spherical Bessel functions,

$$G_j = N r j_\kappa(kr) = N r j_{j+1/2}(kr).$$

F_j is obtained from the second eq. (105′) and the well-known relation

$$j_l'(z) = \frac{l}{z} j_l - j_{l+1} = -\frac{l+1}{z} j_l + j_{l-1}.$$

One finds

$$F_j(r) = N r j_{\kappa-1}(kr) = N r j_{j-1/2}(kr),$$

so that, with $\kappa = j + 1/2$

$$\phi_{\kappa m}^{(V\equiv 0)} = \frac{N}{\sqrt{2}} \left\{ j_{j+1/2}(kr) \varphi_{\kappa m} + \mathrm{i}\, j_{j-1/2}(kr) \varphi_{-\kappa m} \right\}. \tag{108}$$

In deriving the partial wave decomposition of the scattering amplitude we assume, at first, that the potential $V(r)$ decreases at infinity faster than $1/r$, i.e. $\lim_{r\to\infty} rV(r) = 0$. (See discussion in sec. 2.) The modifications due to the long range of the Coulomb potential are considered at the end of this section.

Following eq. (17) we request the solutions to have the following asymptotic form:

$$\phi_{m=1/2} \sim \begin{pmatrix} 1 \\ 0 \end{pmatrix} \mathrm{e}^{\mathrm{i}kz} + \frac{f(\theta, \varphi)}{r} \begin{pmatrix} \mathrm{e}^{-\mathrm{i}\varphi/2} \cos(\theta/2) \\ \mathrm{e}^{\mathrm{i}\varphi/2} \sin(\theta/2) \end{pmatrix} \mathrm{e}^{\mathrm{i}kr}. \tag{109}$$

For the solution at all values of $\boldsymbol{x}$ we write the series expansion

$$\phi_{m=1/2} = \sum_{j=1/2}^{\infty} a_{j1/2}\phi_{j1/2}$$

in terms of the angular momentum eigenstates (107).

The incoming part of ϕ, i.e. the first term of eq. (109), is readily expanded in terms of partial waves

$$\phi_{\text{in}} \equiv \begin{pmatrix}1\\0\end{pmatrix} e^{ikz} = \begin{pmatrix}1\\0\end{pmatrix} \sum_{l=0}^{\infty} i^l j_l(kr)\sqrt{(2l+1)4\pi}\, Y_{l0}$$

$$= \sqrt{4\pi} \sum_{j=1/2}^{\infty} \sum_{l=j-1/2}^{j+1/2} i^l\sqrt{2l+1}\, j_l(kr)\left(l0, \tfrac{1}{2}\tfrac{1}{2} | j\tfrac{1}{2}\right)\varphi_{jl1/2}. \tag{110}$$

Here we have expressed the product of the spin up state $\begin{pmatrix}1\\0\end{pmatrix}$ and of the eigenstate Y_{l0} of orbital angular momentum in terms of states in which l and spin $1/2$ are coupled to total angular momentum $j = l \pm 1/2$. Remembering the definition of the Dirac quantum number κ, we have for positive κ

$$\left.\begin{aligned}\varphi_{\kappa m} &\equiv \varphi_{j=l-1/2\, l=\kappa\, m}\\ \varphi_{-\kappa m} &\equiv \varphi_{j=\bar{l}+1/2\, \bar{l}=\kappa-1\, m}\end{aligned}\right\} \quad (\kappa > 0)$$

This allows us to rewrite eq. (110) by inserting the explicit values of the Clebsch–Gordan coefficients, as given in Table 2.

One finds

$$\phi_{\text{in}} = \sum_{j=1/2}^{\infty} i^{j-3/2}\sqrt{4\pi\left(j+\tfrac{1}{2}\right)}\left\{ j_{j+1/2}(kr)\varphi_{\kappa 1/2} + i\, j_{j-1/2}(kr)\varphi_{-\kappa 1/2}\right\}. \tag{110'}$$

[The reader may check that this is the same as the free solution (108).] For the purpose of reference we note here the well-known asymptotic behaviour of spherical Bessel functions

$$j_l(z) \underset{z\to\infty}{\sim} \frac{1}{z}\sin(z - l\pi/2). \tag{111}$$

Table 2

Explicit expressions for Clebsch–Gordan coefficients $(l\, m_l = m - m_s;\ \tfrac{1}{2} m_s | jm)$.

j	$m_s = +\tfrac{1}{2}$	$-\tfrac{1}{2}$
$l+\tfrac{1}{2}$	$\left(\dfrac{l+\tfrac{1}{2}+m}{2l+1}\right)^{1/2}$	$\left(\dfrac{l+\tfrac{1}{2}-m}{2l+1}\right)^{1/2}$
$l-\tfrac{1}{2}$	$-\left(\dfrac{l+\tfrac{1}{2}-m}{2l+1}\right)^{1/2}$	$\left(\dfrac{l+\tfrac{1}{2}+m}{2l+1}\right)^{1/2}$

For, if we make the following ansatz for the full solution:

$$\phi_{m=1/2}=\frac{1}{kr}\sum_{j=1/2}^{\infty}\sqrt{4\pi(j+1/2)}\,\mathrm{i}^{j-3/2}\mathrm{e}^{\mathrm{i}\eta_j}\{G_j\varphi_{\kappa 1/2}+\mathrm{i}F_j\varphi_{-\kappa 1/2}\},\tag{112}$$

the asymptotic behaviour of the radial functions can be taken to be

$$G_j\sim\sin\left(kr-\frac{j+1/2}{2}\pi+\eta_j\right),\tag{113a}$$

$$F_j\sim\cos\left(kr-\frac{j+1/2}{2}\pi+\eta_j\right).\tag{113b}$$

The asymptotic ansatz (113) makes sure that the *in*coming wave of ϕ, eq. (112), i.e. the piece proportional to e^{-ikr}/r, is indeed equal to the incoming part of ϕ_{in}, eq. (110′). η_j is the phase shift caused by the potential $V(r)$ as compared to the force-free situation where $V\equiv 0$. In order to identify the scattering amplitude $f(\theta,\varphi)$ we must isolate the *out*going spherical wave e^{ikr}/r in the asymptotic expansion of ϕ, eq. (112). We find

$$\phi|_{\text{out}}\sim-\frac{\mathrm{e}^{ikr}}{2ikr}\sqrt{4\pi}\sum_j\sqrt{j+1/2}\,\mathrm{e}^{2i\eta_j}\{\varphi_{\kappa 1/2}-\varphi_{-\kappa 1/2}\}.$$

The asymptotic piece of ϕ_{in}, eqs. (110), which contains the *out*going spherical wave is determined in exactly the same way. Obviously, it is of the same form, but with $\eta_j=0$. Upon comparison of the difference $(\phi-\phi_{\text{in}})_{\text{outg. spher. wave}}$ to the asymptotic form (109) we have

$$f(\theta,\varphi)\begin{pmatrix}\mathrm{e}^{-\mathrm{i}\varphi/2}\cos(\theta/2)\\ \mathrm{e}^{\mathrm{i}\varphi/2}\sin(\theta/2)\end{pmatrix}$$

$$=-\frac{1}{2ik}\sum\sqrt{4\pi(j+1/2)}\,(\mathrm{e}^{2\mathrm{i}\eta_j}-1)\{\varphi_{\kappa 1/2}-\varphi_{-\kappa 1/2}\}.\tag{114}$$

By means of the Clebsch–Gordan coefficients of table 2 we have, with $l=\kappa=j+\frac{1}{2}$,

$$\{\varphi_{\kappa 1/2}-\varphi_{-\kappa 1/2}\}=-\left(\sqrt{\frac{l}{2l+1}}\,Y_{l0}+\sqrt{\frac{l}{2l-1}}\,Y_{l-1,0}\right)\begin{pmatrix}1\\0\end{pmatrix}$$
$$+\left(\sqrt{\frac{l+1}{2l+1}}\,Y_{l1}-\sqrt{\frac{l-1}{2l-1}}\,Y_{l-1,1}\right)\begin{pmatrix}0\\1\end{pmatrix}.$$

Using the definition

$$Y_{lm}=(-)^m\sqrt{\frac{(2l+1)(l-1)!}{4\pi(l+1)!}}\,P_l^m\mathrm{e}^{\mathrm{i}\varphi}\tag{115}$$

and two recursion relations for the associated Legendre functions [R6]

$$P_l^m-xP_{l-1}^m-(l+m-1)\sqrt{1-x^2}\,P_{l-1}^{m-1}=0,$$
$$xP_l^m-(l-m+1)\sqrt{1-x^2}\,P_l^{m-1}-P_{l-1}^m=0,$$
$$(x=\cos\theta)$$

from which we derive

$$P_l^1 - P_{l-1}^1 = l\frac{\sqrt{1-x^2}}{1+x}(P_l + P_{l-1}) = l\,\mathrm{tg}(\theta/2)(P_l + P_{l-1}),$$

one finds

$$\{\varphi_{\kappa 1/2} - \varphi_{-\kappa 1/2}\} = -\sqrt{\frac{l}{4\pi}}(P_l + P_{l-1})\left\{\begin{pmatrix}1\\0\end{pmatrix} + \mathrm{e}^{\mathrm{i}\varphi}\mathrm{tg}(\theta/2)\begin{pmatrix}0\\1\end{pmatrix}\right\}$$

$$= -\sqrt{\frac{l}{4\pi}}\frac{\mathrm{e}^{\mathrm{i}\varphi/2}}{\cos\theta/2}(P_l + P_{l-1})\begin{pmatrix}\mathrm{e}^{-\mathrm{i}\varphi/2}\cos(\theta/2)\\ \mathrm{e}^{\mathrm{i}\varphi/2}\sin(\theta/2)\end{pmatrix};$$

comparing this to eq. (114) and setting $l = j + 1/2$ we obtain at once the final result

$$f(\theta,\varphi) = \frac{1}{2\mathrm{i}k}\frac{\mathrm{e}^{\mathrm{i}\varphi/2}}{\cos(\theta/2)}\sum_{j=1/2}^{\infty}(j+1/2)(\mathrm{e}^{2\mathrm{i}\eta_j} - 1)(P_{j+1/2} + P_{j-1/2}). \tag{116}$$

The phase factor $\mathrm{e}^{\mathrm{i}\varphi/2}$ is the same as the one obtained in Born approximation, eq. (25). As the potential $V(r)$ and hence the scattering process are axially symmetric about the 3-axis we may set $\varphi = 0$ without loss of generality.

Extension to the Coulomb potential. For potentials which do not decrease faster than $1/r$, the asymptotic form (109) is not correct. Very much like in the analogous nonrelativistic situation the phase factor $\mathrm{e}^{\mathrm{i}kr}$ is modified by an additional phase factor which depends on $\ln(2kr)$, so that in eq. (109) we should make the replacement

$$\mathrm{e}^{\mathrm{i}kr} \to \mathrm{e}^{\mathrm{i}(kr + Z\alpha\ln(2kr))}.$$

In the expression (116) the phases η_j, so far, were the scattering phases due to the potential relative to the *force-free* case. In the case of a potential decreasing like $1/r$ a different procedure is indicated: The electrostatic potential created by a nucleus with spherically symmetric charge density $\rho(r)$ is

$$V(r) = -4\pi Ze^2\left\{\frac{1}{r}\int_0^r \rho(r')r'^2\,\mathrm{d}r' + \int_r^{\infty}\rho(r')r'\,\mathrm{d}r'\right\}. \tag{117}$$

As $\rho(r)$ vanishes (or becomes negligibly small) beyond some distance R of the order of the nuclear radius, $V(r)$ approaches the pure $1/r$ potential

$$V_{\mathrm{C}}(r) = -Ze^2/r \tag{118}$$

for $r > R$. Like in the nonrelativistic case the radial Dirac equations (105) can be solved analytically for the case of the potential (118) of point-like charges. In particular, the scattering phases η_j^{C} of this potential can be given explicitly and the scattering amplitude f_{C} be computed from eq. (116). Therefore, the scattering problem for the true potential (117) is solved most economically by computing the *additional* phase shift due to $V(r)$, viz.

$$\delta_j = \eta_j - \eta_j^{\mathrm{C}}, \tag{119}$$

where η_j is the full phase shift of the potential $V(r)$. The construction of the continuum solutions g_κ^{C} and f_κ^{C} for the potential V_{C} is straightforward but tedious and we do not work them out here.*) The solutions which are regular at the origin have the asymptotic behaviour (in the mass zero limit),

$$rg_\kappa^{\mathrm{C}} \sim \sin\left(kr + Z\alpha\ln(2kr) + \delta_\kappa^0\right),$$
$$rf_\kappa^{\mathrm{C}} \sim \cos\left(kr + Z\alpha\ln(2kr) + \delta_\kappa^0\right), \tag{120}$$

where

$$\delta_\kappa^0 \equiv \bar{\eta}_\kappa - \sigma_\kappa - (\gamma_\kappa - 1)\pi/2, \tag{121}$$

with

$$\gamma_\kappa = \sqrt{\kappa^2 - (Z\alpha)^2},$$

$$\bar{\eta}_\kappa(m=0) = -\frac{1}{2}\operatorname{arctg}\frac{Z\alpha}{\gamma_\kappa} - \frac{\pi}{2}\frac{1+\operatorname{sign}\kappa}{2},$$

$$\sigma_\kappa = \arg\Gamma(\gamma_\kappa + \mathrm{i}Z\alpha);$$

comparing this to the general form (113) we see that the Coulomb phase is given by

$$\eta_\kappa^{\mathrm{C}} = \bar{\eta}_\kappa - \sigma_\kappa + (l - \gamma_\kappa + 1)\pi/2 \tag{122}$$

(up to the logarithmic term $Z\alpha\ln 2kr$), or

$$\eta_j^{\mathrm{C}} = \bar{\eta}_j - \sigma_j + \left(j + \tfrac{3}{2} - \gamma_j\right)\pi/2. \tag{122$'$}$$

In eq. (122′) we have written the index j, not κ since we need to consider only positive κ here; eq. (122) holds for all κ, positive and negative. For the sake of simplicity, we have neglected the mass of the electron. If one wishes to retain the mass terms then $\bar{\eta}_\kappa$ of eq. (121) is replaced with**)

$$\bar{\eta}_\kappa(m) = -\tfrac{1}{2}\operatorname{arctg}\frac{y\left(1 + \dfrac{\gamma_\kappa}{\kappa}\dfrac{m}{E}\right)}{\gamma_\kappa - \dfrac{1}{\kappa}y^2\dfrac{m}{E}} - \frac{\pi}{2}\frac{1+\operatorname{sign}\kappa}{2}. \tag{123}$$

All other formulae (120)–(122) remain unchanged provided k is now understood to be the wave number $k = \sqrt{E^2 - m^2}$. η_j^{C} can therefore be obtained analytically from eq. (122). The additional phase shifts δ_j, eq. (119), which are caused by the *difference* of the true potential (117) and the pointlike potential (118) may be obtained as follows: One calculates solutions of the Dirac equations (105′), for a given energy and for $V(r)$ as obtained from eq. (117), which are regular at the origin. As $V(r)$ goes over into $V_{\mathrm{C}}(r)$ at some finite radius R outside the nuclear charge radius, we need the full solution (f, g) only in the inner region $0 \le r \le R$. In the outer region $r \ge R$ the exact solution is a superposition of regular (R) and irregular (I) solutions

*) See e.g. Ref. R20.

**) Arctg defined such that it goes to zero as the argument goes to zero.

Table 3
Coulomb phases and phase shifts, eq. (119) for electron scattering on $^{208}_{82}$Pb. The charge density is described by eq. (125) with $c = 6.6475$ fm, $t = 2.30$ fm.

j	η_j^{C}	δ_j
$\frac{1}{2}$	0.4441	−1.4779
$\frac{3}{2}$	−0.2401	−0.8046
$\frac{5}{2}$	−0.5491	−0.5137
$\frac{7}{2}$	−0.7512	−0.3367
$\frac{9}{2}$	−0.9017	−0.2181
$\frac{11}{2}$	−1.0218	−0.1361
$\frac{13}{2}$	−1.1217	−0.0801
$\frac{15}{2}$	−1.2072	−0.0434
$\frac{17}{2}$	−1.2820	−0.0214
$\frac{19}{2}$	−1.3485	−0.0095
$\frac{21}{2}$	−1.4083	−0.0039

of eqs. (105′) with $V_{\mathrm{C}}(r)$, viz.

$$f_\kappa = af_\kappa^{\mathrm{C,R}} + bf_\kappa^{\mathrm{C,I}},$$
$$g_\kappa = ag_\kappa^{\mathrm{C,R}} + bg_\kappa^{\mathrm{C,I}}. \tag{124}$$

The phase differences δ_j may be obtained by direct comparison of the exact solution to the regular solutions $(f^{\mathrm{C,R}}, g^{\mathrm{C,R}})$ at some $r > R$ at which an asymptotic expansion of these functions is meaningful. In this case, however, $rf^{\mathrm{C,R}}$ and $rg^{\mathrm{C,R}}$ must be expanded beyond the form (120), up to, say, terms of order $1/r^2$. Alternatively, the phases may be obtained from the asymptotic form of eq. (124), expressing them as functions of b/a and η_j^{C} (Ravenhall et al. 1954).

As an example, Table 3 shows the Coulomb phases η_j^{C} for V_{C} with $Z = 82$, as well as the phase shifts δ_j for the Fermi charge density (125) for lead, $Z = 82$, $c = 6.6475$, $t = 2.30$. The electron energy is $k = 300$ MeV. At this energy the nine lowest partial waves are modified appreciably by the deviation of the actual charge density from the point-like charge.

5.6. *Practical analysis of scattering data and information content of partial waves*

In the early stages of this kind of nuclear physics it was customary to analyze elastic scattering data in terms of specific functional forms for the nuclear charge density. An ansatz that was particularly popular is the so-called *Fermi distribution*,

$$\rho(r) = N\frac{1}{1 + \exp((r - c)/z)}, \tag{125}$$

where the normalization factor

$$N = \frac{3}{4\pi c^3}\left\{1+\left(\frac{\pi z}{c}\right)^2 - 6\left(\frac{z}{c}\right)^3 \mathrm{e}^{-c/z}\sum_{n=1}^{\infty}\frac{(-)^n}{n^3}\mathrm{e}^{-nc/z}\right\}^{-1}$$
$$\simeq \frac{3}{4\pi c^3}\frac{1}{1+(\pi z/c)^2} \tag{126}$$

is chosen such as to normalize $\rho(r)$ to one, cf. eq. (2). The parameter c is called the *radius of half-density* because $\rho(r=c)=0.5\rho(r=0)$. The parameter z is a measure for the rate at which the density falls off in the nuclear surface. A good measure for this fall-off is the so-called *surface thickness* t which is defined as follows: Let $\rho(r)$ be a function which decreases monotonically for increasing r. Let $r_{(90)}$ and $r_{(10)}$ be the radii at which the density is 90% and 10% of its value at $r=0$, respectively. Then

$$t := r_{(10)} - r_{(90)}.$$

In the case of the Fermi distribution (125) one has the relation $t = 4z\ln 3 \simeq 4.394z$. The shape of the Fermi density is illustrated by Fig. II.7, curve marked ρ_0.

As a matter of fact, the Fermi density (125) gives a surprisingly good description of charge and matter densities in practically all spherical nuclei except the very lightest ones. It depends on two parameters, c and t, which have the typical values

$$c \simeq 1.1 \times Z^{1/3}\ \mathrm{fm}, \qquad t \simeq 2.2\text{–}2.5\ \mathrm{fm}. \tag{127}$$

There are, of course, deviations from this simple pattern, especially at and close to magic shells but these deviations are never very large. (We shall come back to this, in connection with muonic atoms, below.) Even though a functional form such as the Fermi function (125) may be quite useful as a rough parametrization, it is not adequate for the analysis of precise measurements of the elastic cross section extending over many orders of magnitude, for several reasons: Assuming a specific functional form implies a prejudice about possible shapes of the charge density and, therefore, cannot be the basis of a model-free analysis of the data. Furthermore, such an ansatz contains a finite number of parameters (two in the case of the Fermi distribution (125)) which are then determined by a best fit to the data. What if the data are so numerous and so precise that they contain more information on the charge distribution than what can be described by these parameters? Finally, a specific function such as (125) does not reflect the fact that different parts of $\rho(r)$ have different weights in the angular dependence of cross section. In summary, one would prefer a model independent way of analysis whose final result would be an empirical density function $\rho(r)$, along with an error band that depends on r and reflects the type of experimental input as well as its error bars.

Such methods of analysis which do not rely on specific model densities have been proposed and have been applied successfully to high-precision experiments (Lenz 1969, Friedrich et al. 1972, Sick 1974, Friar et al. 1973, 1975). Figs. 6 and 7 show two typical examples for nuclear charge densities as determined from experiment. Instead of describing these specific methods here we prefer to discuss, in a qualita-

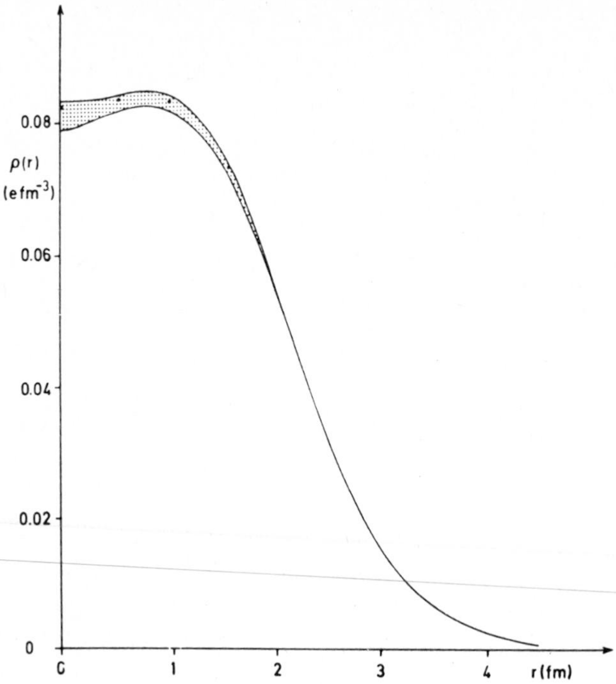

Fig. 6. Quasi model independent determination of the charge density in carbon. Taken from Sick (1974).

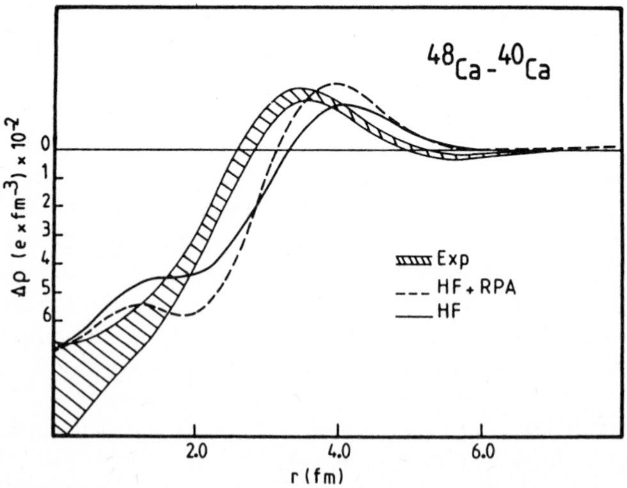

Fig. 7. Difference of charge densities of ^{48}Ca and ^{40}Ca. The shaded band is the result of a quasi model independent determination from elastic electron scattering; its width reflects experimental errors of the cross sections as well as the lack of knowledge of their behaviour at large momentum transfers. Figure taken from Lect. Notes in Phys. 108 (1979), p. 58 (Proceedings of Conf. on Nuclear Physics with Electromagnetic Interactions, Mainz 1979).

tive manner, the physics information carried by the various low, intermediate and high partial waves.

Sensitivity of scattering phases to details of charge density. Obviously, the nature of charge density moments and their number which can be obtained from elastic scattering, depends on:

(i) the primary energy of the electron,
(ii) the range of momentum transfers, and
(iii) the experimental error bars of the differential cross section.

The essential features of such an analysis can be made transparent by studying integral representations for the scattering phases.*) To this end, let us consider two different charge densities $\rho^{(1)}(r)$, $\rho^{(2)}(r)$ which are both normalized to one, as before, but which differ from each other over the domain of the nucleus. Both are supposed to go to zero quickly for $r \gtrsim R$. The corresponding potentials $V^{(1)}(r)$ and $V^{(2)}(r)$, calculated from eq. (117), then differ only for $r \lesssim R$ and both go over into $V_C(r)$ outside the nucleus.

Let $F_j^{(1)}$, $G_j^{(1)}$ and $F_j^{(2)}$, $G_j^{(2)}$ be the solutions of eqs. (105′) for $V^{(1)}$ and $V^{(2)}$, respectively, and let $\delta_j^{(1)}$ and $\delta_j^{(2)}$ be the corresponding phase shifts as defined in eq. (119). From eqs. (105′) and from the properties of the generalized Wronskian (see exercise 6),

$$W(r) = F_j^{(1)}(r)G_j^{(2)}(r) - G_j^{(1)}(r)F_j^{(2)}(r), \tag{128}$$

one derives the relation

$$\sin\left(\delta_j^{(1)} - \delta_j^{(2)}\right) = -\int_0^\infty \mathrm{d}r\left(V^{(1)} - V^{(2)}\right)\left(F_j^{(1)}F_j^{(2)} + G_j^{(1)}G_j^{(2)}\right). \tag{129}$$

The potentials are related to the charge densities through Poisson's equation. Inserting this into eq. (129) and integrating by parts twice, one has the final result

$$\sin\left(\delta_j^{(1)} - \delta_j^{(2)}\right) = 4\pi\int_0^\infty r^2\,\mathrm{d}r\left(\rho^{(1)}(r) - \rho^{(2)}(r)\right)\chi_j(r), \tag{130}$$

where χ_j stands for

$$\chi_j(r) = \int_0^r \frac{\mathrm{d}r'}{r'^2}\int_0^{r'}\mathrm{d}r''\left(F_j^{(1)}F_j^{(2)} + G_j^{(1)}G_j^{(2)}\right). \tag{131}$$

What does eq. (130) tell us about the sensitivity of the phase shifts to the charge density? For answering this question it is useful to consider high, intermediate, and low partial waves separately:

(a) *High partial waves.* Classically speaking, these partial waves correspond to electron trajectories which pass by, far outside, and do not penetrate the nucleus. The phases $\eta_j^{(i)}$ coincide practically with the phases η_j^{C} of a pointlike source. As $\rho^{(1)} \simeq \rho^{(2)}$ these phases do not contain information on the nuclear charge density other than its total charge. (In the example of table 3: $j \gtrsim 19/2$.)

*) See e.g. Elton (1953), Lenz (1969).

(b) *Intermediate partial waves.* These partial waves start penetrating the nucleus somewhat. However, the centrifugal potential terms in eqs. (105′) are still predominant so that the radial functions can be approximated, over the entire nuclear domain, by their power behaviour at the origin, viz.

$$F_j^{(i)} \sim r^{j+1/2}, \qquad G_j^{(i)} \sim r^{j+3/2}, \tag{132}$$

Inserting this into eq. (131) we find that $\chi_j(r)$ may be approximated by a simple power behaviour too, $\chi_j \sim r^{2j+1}$. Eq. (130) then becomes

$$\sin\left(\delta_j^{(1)} - \delta_j^{(2)}\right) \sim 4\pi \int_0^\infty r^2 \mathrm{d}r \left(\rho^{(1)} - \rho^{(2)}\right) r^{2j+1}. \tag{130′}$$

Thus, intermediate partial waves are determined by *even* moments of the charge density. Obviously, these depend primarily on the density at the *nuclear* surface.

(c) *Low partial waves.* The low partial waves, finally, for which the approximation (132) becomes invalid, contain information about the nuclear interior. If these partial waves occur at all, i.e. if they are really distinct from the class b (in fact, this is only the case if the energy is high enough), then they cannot be expressed in terms of simple moments

$$\int_0^\infty r^2 \mathrm{d}r \rho(r) r^{2j+1}$$

and a full partial wave analysis must be carried out.

Summarizing, we may say that the type and quality of information that is obtained from a partial wave analysis of electron scattering is very similar to the information content of the form factor in Born approximation. Therefore, Coulomb distortion, even though important on a quantitative level, is not much more than a technical complication that does not alter the physics of the process.

5.7. *Miscellaneous comments*

We close this somewhat lengthy discussion of electron scattering with a few comments and supplementary remarks, as well as some hints for further reading. The first three supplements concern elastic scattering, the last two inelastic scattering.

Dispersion corrections to elastic scattering. In our treatment of elastic scattering we have assumed the nucleus to be an inert system of stationary charge and current densities. In other terms, we have calculated the fields created by the nuclear ground state at the site of the electron and we have calculated the cross sections from a one-particle wave equation for the electron, in the *external field approximation*. In reality, the nucleus is a dynamical system and has its own internal degrees of freedom which manifest themselves in its rich excitation spectra. These internal degrees of freedom may play a role in elastic scattering if second-order processes of the type sketched symbolically in Fig. 8 become important: In a first step the electron excites the nuclear ground state A to some excited intermediate state A* which deexcites again to the ground state in a second scattering process.

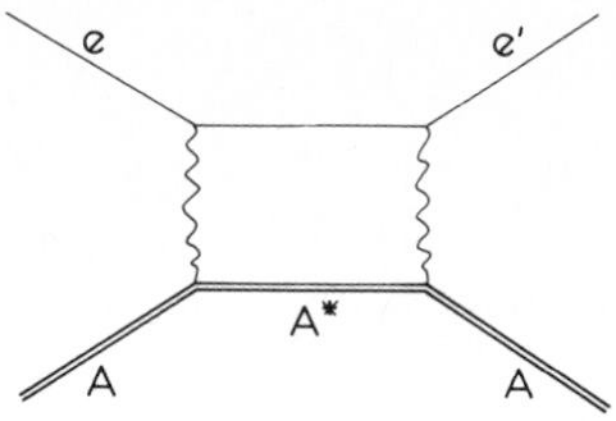

Fig. 8. So-called dispersion corrections to electron scattering: a two-step process in which the nucleus is excited to an intermediate state.

These dispersion corrections are notoriously difficult to calculate in a reliable manner. Fortunately, they are not large and can be neglected in most practical situations.

Elastic scattering on strongly deformed nuclei. So far we have considered only spherically symmetric nuclear charge densities $\rho(r)$. This situation allowed us to separate angular and radial motion of the electron. Obviously, this situation applies only for nuclei with spin zero, $J=0$. For nuclei with $J=1/2$ we have formally the same situation as for the nucleon: Here there is charge scattering on a spherically symmetric charge density as before, plus scattering on a magnetic moment density (M1 scattering). For nuclei with $J \geq 1$ the nuclear charge density also has nonvanishing higher (even) moments, viz. (cf. sec. II.4, eq. (II.101))

$$\begin{aligned} \rho(\boldsymbol{r}) &:= \langle JM=J | \sum_{i=1}^{Z} \delta(\boldsymbol{r}-\boldsymbol{r}_i) | JM=J \rangle \\ &= \rho_0(r) + \sum_{\kappa=1}^{[J]} \sqrt{\frac{4\kappa+1}{16\pi}}\, \rho_{2\kappa}(r) Y_{2\kappa 0}(\theta). \end{aligned} \tag{133}$$

In the case of strongly deformed nuclei the quadrupole term will, in general, be predominant, i.e.

$$\rho(\boldsymbol{r}) \simeq \rho_0(r) + \sqrt{\frac{5}{16\pi}}\, \rho_2(r) Y_{20}(\theta), \tag{134}$$

where $\rho_2(r)$ denotes the radial quadrupole density. The factors in eq. (133) have been chosen such that the nuclear spectroscopic quadrupole moment, by its traditional definition, is given by the second moment of $\rho_2(r)$

$$Q_s = \int_0^\infty \left[\rho_2(r) r^2\right] r^2 \, \mathrm{d}r. \tag{135}$$

While the calculation of the scattering cross section in Born approximation is straightforward (see above), the corresponding partial wave analysis is technically more complicated but still feasible. There may be a difficulty on the experimental side: Such strongly deformed nuclei have rotational excitations of rather low energy, 50 to 200 keV. It then depends on the energy resolution of a given experimental set up whether or not the truly elastic scattering on the nuclear ground state can be distinguished from the excitation cross sections for these low-lying rotator states. Thus, elastic scattering of electrons may not be the ideal tool for investigating the charge density of deformed nuclei. We will see below that in this case muonic atoms offer a more direct approach to this quantity.

Summation of partial wave amplitudes. There is a technical difficulty in summing the partial wave series (116): As it stands this series converges very slowly. The origin of this problem is not difficult to understand. For this purpose let us consider first the scattering on a point charge in the nonrelativistic case. The potential $V_C(r)$ of a point charge introduces a $1/r$ singularity into the Schrödinger equation. The corresponding scattering amplitude which can be derived in closed form, is proportional to $1/\sin^2(\theta/2)$ and hence becomes singular at $\theta = 0$. This singularity can be "smoothened" and the convergence of the partial wave series accelerated by means of the following procedure. Suppose we wish to sum an expression of the form (Ravenhall et al. 1954)

$$F(\theta) = \sum_{l=0}^{\infty} a_l P_l(\cos\theta).$$

If this series converges only very slowly in practice, one may replace it by a so-called reduced series which is defined by

$$(1-\cos\theta)^m F(\theta) = \sum_{l=0}^{\infty} c_l^{(m)} P_l(\cos\theta),$$

and where m is some positive integer. Using standard recursion formulae for Legendre polynomials one shows that the new coefficients $c_l^{(m)}$ can be calculated by means of the recursion relation

$$c_l^{(m+1)} = c_l^{(m)} - \frac{l}{2l-1} c_{l-1}^{(m)} - \frac{l+1}{2l+3} c_{l+1}^{(m)},$$

$$c_l^{(0)} = a_l.$$

The modified series which is obtained after a few iterations converges more rapidly than the original series.

Inelastic scattering and partial wave analysis. The problem of Coulomb distortion in inelastic electron scattering is essentially the same as in elastic scattering. For light nuclei the conceptually simple method of Born approximation may still be adequate (depending, again, on the accuracy of the data). For medium and heavy nuclei it is not. The inelastic form factors of Born approximation, in general, have zeroes in the physical domain of momentum transfers. Therefore, the cross section, if it is dominated by one form factor, again exhibits the typical diffraction pattern of Born approximation. The diffraction zeroes are replaced with minima and are shifted from their initial position if Coulomb distortion is taken into account. Also the absolute values of the cross section can be widely different from what they are in Born approximation, depending on the value of the scattering angle.

The partial wave analysis of inelastic scattering is conceptually similar to the case of elastic scattering but technically much more involved. The first such analyses were carried out in the mid-nineteen-sixties.*) (Griffy et al. 1963, Scheck 1966). A good starting point is the interaction term in the form of eq. (85). It is appropriate to expand the electron wave functions in terms of the central field solutions (104). The initial state $|k\rangle$ is determined by the requirement that in the asymptotic domain it contain the plane wave and incoming spherical waves. Similarly, the final state is the analogous superposition of plane wave and outgoing spherical wave. Inserting these (properly normalized) functions into the matrix element (85) yields a multiple sum of matrix elements of the interaction between states of good angular momentum and parity. This fact allows one to make use of the selection rules due to angular momentum and parity conservation and to perform all angular and spin integrals by means of standard angular momentum algebra. This leaves one with a sum over radial integrals involving the radial functions $f_\kappa(r)$, $g_\kappa(r)$ and some nuclear radial quantity. We know from the analysis of elastic scattering that for $E \simeq 300$ MeV about the nine lowest partial waves penetrate into the nucleus. Obviously, these are the ones which are sensitive to details of the transition charge and current densities. Here, the radial integrals entering the expansion of the scattering amplitude are obtained by numerical integration. For the higher partial waves the eigenfunctions of the point charge may be used and, if one is lucky, the radial integrals can be done analytically (Reynolds et al. 1964). The complexity of such calculations lies in the high number of partial waves which contribute and in the complicated pattern of terms allowed by the selection rules of angular momentum.

6. Muonic atoms—introduction

Like any other long-lived negatively charged particle the muon can be captured in the static Coulomb field of a nucleus and thus can form a hydrogen-like exotic atom. This system has peculiar and unique properties, both regarding its *spatial dimensions* and its dynamical *time structure*, which make it an important tool in exploring electroweak interactions and in probing properties of the nucleus.

This section summarizes first the properties of the muon which are relevant for the subsequent sections of this chapter. We then give a first qualitative picture of the properties of the muonic atoms and their applications. The section closes with a derivation of bound central field solutions in static Coulomb potentials.

Specific and quantitative applications of muonic atoms to quantum electrodynamics and to the investigation of nuclear properties are treated in secs. 7 and 8, respectively.

*) The much simpler case of monopole excitations was treated in Alder et al. (1963).

6.1. Properties of free muons

The muon has all properties of a "heavy electron". It appears in two charge states μ^- and μ^+, which are antiparticles of each other. Its charge is equal to the charge of the electron*); its spin is $1/2$; it carries *lepton number* (see Chap. V). Its *mass* is about 207 times larger than the electron mass. More precisely

$$m_\mu/m_e = 206.768\,297\,(62). \qquad (136)$$

This number is obtained by combining the measured values of
(i) the ratio of magnetic moments $\mu_\mu/\mu_{\rm p}$ of muon and proton;
(ii) the hyperfine splitting $\Delta E = E(F=1) - E(F=0)$ of muonium;
(iii) the anomaly of the g-factor of the muon, $a_\mu = \frac{1}{2}(|g_\mu| - 2)$.**)
[The same combination of data gives very precise information on the equality of the charges of the muon and the electron.]

With $m_{\rm e} = 0.5110034(14)$ MeV, this gives m_μ to about 3 ppm:

$$m_\mu = 105.6593(3)\ \text{MeV}/c^2. \qquad (137)$$

The fact that μ^- carries the charge $Q = -|e|$ means that it has exactly the same coupling to the electromagnetic field as the electron. In Chap. V we shall see that also the weak interactions of muons are exactly the same as those of electrons. Actually, the same statements seem to apply also to the τ-lepton with mass

$$m_\tau = 1784(3)\ \text{MeV}/c^2. \qquad (138)$$

In this sense the interactions of leptons are *universal*: The structure of the coupling terms to the Maxwell field and to the bosons of weak interactions are identically the same for the three kinds of "electrons" $\text{e}^-, \mu^-, \tau^-$. Their coupling strengths (i.e. their electric and weak "charges", respectively) to photons and to weak bosons are the same, respectively.

All *quantitative* differences in physical properties of electrons, muons and τ-leptons (such as anomaly of magnetic dipole moments, scattering cross sections, decay amplitudes) will be due solely to the difference in their *masses*. This can be understood in a qualitative manner as follows. A specific physical situation is always characterized by a typical spatial dimension (examples: Bohr radius $a_{\rm B} = \hbar^2/e^2 m$, Compton wave length $\lambda\!\!\!^- = \hbar/mc$) and typical momenta (examples: momentum in a bound atomic state, momentum transfer in a scattering amplitude) which yield the bulk of the quantity that one wishes to calculate. However, the scale of these characteristic dimensions is set by the mass of the particle or, in some cases, by ratio or difference of masses of different leptons (examples: vacuum polarization due to virtual electron–positron pairs in electronic and in muonic atoms, g-factor anomaly for electrons and for muons). We shall encounter many examples below, in this chapter and in Chap. V.

*)This is known to at least 2 ppm. Cf. the summary by H. Primakoff in "Muon Physics", Ref. R21.

**)See e.g. Scheck (1978).

The *magnetic moment* of the muon relative to the magnetic moment of the proton is known very precisely from measurements in muonium, viz.

$$\mu_{\mu^+}/\mu_p = 3.183\,344\,9(9) \quad (0.3\,\text{ppm}). \tag{139}$$

It is practically a normal Dirac moment,

$$\mu_\mu = \tfrac{1}{2} g_\mu \frac{Q}{2m_\mu} \quad (Q = -|e|), \tag{140}$$

with $|g_\mu| \simeq 2$, the deviation of $|g_\mu|$ from 2, the so-called anomaly a_μ, being predictable on the basis of higher order radiative corrections. This anomaly is defined as $a_\mu = \frac{1}{2}(|g_\mu| - 2)$. The measured value of this anomaly is (Bailey et al. 1979).

$$a_\mu = 1\,165\,924\,(9) \times 10^{-9}. \tag{141}$$

In the same experiment the equality of g_{μ^-} and $-g_{\mu^+}$ which is predicted by the invariance of the muon's interactions under the combined operation *CPT* (charge conjugation *C*, space reflection *P*, and time reversal *T*) has been tested with the result

$$(|g_{\mu^+}| - g_{\mu^-})/g = (0.026 \pm 0.017)\,\text{ppm}. \tag{141'}$$

In contrast to the electron, the muon is unstable. Its primary decay mode is

$$\mu^- \to e^- \bar{\nu}_e \nu_\mu \qquad (\mu^+ \to e^+ \nu_e \bar{\nu}_\mu);$$

its lifetime is

$$\tau_\mu = 2.197\,14(7) \times 10^{-6}\,\text{sec} \tag{142}$$

(see below, Chap. V). This time is very long as compared to typical time scales of electromagnetic processes of muons in a target.

6.2. *Muonic atoms, qualitative discussion*

As the Pauli exclusion principle is not effective between muons and electrons, the trapped muon runs through its Bohr cascade towards the 1s-state, irrespective of the presence of the electronic shells of the host atom. Before we turn to the quantitative analysis of muonic atoms let us first discuss their characteristic spatial dimensions, energy scales and time scales.

Energy scales and spatial dimensions. Since the Bohr radius $a_B = 1/Z\alpha m$ is inversely proportional to the mass of the charged lepton (more precisely: the *reduced* mass of the lepton–nucleus system), the orbits of a muonic atom are smaller by a factor of about 207 (see eq. (136)) than the orbits of the electrons of the host atom. If the main quantum number is smaller than n_0, where

$$n_0^2 a_B(m_\mu) \lesssim a_B(m_e),$$

the muonic orbits ($n, l = n-1$) lie inside the electronic 1s-orbit. This happens for

$$n \lesssim n_0 \simeq \sqrt{m_\mu/m_e} \simeq 14.$$

Therefore, the states with $n \lesssim n_0$ of the muonic atom are essentially hydrogenlike up to screening effects by the electronic shells of the host atom. Screening will be the smaller, the lower the orbit. Thus, for a first qualitative orientation we may use the equations for the nonrelativistic hydrogen atom. The binding energy of a bound state with main quantum number is

$$E_n^{(\mathrm{n.r.})} = -m_\mu (Z\alpha)^2/2n^2. \tag{143}$$

The first relativistic correction to this is of order $(Z\alpha)^4$ but its magnitude relative to (143) is independent of the mass. Thus, the relative importance of relativistic effects in muonic atoms is approximately the same as in electronic atoms. The relevant parameter is $Z\alpha$. For light nuclei these effects will be small, for heavy nuclei such as lead $(Z = 82)$ where $Z\alpha \simeq 0.6$ they will be important. For most estimates and qualitative considerations the formulae of the hydrogen atom will be adequate. The wave functions are [see eq. (II.52)]

$$\psi_{nlm}(\boldsymbol{r}) = \frac{1}{r} y_{nl}(r) Y_{lm}(\theta, \varphi), \tag{144}$$

$$y_{nl}(r) = \sqrt{\frac{(l+n)!}{a_{\mathrm{B}}(n-l-1)!}\,\frac{1}{n(2l+1)!}}\, z^{l+1} e^{-z/2} \times {}_1F_1(-n+l+1; 2l+2; z), \tag{145}$$

where

$$a_{\mathrm{B}} = 1/Z\alpha\mu, \tag{145a}$$

μ being the reduced mass, and z is the dimensionless variable,

$$z = \frac{2}{n a_{\mathrm{B}}} r. \tag{145b}$$

${}_1F_1$ denotes the confluent hypergeometric function [R1].

Let us calculate the Bohr radius for a light, two medium weight and a heavy nucleus, and let us compare it to the nuclear r.m.s. radius:

Nucleus	$\langle r^2\rangle^{1/2}$ [fm]	$a_{\mathrm{B}}(m)$ [fm]
^{4}He $(Z=4)$	3.1	64
^{16}O $(Z=8)$	2.6	32
^{40}Ca $(Z=20)$	3.5	12.8
^{208}Pb $(Z=82)$	5.5	3.12

We see from this comparison that from about $Z = 20$ on up the low muonic orbits start penetrating into the nucleus more and more. Accordingly, the low muonic states must become more and more sensitive, as Z increases, to the finite size of the nucleus and, in particular, to the deviation of the nuclear charge distribution from a point charge. Of course, for large Z the estimates for the binding energies and the radii of low-lying orbits given above become unrealistic. Let us illustrate this by comparing the 2p–1s transition energy in *lead*, for a point charge and for a realistic charge distribution*⁾:

$$\left(E_{2\mathrm{p}1/2}-E_{1\mathrm{s}}\right)_{\substack{\text{point charge}\\(Z=82)}}=(-5.38+20.99)\ \mathrm{MeV}=15.61\ \mathrm{MeV},$$

$$\left(E_{2\mathrm{p}1/2}-E_{1\mathrm{s}}\right)_{\text{finite size}}=(-4.78+10.52)\ \mathrm{MeV}=5.74\ \mathrm{MeV}.$$

We note that the 1s state of this heavy atom shows a much weaker binding than in the pure $1/r$ potential. The 2p state is also shifted upwards, but by a much smaller proportion than the 1s state. At the same time also the radial wave functions are affected by the finite extension of the nuclear charge density. As the states are less bound than for a pointlike charge the radial functions are driven towards larger values of r. Nevertheless, it is still true that the muon in a 1s state of a heavy atom penetrates strongly into the nuclear interior.

Time scales in muonic atoms. The fate of the muon between the moment of its creation, say, from pion decay,

$$\pi^{-}\rightarrow\mu^{-}+\bar{\nu}_{\mu} \tag{146}$$

in a continuum state until it is trapped in some high-lying Bohr orbit of a target atom, i.e. the moderation of the muon from its initial positive energy down to zero kinetic energy through ionization and inelastic scattering processes in the target, is complicated and, in fact, not too well known. For our purposes it will be sufficient to know that these early stages take a relatively short time, of the order of 10^{-10} to 10^{-12} sec. The muon eventually lands in some bound state with quantum numbers (n,l). The question as to what the initial distribution in n and l is, can be (and has been) studied experimentally by looking at the intensities of γ-transitions between these highest states of the cascade. There is, as yet, no satisfactory theory of these initial distributions—a fact that may not seem so surprising if one realizes that the initial (n,l) distribution is a complicated function of the structure and chemical composition of the target material.

The cascade proceeds predominantly through

Auger transitions, i.e. through emission of electrons in the host atom, and

electric dipole γ-radiation.

Auger transitions are important mainly in the upper part of the cascade. They are relatively more important for the lighter elements.**⁾

In the lower part of the cascade the transition energies become large, and E1 γ-transitions quickly take over. The selection rules of Auger and of γ(E1)-transitions are

$$\text{Auger:}\quad l_{\mathrm{f}}=l_{\mathrm{i}}\pm 1,\quad \Delta n=\text{minimal}, \tag{147}$$

$$\text{E1:}\quad l_{\mathrm{f}}=l_{\mathrm{i}}\pm 1,\quad \Delta n=\text{maximal}. \tag{148}$$

*⁾The point charge values are calculated from eq. (162) below. The finite size values are taken from Engfer et al. (1974).

**⁾See Ref. R21, Vol. I, Chap. III.

The selection rule for Δl is a strict one, the rules for Δn are somewhat empirical. They arise from the energy dependence of the rates and from the n-dependence of the transition matrix elements.

The selection rules have the effect of favouring *circular orbits* $(n, l=n-1)$; the lower n, the higher are the relative intensities for transitions between circular orbits. Transitions between inner, non-circular states have comparatively low intensities.

It is not difficult to estimate the time scale of these γ-transitions. The transition probability for an electric dipole transition is given by

$$T(\mathrm{E1}) = 8\pi c \frac{2\alpha}{9}\left(\frac{\Delta E}{\hbar c}\right)^3 \frac{1}{2l_\mathrm{i}+1}(l_\mathrm{f}\|Y_1\|l_\mathrm{i})^2 \langle n_\mathrm{f} l_\mathrm{f}|r|n_\mathrm{i} l_\mathrm{i}\rangle^2. \tag{149}$$

Let us calculate $T(\mathrm{E1})$ for a transition between circular orbits $(n_\mathrm{i}\equiv n, l_\mathrm{i}=n-1)$ $(n_\mathrm{f}=n-1, l_\mathrm{f}=n-2)$ with the wave functions (144). We have

$$(l_\mathrm{f}\|Y_1\|l_\mathrm{i}) = (-)^{l_\mathrm{i}+1}\sqrt{\frac{3}{4\pi}}\sqrt{(2l_\mathrm{i}+1)(2l_\mathrm{i}-1)}\begin{pmatrix} l_\mathrm{i} & 1 & l_\mathrm{i}-1 \\ 0 & 0 & 0 \end{pmatrix}$$
$$= -\sqrt{\frac{3(n-1)}{4\pi}},$$

$$\Delta E \equiv E_n - E_{n-1} = \frac{2n-1}{2n^2(n-1)^2}(Z\alpha)^2\mu c^2,$$

$$\langle n, n-1|r|n-1, n-2\rangle = a_\mathrm{B}\frac{2^{2n+1}n^{n+1}(n-1)^{n+2}}{(2n-1)^{2n}\sqrt{2(2n-1)(n-1)}}.$$

Putting these results together, we find

$$\hbar T(\mathrm{E1}; n\to n-1) = \frac{2^{4n}n^{2n-4}(n-1)^{2n-2}}{3(2n-1)^{4n-1}}\alpha^5\mu c^2 Z^4,$$

or, when expressing μ in terms of the electron mass,

$$T(\mathrm{E1}; n\to n-1) \simeq 5.355\times 10^9 \frac{2^{4n}n^{2n-4}(n-1)^{2n-2}}{(2n-1)^{4n-1}}\times\frac{\mu}{m_\mathrm{e}}Z^4[\mathrm{sec}^{-1}]. \tag{150}$$

The transition probability is proportional to the reduced mass and to the fourth power of the nuclear charge. For the sake of illustration let us calculate the transition time

$$\tau(\mathrm{E1}; n\to n') = 1/T(\mathrm{E1}; n\to n')$$

for the transitions $\{(n=14, l=13)\to(n=13, l=12)\}$ and $\{2\mathrm{p}\to 1\mathrm{s}\}$:

$$\tau(\mathrm{E1}; 14\to 13) \simeq 2.26(1+0.113/A)10^{-7}/Z^4\,\mathrm{sec}, \tag{150a}$$
$$\tau(\mathrm{E1}; 2\to 1) \simeq 7.72(1+0.113/A)10^{-12}/Z^4\,\mathrm{sec}. \tag{150b}$$

[The correction factor stems from the reduced mass $1/\mu \simeq (1/m_\mu)(1+m_\mu/Am_\mathrm{N})$]. These times must be compared with the lifetime (142) of the muon, $\tau \simeq 2.2\times 10^{-6}$

sec. In a very light system such as hydrogen ($Z=1$) the upper part of the cascade is relatively slow, so that many muons will decay during the cascade before they reach the 1s state. In a heavy atom such as lead ($Z=82$), however, the cascade times are scaled down by the factor Z^4. Rough interpolation between $n=14$ and $n=2$ shows that the whole cascade (assuming E1 transitions only) will take about 10^{-14} sec. This time is extremely short as compared to τ_μ.

We conclude: For medium and heavy nuclei the trapping time and the cascade are very short as compared to the muon lifetime. In these atoms the muon behaves exactly like a stable heavy electron. In very light atoms such as muonic hydrogen, however, the upper part of the cascade is affected appreciably by muon decay, at the expense of the intensities of the cascade γ-transitions. These qualitative considerations are confirmed by detailed cascade calculations on computers. A detailed knowledge of the cascade is important in many experimental situations. For instance, if one sets out to study properties of inner states such as the metastable 2s state, these calculations are essential in predicting the relative populations of these states.

After having reached the 1s state the muon either decays, or is captured by the nucleus through the weak interaction process

$$\mu^- + (Z, A) \rightarrow (Z-1, A) + \nu_\mu.$$

The capture rate Γ_{cap}, roughly, increases like Z^4. In light elements decay width and capture width are of similar magnitude. However, as Z increases, the capture reaction becomes predominant. In the heavy atoms, the lifetime of the muon is reduced, due to capture, to about 10^{-7} seconds.

These time scales of the muonic atom are illustrated in Fig. 9. They should be compared to typical time scales in the electronic shell as well as in the nuclear excitation spectrum. As to the former, we show, as an example, the lifetime of a hole state in the K-shell. As may be seen from the figure this atomic lifetime is comparable to cascade times of the muon. One consequence is this: If the muon, in the upper part of the cascade, has created a hole in e.g. the K-shell, it will make the rest of its cascade in presence of an incomplete electronic shell. This may be relevant in precision measurements of muonic transitions if the screening of the nuclear charge by the electrons has to be taken into account.

Regarding the nuclear excitation spectrum it can happen that due to accidental (near) degeneracy between muonic transition energies and the energies of certain nuclear excited states, the nucleus remains in an excited state when the muon has reached the 1s orbit. Comparison of typical lifetimes of such nuclear states with the lifetime $\tau(1s)$ of the muon in its 1s state (due to free decay and capture) shows a remarkable fact: The nuclear lifetimes are generally much shorter than $\tau(1s)$. Therefore, the nucleus can return to its ground state, through emission of a γ-ray, *while the muon remains in the 1s orbit*. This offers the unique possibility of observing a nuclear transition in presence of the muon in the 1s orbit which, as we know, penetrates strongly into the interior of the nucleus. The additional charge $-e$ leads to *isomer shifts*, the interaction of the nuclear magnetic moment with the muon's magnetic moment to *magnetic hyperfine structure*.

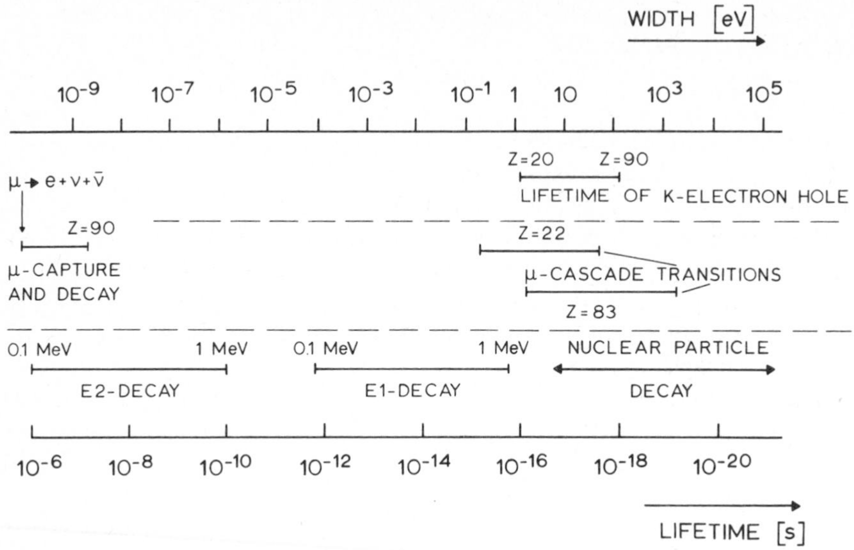

Fig. 9. Comparison of various lifetimes which are relevant for the dynamics in a muonic atom. The lower part of the figure shows typical times for nuclear γ- and particle decays. Taken from Ref. R21, Vol. I, Chap. III.

On the basis of these qualitative considerations we may group the information obtainable from muonic atoms according to the spatial extension of the orbits in question.

(a) For *intermediate and heavy elements* we distinguish

(a.1) *High-lying orbits*: These are the ones which overlap strongly with the electronic cloud. These states are affected in an essential way by the state of the host atom and by the chemical composition and physical structure of the target.

(a.2) *Very low orbits*: As the muon moves well inside the lowest electronic shell, screening effects are very small and often negligible. These orbits penetrate into the nucleus and, therefore, are sensitive to the spatial structure of nuclear charge, magnetization and current densities. As the transition energies are comparable to nuclear excitation energies and as the overlap with nuclear states is large, dynamical mixing effects between muonic and nuclear states can occur. Also, instead of emitting γ-rays the muon can transfer its energy to the nucleus which then decays via fission or via emission of neutrons. These radiationless transitions are especially important in heavy elements. Muon induced fission is an important tool for the study of fission in transuranium elements (fission barriers, fission isomers).

(a.3) *Intermediate orbits* ($3 \lesssim n \lesssim 6$): These are the ones which are the most hydrogenlike. Indeed, effects due to the finite size of the nucleus are small and can be calculated to a high degree of accuracy. Likewise, the screening effects due to the electronic cloud are small and under good control. Energies and wave functions of these states can be calculated to very high precision. These orbits are ideally suited

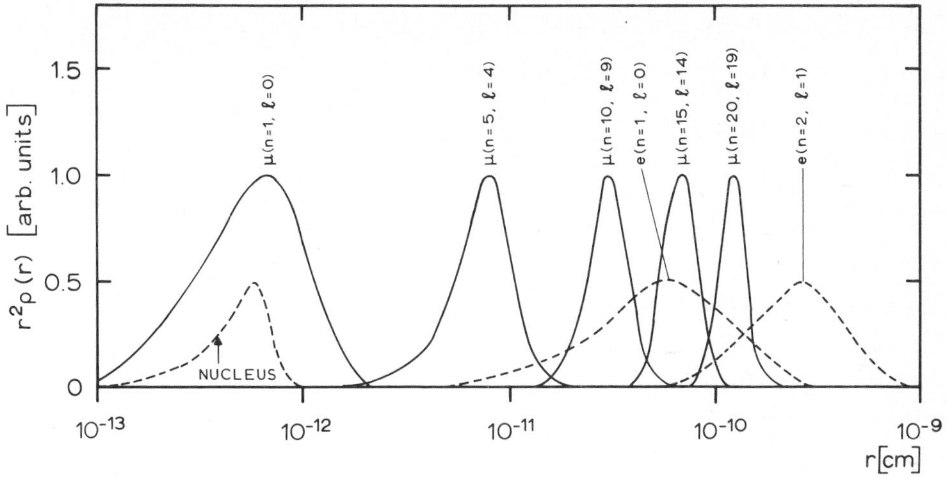

Fig. 10. Spatial structure of a heavy muonic atom. The scale in r (abscissa) is logarithmic; the densities are not normalized. Taken from Ref. R21, Vol. I, Chap. III.

for model-independent measurements of *static nuclear moments* (especially electric quadrupole and hexadecapole moments), and of *radiative corrections*.

Fig. 10 shows the densities $R_{nl}^2(r) \cdot r^2$ of selected muonic circular orbits in a heavy element (bismuth, $Z = 83$), in comparison to the nuclear charge density and to the density of K and L electrons. The figure illustrates quite clearly the three groups of orbits discussed above.

(b) In *very light elements* (such as hydrogen and helium) we have essentially only orbits of type (a.3). The electrons of the host atom are completely stripped off by Auger effect so that the muon sees the bare charge of the nucleus. On the nuclear side, the muonic orbit radii are such that penetration effects are small. Thus, the study of energies of very light muonic atoms concerns primarily tests of radiative corrections as predicted by quantum electrodynamics.

6.3. Dirac bound states in a central field

In a quantitative analysis of data on muonic atoms (i.e. transition energies and X-ray intensities) we need the exact bound state solutions of the Dirac equation. In the case of a spherically symmetric potential (central field) the total angular momentum $\boldsymbol{j} = \boldsymbol{l} + \boldsymbol{s}$ commutes with the energy and can be chosen diagonal. The problem is then reduced to the calculation of the radial wave functions $f_\kappa(r)$ and $g_\kappa(r)$ of the central field solutions

$$\Phi_{\kappa m}(\boldsymbol{r}) = \begin{pmatrix} g_\kappa(r)\varphi_{\kappa m} \\ \mathrm{i} f_\kappa(r)\varphi_{-\kappa m} \end{pmatrix}. \tag{151}$$

We recall the meaning of the symbols in eq. (151):

$$\kappa = \pm 1, \pm 2, \pm 3, \ldots \quad \text{(Dirac quantum number)},$$

$$j = |\kappa| - \tfrac{1}{2} \quad \text{and} \quad \begin{cases} l = \kappa & \text{for } \kappa > 0, \\ l = -\kappa - 1 & \text{for } \kappa < 0, \end{cases}$$

$$\varphi_{\kappa m} \equiv \varphi_{jlm} = \sum_{m_l, m_s} (l m_l, \tfrac{1}{2} m_s | jm) Y_{l m_l} \chi_{m_s}.$$

$f(r)$ and $g(r)$ satisfy a system of first-order differential equations, which contains the spherically symmetric potential $V(r)$, viz.

$$\frac{\mathrm{d} f_\kappa}{\mathrm{d} r} = \frac{\kappa - 1}{r} f_\kappa - \{E - V(r) - m\} g_\kappa,$$

$$\frac{\mathrm{d} g_\kappa}{\mathrm{d} r} = -\frac{\kappa + 1}{r} g_\kappa + \{E - V(r) + m\} f_\kappa. \tag{152}$$

g_κ is called the "large" component, f_κ the "small" component: in the nonrelativistic limit $m - |E| \ll 1$ f_κ vanishes, whilst g_κ goes over into the corresponding Schrödinger wave function (see below).

6.3.1. *Case (a): Potential of a point-like charge*

As nuclei are well confined systems the electrostatic potential eq. (117) around any nucleus approaches rapidly the potential of a point charge, $V_c = -Z\alpha/r$, outside the nuclear radius. Therefore, this potential is the reference case to which the actual potential should be compared. The bound states of the true potential are shifted with respect to the case of the point charge. The corresponding wave functions are more or less distorted Coulombic wave functions. Information about the true potential is primarily contained in the shifts of the energy levels, to some extent also in the wave functions and in observables derived from them.

Technically, the bound state problem is different for the two cases and the potential of the point charge must be considered separately. The reason for this technical difference is easy to understand: The true potential of an extended charge distribution is regular at the origin, $r = 0$. (Consider, for instance, the homogeneous density, for which $V(r) \sim C_0 + C_1 r^2$ is parabolic.) $V_c(r)$, on the contrary, is singular at $r = 0$. In any relativistic wave equation, both V_c and its square enter. This is obvious in the Klein–Gordon equation (II.36) where the term $(E - V_c(r))^2$ is added to the kinetic energy. In the case of Dirac spinors we know that each component also satisfies the Klein–Gordon equation. Alternately, this may be recovered directly from eqs. (152) by deriving from them uncoupled second-order equations for $f(r)$ and $g(r)$ separately (so-called "iterated form" of the Dirac equation).

In either case, the term $V_c^2(r)$, being proportional to $1/r^2$, has the same type of singularity at the origin as the centrifugal potential $l(l+1)/r^2$. This singularity, as is well-known, determines the behaviour of the radial wave functions at $r = 0$. Thus, for a regular potential we expect the standard behaviour r^l or r^{-l-1}, while for the $1/r$ potential the characteristic exponent is modified by terms of order $Z\alpha$. At infinity, $r \to \infty$, on the other hand, the two cases are obviously the same. So only the solutions inside a suitably defined matching radius R where the two potentials differ substantially, must be derived separately. After this long introduction we now turn to the derivation of the bound Coulomb states.

For very large r the differential equations (152) give the approximate equation

$$\begin{aligned} g''(r) - (m^2 - E^2) g(r) &\simeq 0 \\ f'(r) &\simeq -(E-m) g(r) \end{aligned} \qquad \text{(large } r\text{)}.$$

For bound states $E < m$ and $\lambda := \sqrt{m^2 - E^2}$ is real positive. Thus $g(r)$ behaves like $e^{-\lambda r}$, $f(r)$ like $\sqrt{(m-E)/(E+m)}\, g(r)$. For convenience we take out a factor $1/r$, in f and g, and we make the ansatz

$$\begin{aligned} g(r) &= \frac{1}{r} e^{-\lambda r} \sqrt{E+m}\, \{ y_1(r) + y_2(r) \}, \\ f(r) &= \frac{1}{r} e^{-\lambda r} \sqrt{m-E}\, \{ y_1(r) - y_2(r) \}. \end{aligned} \tag{153}$$

The asymptotic behaviour of $y_1(r)$ and $y_2(r)$ must be such that the exponential factor $e^{-\lambda r}$ is not compensated. Let $x := 2\lambda r$, then y_1 and y_2 satisfy the system

$$\begin{aligned} \frac{\mathrm{d} y_1}{\mathrm{d} x} &= \left(1 - \frac{Z\alpha E}{\lambda x}\right) y_1 - \left(\frac{\kappa}{x} + \frac{Z\alpha m}{\lambda x}\right) y_2, \\ \frac{\mathrm{d} y_2}{\mathrm{d} x} &= \left(-\frac{\kappa}{x} + \frac{Z\alpha m}{\lambda x}\right) y_1 + \frac{Z\alpha E}{\lambda x} y_2. \end{aligned} \tag{154}$$

The behaviour at $r = 0$ remains to be determined, keeping in mind the remarks made before. For that purpose, we write

$$y_{1,2}(x) = x^{\gamma} \phi_{1,2}(x), \tag{155}$$

with $\phi_{1,2}(0) \neq 0$ but finite and determine γ from eqs. (154): For $x = 0$ we obtain the linear system

$$\begin{aligned} \gamma \phi_1(0) &= -\frac{Z\alpha E}{\lambda} \phi_1(0) - \left(\kappa + \frac{Z\alpha m}{\lambda}\right) \phi_2(0), \\ \gamma \phi_2(0) &= \left(-\kappa + \frac{Z\alpha m}{\lambda}\right) \phi_1(0) + \frac{Z\alpha E}{\lambda} \phi_2(0). \end{aligned}$$

This homogeneous system has a nontrivial solution only if the determinant of the coefficient matrix

$$\begin{pmatrix} \gamma + Z\alpha E/\lambda & \kappa + Z\alpha m/\lambda \\ \kappa - Z\alpha m/\lambda & \gamma - Z\alpha E/\lambda \end{pmatrix}$$

vanishes. This gives $\gamma^2 = \kappa^2 - (Z\alpha)^2$. The solutions regular at the origin require the positive square root,

$$\gamma = \sqrt{\kappa^2 - (Z\alpha)^2}. \tag{156}$$

With the ansatz (155), where γ is given by eq. (156), one derives easily the differential equations satisfied by $\phi_1(x)$ and $\phi_2(x)$, viz.

$$\frac{d\phi_1}{dx} = \left\{1 - \left(\gamma + \frac{Z\alpha E}{\lambda}\right)\frac{1}{x}\right\}\phi_1 - \left(\kappa + \frac{Z\alpha m}{\lambda}\right)\frac{1}{x}\phi_2,$$

$$\frac{d\phi_2}{dx} = -\left(\kappa - \frac{Z\alpha m}{\lambda}\right)\frac{1}{x}\phi_1 - \left(\gamma - \frac{Z\alpha E}{\lambda}\right)\frac{1}{x}\phi_2. \tag{157}$$

The functions $\phi_i(x)$ represent what remains of the radial functions f and g after we have taken out the characteristic exponent (x^{γ}) near the origin and the exponential factor ($e^{-x/2}$) at infinity. Our experience with analogous problems in nonrelativistic quantum mechanics suggests that $\phi_i(x)$ are simple polynomials (they must be orthogonal for fixed κ). In fact, one can show that they can be written in terms of the well-known confluent hypergeometric function. One way of seeing this is this: Starting from eqs. (157) derive a second-order differential equation for ϕ_2 (or ϕ_1) alone. One finds

$$x\frac{d^2\phi_2}{dx^2} + (2\gamma + 1 - x)\frac{d\phi_2}{dx} - \left(\gamma - \frac{Z\alpha E}{\lambda}\right)\phi_2 = 0. \tag{158}$$

This is, indeed, Kummer's equation with $a \equiv \gamma - Z\alpha E/\lambda$, $b \equiv 2\gamma + 1$. The solution with the required properties is [R1, R22]

$$\phi_2(x) = {}_1F_1\left(\gamma - \frac{Z\alpha E}{\lambda}; 2\gamma + 1; x\right). \tag{159a}$$

ϕ_1 is found from the second equation (157) and the recurrence relation

$$x\,{}_1F_1'(a; b; x) + a\,{}_1F_1(a; b; x) = a\,{}_1F_1(a + 1; b; x)$$

for the confluent hypergeometric function. Thus

$$\phi_1(x) = \frac{Z\alpha E/\lambda - \gamma}{\kappa - Z\alpha m/\lambda}\,{}_1F_1(1 + \gamma - Z\alpha E/\lambda; 2\gamma + 1; x). \tag{159b}$$

The asymptotic behaviour of ${}_1F_1$ is, to leading order in $1/x$,

$${}_1F_1(a; b; x) \sim \frac{\Gamma(b)}{\Gamma(b-a)}(-x)^{-a} + \frac{\Gamma(b)}{\Gamma(a)}e^x x^{a-b}. \tag{160}$$

Obviously, the second term of this must vanish if we do not wish to destroy the exponential decrease of the bound state solutions, i.e. the factor $e^{-x/2}$ in eqs. (153). This can only be achieved if the parameter a is zero or a negative integer because in that case $1/\Gamma(a)$ vanishes. When applied to ϕ_2, this gives the condition

$$Z\alpha E/\lambda - \gamma = n', \tag{161}$$

with

$$\lambda = \sqrt{m^2 - E^2}, \qquad n' = 0, 1, 2, \ldots .$$

Does the same condition make also $\phi_1(x)$, eq. (159b), remain regular at infinity? For $n' \geq 1$ this is obvious. For $n' = 0$, a little more care is necessary: The function ${}_1F_1(1; 2\gamma + 1; x)$ is not regular but the factor $Z\alpha E/\lambda - \gamma = 0$ in front of it makes ϕ_1

vanish, provided the denominator $(\kappa - Z\alpha m/\lambda)$ does not vanish. This is what has to be checked. With $\gamma = Z\alpha E/\lambda$ we have

$$\kappa^2 = \gamma^2 + (Z\alpha)^2 = (Z\alpha)^2 \frac{\lambda^2 + E^2}{\lambda^2} = \left(Z\alpha \frac{m}{\lambda}\right)^2$$

and therefore $\kappa = \pm Z\alpha m/\lambda$. The solution $\kappa = +Z\alpha m/\lambda$ must indeed be excluded, whilst $\kappa = -Z\alpha m/\lambda$ is acceptable. Therefore, for $n' = 0$ only negative κ is allowed.

For convenience, we set $n' = n - |\kappa|$ with $n = 1, 2, \ldots$. Our results can then be summarized as follows. From eq. (161) we find

$$E_{n|\kappa|} = m \left\{ 1 + \left(\frac{Z\alpha}{n - |\kappa| + \sqrt{\kappa^2 - (Z\alpha)^2}} \right)^2 \right\}^{-1/2}, \tag{162}$$

where the quantum numbers n, κ and $|\kappa| - 1/2$ assume the following values:

$$\begin{aligned} &n = 1, 2, \ldots, \\ &\kappa = \pm 1, \pm 2, \ldots, \pm (n-1), -n, \\ &j = |\kappa| - \tfrac{1}{2} = \tfrac{1}{2}, \tfrac{3}{2}, \ldots, n - \tfrac{1}{2}. \end{aligned} \tag{163}$$

$\kappa = +n$, as we said above, is excluded. n is the familiar main quantum number of the nonrelativistic hydrogen atom. This can be seen, for instance, by expanding the energy eigenvalues (162) in terms of Z,

$$E_{n|\kappa|} \simeq m \left\{ 1 - \frac{(Z\alpha)^2}{2n^2} - \frac{(Z\alpha)^4}{2n^4} \left(\frac{n}{|\kappa|} - \frac{3}{4} \right) \right\}, \tag{164}$$

the first term of which is the rest mass, whilst the second gives the binding energy (143) of the nonrelativistic hydrogen atom. The third term is the first relativistic correction. This term is independent of the mass (relative to the others) but depends on the angular momentum $j = |\kappa| - \frac{1}{2}$. For constant n it is relatively more important for small values of j than for large values.

As may be seen from the exact formula (162) the dynamical l-degeneracy of the nonrelativistic hydrogen atom is almost completely lifted. Only energy eigenvalues of equal n and $|\kappa|$ are degenerate. As $|\kappa|$ can be positive and negative, except for the largest value of κ where $\kappa = -n$, all j-values except for the highest $j = n - \frac{1}{2}$ have a *two*fold dynamical degeneracy, in addition to the usual directional degeneracy in m_j, the magnetic quantum number. Thus, the $2s_{1/2}$ and $2p_{1/2}$ states are degenerate, the $3s_{1/2}$ and $3p_{1/2}$ states, the $3p_{3/2}$ and $3d_{3/2}$ states, and so on.

This remaining degeneracy of the relativistic atom is lifted eventually by radiative corrections (Lamb shift).

The eigenfunctions are given by our eqs. (153), (155) and (159). The normalization to 1 is best performed by making use of well-known integrals involving confluent hypergeometric functions, exponentials and powers.

The result is

$$\begin{aligned} g_{n\kappa}(r) &= 2\lambda N(n,\kappa)\sqrt{m+E}\, x^{\gamma-1} e^{-x/2} \\ &\quad \times \Big\{ -(n-|\kappa|)\, {}_1F_1(-n+|\kappa|+1; 2\gamma+1; x) \\ &\qquad + \left(\frac{Z\alpha m}{\lambda} - \kappa\right) {}_1F_1(-n+|\kappa|; 2\gamma+1; x) \Big\} \\ f_{n\kappa}(r) &= -2\lambda N(n,\kappa)\sqrt{m-E}\, x^{\gamma-1} e^{-x/2} \\ &\quad \times \Big\{ (n-|\kappa|)\, {}_1F_1(-n+|\kappa|+1; 2\gamma+1; x) \\ &\qquad + \left(\frac{Z\alpha m}{\lambda} - \kappa\right) {}_1F_1(-n+|\kappa|; 2\gamma+1; x) \Big\}. \end{aligned} \tag{165}$$

The normalization constant $N(n, \kappa)$ is given by

$$N(n,\kappa) = \frac{\lambda}{m} \frac{1}{\Gamma(2\gamma+1)} \left\{ \frac{\Gamma(2\gamma+n-|\kappa|+1)}{2Z\alpha(Z\alpha m/\lambda - \kappa)\Gamma(n-|\kappa|+1)} \right\}^{1/2}. \tag{166a}$$

As before,

$$x = 2\lambda r, \tag{166b}$$

$$\lambda = \sqrt{m^2 - E_{n|\kappa|}^2} = \frac{Z\alpha m}{\sqrt{n^2 - 2(n-|\kappa|)(|\kappa|-\gamma)}}, \tag{166c}$$

$$\gamma = \sqrt{\kappa^2 - (Z\alpha)^2}. \tag{166d}$$

Strictly speaking, the orbital angular momentum l is not a good quantum number. However, in the limit of weakly relativistic motion, i.e. for $Z\alpha \ll 1$, $\sqrt{m-E} = \mathcal{O}(Z\alpha)$ is small compared to $\sqrt{m+E} \simeq 2m$. This shows that the wave function $g_{n\kappa}(r)$ is large compared to the wave function $f_{n\kappa}(r)$. Thus, the upper component in $\Phi_{n\kappa m}$, eq. (151), is large compared to the lower one, and its angular momentum l can be used to label the state, even though the lower component carries a different angular momentum $\bar{l} = l \pm 1$. The nomenclature is approximate and refers to the corresponding nonrelativistic situation. As an example let us consider all states with $n = 2$. Here we have

$$n = 2,\ \kappa = -1,\ j = \tfrac{1}{2}: \quad \begin{pmatrix} l = 0 \\ \bar{l} = 1 \end{pmatrix} \text{ “2s}_{1/2}\text{-state”},$$

$$n = 2,\ \kappa = +1,\ j = \tfrac{1}{2}: \quad \begin{pmatrix} l = 1 \\ \bar{l} = 0 \end{pmatrix} \text{ “2p}_{1/2}\text{-state”},$$

$$n = 2,\ \kappa = -2,\ j = \tfrac{3}{2}: \quad \begin{pmatrix} l = 1 \\ \bar{l} = 2 \end{pmatrix} \text{ “2p}_{3/2}\text{-state”}.$$

The example shows that in the relativistic case (i.e. $Z\alpha$ not small compared to 1), the $2p_{1/2}$-state is a closer parent to the $2s_{1/2}$ than to the $2p_{3/2}$. We expect relativistic effects in the $2p_{1/2}$ to be more important than in the $2p_{3/2}$.

The limit to the purely nonrelativistic case can be verified on the expressions (165) for the wave functions. As

$$\lambda \simeq \frac{Z\alpha m}{n}\left(1+\frac{(Z\alpha)^2}{2n^2}\frac{n-|\kappa|}{|\kappa|}\right),$$

this means calculating the wave functions to lowest non-vanishing order in $Z\alpha$. In this limit $m-E\simeq 0$ and $f_{n\kappa}(r)\simeq 0$. For $\kappa>0$ we have $l=\kappa$,

$$x \simeq \frac{2Z\alpha m}{n}r = \frac{2r}{na_{\mathrm{B}}} =: z,$$

and

$$g_{n\kappa=l} \simeq 2\frac{Z\alpha m}{n}\frac{Z\alpha}{n}\frac{1}{\Gamma(2l+1)}\sqrt{\frac{\Gamma(n+l+1)}{2Z\alpha(n-l)\Gamma(n-l+1)}}\sqrt{2m}\,z^{l-1}\mathrm{e}^{-z/2}$$
$$\times(n-l)\{{}_1F_1(-n+l;2l+1;z)-{}_1F_1(-n+l+1;2l+1;z)\},$$

which equation, by means of the recurrence relation

$$b\{{}_1F_1(a;b;z)-{}_1F_1(a-1;b;z)\}=z\,{}_1F_1(a;b+1;z) \tag{167}$$

goes over into the nonrelativistic wave function $(1/r)y_{nl}(r)$, eq. (145). For negative κ, $\bar{l}=-\kappa-1=|\kappa|-1$, $g_{n\kappa}(r)$ must go over into $(1/r)y_{n,\bar{l}}(r)$, since in the nonrelativistic limit the states $(n,\ j=l+\frac{1}{2})$ and $(n,\ j=l-\frac{1}{2})$ have the same radial wave function. That this is indeed so may be verified from eq. (165), and the recurrence relation

$$(1+a-b)\,{}_1F_1(a;b;z)-a\,{}_1F_1(a+1;b;z)+(b-1)\,{}_1F_1(a;b-1;z)=0 \tag{168}$$

(see exercise 8).

6.3.2. *Case (b): Potential of a spherically symmetric charge distribution of finite size*

For simplicity we consider a *spherically symmetric* charge distribution $\rho(r)$ of finite size. (The more general case of an arbitrary density $\rho(\boldsymbol{r})$ is dealt with in sec. 8, below.) Unlike the case of a point charge $\rho(r)$ is supposed to be regular at the origin and to admit a Taylor expansion around $r=0$,

$$\rho(r)=\rho_0+\rho_1 r+\frac{1}{2!}\rho_2 r^2+\mathcal{O}(r^3). \tag{169}$$

Inserting this series into the formula (117) for the potential this yields a similar expansion for $V(r)$, viz.

$$V(r)=-4\pi Z\alpha^2\left\{\int_0^\infty \rho(r')r'\mathrm{d}r'-\tfrac{1}{6}\rho_0 r^2-\tfrac{1}{12}\rho_1 r^3-\tfrac{1}{40}\rho_2 r^4+\mathcal{O}(r^5)\right\}. \tag{170}$$

This shows that $V(r)$ behaves like a parabola, $V_0+V_1r^2$, close to the origin. As a consequence the behaviour of the radial solutions $f(r)$ and $g(r)$ near the origin is determined entirely by the centrifugal potential but not by the electrostatic potential.

For $\kappa > 0$ the upper component of the spinor (151) carries the orbital angular momentum $l = \kappa$, whilst the lower component carries $\bar{l} = \kappa - 1$. We expect, therefore, the solutions regular at the origin to behave according to

$$\kappa > 0 \quad \begin{cases} g_\kappa(r) \sim r^\kappa \\ f_\kappa(r) \sim r^{\kappa-1}. \end{cases} \tag{171a}$$

Similarly, for $\kappa < 0$, the upper component has $l = -\kappa - 1$, the lower component has $\bar{l} = -\kappa$, so that we expect

$$\kappa < 0 \quad \begin{cases} g_\kappa(r) \sim r^{-\kappa-1} \\ f_\kappa(r) \sim r^{-\kappa}. \end{cases} \tag{171b}$$

It is not difficult to prove these assertions. The second-order differential equations for $f(r)$ and $g(r)$ alone, which one derives from the system (152), contain the centrifugal terms $\kappa(\kappa-1)/r^2$ and $\kappa(\kappa+1)/r^2$, respectively. The characteristic exponents α and β in the ansatz $f(r) = r^\alpha \Sigma a_n r^n$ and $g(r) = r^\beta \Sigma b_n r^n$ are found to satisfy the equations

$$\alpha(\alpha+1) = \kappa(\kappa-1), \qquad \beta(\beta+1) = \kappa(\kappa+1),$$

whose solutions are $\alpha_1 = \kappa - 1$, $\alpha_2 = -\kappa$ and $\beta_1 = \kappa$, $\beta_2 = -\kappa - 1$. In fact, the two cases of positive and negative κ, can be written in a particularly simple and compact form if we introduce the definitions

$$\kappa > 0 \quad \begin{cases} g_\kappa(r) = r^\kappa G_\kappa(r), \\ f_\kappa(r) = r^{\kappa-1} F_\kappa(r), \end{cases} \tag{171c}$$

$$\kappa < 0 \quad \begin{cases} g_\kappa(r) = r^{-\kappa-1} F_\kappa(r), \\ f_\kappa(r) = -r^{-\kappa} G_\kappa(r). \end{cases} \tag{171d}$$

The functions F_κ and G_κ obey the system of first-order differential equations

$$\frac{\mathrm{d}F_\kappa}{\mathrm{d}r} = \{V(r) - E + m \operatorname{sign} \kappa\} r G_\kappa,$$

$$\frac{\mathrm{d}G_\kappa}{\mathrm{d}r} = -\frac{1}{r}\{(2|\kappa|+1)G_\kappa + [V(r) - E - m \operatorname{sign} \kappa] F_\kappa\}, \tag{172}$$

with initial conditions

$$F_\kappa(0) = a_0 \neq 0, \qquad G_\kappa(0) = -\frac{V(0) - E - m \operatorname{sign} \kappa}{2|\kappa|+1} a_0,$$

$$\frac{\mathrm{d}F_\kappa}{\mathrm{d}r}(0) = \frac{\mathrm{d}G_\kappa}{\mathrm{d}r}(0) = 0. \tag{173}$$

The system (172) with the initial conditions (173) is well adapted for numerical integration of the wave functions F and G, from which f and g are then obtained by means of (171c) or (171d), respectively. These solutions, which by construction are regular at the origin, must also be regular at infinity. This condition fixes the eigenvalues $E_{n\kappa}$ for given κ. In order to achieve this, it is convenient to use the following trick.

The density $\rho(r)$ has a finite extension, i.e. beyond a certain radius R_M the density is zero or negligibly small. (In case of nuclei R_M is typically 2 to 3 times the nuclear radius.) Therefore, for $r \gtrsim R_M$ the potential $V(r)$ is practically indistinguishable from $V_c = -Z\alpha/r$, the potential of a point-like charge, and the true wave functions (171c, d) must be linear combinations of two linearly independent solutions of the system of differential equations treated above, case (a). One such set of independent solutions for V_c could be the solution (165), regular at the origin, together with another solution which is singular at the origin.

The trick now consists in constructing that specific linear combination which, for *arbitrary* energy E, decreases exponentially at infinity (i.e. is regular at infinity), irrespective of its behaviour at $r=0$. Call this solution (f_∞, g_∞). It then suffices to vary E until the inner solutions f_0, g_0 which are regular at $r=0$ and which are obtained by numerical integration, match the outer solution continuously at the point $r=R_M$, viz.

$$[f_\infty(r)g_0(r) - g_\infty(r)f_0(r)]_{r=R_M} = 0. \tag{174}$$

For the construction of (f_∞, g_∞) we return to eq. (158). It is not difficult to verify that a solution independent of the regular one $\phi_2^R(x) = {}_1F_1(a; b; x)$, where $a = \gamma - Z\alpha E/\lambda$, $b = 2\gamma + 1$, is this:

$$\begin{aligned}\phi_2^I(x) &= x^{1-b}{}_1F_1(1+a-b; 2-b; x)\\ &= x^{-2\gamma}{}_1F_1(-\gamma - Z\alpha E/\lambda; -2\gamma+1; x).\end{aligned}$$

According to the defining eqs. (153) and (155) ϕ^B and ϕ^I enter into f and g with the factor $x^\gamma e^{-x/2}$. Thus, rf and rg are linear combinations of the functions

$$M(\varepsilon, s; x) = x^s e^{-x/2}{}_1F_1(s-\varepsilon; 2s+1; x), \tag{175}$$

where $\varepsilon := Z\alpha E/\lambda$ and $s = \pm\gamma$.

The asymptotic behaviour of $M(\varepsilon, s, x)$ is determined by the asymptotic form (160) of the confluent hypergeometric function. For real and positive x the second term in eq. (160) dominates and we have

$$M(\varepsilon, \pm\gamma; x) \underset{x\to\infty}{\sim} \frac{\Gamma(\pm 2\gamma+1)}{\Gamma(\pm\gamma-\varepsilon)} x^{-\varepsilon-1} e^{x/2}. \tag{176}$$

We now combine the two solutions for positive and negative γ in such a way as to cancel out the exponentially increasing term (176). Noting that $\Gamma(\pm 2\gamma+1) = \pm 2\gamma\Gamma(\pm 2\gamma)$, this is achieved by taking the combination

$$W(\varepsilon, s; x) := \frac{\Gamma(-2\gamma)}{\Gamma(-\gamma-\varepsilon)} M(\varepsilon, \gamma; x) + \frac{\Gamma(2\gamma)}{\Gamma(\gamma-\varepsilon)} M(\varepsilon, -\gamma; x). \tag{177}$$

This specific linear combination is regular at infinity for all values $\varepsilon = Z\alpha E/\lambda$.*)

*) W is a Whittaker function, except for an extra factor $x^{1/2}$ on the right-hand side of eq. (177).

Repeating some of the steps of case (a) it is straightforward to construct the radial solution (f_∞, g_∞) from (177). One finds, up to arbitrary normalization,

$$rf_\infty(r) = \sqrt{m - E}\left\{\left(\kappa + \frac{Z\alpha m}{\lambda}\right)W(\varepsilon - 1, \gamma; x) - W(\varepsilon, \gamma; x)\right\}, \tag{178a}$$

$$rg_\infty(r) = \sqrt{m + E}\left\{\left(\kappa + \frac{Z\alpha m}{\lambda}\right)W(\varepsilon - 1, \gamma; x) + W(\varepsilon, \gamma; x)\right\}. \tag{178b}$$

Note that these solutions are indeed regular at infinity for any value of the energy E. (This still remains true when E becomes complex. This occurs if the Dirac equation contains a complex optical potential.)

Remarks: (i) The functions (178) are, in general, not regular at the origin because of the factor $x^{-\gamma}$ in the second term of eq. (177). (ii) If one imposes regularity at the origin, too, this term must vanish and one recovers the eigenvalue condition of the previous case (a), cf. eq. (161).

7. Muonic atoms and quantum electrodynamics

In a sense, vacuum polarization is the simplest and most fundamental radiative effect in the quantized theory of photons in interaction with matter. What is this effect and why is it fundamental? In the quantized Maxwell theory the electromagnetic forces which act between charged particles are described by the exchange of photons between these particles. The exchanged photon, through its quantum nature, can go over into all possible intermediate states which are allowed by the laws of conservation of the theory. These intermediate states can annihilate and go over into the same photon state again, as sketched in Fig. 11. In this diagram the hatched loop stands for the sum of all many particle and photon states which are allowed by the rules of the theory. The net effect of these insertions is a modification of the photon propagator or, in other words, of the classical forces between charged matter particles. The range of these modifications, in coordinate space, is a function of the masses of the particles in the intermediate states. The phenomenon occurs in any local gauge theory (the photon being replaced by the vector gauge bosons of the

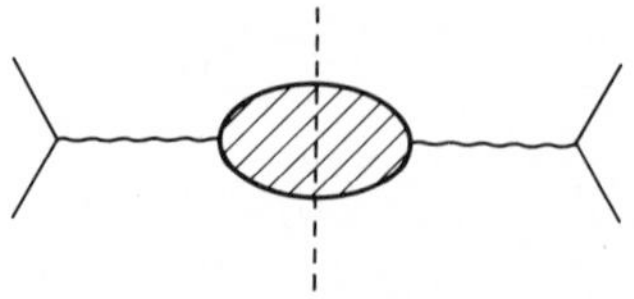

Fig. 11. Vacuum polarization in quantum electrodynamics leading to modification of the photon propagator.

theory), and it reflects fundamental properties of the theory. This can be understood qualitatively by cutting the diagram in Fig. 11 as indicated by the broken line: If we change the external momenta such that the particle lines at the left and at the right of this figure represent incoming and outgoing particle-antiparticle pairs respectively, then the cut diagrams represent the total pair annihilation cross sections.

In quantum electrodynamics (QED), to lowest order in the fine structure constant α, vacuum polarization is represented by the virtual creation and re-annihilation of all possible pairs of one fermion and its antifermion, as shown in Fig. 12(a). This diagram (and likewise all diagrams of higher order in α), when added to the single photon exchange, leads to a modification of the photon propagator and has two basic effects: The first effect is an infinite, logarithmically divergent, contribution which, however, is the same in any diagram where the photon couples to a given particle of charge e_0. The presence of this divergent term indicates that the *bare* charge e_0 (i.e. the coupling constant appearing in the original Lagrangian), is renormalized, by an infinite amount, to the *physical* charge e. The infinity can be circumvented formally by a redefinition of the coupling constant which means replacing the bare charge e_0 everywhere by the physical charge e. This prescription is formal because e_0 is necessarily infinite and yet, the replacement is to be made, order by order, to the order α^n of perturbation theory to which we work, as if everything were finite. As the theory does not allow to predict the magnitude of the charge e, this first and divergent effect of vacuum polarization is not observable.

In contrast to this, the remainder of vacuum polarization is finite and unique, and does have observable consequences. This second effect can be calculated uniquely, in successive orders of α, and can be confronted with precision measurements, as a test of QED. The physical effect of (finite) vacuum polarization is best understood in the case of an external, electrostatic potential. In this case vacuum polarization leads to

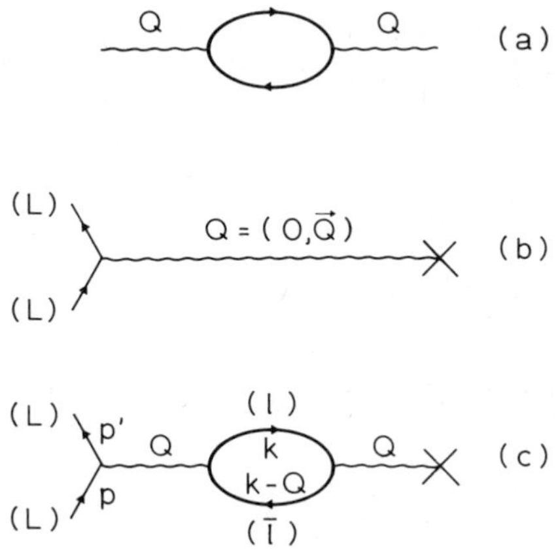

Fig. 12. (a) Modification of a photon propagator to order α, due to a loop of virtual charged particles (electron, muon, etc.). (b) Interaction of a charged lepton (L) with an external potential (represented by the cross). (c) Modification of the interaction due to diagram (b) through lowest-order vacuum polarization.

a distortion of the given potential, over a distance which is characterized by the Compton wave length $\lambda\!\!\!^{-}(i)=\hbar/m_i c$ of the particle that runs along the loop in the diagram of Fig. 12(a). Clearly, the longest range in this distortion effect is due to the lightest particle, the electron. Thus, vacuum polarization due to virtual *electron–positron pairs* is expected to distort the original potential over a distance of the order

$$\lambda\!\!\!^{-}(\mathrm{e}) \simeq 386 \text{ fm}. \tag{179}$$

If we wish to see the quantitative importance of this effect in atoms we must check whether the orbit radii are large as compared to $\lambda\!\!\!^{-}(\mathrm{e})$ or whether they are comparable to or smaller than $\lambda\!\!\!^{-}(\mathrm{e})$. Hydrogen and muonium ($\mu^+\mathrm{e}^-$) have Bohr radii $\sim 1/\alpha m \simeq 5\times 10^4$ fm and belong to the first class. In those "dilute" systems vacuum polarization is a small effect as compared to other radiative corrections due to vertex modification and to the anomalous magnetic moments. In contrast to these, muonic atoms have sizes which are indeed comparable to $\lambda\!\!\!^{-}(\mathrm{e})$, eq. (179). As a consequence, vacuum polarization is the predominant radiative correction in muonic atoms. As far as tests of QED are concerned, weakly bound systems, such as ordinary light atoms and muonium, and muonic atoms are complementary. They test different and complementary predictions of QED. We discuss first the finite and observable parts of vacuum polarization, in lowest order. We then say a little more about higher orders and give some characteristic examples. The discussion of muonium (which is an example for the complementary situation, similar to hydrogen) can be found e.g. in Scheck (1978).

7.1. *Observable part of vacuum polarization to order $O(\alpha)$*

For the sake of simplicity let us consider the case where the photon line in Fig. 12(a) represents the interaction of a fermion (L) with an *external* electrostatic potential of a *point charge c*. This potential is represented by a cross in Fig. 12(b) and (c). According to standard Feynman rules the translation of this cross and the photon line into a formula is

$$\tilde{A}_\mu(\boldsymbol{Q}^2)=\frac{c}{(2\pi)^3}\delta_{\mu 0}\frac{1}{\boldsymbol{Q}^2}, \tag{180}$$

which is the Fourier transform of the usual $1/r$-potential*)

$$\tilde{A}_\mu(\boldsymbol{Q}^2)=\frac{1}{(2\pi)^3}\int \mathrm{d}^3Q\,\mathrm{e}^{-\mathrm{i}\boldsymbol{Q}\boldsymbol{x}}\frac{c}{4\pi|\boldsymbol{x}|}\delta_{\mu 0}. \tag{181′}$$

Consider now the vacuum loop of a charged lepton ℓ illustrated by Fig. 12(c).

*) In the standard formulation of Feynman rules one chooses natural units so that $e^2/4\pi \equiv \alpha$.

According to the Feynman rules the sum of diagrams b and c is given by

$$M = 2\pi i e_0 \overline{u_{\rm L}(p')}\gamma_\mu u_{\rm L}(p)$$
$$\times\left[g^{\mu\nu} + \frac{1}{Q^2}\frac{ie_0^2}{(2\pi)^4}\int d^4k\,\mathrm{Sp}\left\{\gamma^\mu \frac{1}{\not k - m_\ell + i\varepsilon}\gamma^\nu \frac{1}{\not k - \not Q - m_\ell + i\varepsilon}\right\}\tilde{A}_\nu\right], \tag{181}$$

where $Q = p - p'$. In the specific case of elastic scattering of the fermion L in an external potential we have $p^0 = p'^0$, so that $Q^2 = -\boldsymbol{Q}^2$. In a more general diagram Q is the momentum carried by the photon lines entering and leaving the closed fermion loop (see Fig. 12(a)). In any such case the vacuum polarization loop of order e^2 is represented by the tensor

$$\Pi^{\mu\nu}(Q) = \frac{ie_0^2}{(2\pi)^4}\int d^4k\,\mathrm{Sp}\left\{\gamma^\mu \frac{1}{\not k - m_\ell + i\varepsilon}\gamma^\nu \frac{1}{\not k - \not Q - m_\ell + i\varepsilon}\right\}. \tag{182}$$

As it stands this integral is divergent and we must be very careful in performing algebraic manipulations on it. It is well-defined and finite only in a regularized form of QED. Methods of regularization are described in textbooks on quantum electrodynamics. They are essential in identifying the precise nature of the singularities of divergent quantities and in isolating these from the finite parts. If the regularization respects Lorentz invariance and gauge invariance of the theory, these latter, finite parts are unique.

In what follows we assume that the tensor $\Pi^{\mu\nu}(Q)$ is already regularized. For instance, using the method of Pauli and Villars, one finds

$$\Pi^{\mu\nu}_{\rm reg}(Q) = (Q^\mu Q^\nu - Q^2 g^{\mu\nu})\left[C + \Pi(Q^2)\right], \tag{183}$$

where the first term depends on a fictitious regulator mass M,

$$C = \frac{\alpha_0}{3\pi}\ln\left(\frac{M}{m_\ell}\right)^2 \tag{184}$$

(and hence is logarithmically divergent), whilst the second term

$$\Pi(Q^2) = -\frac{2\alpha}{\pi}\int_0^1 dz\, z(1-z)\ln\left(\frac{m_\ell^2 - Q^2 z(1-z)}{m_\ell^2 - i\varepsilon}\right) \tag{185}$$

is finite and independent of the method of regularization used. It can be shown that the logarithmic divergence (184) represents a formal renormalization of the charge and that it can be absorbed if the bare charge is replaced by the physical charge,*)

$$e = e_0\sqrt{1 - \frac{\alpha}{3\pi}\ln\left(\frac{M}{m}\right)^2}\,. \tag{186}$$

Finally, we note that the specific covariant form of $\Pi^{\mu\nu}$, eq. (183), is a consequence

*) As we work in second order here, it is consistent to insert $\alpha = e^2/4\pi$, not $\alpha_0 = e_0^2/4\pi$ into (185) and the square root in (186).

of gauge invariance which requires

$$Q_\mu \Pi^{\mu\nu}_{\text{reg}}(Q) = 0 = \Pi^{\mu\nu}_{\text{reg}}(Q) Q_\nu .$$

In order to understand the physical content of the finite part $\Pi(Q^2)$ let us transform the integral (185) somewhat so that it may easily be transformed to coordinate space, by means of Fourier transformation. Let $z = (1-y)/2$ and, therefore, $1 - z = (1+y)/2$. Then

$$\Pi(Q^2) = -\frac{\alpha}{2\pi}\int_0^1 \mathrm{d}y\,(1-y^2)\ln\left\{1 - \frac{Q^2}{m^2 - \mathrm{i}\varepsilon}\,\frac{1-y^2}{4}\right\} \quad (m \equiv m_\ell).$$

By partial integration we can get rid of the logarithm and obtain

$$\Pi(Q^2) = \frac{\alpha}{\pi} Q^2 \int_0^1 \mathrm{d}y \frac{y^2(1-y^2/3)}{4m^2 - Q^2(1-y)^2 - \mathrm{i}\varepsilon}. \tag{187}$$

An equivalent representation is obtained by means of the substitution

$$\kappa^2 := \frac{4m^2}{1-y^2},$$

i.e.

$$y^2 = 1 - \frac{4m^2}{\kappa^2}, \qquad \mathrm{d}\kappa^2 = \frac{\kappa^4}{2m^2} y\,\mathrm{d}y,$$

which gives

$$\Pi(Q^2) = \frac{\alpha Q^2}{3\pi}\int_{4m^2}^{\infty} \mathrm{d}\kappa^2 \frac{(1+2m^2/\kappa^2)\sqrt{1-4m^2/\kappa^2}}{\kappa^2(\kappa^2 - Q^2 - \mathrm{i}\varepsilon)}. \tag{188}$$

Let us now return to the example of scattering in an external electrostatic potential. Inserting the result (188) into the amplitude (181), we obtain after having renormalized the charge (to the order at which we work),

$$\begin{aligned} M &= 2\pi \mathrm{i} e \overline{u_{\mathrm{L}}(p')}\gamma_\mu u_{\mathrm{L}}(p)\left[g^{\mu\nu} + \frac{1}{Q^2}(Q^\mu Q^\nu - Q^2 g^{\mu\nu})\Pi(Q^2)\right]\tilde{A}_\nu \\ &= 2\pi \mathrm{i} e \overline{u_{\mathrm{L}}(p')}\gamma^\nu u_{\mathrm{L}}(p)\left[1 - \Pi(Q^2)\right]\tilde{A}_\nu . \end{aligned}$$

In this example $Q^2 = -\boldsymbol{Q}^2$, $\tilde{A}_\nu$ is given by eq. (180), $\Pi(Q^2 = -\boldsymbol{Q}^2)$ by eq. (188). The factor $[1-\Pi]A$ can be read as a modified external potential and may easily be transformed to coordinate space. Let

$$\tilde{V}(\boldsymbol{Q}^2) := -e\left[1 - \Pi(-\boldsymbol{Q}^2)\right]\tilde{A}_0(\boldsymbol{Q}^2). \tag{189}$$

The potential in coordinate space is the Fourier transform of $\tilde{V}$,

$$V(\boldsymbol{x}) = \frac{C}{(2\pi)^3}\int \mathrm{d}^3\boldsymbol{Q}\,\mathrm{e}^{\mathrm{i}\boldsymbol{Q}\boldsymbol{x}}\left\{\frac{1}{\boldsymbol{Q}^2} + \frac{\alpha}{3\pi}\int_{4m^2}^{\infty}\mathrm{d}\kappa^2 \frac{(1+2m^2/\kappa^2)\sqrt{1-4m^2/\kappa^2}}{\kappa^2(\kappa^2 + \boldsymbol{Q}^2 - \mathrm{i}\varepsilon)}\right\}, \tag{189'}$$

where we set $-ce = C$.

The integrals over $\boldsymbol{Q}$ can be performed by means of the formula

$$\frac{1}{(2\pi)^3}\int \mathrm{d}^3 Q \frac{\mathrm{e}^{\mathrm{i}\boldsymbol{Q}\boldsymbol{x}}}{\boldsymbol{Q}^2+a^2} = \mathrm{e}^{-ar}/4\pi r \quad (r=|\boldsymbol{x}|),$$

taking $a = 0$ in the first term, $a = \kappa$ in the second term of eq. (189′). This gives

$$V(r) = \frac{C}{4\pi}\left\{\frac{1}{r} + \frac{\alpha}{3\pi}\int_{4m^2}^{\infty} \mathrm{d}\kappa^2 \frac{(1+2m^2/\kappa^2)\sqrt{1-4m^2/\kappa^2}}{\kappa^2}\frac{\mathrm{e}^{-\kappa r}}{r}\right\}. \tag{190}$$

Thus, the original $1/r$ potential is modified by a superposition of Yukawa terms with ranges greater or equal $1/2m = \frac{1}{2}\lambda\!\!\!^{-}(\mathrm{e})$.

For practical use one may replace the variable κ^2 by a dimensionless integration variable, viz. $\kappa^2 = 4m^2x^2$, so that (Uehling 1935)

$$V(r) = \frac{C}{4\pi r}\left\{1 + \frac{2\alpha}{3\pi}\int_{1}^{\infty} \mathrm{d}x\, \mathrm{e}^{-2mxr}\left(1+\frac{1}{2x^2}\right)\frac{\sqrt{x^2-1}}{x^2}\right\}. \tag{190′}$$

Remember that c is the external charge. For instance, if this is a proton then $c = |e|$; if it is a point-like nucleus then $c = +Z|e|$. (The case of an extended charge distribution is treated below.)

There are two limiting situations where it is easy to estimate the integral in eq. (190′). If $rm \gg 1$, i.e. if r is very large compared to $\lambda\!\!\!^{-}(\mathrm{e})$, the integrand is large only close to the lower limit of the integral. We approximate

$$\left(1+\frac{1}{2x^2}\right)\frac{\sqrt{x^2-1}}{x^2} \simeq \frac{3}{2}\sqrt{2(x-1)}$$

near $x = 1$ and substitute $x - 1 = u$, so that

$$\int_{1}^{\infty} \mathrm{e}^{-2mxr}\left(1+\frac{1}{2x^2}\right)\frac{\sqrt{x^2-1}}{x} \simeq \mathrm{e}^{-2mr}\int_{0}^{\infty} \mathrm{d}u\, \mathrm{e}^{-2mru}\tfrac{3}{2}\sqrt{2u}$$
$$= \frac{3\sqrt{\pi}}{8}\frac{\mathrm{e}^{-2mr}}{(mr)^{3/2}},$$

and therefore

$$V(r) \simeq \frac{C}{4\pi r}\left\{1 + \frac{\alpha}{4\sqrt{\pi}}\frac{\mathrm{e}^{-2mr}}{(mr)^{3/2}}\right\}, \quad r \gg \lambda\!\!\!^{-}(\mathrm{e}). \tag{191a}$$

This limit is relevant in normal atoms whose orbit radii are indeed large as compared to $\lambda\!\!\!^{-}(\mathrm{e})$. Because of the exponential the correction term is very small. This is in contrast to muonic atoms whose orbit radii are comparable with or smaller than $\lambda\!\!\!^{-}(\mathrm{e})$. Here vacuum polarization is a large effect.

Similarly, if $rm \ll 1$ i.e. if $r \ll \lambda\!\!\!^{-}(\mathrm{e})$ the integral (190′) can be solved approximately, too, and is found to be (Blomqvist 1972)

$$V(r) \simeq \frac{C}{4\pi r}\left\{1 - \frac{2\alpha}{3\pi}(\ln(mr) + C_{\mathrm{E}} + 5/6)\right\}, \tag{191b}$$

where $C_{\mathrm{E}} = 0.577216$ (Euler's constant).

7.2. Illustration and interpretation of vacuum polarization of order $\alpha Z\alpha$

Fig. 13 illustrates the $1/r$ potential and the vacuum polarization potential of eq. (190′): In order to get rid of dimensional quantities and of the numerical factors in front of the integral, we have multiplied $V(r)$, eq. (190′), by $(-6\pi^2/\alpha Cm)$. Thus the upper curve represents the reduced vacuum polarization potential

$$v_{\text{vacpol}}(u) = -\frac{1}{u}\int_1^\infty \mathrm{d}x\, \mathrm{e}^{-2ux}\left(1+\frac{1}{2x^2}\right)\frac{\sqrt{x^2-1}}{x^2} \tag{192}$$

as a function of $u = rm \mathrel{\hat{=}} r/\lambda\!\!\!^{-}(\mathrm{e})$. The figure shows that v_{vacpol} is small compared to the uncorrected $1/r$ potential for distances $r \gtrsim \lambda\!\!\!^{-}(\mathrm{e})$, but becomes very strong at small distances $r \ll \lambda\!\!\!^{-}(\mathrm{e})$. Furthermore, the potential due to vacuum polarization has the same sign as the $1/r$ potential everywhere. Thus, in an attractive $1/r$ potential, vacuum polarization leads to even more attraction. This result contradicts naive expectations if we think of vacuum polarization in analogy to electric polarizability of ordinary matter. Indeed, for a positive point charge, we would expect the virtual positrons to be pushed away from the origin and the virtual electrons to be pulled towards the origin. As the total induced charge is zero, this would mean that the original positive point charge effectively is smeared out over a certain region of space. However, this would lead to an effective screening of the charge and hence to a *reduction* of its field, not to the increase seen in the figure.

This result becomes even more puzzling if we consider the induced charge density $\rho_{\text{Pol}}(r)$ pertaining to the vacuum polarization potential: $\rho_{\text{Pol}}(r) = -\Delta V_{\text{Pol}}(r)$ (for technical reasons we introduce a convergence factor into the integral),

$$V_{\text{Pol}}(r) = \frac{C}{4\pi r}\frac{2\alpha}{3\pi}\lim_{M\to\infty}\int_1^\infty \mathrm{d}x\, \mathrm{e}^{-(m/M)x}\mathrm{e}^{-2mrx}\left(1+\frac{1}{2x^2}\right)\frac{\sqrt{x^2-1}}{x^2}.$$

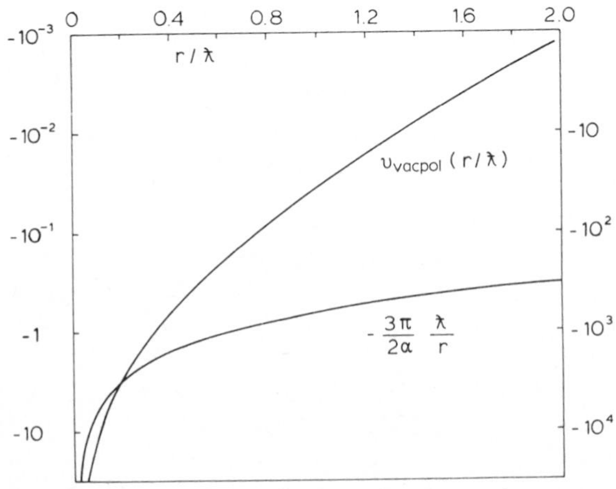

Fig. 13. Potential due to vacuum polarization, in dimensionless form, as a function of $r/\lambda\!\!\!^{-}$ where $\lambda\!\!\!^{-}$ is the Compton wave length of the virtual particle in the loop of Fig. 12(a). Also shown is the $1/r$ potential, scaled by the same factor as the former. The right-hand scale holds for the latter.

Using the well-known formula

$$(\Delta - \kappa^2)\frac{e^{-\kappa r}}{r} = -4\pi\delta(\boldsymbol{r})$$

and taking $\kappa = 2mx$, one finds

$$\rho_{\text{Pol}}(r) = \frac{C}{4\pi}\frac{2\alpha}{3\pi}\lim_{M\to\infty}\left\{4\pi\delta(\boldsymbol{r})\int_1^\infty \mathrm{d}x\, \mathrm{e}^{-(m/M)x}\left(1+\frac{1}{2x^2}\right)\frac{\sqrt{x^2-1}}{x^2}\right.$$
$$\left. - \frac{4m^2}{r}\int_1^\infty \mathrm{d}x\, \mathrm{e}^{-(m/M)x}\mathrm{e}^{-2mrx}\left(1+\frac{1}{2x^2}\right)\sqrt{x^2-1}\right\}. \tag{193}$$

(The convergence factor is needed in the first term of this expression because the integral diverges logarithmically; it is irrelevant in the second term.) This polarization charge density has rather curious properties: For finite argument $r \neq 0$ it has the same sign everywhere. At $r = 0$ it has two singularities, a δ-distribution with a linearly divergent coefficient

$$4\pi\delta(\boldsymbol{r})\ln(M/m) \tag{194}$$

and a $1/r^3$ pole which comes from the second term (see below eq. (197a)). Yet, the integral of ρ_{Pol} over all space vanishes, as it should. This is seen by making use of the formula

$$\int_0^\infty \frac{\mathrm{e}^{-\kappa r}}{r} r^2\,\mathrm{d}r = 1/\kappa^2$$

with $\kappa = 2mx$:

$$\int \rho_{\text{Pol}}(r)\,\mathrm{d}^3r = 4\pi\frac{C}{4\pi}\frac{2\alpha}{3\pi}\lim_{M\to\infty}\left\{\int_1^\infty \mathrm{d}x\, \mathrm{e}^{-(m/M)x}\left(1+\frac{1}{2x^2}\right)\frac{\sqrt{x^2-1}}{x^2}\right.$$
$$\left. - \int_1^\infty \mathrm{d}x\, \mathrm{e}^{-(m/M)x}\left(1+\frac{1}{2x^2}\right)\frac{\sqrt{x^2-1}}{x^2}\right\} = 0. \tag{195}$$

Fig. 14 shows the polarization charge density for $r \neq 0$. Again, in order to get rid of dimensional quantities we have plotted the function

$$f(u) = \left(-\frac{6\pi^2}{\alpha C m^3}\right)\rho_{\text{Pol}}(r)$$
$$= \frac{4}{u}\int_0^\infty \mathrm{d}x\, \mathrm{e}^{-2ux}\left(1+\frac{1}{2x^2}\right)\sqrt{x^2-1} \quad (r \neq 0), \tag{196}$$

where $u = mr \mathrel{\hat{=}} r/\lambda\!\!\!^{-}(\mathrm{e})$, as a function of this variable. The connection to the function (192) is

$$\frac{1}{u^2}\frac{\mathrm{d}}{\mathrm{d}u}\left(u^2\frac{\mathrm{d}v(u)}{\mathrm{d}u}\right) = -f(u), \quad u \neq 0.$$

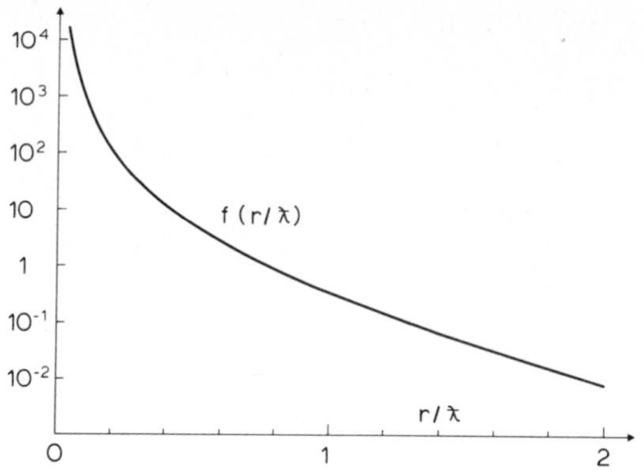

Fig. 14. Induced polarization charge density, eq. (196), in dimensionless form and as a function of $r/\lambdabar$. Note that this density is positive (attractive) for all finite r, but is singular at $r=0$ so that its integral over all space vanishes.

From the approximate expressions (191a, b) for the potential at large and small r, respectively, we derive the corresponding limiting behaviour of the polarization density (196), viz. ($u = mr \hat{=} r/\lambdabar(\mathrm{e})$)

$$r \ll \lambdabar \quad f(u) \simeq 1/u^3, \tag{197a}$$

$$r \gg \lambdabar \quad f(u) \simeq \frac{3\sqrt{\pi}}{8} u^{-5/2} \mathrm{e}^{-2u}. \tag{197b}$$

Clearly, the polarization charge density is not very *anschaulich*. It has the expected property (195) but it is qualitatively different from a polarization density in ordinary matter. The singular term (194) at $r=0$ is reminiscent of the charge renormalization (184). The only semi-physical statement one may make is this: When a test charge or a photon probes the field of our point charge at large distances, i.e. at low momentum transfers, then it sees what is called the *physical* charge. The more it approaches the point charge (i.e. the larger the momentum transfers), the more the test particle sees of the *bare* charge. As the bare charge is larger than the physical charge (see eq. (186)), the test particle sees an enhanced field at short distances. The closer it comes the more enhanced the field. Unfortunately, the bare charge is infinite. The phenomenon of vacuum polarization evades simple analogies to classical polarization phenomena because infinite charge renormalization is not *anschaulich*.

7.3. *Radiative corrections in muonic atoms*

The polarization potential of order $O(\alpha C)$, eq. (190′), which we have discussed so extensively, yields the dominant contribution to radiative corrections to the energies of muonic atoms. In this section we give some examples, for the sake of illustration. We then discuss other radiative corrections such as the Lamb shift and vacuum polarization of higher order.

It is not difficult to generalize the result (190′) for a point-like charge to the case of the extended charge density of a nucleus. Here $C = -Ze^2 = -4\pi Z\alpha$ and

$$V(\boldsymbol{r}) = -Z\alpha \int \mathrm{d}^3 r' \frac{\rho(\boldsymbol{r}')}{|\boldsymbol{r}-\boldsymbol{r}'|} \times \left\{1 + \frac{2\alpha}{3\pi}\int_1^\infty \mathrm{e}^{-2m|\boldsymbol{r}-\boldsymbol{r}'|x}\left(1+\frac{1}{2x^2}\right)\frac{\sqrt{x^2-1}}{x^2}\mathrm{d}x\right\}, \qquad (198)$$

where $\rho(\boldsymbol{r})$ is the charge density normalized to one.

Clearly, the singular properties of the polarization potential and charge density remain unchanged by the integration over the finite charge density.

If the charge density is spherically symmetric the expression (197) simplifies somewhat. The first term on the r.h.s. goes over into the uncorrected spherical potential (117). The integral over angular variables in the polarization potential can be done by elementary means (exercise!) giving

$$V_{\text{vacpol}}(r) = -Z\alpha \frac{2\alpha}{3m}\int_0^\infty \mathrm{d}r' \frac{r'}{r}\rho(r')\{I(|r-r'|) - I(r+r')\} \qquad (199)$$

with

$$I(z) = \int_1^\infty \mathrm{e}^{-2mzx}\left(1+\frac{1}{2x^2}\right)\frac{\sqrt{x^2-1}}{x^3}\mathrm{d}x.$$

In a deformed nucleus, the multipole expansion (133) of the charge density should be inserted into eq. (198). (We see from this, in particular, that vacuum polarization will contribute to electric quadrupole hyperfine structure as well.)

Let us now illustrate the importance of vacuum polarization of order $O(\alpha Z\alpha)$ by a few practical examples and let us compare this correction to the remaining radiative corrections.

In the second column of Table 4 we give the transition energy $2p_{1/2}$–$1s_{1/2}$ in four typical atoms. The third column shows the order $\alpha Z\alpha$ vacuum polarization whilst the fourth and fifth columns show vacuum polarization corrections of higher order and the remainder of the Lamb shift, respectively. Clearly, the $O(\alpha Z\alpha)$ term is the dominant correction in all cases.

Table 4
Realistic $2p_{1/2}$–$1s_{1/2}$ transition energies and radiative corrections in muonic atoms. (The transition energies contain all corrections.) All energies in keV. Numbers taken from Engfer et al. (1974).

Nucleus	Trans. energy (keV)	Vac. pol. $\alpha Z\alpha$ (keV)	Vac. pol., higher orders (keV)	Lamb shift (keV)
$^{12}_{6}$C	75.25	0.372	0.002	−0.006
$^{\text{nat}}_{20}$Ca	783.79	6.049	0.044	−0.208
$^{116}_{50}$Sn	3418.99	25.455	0.109	−1.548
$^{208}_{82}$Pb	5778.01	34.804	−0.106	−2.683

Table 5
Lamb shifts in muonic hydrogen and helium. All energies in meV ($=10^{-3}$ eV). Calculated values from Borie et al. (1982). Experimental values from G. Carboni et al. Nucl. Phys. A278(1977)381.

	$^1_1$H	$^4_2$He		
Lamb shift	$2p_{3/2}-2s_{1/2}$	$2p_{1/2}-2s_{1/2}$	$2p_{3/2}-2s_{1/2}$	$3d_{3/2}-3p_{3/2}$
fine structure	8.4	0	145.7	0
vac. pol. $\alpha Z\alpha$	204.9	1665.8	1666.1	110.560
vac. pol., higher	1.5	12.0	12.0	0.925
vertex correction	-0.6	-11.1	-10.8	-0.069
recoil	~ 0	-0.2	-0.2	0.005
finite size of nucleus	-3.4 ± 0.1	$-103\ \langle r^2 \rangle$[a)]	$-103\ \langle r^2 \rangle$[a)]	~ 0
polarizability	$\sim 2 \times 10^{-2}$	3.1 ± 0.6	3.1 ± 0.6	~ 0
total theoretical	210.8 ± 0.1	1380.9(4.2)	1527.2(4.2)	111.42
experiment		1381.3(0.5)	1527.5(0.3)	

[a)] $\langle r^2 \rangle^{1/2} = 1.674 \pm 0.012$ fm.

Table 5 shows the radiative corrections for muonic hydrogen and helium 4. The second column contains the radiative and other contributions to the energy splitting of the $2p_{3/2}$ and $2s_{1/2}$ states in hydrogen. The origin of individual contributions is indicated in the first column. The correction labeled "polarizability" is explained in the next section. The third and fourth column give the details of the Lamb shift in the $n=2$ states of muonic ^{4}He. The last column, finally, shows an example for the Lamb shift in the $n=3$ levels of helium.

The bottom line shows the experimental results obtained for the $n=2$ system in helium. The agreement with the theoretical predictions is excellent. However, the uncertainty on the theoretical numbers is about a factor of ten larger than the experiment error bar. As may be seen from the table, this uncertainty stems almost entirely from the finite size correction, i.e. from the experimental error bar of the r.m.s. radius of ^{4}He. The polarizability shift is also sizeable but is believed to be calculable to the accuracy indicated in the table, at least. In this respect, the $n=3$ states in helium and the Lamb shift in hydrogen are somewhat clearer tests of radiative corrections as both the finite size and polarizability corrections are small. So far, these shifts have not been measured as yet.

In the examples shown in tables 4 and 5 the corrections due to vacuum polarization of order higher than $\alpha Z\alpha$ are due primarily to

(i) the terms of order $\alpha^2 Z\alpha$ which are depicted in Fig. 15(a),

(ii) terms of order $\alpha(Z\alpha)^3$ (more generally $\alpha(Z\alpha)^{2n+1}$) which are illustrated by Fig. 15(b). These latter terms, even though they are proportional to α^4, appear enhanced in heavy nuclei because $(Z\alpha)$ is not small as compared to one anymore. For example, in lead we have $Z\alpha \simeq 0.6$. Here the correction to 2p–1s transition

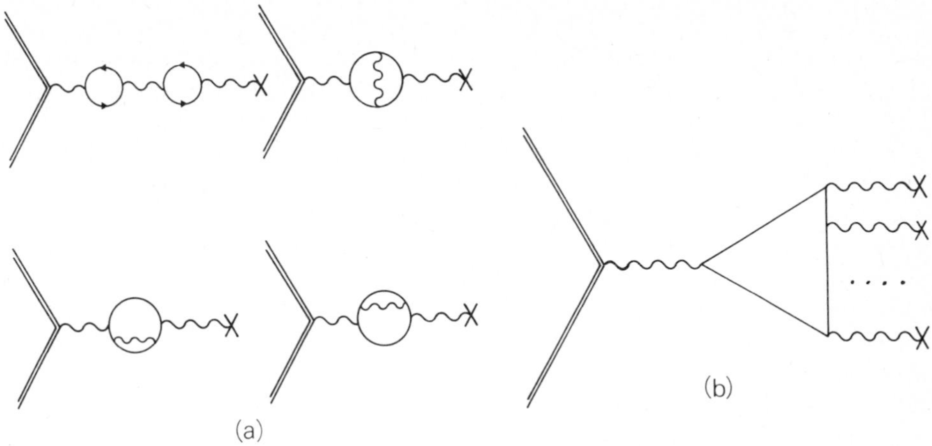

Fig. 15. (a) Vacuum polarization to order $\alpha^2(Z\alpha)$, Z being the charge of the external potential. The double line represents the external, bound muon. (b) Vacuum polarization of order $\alpha(Z\alpha)^{2n+1}$ for $n = 1$ and higher.

energy due to diagram b) is as large as 40 eV. Other corrections such as vacuum polarization due to a virtual *muon* loop are found to be very small.

Finally, the remaining radiative corrections, i.e. the Lamb shift and the anomalous magnetic moment interaction, can be represented by additional potentials in the muon's wave equation. The details and references to the literature are found in Ref. R21, vol. I, Chapter III.

We may summarize this section by saying that vacuum polarization of order $\alpha Z\alpha$ is tested at a level of a few parts per million. The higher order terms of vacuum polarization are tested to about 20%. This provides another piece of evidence for the success of quantum electrodynamics, which is complementary to the very impressive classical tests of QED in electronic atoms and in the g-factor anomaly of electron and muon.

8. Muonic atoms and nuclear structure

The study of spectra and X-ray intensities of muonic atoms provides very precise information about electric and magnetic properties of nuclear ground states and, to a lesser extent, also of nuclear excited states. This is so because these properties can be measured very accurately and because the analysis in terms of nuclear properties is well-defined and conceptually simple. We do not give a complete survey of the subject but rather concentrate on a few examples that illustrate the physics and demonstrate the power of the method. For more detailed analysis we refer to the monographs and reviews on muonic atoms (R21 Chap. III, Borie et al. 1982 and references therein).

8.1. Monopole charge densities of nuclear ground states

The low-lying orbits of muonic atoms are affected by the finite size of the nuclear charge distribution. In the order of decreasing sensitivity these are the states: $1s_{1/2}$, $2s_{1/2}$, $2p_{1/2}$, $2p_{3/2}$, etc. As the muonic cascade runs primarily through the circular orbits it is clear that information on the static charge distribution is obtained mostly from the (2p–1s) and (3d–2p) transition energies. In order to see the sensitivity of energies of bound states to details of the charge distribution let us perform a simple calculation. Write the radial equations (152), once for the point charge and once for the finite charge distribution. Let V^0, E^0, f_κ^0, g_κ^0 be the potential, the energy and the radial wave functions, respectively, in the former case, whilst the same quantities in the latter case shall be written without the superscript 0. Multiplying the first of eqs. (152) by g_κ^0, the second by $(-f_\kappa^0)$, then doing the same with the role of (V^0, E^0, f^0, g^0) and (V, E, f, g) exchanged, adding and subtracting the four equations obtained in this way, one finds (exercise 10)

$$\frac{\mathrm{d}}{\mathrm{d}r}\{f_\kappa^0 g_\kappa - f_\kappa g_\kappa^0\} = -\frac{2}{r}\{f_\kappa^0 g_\kappa - f_\kappa g_\kappa^0\} - \{(V-V^0)-(E_\kappa - E_\kappa^0)\}\{(f_\kappa^0 f_\kappa + g_\kappa^0 g_\kappa)\}. \tag{200}$$

The l.h.s. and the first term on the r.h.s. can be combined to $(1/r^2)(\mathrm{d}/\mathrm{d}r)(r^2\{\ldots\})$. The integral over the derivative vanishes, when taken from zero to infinity,

$$\int_0^\infty \frac{1}{r^2}\frac{\mathrm{d}}{\mathrm{d}r}(r^2\{f_\kappa^0 g_\kappa - f_\kappa g_\kappa^0\})r^2\,\mathrm{d}r = 0,$$

because the radial functions of bound states are regular at the origin and vanish at infinity. Therefore the energy shift $E - E^0$ is related to the difference in the potentials $V - V^0 = V + Z\alpha/r$ by

$$(E_\kappa - E_\kappa^0)\int_0^\infty \{f_\kappa^0 f_\kappa + g_\kappa^0 g_\kappa\} r^2\,\mathrm{d}r = \int_0^\infty \left(V(r) + \frac{Z\alpha}{r}\right)\{f_\kappa^0 f_\kappa + g_\kappa^0 g_\kappa\} r^2\,\mathrm{d}r. \tag{201}$$

If E does not differ too much from E^0, the functions f and g on the l.h.s. may be replaced by f^0 and g^0. This replacement is not reliable on the r.h.s. of eq. (201) because for small r, $V(r)$ differs appreciately from $V^0 = Z\alpha/r$. Thus,

$$E_\kappa - E_\kappa^0 \simeq \int_0^\infty \left(V(r) + \frac{Z\alpha}{r}\right)\{f_\kappa^0 f_\kappa + g_\kappa^0 g_\kappa\} r^2\,\mathrm{d}r. \tag{202}$$

Clearly, the integrand vanishes outside the nucleus. If the muon penetrates but little into the nucleus, we may be allowed to replace the wave functions by their power behaviour near the origin, cf. eqs. (165) and (171). If we keep only the lowest power in the curly brackets of eq. (202), then

$$E_\kappa - E_\kappa^0 \propto \int_0^\infty \left(V(r) + \frac{Z\alpha}{r}\right) r^{|\kappa|+\gamma}\,\mathrm{d}r. \tag{203a}$$

By applying partial integration twice to this integral and making use of Poisson's equation we find that the energy shift is proportional to a noninteger moment of the nuclear charge density,

$$E_\kappa - E_\kappa^0 \propto \int_0^\infty r^{\gamma+|\kappa|}\rho(r)r^2\,\mathrm{d}r \tag{203b}$$

(see also exercise 11). It is true that the (2p–1s) transition in light muonic atoms does indeed determine the moment $\langle r^{\gamma+1}\rangle \simeq \langle r^2\rangle$ of the nuclear density. However, in medium-weight and heavy nuclei the approximations leading to eqs. (203) are not reliable and a better analysis of the energies must be invoked. For example, one may assume a specific, analytical form for the charge density containing a set of free parameters. One calculates the potential (117) and, in a second step, the energies of the muonic bound states by integration of eqs. (172). The parameters in the density are varied until the experimental transition energies are well reproduced. This yields a model density from which various moments may be calculated. We note, however, that the—largely model independent—analysis of electron scattering data that we mentioned above, can also be applied to muonic atoms, or to electron scattering and muonic atoms combined.

The approximations (203) fail for larger values of the nuclear charge because the muon penetrates the nucleus. As a consequence the muonic wave functions can no longer be replaced by their power behaviour at the origin. Their exponential fall-off for larger r is already felt within the nuclear domain. This makes understandable the empirical finding that muonic transitions determine in essence generalized moments of the type of eq. (II.59) (here of the *charge* density),*⁾

$$\langle r^\sigma e^{-\alpha r}\rangle := 4\pi\int_0^\infty \rho(r)r^\sigma e^{-\alpha r}r^2\,\mathrm{d}r. \tag{204}$$

σ is generally found to be close to $(2l+2)$, whilst α varies in the range of 0.04/fm to 0.15/fm. Actual values of these parameters have been determined all through the periodic table and can be found in the article by Engfer et al. (1974).

8.2. *Resonance excitation: The example of atoms with strongly deformed nuclei*

We noted above that muonic transition energies in heavy atoms are comparable to typical nuclear transition energies. Therefore, the muonic spectrum is often found to be in resonance with the nuclear spectrum. We consider here the simple but characteristic example of an atom with a strongly deformed nucleus. In order to simplify matters let us assume that the nucleus has an excited rotator state with spin and parity 2^+ such that its energy difference to the ground state 0^+,

$$\Delta_\mathrm{n} := E_\mathrm{n}(2^+) - E_\mathrm{n}(0^+) \tag{205}$$

is of the same order as the muonic fine structure in the 2p-states,

$$\Delta_\mu := E_\mu(2\mathrm{p}_{3/2}) - E_\mu(2\mathrm{p}_{1/2}). \tag{206}$$

This assumption is indeed realistic because in the rare earths, for instance, $\Delta_\mathrm{n} \simeq$

*⁾See the addendum by Barrett to Chap. III of Ref. R21.

100–200 keV, and $\Delta_\mu \simeq$ 100–200 keV. The whole system, nucleus plus muon, then has two nearly degenerate states with total angular momentum $F=\frac{3}{2}$,

$$|1\rangle := |(0^+;2p_{3/2})\,F=\tfrac{3}{2}\rangle, \tag{207a}$$

$$|2\rangle := |(2^+;2p_{1/2})\,F=\tfrac{3}{2}\rangle, \tag{207b}$$

which interact via the electric quadrupole term in the Coulomb interaction

$$\sum_{i=1}^{Z} \frac{-e^2}{|\boldsymbol{r}_i-\boldsymbol{r}_\mu|} \to H_{E2} = -e^2\frac{4\pi}{5}\sum_i \frac{r_<^2}{r_>^3}\sum_{m=-2}^{+2} Y_{2m}^*(\hat{\boldsymbol{r}}_i)Y_{2m}(\hat{\boldsymbol{r}}_\mu). \tag{208}$$

It is a known property of the nuclear rotator model that the nondiagonal matrix element of the interaction (208) between the states 0^+ and 2^+ is large and comparable to Δ_n. Therefore the states (207) cannot be eigenstates of the Hamiltonian that describes the combined system. Instead the eigenstates will be (cf. exercise 13),

$$|\phi_1\rangle = |1\rangle\cos\alpha + |2\rangle\sin\alpha, \tag{209a}$$

$$|\phi_2\rangle = -|1\rangle\sin\alpha + |2\rangle\cos\alpha, \tag{209b}$$

with $\cos\alpha$ a known function of Δ_n, Δ_μ, and the matrix elements of H_{E2}, eq. (208). Exercise 13 shows that in case of near degeneracy α comes close to 45°, the case of maximal mixing. Let us then consider the consequences of this dynamical mixing.

The muonic (2p–1s) transitions take place in about 10^{-17} sec, whereas the nuclear (2^+–0^+) transition takes about 10^{-9} to 10^{-11} sec. Therefore, the component $|1\rangle$ decays as follows

$$|1\rangle = |(0^+;2p_{3/2})\,F=\tfrac{3}{2}\rangle \xrightarrow[10^{-17}\text{sec}]{} |(0^+;1s_{1/2})\,F=\tfrac{1}{2}\rangle, \tag{210a}$$

whilst the component $|2\rangle$ decays in two steps, one fast and one slow,

$$|2\rangle = |(2^+;2p_{1/2})\,F=\tfrac{3}{2}\rangle \xrightarrow[10^{-17}\text{ sec}]{} |(2^+;1s_{1/2})\,F=\tfrac{3}{2}\text{ or }\tfrac{5}{2}\rangle$$

$$\xrightarrow[10^{-9}\text{–}10^{-11}\text{ sec}]{} |(0^+;1s_{1/2})\,F=\tfrac{1}{2}\rangle. \tag{210b}$$

During the upper part of the muon's cascade the nucleus remains in its ground state 0^+. Therefore, of the (nearly) degenerate states (209) the state $|\phi_1\rangle$ will have a population proportional to $\cos^2\alpha$, whilst the population of state $|\phi_2\rangle$ is proportional to $\sin^2\alpha$. The relative intensity of branch (210b) as compared to branch (210a) is then

$$\frac{2\sin^2\alpha\cos^2\alpha}{\sin^4\alpha+\cos^4\alpha}. \tag{211}$$

This means that the intensity of the ($2p_{1/2}-1s_{1/2}$) transition is enhanced and the intensity of the ($2p_{3/2}-1s_{1/2}$) transition is weakened as compared to their values without mixing ($\alpha=0$). In other terms, the intensity ratio

$$I(2p_{3/2}-1s_{1/2})/I(2p_{1/2}-1s_{1/2})$$

is perturbed and becomes smaller than for cases without nuclear resonance. Further-

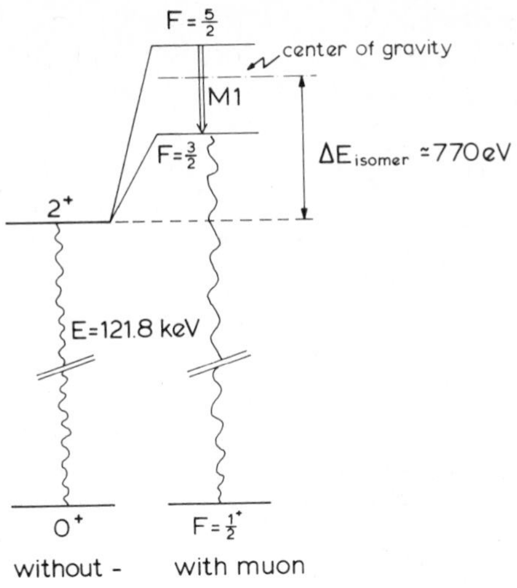

Fig. 16. The ground state 0^+ and first excited, rotational state 2^+ in ^{152}Sm with and without a muon in the $1s_{1/2}$-state. M1 indicates a conversion transition which depopulates the state $F=\frac{5}{2}$ due to Auger transitions.

more, while the muon has reached the atomic ground state there is a nonvanishing probability to find the nucleus in its excited 2^+-state. This implies that the nucleus will make its transition back to the ground state in presence of the muon in its atomic 1s-state. This, in turn, has two consequences:

(i) The nuclear 2^+-state and the muonic $1s_{1/2}$-state have magnetic coupling which leads to hyperfine splitting into two states of the coupled system with $F=\frac{5}{2}$ and $F=\frac{3}{2}$, respectively, cf. branch (210b).

(ii) The charge density of the muon gives rise to an additional electrostatic energy which appears as a shift of the center of gravity of the two excited states above versus the ground state. The shift is called *isomer shift* and is given by the expression

$$\Delta E_{\text{isomer}} = -Ze^2 \int d^3 r_n \int d^3 r_\mu \frac{\rho_{1s}(\boldsymbol{r}_\mu)}{|\boldsymbol{r}_\mu - \boldsymbol{r}_n|} \{\rho_{2^+}(\boldsymbol{r}_n) - \rho_{0^+}(\boldsymbol{r}_n)\}. \tag{212}$$

All these effects were indeed observed in numerous muonic atoms. An example is given in Fig. 16. A compilation of measured isomer shifts can be found in Engfer et al. (1974).

8.3. *Hyperfine structure due to magnetic dipole and electric quadrupole interactions*

Generally speaking, magnetic dipole and electric quadrupole hyperfine interactions are strongly enhanced in muonic atoms as compared to electronic atoms: For pointlike sources both have the radial dependence r^{-3}. As muonic orbits are closer to the nucleus than electronic orbits one expects an enhancement factor of roughly

$(m_\mu/m_e)^3$. Whilst this is indeed true for electric quadrupole couplings, the enhancement factor in the case of magnetic interactions is reduced to $(m_\mu/m_e)^2$ because the muon's magnetic moment is smaller than the electron's moment by the factor m_e/m_μ.

For low orbits, clearly, there are strong penetration effects and the effective splittings are reduced as compared to the case of pointlike sources. In this case the muon effectively probes the *spatial* distribution of magnetic dipole density and of electric quadrupole density (the function $\rho_2(r)$ in eq. (134)). The magnetic and electric hyperfine splittings are given by integrals of these densities weighted with the muonic density.

We do not develop the theory of hyperfine interactions here. We refer the reader to the literature for more details*) but quote a few examples for both cases.

The magnetic interaction energy of a state with nuclear spin J and muonic angular momentum j, coupled to total angular momentum F, is given by

$$W_{\mathrm{M1}}(J, n\kappa; F) = \frac{1}{2Ij}\{F(F+1) - J(J+1) - j(j+1)\} \times \{A_l(J, n\kappa) + A_s(J, n\kappa)\} \tag{213}$$

A_l contains the contribution of the nuclear orbital magnetization density, A_s the contribution of the spin magnetization, both weighted with the muonic radial wave functions and the typical dipole interaction terms. As an example, we quote the case of muonic $^{209}_{83}\mathrm{Bi}$ whose ground state has spin/parity $\frac{9}{2}^-$. For a pointlike nucleus we would obtain

$$A_l(\text{point nucleus}) = 3.60\ \mathrm{keV}, \qquad A_s(\text{point nucleus}) = -1.68\ \mathrm{keV},$$

whilst for the extended nucleus one finds

$$A_l = 2.12\ \mathrm{keV}, \qquad A_s = -0.474\ \mathrm{keV}.$$

These results show indeed a considerable reduction effect due to finite size. It is particularly noteworthy that the spin magnetic coupling is more affected than the orbital contribution,

$$\frac{A_s(\text{pt. nucl.}) - A_s}{A_s(\text{pt. nucl.})} = 0.72, \qquad \frac{A_l(\text{pt. nucl.}) - A_l}{A_l(\text{pt. nucl.})} = 0.41.$$

The electric quadrupole energy of state with $J \geq 1$ and $j \geq \frac{3}{2}$, coupled to total angular momentum F, is given by

$$W_{\mathrm{E2}}(J, n\kappa; F) = \frac{3X(X-1) - 4J(J+1)j(j+1)}{2j(2j-1)J(2J-1)} A_2(J, n\kappa), \tag{214}$$

with

$$X = J(J+1) + j(j+1) - F(F+1), \tag{215}$$

where A_2 is the quadrupole hyperfine constant which depends on an integral over the

*) See e.g. Chap. III of Ref. R21.

quadrupole charge density $\rho_2(r)$ weighted with the muonic wave functions and the $r_<^2/r_>^3$ interaction term. The reduction effects due to finite size are important in the low orbits. For example in $^{209}_{83}\mathrm{Bi}_\mu$ one finds for

$$A_2^0\left(\tfrac{9}{2}^-;2\mathrm{p}_{3/2}\right) = -5.90\ \mathrm{keV}\quad \text{(point nucleus)},$$

$$A_2\left(\tfrac{9}{2}^-;2\mathrm{p}_{3/2}\right) = -2.74\ \mathrm{keV}\quad \text{(finite size nucleus)}.$$

Whilst the magnetic interaction is relatively weak and can always be treated in perturbation theory, the electric quadrupole interaction is strong and has strong nondiagonal matrix elements too. Therefore the E2 interaction usually involves a certain number of states (the muonic fine structure doublet plus, possibly, a few excited nuclear states) and must be diagonalized in a finite dimensional space of neighbouring configurations.

A case of special interest is the one of intermediate muonic orbits in atoms with a nucleus having spin $J \geq 1$. In this case the penetration effects are small and A_2 is proportional to

$$Q_s = \int_0^\infty \rho_2(r) r^4 \mathrm{d}r, \tag{216}$$

the spectroscopic quadrupole moment of the nucleus. Therefore, the muonic E2-hyperfine structure provides a direct and model-independent method of determining spectroscopic quadrupole moments. In a few test cases in the rare earths accuracies of 6×10^{-3} were achieved. To our knowledge this is the most accurate, model-free method of measuring nuclear quadrupole moments.

References

Alder, K., A. Bohr, T. Huus, B. Mottelson and A. Winther, 1956, *Rev. Mod. Phys.* 28, 432.

Alder, K. and T.H. Schucan, 1963, *Nucl. Phys.* 42, 498.

Bailey, J., K. Borer, F. Combley, H. Drumm, C. Eck, F.J.M. Farley, J.H. Field, W. Flegel, P.M. Hattersley, F. Krienen, F. Lange, G. Lebée, E. McMillan, G. Petrucci, E. Picasso, O. Runolfsson, W. von Rüden, R.W. Williams and S. Wojcicki, 1979, *Nucl. Phys.* B150, 1.

Blomqvist, J., 1972, *Nucl. Phys.* B48, 95.

Borie, E.F. and G.A. Rinker, 1982, *Rev. Mod. Phys.* 54, 67.

Borkowski, F., G.G. Simon, V.H. Walther and R.D. Wendling, 1975, *Nucl. Phys.* B93, 461.

Borkowski, F., G. Höhler, E. Pietarinen, I. Sabba-Stefanescu, G.G. Simon, V.H. Walther and R.D. Wendling, 1976, *Nucl. Phys.* B114, 505.

De Forest, T. and J.D. Walecka, 1966, *Adv. in Phys.* 15, 1.

Donnelly, T.W. and J.D. Walecka, 1975, *Ann. Rev. Nucl. Sci.* 25, 329.

Elton, L.R.B., 1953, *Proc. Roy. Soc.* (London) A66, 806.

Engfer, R., H. Schneuwly, J.L. Vuilleumier, H.K. Walter and A. Zehnder, 1974, *Atomic and Nuclear Data Tables* 14, 509.

Friedrich, J. and F. Lenz, 1972, *Nucl. Phys.* A183, 523.

Friar, J.L. and J.W. Negele, 1973, *Nucl. Phys.* A212, 93.

Griffy, T.A., D.S. Onley, J.T. Reynolds and L.C. Biedenharn, 1963, *Phys. Rev.* 128, 833 and 129, 1698.

Lenz, F., 1969, *Zeit. Physik* 222, 491.

Mott, N.F., 1929, *Proc. Roy. Soc.* (London) A124, 429.

Ravenhall, D.G., D.R. Yennie and R.N. Wilson, 1954, *Phys. Rev.* 95, 500.

Reynolds, J.T., D.S. Onley and L.C. Biedenharn, 1964, *J. Math. Phys.* 5, 411.
Rosenbluth, M.N., 1950, *Phys. Rev.* 79, 615.
Scheck, F., 1978, *Phys. Reports* 44, 187.
Scheck, F., 1966, *Nucl. Phys.* 77, 577.
Sick, I., 1974, *Nucl. Phys.* A218, 509 and *Phys. Lett.* 53B, 15.
Simon, G.G., Ch. Schmitt, F. Borkowski and V.H. Walther, 1980, *Nucl. Phys.* A333, 381.
Uehling, E.A., 1935, *Phys. Rev.* 48, 55.

Exercises

1. Show that eq. (17b) can be transformed into eq. (17a) by means of a rotation and a reflection. If the interaction is invariant with respect to these operations, the scattering amplitudes must be the same.

2. Derive the cross section, eqs. (26) and (35), for electron scattering off a spin-zero target, on the basis of Feynman rules.

3. Prove eq. (41) starting from eq. (39). *Hint:* Consider first the case of an infinitesimal translation. Make then use of

$$e^x = \lim_{n\to\infty} (1 + x/n)^n .$$

4. Prove relation (93) and use this relation to calculate the interaction term (92).

5. Derive the elastic form factor for a homogenous charge density (102). Calculate the cross section and discuss the result.

6. Prove the representation (129). *Hints:* Derive first an equation for $(\mathrm{d}/\mathrm{d}r)W(r)$, with $W(r)$ as defined in eq. (128), by combining the differential equations (105′) for the two potentials. Calculate $W(r = R_\infty)$ at some large radius, making use of the asymptotic forms of the radial functions, and let R_∞ go to infinity.

7. An unstable particle of lifetime τ is produced with a given energy E on a fixed target in the laboratory. Over which length can one reasonably hope to transport a beam of such particles? Consider the example of muons and pions at kinetic energies between 10 MeV and 200 MeV.

8. For $\kappa < 0$, $l = -\kappa - 1$, show that $rg_{n\kappa}(r) \simeq y_{nl}(r)$. This means that the states $(n, \kappa, j = -\kappa - \frac{1}{2} = l + \frac{1}{2})$ and $(n, \kappa' = -\kappa - 1, j = -\kappa - \frac{3}{2} = l - \frac{1}{2})$ have the same radial wave function in the nonrelativistic limit.

9. For circular orbits compare the fine structure splitting as predicted by eq. (162) with the one obtained in first-order perturbation theory from

$$V_{ls} = \frac{1}{2m^2}\frac{1}{r}\frac{\mathrm{d}V}{\mathrm{d}r}\boldsymbol{l}\cdot\boldsymbol{s}$$

and nonrelativistic wave functions.

10. Carry out the derivation leading to eq. (200).

11. In a muonic atom consider two potentials V_1 and V_2 which differ only over the nuclear domain. Expand the muonic density according to $|\Psi|^2 \simeq a + br^2$ and show that the eigenvalues of the energy for V_1 and V_2 differ approximately by

$$\Delta E \simeq Ze^2\frac{2\pi}{3}a\left\{\Delta\langle r^2\rangle + \frac{3}{10}\frac{b}{a}\Delta\langle r^4\rangle\right\},$$

where $\Delta\langle r^{2n}\rangle$ is the difference of the corresponding nuclear charge moments. Derive an approximate formula for $E_{2p_{1/2}} - E_{2s_{1/2}}$ as compared to the value of this difference in the case of the point charge.

12. Consider eq. (202) for two potentials $V_1(r)$ and $V_2(r)$ which do not differ much and which are regular at the origin. How would you estimate the difference $E_\kappa^1 - E_\kappa^2$ of the corresponding bound state energies? Convert the integral to an integral over the nuclear charge density.

13. Let $\Delta_{ik} := \langle i|H_{E2}|k\rangle$. Determine the eigenstates (209) and the corresponding energy eigenvalues of the Hamiltonian of the combined system. Discuss the cases (a) $\Delta_n = \Delta_\mu$, $\Delta_{11} \simeq \Delta_{22}$; (b) $\Delta_{12} \gg \Delta_n, \Delta_\mu, |\Delta_{11} - \Delta_{22}|$.

14. Prove the following identity, starting from eqs. (III.62 – 62′):

$$(p+p')_\alpha \overline{u(p')}\, u(p) = \overline{u(p')}\left\{2M\gamma_\alpha + i\sigma_{\alpha\beta}(p-p')^\beta\right\} u(p).$$

Chapter V

WEAK INTERACTIONS

This chapter gives an introduction to the phenomenology and the theory of weak interaction processes involving leptons and hadrons. In sections 1 and 2 we collect the most prominent and characteristic properties of weak interactions as they follow from the analysis of a set of key experiments, old and recent. The following sections 3 to 5 deal with the elements of non-Abelian local gauge theories in general, and with the unified theory of electroweak interactions of Glashow, Salam and Weinberg (GSW), in particular.

The GSW theory is a *minimal* theory in the sense that it incorporates all known phenomenology and successfully describes all known electromagnetic and weak interaction processes, including the masses and properties of the recently discovered W- and Z^0-bosons. However, there are good reasons to expect that this minimal model is only an approximation to some, more global, theory of leptons and hadrons. The underlying theory could be either a larger local gauge theory of all interactions, including the strong interactions and gravitation, or a theory of more elementary objects which would, in some sense, appear to be the constituents of the known leptons and quarks. In any event, the minimal GSW theory needs to be tested in precision experiments at all available energies. Sections 6 and 8 of this chapter are devoted to a discussion of such tests and to specific possibilities of searching for possible deviations from its predictions.

Weak interaction processes involving hadrons have another important aspect: Assuming the structure of the weak couplings to be known and to be given by the typical vector and axial vector current couplings, the matrix elements of such currents, taken between hadronic states, can be used to extract information on strong interaction quantities. This is analogous to the case of purely electromagnetic interactions such as electron scattering which allow to determine hadronic form factors, as described in Chap. IV. This aspect of weak interactions at medium energies is dealt with in our discussion of various semileptonic decays of hadrons in several sections below.

1. Phenomenological aspects of weak interactions

The theory of electroweak interactions of *leptons* can be formulated in terms of elementary fermion fields and in terms of interaction Lagrangians whose solutions can be constructed in straightforward perturbation theory. Weak processes involving *hadrons* are complicated by the fact that any hadronic matrix element of weak interaction vertices is renormalized by strong interactions and is not readily calculable in a perturbation series. In the first place, however, it is not so much the absolute magnitude of weak hadronic matrix elements that matters, but rather their Lorentz-structure, their quantum numbers and selection rules.

The Lorentz structure and form factor decomposition of the hadronic matrix element of a given Lorentz tensor operator $T_{\alpha\beta\ldots}$

$$\langle A|T_{\alpha\beta\ldots}|B\rangle$$

is well defined as soon as the spins and the intrinsic parities of the hadronic states A and B are known. Let us consider some examples: The matrix element of a *vector* current between a one-pion state and the vacuum must vanish,

$$\langle 0|v_\alpha(x)|\pi(q)\rangle = 0, \tag{1}$$

independently of what that vector current is and of what the internal structure of the one-pion state is. This is so because under rotations

$$\begin{array}{lll} v_0 & \text{behaves like an object with spin/parity} & 0^+, \\ \boldsymbol{v} & \text{behaves like an object with spin/parity} & 1^-. \end{array} \tag{2}$$

The vacuum state carries total angular momentum zero and is assigned positive parity. The pion has no spin and carries negative intrinsic parity. Its orbital state can be decomposed into states of good orbital angular momentum $l = 0, 1, 2, 3, \ldots$ whose parity is $(-)^l$. Combining the intrinsic and orbital properties one sees that the states of good total angular momentum and parity of the pion are

$$(J^P)_\pi = 0^-, 1^+, 2^-, \ldots . \tag{3}$$

Obviously, none of these possibilities, when combined with the properties (2) of the operator v^α, can yield the $(J^P)_{\text{vac}} = 0^+$ of the vacuum state. Hence the matrix element (1) vanishes.

On the other hand, if instead of a vector operator we consider the same matrix element of an axial vector operator,

$$\langle 0|a_\alpha(x)|\pi(q)\rangle, \tag{4}$$

then this matrix element need not be zero because under rotations

$$a_\alpha \text{ behaves like } \{0^-, 1^+\}, \tag{5}$$

which can indeed be coupled with $(J^P)_\pi$ to form the vacuum state 0^+. If the pion state is a plane wave with four momentum q we can translate the operator $a_\alpha(x)$ to $x = 0$, or some other fixed point in space–time. From eq. (IV.42) we have

$$\langle 0|a_\alpha(x)|\pi(q)\rangle = \mathrm{e}^{-\mathrm{i}xq}\langle 0|a_\alpha(0)|\pi(q)\rangle. \tag{4'}$$

Whether or not this quantity is different from zero depends now on the internal quantum numbers of a_α. The current $a_\alpha(x)$ must contain a piece proportional to an interpolating pion field, i.e. a field that can indeed create or annihilate a one-pion state. This field, therefore, must carry the quantum numbers of isospin, strangeness, baryon number, etc. of one-pion states. If all this is fulfilled, then the matrix element (4) must be a *vector* itself. Since q is the only vector available the covariant decomposition of eq. (4) must be

$$\langle 0|a_\alpha(0)|\pi(q)\rangle = F_\pi(q^2)q_\alpha \frac{1}{(2\pi)^{3/2}}. \tag{6}$$

However, $q^2 = m_\pi^2$, so the invariant form factor F_π, in fact, is a constant. Finally, from the behaviour of the pion state and of $a_\alpha(x)$ under time reversal T and charge conjugation one can show that F_π is pure imaginary, $F_\pi = \mathrm{i} f_\pi$ with f_π real. This will be shown below, in sec. 7.

The selection rules which follow from charge conjugation C, G-parity and the internal quantum numbers of the hadronic states and the relevant tensor operators can be kept track of in a transparent manner by writing the weak interaction covariants in terms of elementary quark fields and by making use of the constituent quark model for mesons and baryons.*)

Therefore we start our discussion of the phenomenology by a summary of the properties of the leptons and of the quark families, for the sake of reference in the following sections.

1.1. *Basic properties of leptons and quarks*

1.1.1. *The lepton families*

Three families of leptons are known to us: the electron e and its neutrino partner ν_e, the muon μ and its neutrino ν_μ, and the τ-lepton (presumably) accompanied by still another neutrino ν_τ:

$$\begin{pmatrix} \nu_e \\ e^- \end{pmatrix} \quad \begin{pmatrix} \nu_\mu \\ \mu^- \end{pmatrix} \quad \begin{pmatrix} \nu_\tau \\ \tau^- \end{pmatrix}, \tag{7}$$

The masses of the charged particles of these families are

$$m_e = 0.511\,003\,4(14)\ \mathrm{MeV}/c^2, \tag{8a}$$

$$m_\mu = 105.659\,3(3)\ \mathrm{MeV}/c^2, \tag{8b}$$

$$m_\tau = 1784.2(3.2)\ \mathrm{MeV}/c^2. \tag{8c}$$

Actually, the mass *ratio* of the muon to the electron mass is known to a still higher accuracy than that of m_e, from the measurement of the muon's magnetic moment and of its g-factor anomaly, viz. (Scheck 1978, Heil et al. 1984)

$$m_\mu/m_e = 206.768\,297(62). \tag{9}$$

The neutrinos are believed to be massless. However, the experimental upper limits for $m(\nu_\mu)$ and $m(\nu_\tau)$ are not very stringent and there is indeed the possibility for them to have nonzero and measurable masses. For example, for the muon neutrino we know only that $m(\nu_\mu) < \sim 0.5\ \mathrm{MeV}/c^2$. As to the mass of the electron neutrino, there are indications (from the electron spectrum in the beta decay of tritium) that it

*) For an introduction see Ref. R13.

is nonzero and of the order of 30 eV/c^2. For most of what we discuss in the following sections the neutrinos can be assumed to be rigorously massless simply because $m(\nu_e)$ and $m(\nu_\mu)$ are definitely much smaller than the energy release in the decay processes that we consider.

To the best of our knowledge, neutrinos seem to occur in states of definite helicity (cf. sec. 8.3 of Chap. III), with ν_f having negative helicity, $\bar{\nu}_f$ having positive helicity,

$$h(\nu_f) = -1, \qquad h(\bar{\nu}_f) = +1, \tag{10}$$

The density matrix describing the choice (10) is given in eq. (III.137) above. Their *charged* partners, to the contrary, can occur in any polarization state of a spin $-1/2$ particle. Even in kinematical situations where their mass can be neglected they can couple to other particles in states of positive and negative helicity: For instance, a vertex (eeγ) of two external electrons to a real or virtual photon must involve states of positive helicity and states of negative helicity, with equal weight, because electromagnetic couplings conserve parity.

Each of the three lepton families (7) seems to carry its own additive quantum number L_f which is the only distinctive characteristic for the family (f, ν_f) and which is strictly conserved in all electromagnetic and weak reactions involving leptons. The eigenvalues of these *lepton family numbers* $L_f = L_e$, L_μ, or L_τ are assigned as follows:

$$L_{f'}(f^-) = \delta_{ff'}, \qquad L_{f'}(\nu_f) = \delta_{ff'}, \tag{11a}$$

so that for the antiparticles

$$L_{f'}(f^+) = L_{f'}(\bar{\nu}_f) = -\delta_{ff'}. \tag{11b}$$

The total lepton number L of a given particle ℓ, which is then also strictly conserved, is given by the sum

$$L = \sum_{f=e,\mu,\tau} L_f(\ell). \tag{12}$$

The dynamical origin of these conservation laws is not known; they are fulfilled to a very high degree of accuracy. For instance, processes like

$$\mu^+ \to e^+ + \gamma,$$

$$\mu^+ \to e^+ e^- e^+,$$

$$K^+ \to \pi^+ e^+ \mu^-,$$

$$K^+ \to \pi^+ e^- \mu^+,$$

in which L is conserved but L_e and L_μ are not, have never been observed. They are known to be suppressed by at least nine to ten orders of magnitude in branching ratio as compared to the main decay modes of the parent particle.

1.1.2. The quark families

Except for some exotic states, all hadrons are believed to be composites of quarks and antiquarks. The low-lying mesons, for instance, should be predominantly states of one quark and one antiquark with definite relative orbital angular momentum L, with spin S and total angular momentum $J = L + S, L + S - 1, \ldots, |L - S|$,

$$(q\bar{q})_{L,S\to J}.$$

The spin of the meson is J, its intrinsic parity is $P = (-)^{L+1}$. If it is an eigenstate of charge conjugation C, then $C = (-)^{L+S}$. If $P = (-)^J$ one says that the meson has *natural* parity. In this case, and if the meson is an eigenstate of C, then one shows easily that it must have $P = C$. This is a typical prediction of the quark model.

The baryons are described by states of three quarks, with or without resulting internal angular momentum L and coupled to a resulting spin S, so that the spin of the particle J is the vector sum of $\boldsymbol{L}$ and $\boldsymbol{S}$.

The quarks carry *flavour* quantum numbers which add up to the flavour properties of the physical hadrons, such as baryon number B, isospin I, strangeness S, charm c etc. All quarks are assigned baryon number $B = \frac{1}{3}$ so that any state with three quarks (qqq) has indeed $B = +1$. Since baryon number is an additive, charge-like quantum number, and since quarks are fermions, a single antiquark $\bar{q}$ has $B = -\frac{1}{3}$, a state of three antiquarks $(\bar{q}\bar{q}\bar{q})$ has $B = -1$, and the meson states $(q\bar{q})$ have $B = 0$. [Baryon number conservation is an empirical conservation law which is derived from the observation that in all known reactions or decay processes the number of fermionic hadrons—with the correct counting of particles and antiparticles—is conserved. For instance, the stability of the proton and the stability of the hydrogen atom are very good indications of this conservation law. Note, however, that grand unified theories of *all* interactions as well as the cosmology of the big bang would prefer not to have baryon number as an exact conservation law. In these theories the proton is expected to be very long-lived but not absolutely stable. The present lower limit on the proton's lifetime is of the order of 10^{31} years.]

The quarks are denoted u ("up"), d ("down"), s ("strange"), c ("charmed"), b ("bottom" or "beauty"), t ("top" or "truth"). In the older literature on the constituent quark model, u, d, s are sometimes also denoted p, n, λ, because, except for baryon number and charge, they have the same properties as the physical states: proton, neutron and lambda $\Lambda(1116\ \mathrm{MeV}/c^2)$. The quark s is the carrier of strangeness S, i.e. it is assigned strangeness -1 while all other quarks have $S = 0$,

$$S(s) = -1, \quad S(\bar{s}) = +1, \quad S(q) = 0 \quad \forall q \neq s.$$

Similarly, c is the carrier of charm, b the carrier of beauty, etc., all of which, like the strangeness, are additive quantum numbers.

In contrast to these, isospin is an internal spectrum symmetry which has the structure of SU(2). Thus, as for angular momentum, the physical hadrons are eigenstates of total isospin $\boldsymbol{I}^{s^2}$, with the same eigenvalue I^s in a given mass-degenerate multiplet. The members of such a multiplet are classified according to I_3^s, the projection of I^s onto the 3-axis in isospin space, with $I_3^s = -I^s, -I^s + 1, \ldots, I^s$, as usual. Let us call this symmetry the *strong interaction isospin*, in contrast to the weak interaction isospin to be defined below. This symmetry is indeed an almost exact spectrum symmetry on the level of hadron physics, as may be seen by observing, for instance, that the difference of the proton and the neutron masses

$$m_n - m_p = -1.293\,43(4)\ \mathrm{MeV}/c^2 \tag{13}$$

is very small as compared to their absolute values and, similarly, that

$$m_{\pi^\pm} - m_{\pi^0} = 4.604\,3(37)\ \text{MeV}/c^2 \tag{14}$$

amounts to only about 3.3% of m_π. The strong isospin quantum numbers are carried by the u and d quarks: Both have $I^s = \frac{1}{2}$, with $I_3^s(\text{u}) = +\frac{1}{2}$, $I_3^s(\text{d}) = -\frac{1}{2}$. The electric charges finally are

$$Q = +\tfrac{2}{3} \quad \text{for u, c, t,} \tag{15a}$$

$$Q = -\tfrac{1}{3} \quad \text{for d, s, b.} \tag{15b}$$

Quarks also carry *colour* quantum numbers with respect to the local gauge theory $SU(3)_c$ which describes the strong interactions. In contrast to the flavour symmetries mentioned above and in contrast to the local gauge theory of electroweak interactions, this symmetry remains unbroken at all levels of the theory. Electroweak interactions couple only to the flavour degrees of freedom and are indifferent to colour. In other terms, even though each quark flavour state carries an additional SU(3)-colour index,

$$\text{u}^i, \text{d}^k, \text{s}^l, \ldots, \quad i, k, l, \ldots = 1, 2, 3 \in SU(3)_c, \tag{16}$$

the electromagnetic and weak coupling constants of the three colour states of the up-quark u^1, u^2, and u^3 are the same (and likewise for d^1, d^2, d^3, etc.). Therefore, in most of our discussion of electromagnetic and weak interactions we shall suppress the SU(3) index describing the colour degrees of freedom. (See however sec. 4.5.2 below.)

It is not difficult to construct the wave functions of the mesons in flavour space. Denoting the spin projections $m_S = \pm\frac{1}{2}$ by arrows pointing up or down, respectively, we have for instance

$$\text{K}^+ = \frac{1}{\sqrt{2}}(\text{u}\uparrow\bar{\text{s}}\downarrow + \text{u}\downarrow\bar{\text{s}}\uparrow), \tag{17a}$$

$$\pi^+ = \frac{1}{\sqrt{2}}(\text{u}\uparrow\bar{\text{d}}\downarrow + \text{u}\downarrow\bar{\text{d}}\uparrow), \tag{17b}$$

$$\pi^0 = \frac{1}{\sqrt{4}}(\text{u}\uparrow\bar{\text{u}}\downarrow + \text{u}\downarrow\bar{\text{u}}\uparrow - \text{d}\uparrow\bar{\text{d}}\downarrow - \text{d}\downarrow\bar{\text{d}}\uparrow), \tag{17c}$$

where $\text{q}\bar{\text{q}}$, as written in eqs. (17), means a quark and an antiquark in a relative s-state, coupled to spin 0 and to appropriate isospin $\frac{1}{2}$ for the kaon, 1 for the pion. Note that we have coupled the spins and isospins according to the scheme

$$(-)^{j-m'}(jm, j\ -m'|JM)(\text{q})_{jm}(\bar{\text{q}})_{jm'}$$

characteristic for the couplings of particles and antiparticles (or particles and "holes").

For the baryons, the construction of the flavour wave functions is well-defined if we require that (i) the three quarks be in a relative orbital s-state, and, (ii) the wave function in spin and flavour degrees of freedom be totally *symmetric*. For example, a Δ in the state with charge $Q = +2$ and maximal spin projection $m_s = +\frac{3}{2}$ is described by

$$\Delta^{++}_{m_s=\frac{3}{2}} = \mathrm{u}\uparrow\mathrm{u}\uparrow\mathrm{u}\uparrow. \tag{18}$$

Likewise, the $\Lambda(1116)$, having isospin zero, must contain a pair (ud − du). As the state is required to be symmetric in all quarks, this pair must also be in a state of spin zero. Finally, the strangeness −1 of Λ must be carried by an s-quark. Putting these facts together we have

$$\Lambda_{m_s} = \frac{1}{\sqrt{12}} \sum_{\pi} (\mathrm{u}\uparrow\mathrm{d}\downarrow - \mathrm{u}\downarrow\mathrm{d}\uparrow)\mathrm{s}_{m_s}, \tag{19a}$$

where the sum must be taken over all six permutations of the symbols u, d, s. The proton and neutron states are obtained from (19a) by replacing simply s by u or d, respectively. The normalization constant changes, however, because, with s replaced by u or d, six permutations out of the twelve terms in eq. (19a) are pairwise equal. One obtains

$$\mathrm{p}_{m_s} = \frac{1}{\sqrt{18}} \sum_{\pi} (\mathrm{u}\uparrow\mathrm{d}\downarrow - \mathrm{u}\downarrow\mathrm{d}\uparrow)\mathrm{u}_{m_s}, \tag{19b}$$

$$\mathrm{n}_{m_s} = \frac{1}{\sqrt{18}} \sum_{\pi} (\mathrm{u}\uparrow\mathrm{d}\downarrow - \mathrm{u}\downarrow\mathrm{d}\uparrow)\mathrm{d}_{m_s}. \tag{19c}$$

It is easy to verify that the states (17)–(19) reproduce the correct spin, parity and internal quantum numbers of the corresponding mesons and baryons. However, for the baryons there remains a problem: if the decomposition of baryon states in terms of quark states is to make sense at all, we would expect the *radial* wave functions of these states to be *symmetric*, too. [Unless the potential has a weird radial dependence, ground states in quantum mechanics have symmetric spatial wave functions.] This obvious clash with Fermi–Dirac statistics is repaired by introducing the colour degrees of freedom: Indeed, it is assumed that all physical hadron states are singlets with respect to $\mathrm{SU}(3)_c$. For the baryons this means that each product $(\mathrm{q}^1\mathrm{q}^2\mathrm{q}^3)$ in eqs. (18) and (19) is to be replaced by

$$\frac{1}{\sqrt{6}} \sum_{ikl=1}^{3} \varepsilon_{ikl}\mathrm{q}^{1i}\mathrm{q}^{2k}\mathrm{q}^{3l}. \tag{20}$$

This is a singlet with respect to $\mathrm{SU}(3)_c$. Obviously, it is antisymmetric in the three quarks so that the complete baryon wave functions with symmetric orbital functions are indeed antisymmetric.

1.2. Empirical information on weak interactions

In this section we summarize some characteristic properties of the weak interactions. In quoting experimental information on these properties we do not follow the historical development. Instead, we mention those experiments which illustrate in the simplest and most transparent manner the point under discussion.

1.2.1. Range of weak interactions

All purely leptonic and semileptonic weak scattering processes or decays that have been observed in the laboratory up to now, involve four external fermionic particles, leptons and/or quarks. For purely leptonic processes such as

$$\mu^- \to e^- \bar{\nu}_e \nu_\mu,$$

$$\nu_\mu + e^- \to e^- + \nu_\mu,$$

or semileptonic processes involving baryons, such as

$$n \to p e^- \nu_e,$$

$$\bar{\nu}_\mu + p \to n + \mu^+.$$

this is evident. For semileptonic decays of mesons, such as

$$\pi^+ \to \mu^+ \nu_\mu,$$

$$\pi^+ \to \pi^0 e^+ \nu_e,$$

$$K^+ \to \pi^0 \mu^+ \nu_\mu,$$

this is seen most easily by considering these processes in the framework of the quark model, as discussed above. For instance, in the decay $\pi^+ \to \mu^+ \nu_\mu$, the fundamental four-fermion process is

$$u + \bar{d} \to \mu^+ + \nu_\mu.$$

The quark and the antiquark of the pion annihilate at one vertex, the lepton pair is created at another vertex. Thus a typical weak amplitude contains *four* fermionic field operators.

There is clear experimental evidence that the weak interactions are of very short range. All known weak amplitudes at low and intermediate energies contain essentially only s- and p-waves in their partial wave decomposition. This indicates that the interaction is very close to being a *contact* interaction, effective only when the four particles are all at the same point of space and time. Indeed, if the external particles had spin zero, then a contact interaction would yield only s-wave amplitudes. As they carry spin one half, such an interaction can also yield p-wave amplitudes because of spin-orbit or spin-momentum coupling.

As an example, let us consider the process

$$\nu_\mu + e \to e + \nu_\mu. \tag{21}$$

Suppose, this process is due to the exchange of a heavy photon-like boson whose mass M_B is large as compared to the invariant momentum transfer, $M_B^2 \gg |q^2|$. In this case, its propagator can be approximated

$$\frac{-g^{\alpha\beta} + q^\alpha q^\beta / M_B^2}{q^2 - M_B^2} \simeq \frac{g^{\alpha\beta}}{M_B^2}. \tag{22}$$

We can make use of eq. (IV.56) provided we replace the photon propagator $1/q^2$ by

$1/M_B^2$ and replace $4\pi\alpha$ by $g_e \cdot g_\mu$, the product of the coupling constants of the boson to the lepton pairs (e, e) and (ν_μ, ν_μ) [we assume these to be equal, up to factors of order unity]. Let E^* be the neutrino energy in the c.m. system, z^* the cosine of the c.m. scattering angle, and assume $E^* \gg m_e$. From eq. (IV.56) with $M \equiv m_e$, $m \equiv m_\nu = 0$, we have

$$s = 2E^{*2} + m^2 + 2E^*\sqrt{m^2 + E^{*2}} \simeq 4E^{*2},$$
$$t = -2E^{*2}(1 - z^*).$$

Setting $F_1(t) \equiv 1$, $F_2(t) \equiv 0$, eq. (IV. 56) gives

$$\left(\frac{d\sigma}{d\Omega}\right)_{c.m.} \simeq \left(\frac{g^2}{M_B^2}\right)^2 \frac{E^{*2}}{4\pi^2}\left\{1 + \frac{t}{s} + \frac{1}{2}\left(\frac{t}{s}\right)^2\right\}$$
$$= \left(\frac{g^2}{M_B^2}\right)^2 \frac{E^{*2}}{4\pi^2}\left\{1 - \tfrac{1}{2}(1 - z^*) + \tfrac{1}{8}(1 - z^*)^2\right\}.$$

If this expression is integrated over all angles one obtains an expression for the total elastic cross section of the form

$$\sigma = \int d\Omega \left(\frac{d\sigma}{d\Omega}\right)_{c.m.} = \text{const. } E^{*2} \simeq \text{const. } \frac{s}{4}, \tag{23a}$$

i.e. proportional to the square of the c.m. energy of the neutrino. When expressed in the laboratory system where $s = m^2 + 2mE_\nu \simeq 2mE_\nu$ this means that the cross section is proportional to the laboratory energy of the neutrino,

$$\sigma = \text{const. } \tfrac{1}{2}m(E_\nu)_{\text{lab}}. \tag{23b}$$

This linear increase of neutrino cross sections is indeed verified experimentally up to laboratory energies of the order several 100 GeV. It is found to hold also for neutrino reactions such as $\nu_\mu + N \to N' + \mu$ etc. Up to energies of this order of magnitude the weak interactions effectively behave like contact interactions of the form

$$a_\Gamma \frac{g^2}{M_B^2} \overline{\Psi(1)}\, \Gamma_\alpha \Psi(2) \overline{\Psi(3)}\, \Gamma^\alpha \Psi(4), \tag{24}$$

where Γ_α denotes certain Dirac matrices such as γ^α, $\gamma^\alpha\gamma_5$ etc. and a_Γ is a dimensionless constant of order unity. If this is so then the widths Γ of weakly decaying states are proportional to $(a_\Gamma g^2/M_B^2)^2$ which has the dimension E^{-4}. As the decay width has the dimension of an energy, for dimensional reasons, we expect it to be proportional to the fifth power of the released energy. Except for some special cases (such as the decay $\pi \to e\nu_e$) in which there are additional hindrance factors due to angular momentum conservation, this is indeed what one finds empirically. The most striking proof is the great variety of lifetimes in nuclear β-decay which cover many orders of magnitude, from fractions of seconds to many thousands of years.

As an example, let us compare the lifetimes of the muon and of the neutron in the processes

$$\mu \to \mathrm{e}\,\nu\bar{\nu},$$

$$\mathrm{n} \to \mathrm{p}\,\mathrm{e}\bar{\nu}.$$

In the first case the energy release is $(\Delta E)_\mu \simeq m_\mu/2$, in the second case it is $(\Delta E)_\mathrm{n} \simeq m_\mathrm{n} - m_\mathrm{p}$. If the widths are proportional to $(\Delta E)^5$, the lifetimes are proportional to $(\Delta E)^{-5}$. The measured lifetimes are

$$\tau_\mu = 2.2 \times 10^{-6}\ \text{sec},$$

$$\tau_\mathrm{n} = 917\ \text{sec},$$

their ratio being $\tau_\mu/\tau_\mathrm{n} = 2.4 \times 10^{-9}$. This is to be compared to

$$\left(\frac{m_\mathrm{n} - m_\mathrm{p}}{m_\mu/2}\right)^5 = 8.7 \times 10^{-9}.$$

Additional remarks. More detailed expressions for neutrino cross sections as well as their absolute magnitude will be given below. We note, in particular, that the behaviour (23) cannot hold indefinitely since from a certain energy on it contradicts unitarity. At which energy this will happen is worked out below. Finally, the quantitative analysis will show that the actual range of weak interactions is of the order of 10^{-16} cm.

1.2.2. Charged current and neutral current vertices

We distinguish two classes of weak interactions:

(i) The *charged current* (CC) interactions which are described by the exchange of heavy, *charged* bosons $\mathrm{W}^\pm$. They are characterized by vertices of the kind shown in Figs. 1a)–c) at which either a lepton state f is converted into a neutrino state ν_f of the same lepton number L_f, or a quark of charge Q is converted into a quark of charge $Q \pm 1$. Examples are

$$\nu_\mu + \mathrm{e}^- \to \mu^- + \nu_\mathrm{e},$$

where ν_μ is converted to μ^-, through emission of a virtual W^-, and where e^- is converted to ν_e, through absorption of the same W^-. Similarly, in the reaction

$$\bar{\nu}_\mathrm{e} + \mathrm{p} \to \mathrm{n} + \mathrm{e}^+$$

the $\bar{\nu}_\mathrm{e}$ is converted into a e^+, a u-quark into a d-quark. Clearly, the incoming (outgoing) fermion can also be replaced by the corresponding outgoing (incoming) antifermion. In these cases pairs of the type $(\mathrm{f}^-, \bar{\nu}_\mathrm{f})$, $(\mathrm{f}^+, \nu_\mathrm{f})$, $(\mathrm{u}, \bar{\mathrm{d}})$, $(\mathrm{u}, \bar{\mathrm{s}})$ etc. are created or annihilated with simultaneous absorption or emission of a W^+ or W^-.

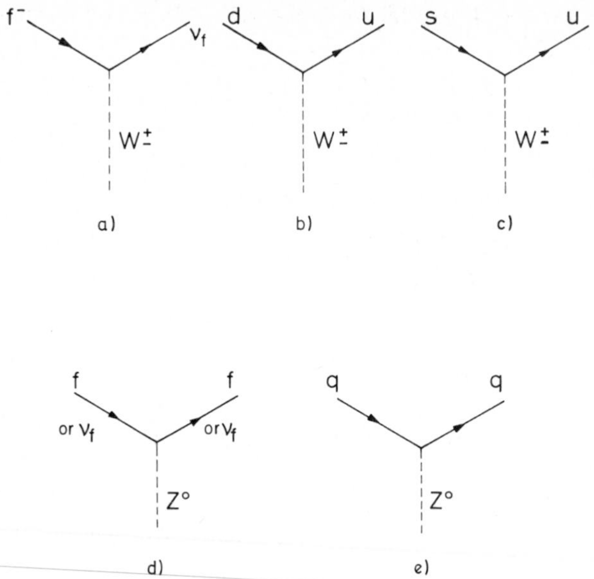

Fig. 1. Charged current vertices (a–c) and neutral current vertices (d, e) describing the coupling of a fermion pair (quarks or leptons) to vector bosons $W^{\pm}$ and Z^0, respectively.

Examples are

$\mu^- \to e^- \bar{\nu}_e \nu_\mu$	where $\mu^- \to \nu_\mu W^-$,	$W^- \to e^- \bar{\nu}_e$,
$\Lambda \to p\mu^- \bar{\nu}_\mu$	where $s \to uW^-$,	$W^- \to \mu^- \bar{\nu}_\mu$,
$\pi^+ \to \mu^+ \nu_\mu$	where $u\bar{d} \to W^+$,	$W^+ \to \mu^+ \nu_\mu$,
$K^- \to \mu^- \bar{\nu}_\mu$	where $\bar{u}s \to W^-$,	$W^- \to \mu^- \bar{\nu}_\mu$.

(ii) The *neutral current* (NC) *interactions* which we describe by the exchange of one (or, possibly, several) neutral boson(s) Z^0. Typical vertices are shown in Figs. 1d) and e). These interactions are somewhat analogous to electromagnetic interactions with the exchange of a photon, except for the following differences: Unlike the photon which is massless, the Z^0 boson is massive and very heavy; the couplings at vertices of the kind shown in Fig. 1 involve simultaneously vector and axial vector currents; neutrinos which have zero electric charge, have nonvanishing couplings to Z^0-bosons. Because the Z^0 is electrically neutral, amplitudes due to exchange of virtual Z^0 can interfere with amplitudes due to photon exchange. This is discussed in more detail below.

Examples of pure NC reactions are

$$\nu_\mu + e^- \to e^- + \nu_\mu,$$

$$\bar{\nu}_\mu + p \to p + \bar{\nu}_\mu,$$

whilst the interference of NC couplings with electromagnetic vertices can be detected in parity violating spin-momentum correlations.

1.2.3. Parity violation, vector and axial vector currents

It is well-known that the strong and electromagnetic interactions are both invariant under the operation of space inversion (parity). In fact, the parity transformation is defined in such a way that it leaves invariant the Lagrangian of strong and of electromagnetic interactions. The weak interactions, however, are not invariant under this parity operation, defined with respect to strong and electromagnetic interactions. Parity violation in weak interactions manifests itself, for instance, in nonvanishing, observable correlations between a spin $\boldsymbol{S}$ and a momentum $\boldsymbol{q}$, $\boldsymbol{S}\cdot\boldsymbol{q}$, in reactions and decay processes where there is no such correlation in the initial state. We give some examples:

(i) In charged pion decay, $\pi^- \to \mu^- \bar{\nu}_\mu$, experiment shows that the muon is fully polarized along its momentum. A measurement based on polarization studies in muonic atoms (formed with muons from pion decay), as well as measurements of polarization quantities in muon capture, show that the longitudinal polarization $P_\ell \equiv (\boldsymbol{\sigma}\cdot\boldsymbol{q})/q$ is 1 within about 10% (Abela et al. 1982, Roesch et al. 1982).

(ii) The electron (positron) from μ^- (μ^+) decay $\mu \to e\nu\bar{\nu}$ is found to be fully polarized along its momentum, even if the initial muon is unpolarized. The latest result from μ^+-decay (Corriveau et al. 1981) is

$$P_\ell = 1.01 \pm 0.06. \tag{25}$$

(iii) In the decay of polarized muons the electron is emitted anisotropically with respect to the muon spin direction. If x denotes the electron energy E in units of its maximum W, $x = E/W$, and if θ is the angle between the muon spin and the electron momentum, the double differential decay probability is proportional to (cf. eq. (282) below, summed over the electron spin)

$$\frac{\mathrm{d}^2\Gamma\left(\mu^- \to e^- \bar{\nu}_e \nu_\mu\right)}{\mathrm{d}(\cos\theta)\,\mathrm{d}x} \propto x^2\{(3-2x) - \xi(2x-1)\cos\theta\}. \tag{26}$$

Thus, the parameter ξ measures the correlation between the muon spin and the electron momentum. Experiment gives the value (Akhmanov et al. 1967)

$$\xi = 0.975 \pm 0.015. \tag{27}$$

Thus, for x close to 1, the electron is emitted preferentially in the direction opposite to the muon polarization ζ_μ, but is never emitted along the direction ζ_μ. This is in agreement with the observation (25) and with the fact that neutrinos are left-handed, antineutrinos are right-handed: For $x = 1$, i.e. maximal energy, the neutrinos must go out with parallel momenta but opposite to the electron momentum. This is illustrated by Fig. 2 which holds for $\theta = 180°$ and which shows that conservation of angular momentum favours this emission angle but forbids $\theta = 0$.

(iv) The most precise polarization data comes from nuclear β-decay. For instance, the longitudinal polarization of β-particles from Gamow–Teller transitions is found

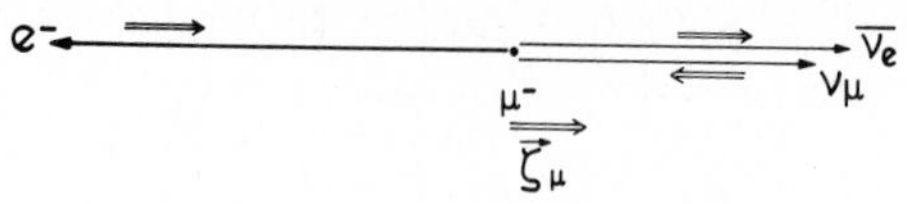

Fig. 2. Decay of negative muon at maximal energy of the decay electron at $\theta = 180°$. The neutrino spins compensate, the electron takes over the spin projection ζ_μ of the decay muon.

to be (Koks et al. 1976)

$$P_\ell = (1.001 \pm 0.008)\, v/c. \tag{28}$$

All of these examples show not only that parity is violated, but that it is violated in a *maximal* way: the longitudinal polarizations and correlation parameters assume their maximal theoretical value. This is true for leptonic CC interactions. For hadrons, maximal parity violation is sometimes hidden and attenuated by strong interaction effects. The intermediate bosons $W^\pm$ and Z^0, very much like the photon, carry spin 1. In contrast to the photon, however, they couple to Lorentz vector as well as axial vector currents. For a long time it was not possible to disprove the alternative possibility that the CC weak interactions were due exclusively to the exchange of bosons with spin 0 and 2, giving rise to Lorentz scalar, pseudoscalar and tensor, pseudotensor couplings, respectively. We now have clear evidence that such a pure situation is not realized in the weak interactions. There may be several bosons with spin 1, of the kind of the W. There may even be, in addition to these, bosons with spin 0 and, possibly, bosons with spin 2. However, the interactions due to the exchange of W-bosons must be predominant over the spin 0 or spin 2 exchanges. The evidence for this comes from an experiment which measures the transfer of leptonic helicity in the following inclusive reaction of $\overline{\nu_\mu}$ on iron (Jonker et al. 1979, 1983)

$$\overline{\nu_\mu} + \mathrm{Fe} \to \mathrm{X} + \mu^+, \tag{29}$$

in which X stands for any final state that can be reached in this reaction. This needs some explanation: A pair of external fermions f_1, f_2 couples to an intermediate boson B with spin J, via a covariant of the form

$$\overline{\Psi_{f_1}(x)}\, \Gamma^{(J)} \Psi_{f_2}(x). \tag{30}$$

If the boson has spin zero $J = 0$, the matrix $\Gamma^{(0)}$ is a linear combination of the unit matrix and of γ_5,

$$\Gamma^{(J=0)} = x_0 \mathbb{1} + y_0 \gamma_5. \tag{31}$$

This can also be written as follows:

$$\begin{aligned}\Gamma^{(J=0)} &= (x_0 + y_0)\tfrac{1}{2}(\mathbb{1} + \gamma_5) + (x_0 - y_0)\tfrac{1}{2}(\mathbb{1} - \gamma_5)\\ &= P_+(x_0 + y_0)P_+ + P_-(x_0 - y_0)P_-,\end{aligned} \tag{31$'$}$$

where we have denoted the projection operators (III.54) by P_+, P_- and have made use of their property of being idempotent. In the experiment (29) the muon is produced at an average, squared momentum around 4 GeV^2. As this is large as compared to m_μ^2 the muon mass may be neglected in the analysis. However, the matrices P_+, P_-, when applied to states of *massless* external fermions, project onto eigenstates of helicity. Using, for example, the standard representation we have

$$P_\pm = \frac{1}{2}(\mathbb{1}_4 \pm \gamma_5) = \frac{1}{2}\begin{pmatrix} \mathbb{1} & \pm\mathbb{1} \\ \pm\mathbb{1} & \mathbb{1} \end{pmatrix} \tag{32}$$

and, from the explicit spinor solutions (III.68–69),

$$P_\pm u(\boldsymbol{p}) = \frac{1}{2}\sqrt{p^0}\begin{pmatrix} (\mathbb{1} \pm h)\chi^{(r)} \\ (\pm\mathbb{1} + h)\chi^{(r)} \end{pmatrix}, \tag{33a}$$

$$P_\pm v(\boldsymbol{p}) = \frac{1}{2}\sqrt{p^0}\begin{pmatrix} \pm(\mathbb{1} \mp h)\chi^{(s)} \\ (\mathbb{1} \mp h)\chi^{(s)} \end{pmatrix}, \tag{33b}$$

where $h = (\boldsymbol{\sigma}\cdot\boldsymbol{p})/|\boldsymbol{p}|$ and $p^0 = |\boldsymbol{p}|$. [In deriving eq. (33b) one must remember that in the spinor $v(\boldsymbol{p})$ the Pauli spinor $\chi^{(s)} = \binom{0}{1}$ represents "spin up".] Thus, P_+ projects onto incoming *particle* states with *positive* helicity, and onto incoming *antiparticle* states with *negative* helicity. Likewise, P_- projects onto particle states with negative helicity and onto antiparticle states with positive helicity.

For outgoing states the same statements hold with the sign of the helicity reversed. We prove this by way of an example

$$\overline{u(\boldsymbol{p})}\,P_+ = \tfrac{1}{2}u^+(\boldsymbol{p})\gamma^0(\mathbb{1} + \gamma_5) = \tfrac{1}{2}u^+(\boldsymbol{p})(\mathbb{1} - \gamma_5)\gamma^0$$
$$= \tfrac{1}{2}((\mathbb{1} - \gamma_5)u(\boldsymbol{p}))^\dagger\gamma^0 = (P_-u(\boldsymbol{p}))^\dagger\gamma^0.$$

The action of the operator $\Gamma^{(0)}$, eq. (31′), on the helicities at a vertex with, say, an incoming neutrino ν_f and an outgoing charged lepton f^- is illustrated by Figs. 3a) and b). Specifically, we note that if the incoming neutrino is fully left-handed, the

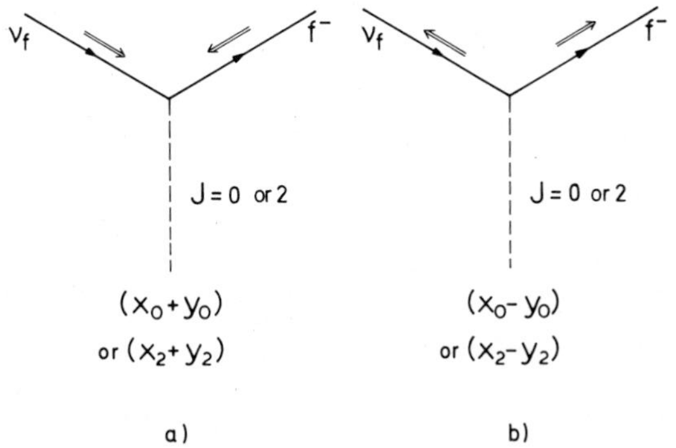

Fig. 3. Pattern of helicity transfer at a vertex of two massless fermions which couple to a boson carrying spin 0 or 2.

outgoing charged lepton must be fully right-handed (in the limit of neglecting its mass m_f).

It is not difficult to see that the same situation applies to the case of an intermediate boson with spin 2. In this case

$$\begin{aligned}\Gamma^{(J=2)} &= x_2\sigma^{\alpha\beta} + y_2\sigma^{\alpha\beta}\gamma_5 \\ &= (x_2+y_2)\sigma^{\alpha\beta}P_+ + (x_2-y_2)\sigma^{\alpha\beta}P_-, \end{aligned} \tag{34}$$

where $\sigma^{\alpha\beta} = (i/2)(\gamma^\alpha\gamma^\beta - \gamma^\beta\gamma^\alpha)$. As both P_+ and P_- commute with a product of two γ-matrices, and hence with $\sigma^{\alpha\beta}$, this is equal to

$$\Gamma^{(2)} = P_+(x_2+y_2)\sigma^{\alpha\beta}P_+ + P_-(x_2-y_2)\sigma^{\alpha\beta}P_-. \tag{34$'$}$$

$\Gamma^{(2)}$ has the same structure as $\Gamma^{(0)}$, eq. (31′), and, therefore, the transfer of helicity at a vertex with a boson of spin 2 is the same as for the case of spin 0, see Fig. 3.

The situation is different for the case of an intermediate boson with spin 1. The covariant (30) is now a linear combination of vector and axial vector terms, so that the matrix $\Gamma^{(1)}$ has the form

$$\begin{aligned}\Gamma^{(1)} &= x_1\gamma^\alpha + y_1\gamma^\alpha\gamma_5 \\ &= (x_1+y_1)\gamma^\alpha P_+ + (x_1-y_1)\gamma^\alpha P_-. \end{aligned} \tag{35}$$

Again, we replace P_+ and P_- by their squares $(P_\pm)^2$ and then move one factor to the left in both terms of eq. (35). However, as γ_5 anticommutes with every γ^α, cf. eq. (III.58), one has now

$$\gamma^\alpha P_+ = P_-\gamma^\alpha, \qquad \gamma^\alpha P_- = P_+\gamma^\alpha,$$

so that

$$\Gamma^{(1)} = P_-(x_1+y_1)\gamma^\alpha P_+ + P_+(x_1-y_1)\gamma^\alpha P_-. \tag{35$'$}$$

Upon comparison with eqs. (31′) and (34′) one sees that the transfer of helicity is different from the two previous cases, as indicated in Fig. 4. In particular, if the incoming neutrino is fully left-handed, the outgoing (approximately massless) lepton is also fully left-handed.

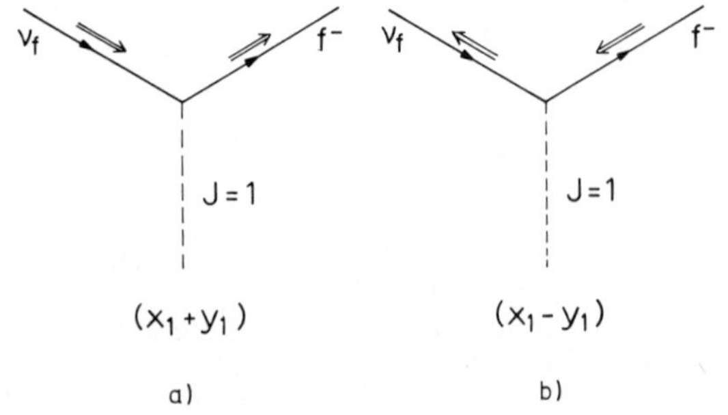

Fig. 4. Pattern of helicity transfer at a vertex of two massless fermions which couple to a boson with spin 1.

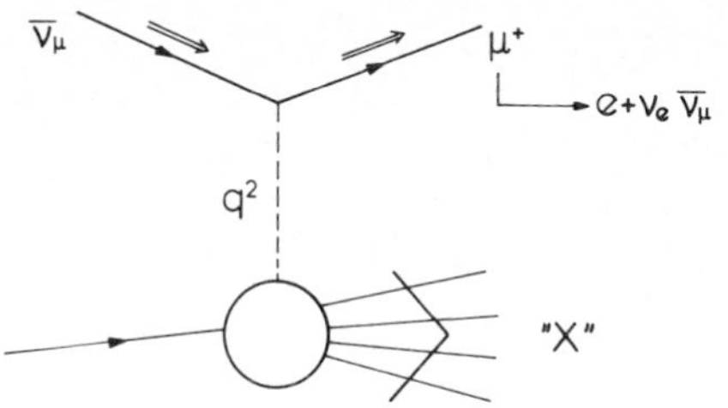

Fig. 5. Helicity transfer from the incoming muonic antineutrino to the positive muon in the inclusive charged current interaction (29). The longitudinal polarization of the outgoing μ^+ is determined from the decay asymmetry in its decay.

Equipped with this knowledge let us now analyze the results of reaction (29). The incoming $\overline{\nu}_\mu$ beam originates primarily from π^- and K^- decays and, therefore, is polarized *along* its momentum: $h(\overline{\nu}_\mu)$ is known to be $+1$ within, say, 10%, from the experiments mentioned above. In the experiment the average polarization of the outgoing μ^+ was measured. Integrating eq. (26) over the variable x from 0 to 1, and reversing the sign of the correlation term in $\cos\theta$ (eq. (26) holds for μ^--decay), one obtains

$$\int_0^1 \mathrm{d}x \frac{\mathrm{d}^2\Gamma(\mu^+ \to e^+\nu_e\overline{\nu}_\mu)}{\mathrm{d}(\cos\theta)\,\mathrm{d}x} \propto \tfrac{1}{2}\{1 + \tfrac{1}{3}\xi\cos\theta\}. \tag{36}$$

From the measured asymmetry (36) it was deduced that the μ^+ has a longitudinal polarization of $+1$, within an error of about 20%. Thus, the helicity transfer was found to be predominantly of the type shown in Fig. 5. This important result tells us that the reaction must be due predominantly to exchanges of spin 1.

1.2.4. Helicity transfer at weak and electromagnetic vertices: additional remarks

In the context of the discussion of helicity transfer in weak CC and NC reactions we wish to add some further remarks of general interest.

(i) Our analysis holds strictly only for massless external fermions. If the mass of the charged leptons is taken into account, then the helicities ± 1 are to be replaced by longitudinal polarizations $\pm v/c$, respectively. In many situations, very much like in the case discussed above, the kinematical situation is such that the mass may be neglected to a good approximation.

(ii) The pattern of helicity transfer depends only on the spin of the exchanged boson, but not on the parameters x_i and y_i in eqs. (31), (34) and (35). Therefore, the situations depicted in Figs. 3 and 4 are independent of whether or not there is parity violation in the interaction. As an example, consider eq. (35) with $x_1 = 1, y_1 = 0$. The covariant (30) is then a Lorentz vector $\overline{\Psi_{f_1}(x)}\,\gamma^\alpha \Psi_{f_2}(x)$ and may represent, for $f_1 = f_2$, the electromagnetic current coupling to a photon line. The transfer of helicity is still as described above. For example, in electron scattering at high energies the helicity of the incoming electron is transferred to the outgoing electron according to the same pattern as shown in Figs. 4. The two situations have the same coupling constants and, in case the incoming electron is unpolarized, they occur with equal weights. There is no helicity flip. This is a result that we had found already in sec. IV.2, in the context of a somewhat different approach.

(iii) Consider the decay of a particle with spin zero and mass M into two leptons with masses m_1 and m_2. Examples are

$$\begin{aligned} &\eta^0(549) \to \mu^+\mu^-, &&\eta^0(549) \to e^+e^-,\\ &\pi^0 \to e^+e^-,\\ &\pi^+ \to \mu^+\nu_\mu, &&\pi^+ \to e^+\nu_e,\\ &K^+ \to \mu^+\nu_\mu, &&K^+ \to e^+\nu_e. \end{aligned} \tag{37}$$

In the rest system of the decaying meson the two leptons are emitted with equal and opposite 3-momentum. Let the direction of this c.m. momentum be the 3-axis. The total angular momentum J as well as its projection J_3 onto that axis are conserved. As the outgoing particles are described by plane waves propagating parallel to the 3-axis, the projection l_3 of the relative *orbital* angular momentum vanishes. Conservation of J_3 then means that the projection S_3 of the total spin is also conserved. As $J_3 = 0$ before the decay, the two leptons must emerge with the *same* longitudinal polarization or helicity.

Suppose now that the interaction responsible for the decay is due to the exchange of bosons with spin 1 ($W^\pm$ or photons, respectively) and that the leptons in the final state are massless. From the analysis given above we know that they must be emitted with *opposite* helicity cf. Figs. 4. As this is in conflict with the conservation of angular momentum we conclude:

A state of spin zero cannot decay into two massless leptons if the interaction is due to bosons with spin 1.

In reality, the charged leptons f^- are not massless and the decays (37) are not strictly forbidden. For a massive particle the projection operators $P_\pm$, when acting on the particle's spinor wave function, yield nonvanishing components of either longitudinal polarization. However, the component required by conservation of angular momentum in the decays (37) is found to be proportional to m_f, the mass of the charged lepton. For example, if the $W^\pm$ boson couples with equal strength to (e, ν_e) and to (μ, ν_μ), the ratio of the decay width $\pi \to e\nu_e$ and $\pi \to \mu\nu_\mu$ is found to be

$$\begin{aligned} R := \frac{\Gamma(\pi^+ \to e^+\nu_e)}{\Gamma(\pi^+ \to \mu^+\nu_\mu)} &= \frac{m_e^2}{m_\mu^2}\left(\frac{1-(m_e/m_\pi)}{1-(m_\mu/m_\pi)}\right)^2\\ &= 2.34 \times 10^{-5} \cdot 5.49 = 1.28 \times 10^{-4}. \end{aligned} \tag{38}$$

The second factor in eq. (38) is the squared ratio of the c.m. momenta in the two decays. It is larger than 1 reflecting the fact that from kinematics alone, the decay into the lighter fermion is preferred over the heavier one. The first factor $(m_e/m_\mu)^2$ is the suppression factor discussed above. It is due to the conflict between the helicity transfer as required by the exchange of a boson with spin 1 and conservation of angular momentum. We emphasize, in particular, that this result does not depend on the specific form of $\Gamma^{(1)}$, eq. (35), and holds for any choice of the parameters x_1, y_1.

The result (38) receives a small additional contribution from radiative corrections, so that the theoretical prediction is

$$R_{\text{theor}} = 1.233 \times 10^{-4}, \tag{39}$$

to be compared with the experimental value

$$R_{\text{exp}} = (1.232 \pm 0.012) \times 10^{-4}. \tag{40}$$

1.2.5. Universality of weak interactions and some properties of semileptonic processes

There is good empirical evidence for the hypothesis that the weak interactions of leptons and of hadrons are universal in the following sense: All weak interactions are mediated by heavy, charged or neutral bosons carrying spin 1. External fermions couple to them via Lorentz vector and axial vector currents with coupling constants which are the same for leptons and quarks except for effects of state mixing and possible renormalization effects from strong interactions in the quark sector. Up to these modifications in the hadronic sector, a precise definition of which will be given below, semileptonic processes are governed by the same effective interaction. The decay widths for, say, a Fermi transition in a nucleus, for pion β-decay $\pi^+ \to \pi^0 e^+ \nu_e$, and for muon decay, are all determined by the same effective coupling constant. The exact relationship between charged interaction and neutral interaction coupling constants for leptons and quarks will be formulated in some detail below, in discussing the GSW unified theory.

Nevertheless, a few empirical properties of semileptonic weak processes can be formulated already at this point without the formalism of a unified gauge theory at hand.

A semileptonic process, by definition, involves both hadronic and leptonic states. The empirical selection rules and regularities that we now discuss, always refer to the hadronic part of the process under consideration. Regarding the additive quantum numbers of hadrons, the weak interactions neither conserve strangeness S, nor charm, nor any of the other, new quantum numbers associated with the b-, t-quarks. Furthermore, they are not scalars with respect to isospin: the charged weak currents which do not change the strangeness, $\Delta S = 0$, carry isospin 1, as demonstrated, for instance, by the decay modes

$$\pi^+ \to \mu^+ \nu_\mu, \qquad \pi^+ \to \pi^0 e^+ \nu_e. \tag{41}$$

Weak neutral currents with $\Delta S = 0$, contain both isoscalar and isovector pieces. If the strangeness is changed, $\Delta S \neq 0$, the weak current carries isospin 1/2, as is clear from the existence of the kaonic decay modes, analogous to (41),

$$K^+ \to \mu^+ \nu_\mu, \qquad K^+ \to \pi^0 e^+ \nu_e. \tag{42}$$

A most remarkable property of such interactions with $\Delta S \neq 0$ is this: the change in strangeness is always equal to the change of electric charge in the hadronic states,

$$\Delta S = (\Delta Q)_{\text{hadrons}}. \tag{43}$$

As an example, the decays

$$\Sigma^+ \to \mathrm{n f}^+ \nu_\mathrm{f},$$

where $\Delta S = -1$ but $(\Delta Q)_{\text{hadrons}} = +1$, have never been observed, in contrast to

$$\Sigma^- \to \mathrm{n f}^- \overline{\nu}_\mathrm{f},$$

which are allowed by the selection rule (43) and which have indeed been seen.

The selection rule (43) implies, in particular, that there are no strangeness changing processes with $\Delta Q = 0$, or, in other terms, that hadronic *neutral currents do not change the strangeness*. For example, the decay processes

$$\mathrm{K}^0_\mathrm{L} \to \mu^+\mu^-, \qquad \mathrm{K}^+ \to \pi^+ \mathrm{e}^+\mathrm{e}^- \tag{44}$$

are found to be strongly suppressed as compared to the other kaonic decay modes. The experimental branching ratios are

$$R\left(\mathrm{K}^0_\mathrm{L} \to \mu^+\mu^-/\mathrm{K}^0_\mathrm{L} \to \text{all}\right) = (9.1 \pm 1.9) \times 10^{-9}, \tag{45a}$$

$$R(\mathrm{K}^+ \to \pi^+\mathrm{e}^+\mathrm{e}^-/\mathrm{K}^+ \to \text{all}) = (2.6 \pm 0.5) \times 10^{-7}. \tag{45b}$$

[A more detailed analysis shows that the experimental results (45) can, in fact, be described by processes of higher than first order.]

2. Vector and axial vector covariants; effective Lagrangians with V and A couplings

In this section we analyze the covariants (30) for the case of vector and axial vectors in more detail. We then construct the general effective four-fermion Lagrangian for the case of vector (V) and axial vextor (A) couplings, i.e. the limit of the interaction due to exchange of bosons with spin 1, taken at momentum transfers which are small as compared to the boson masses. The behaviour of this Lagrangian under parity P, charge conjugation C, and time reversal T is investigated. The full fermion–boson interaction that leads to the effective Lagrangians will be derived in secs. 3 and 4 below, whilst the case of more general covariants is deferred to the discussion of possible deviations from V and A couplings, in sec. 6.

2.1. Vectors and axial vectors

Let us start by recalling the definition of $\overline{\Psi(x)}$ in eq. (III.55), viz.

$$\overline{\Psi(x)} = \left(\chi^{*b}(x)\phi^*{}_B(x)\right) = \left(\phi^{*a}(x)\chi^*{}_A(x)\right)\gamma^0, \tag{46}$$

where we had defined

$$\phi^{*a}(x) \coloneqq \phi^*{}_A(x)(\hat{\sigma}^0)^{Aa}, \tag{46a}$$

$$\chi^*{}_A(x) \coloneqq \chi^{*a}(x)(\sigma^0)_{aA}, \tag{46b}$$

and γ^0 was given by eq. (III.49′). In the four-component formalism eq. (46) is often written, somewhat inaccurately, as

$$\overline{\Psi(x)} = \Psi^\dagger(x)\gamma^0 \tag{47}$$

without explicit reference to the class of representations for the matrices of γ^μ that is being used. Clearly, in the high-energy representation we may write this relation, provided we keep in mind the correct position of indices in eq. (46). If we wish to use any other representation which is obtained from the high-energy representation via a linear, nonsingular transformation S, then the relation (47) can only be maintained if S is *unitary*. This is seen very easily by observing that with $\Psi' = S\Psi$ and $\gamma'^\mu = S\gamma^\mu S^{-1}$, eq. (47) transforms as follows

$$\overline{\Psi(x)} = \Psi'^\dagger(S^{-1})^\dagger S^{-1}\gamma'^0 S.$$

This is equal to $\overline{\Psi'(x)}\,S$ only if S is unitary, in which case $\overline{\Psi'(x)}\,\Psi'(x) = \overline{\Psi(x)}\,\Psi(x)$.

In Chap. III, sec. 3, we showed that under a proper, orthochronous Lorentz transformation $x' = \Lambda x$, $\Lambda \in L_+^\uparrow$ the matrices $\sigma^\mu\partial_\mu$ and $\hat{\sigma}^\mu\partial_\mu$ behave as follows:

$$(\sigma^\alpha\partial'_\alpha) = A(\Lambda)(\sigma^\beta\partial_\beta)A^\dagger(\Lambda), \tag{48a}$$

$$(\hat{\sigma}^\alpha\partial'_\alpha) = \hat{A}(\Lambda)(\hat{\sigma}^\beta\partial_\beta)A^{-1}(\Lambda). \tag{48b}$$

As the derivative transforms according to

$$\partial'_\alpha = \Lambda_\alpha{}^\beta\partial_\beta,$$

it is clear that both

$$\chi^{*a}(x)(\sigma^\alpha)_{aB}\chi^B(x) \quad \text{and} \quad \phi^*{}_A(x)(\hat{\sigma}^\alpha)^{Ab}\phi_b(x) \tag{49}$$

transform like contravariant vectors with respect to $L_+^\uparrow$, i.e.

$$\chi^{*a}(\Lambda x)(\sigma^\alpha)_{aB}\chi^B(\Lambda x) = \Lambda^\alpha{}_\beta\chi^{*c}(x)(\sigma^\beta)_{cD}\chi^D(x),$$

and similarly for the second covariant. Let us then see how these vectors transform under the parity operation P. From eq. (III.75) we know that under P

$$\phi_a(x) \underset{P}{\rightarrow} (\sigma^0)_{aB}\chi^B(Px),$$

$$\chi^A(x) \underset{P}{\rightarrow} (\hat{\sigma}^0)^{Ab}\phi_b(Px). \tag{50}$$

Inserting this into the expressions (49) and observing that

$$\hat{\sigma}^0\sigma^\alpha\hat{\sigma}^0 = \begin{pmatrix} \hat{\sigma}^0 \\ -\hat{\sigma}^i \end{pmatrix},$$

$$\sigma^0\hat{\sigma}^\alpha\sigma^0 = \begin{pmatrix} \sigma^0 \\ -\sigma^i \end{pmatrix},$$

we find that the sum

$$\chi^*\sigma^\alpha\chi + \phi^*\hat{\sigma}^\alpha\phi =: v^\alpha \tag{51}$$

behaves like a Lorentz *vector*, i.e.

$$v^\alpha(x) \underset{P}{\rightarrow} (-)^{1+\delta_{\alpha 0}} v^\alpha(Px), \tag{52}$$

while the difference

$$-\chi^* \sigma^\alpha \chi + \phi^* \hat{\sigma}^\alpha \phi =: a^\alpha \tag{53}$$

behaves like an *axial vector*, i.e.

$$a^\alpha(x) \underset{P}{\rightarrow} (-)^{\delta_{\alpha 0}} a^\alpha(Px). \tag{54}$$

It is easy to verify that the quantities v^α and a^α, when expressed in terms of Dirac spinors, are given by

$$v^\alpha(x) = \overline{\Psi(x)}\,\gamma^\alpha \Psi(x), \tag{51'}$$

$$a^\alpha(x) = \overline{\Psi(x)}\,\gamma^\alpha \gamma_5 \Psi(x). \tag{53'}$$

Having derived the behaviour of the currents (51) and (53) under proper, orthochronous Lorentz transformations and parity, let us now investigate how they transform under charge conjugation and time reversal.

The transformation properties of spinors under C was derived in sec. III.5, viz.

$$\phi_a(x) \underset{C}{\rightarrow} \chi_a^*(x),$$

$$\chi^A(x) \underset{C}{\rightarrow} -\phi^{*A}(x),$$

and, therefore, $\chi^{*a} \rightarrow \phi^a$, $\phi_A^* \rightarrow -\chi_A$. Thus

$$\chi^{*a}(\sigma^\alpha)_{aB}\chi^B \rightarrow -\phi^a(\sigma^\alpha)_{aB}\phi^{*B} = +\phi_a(\hat{\sigma}^{\alpha *})^{aB}\phi_B^* = -\phi_B^*(\hat{\sigma}^\alpha)^{Ba}\phi_a, \tag{55a}$$

where we have used rel. (III.38) and have made use of the fact that ϕ_a and ϕ_B^* anticommute and that the matrices $\hat{\sigma}^\alpha$ are unitary. Similarly,

$$\phi_A^*(\hat{\sigma}^\alpha)^{Ab}\phi_b \rightarrow -\chi_A(\hat{\sigma}^\alpha)^{Ab}\chi_b^* = +\chi^A(\sigma^{\alpha *})_{Ab}\chi^{*b} = -\chi^{*b}(\sigma^\alpha)_{bA}\chi^A. \tag{55b}$$

This implies that v^α is odd, a^α is even under charge conjugation. We note, in particular, that the field operators $\overline{\Psi(x)}$ and $\Psi(x)$ in eqs. (53′), (51′) are interchanged in taking the charge conjugate of v^α and a^α.

Time reversal, finally, has the following effect:

$$\phi_a(x) \underset{T}{\rightarrow} \phi_a^*(Tx) = (\sigma^0)_{aB}\varepsilon^{BD}\phi_D^*(Tx),$$

$$\chi^A(x) \underset{T}{\rightarrow} \chi^{*A}(Tx) = (\hat{\sigma}^0)^{Ab}\varepsilon_{bd}\chi^{*d}(Tx),$$

so that

$$\chi^{*a}(x) \rightarrow \chi^B(Tx)\varepsilon_{BD}(\hat{\sigma}^0)^{Da},$$

$$\phi^*{}_A(x) \rightarrow \phi_b(Tx)\varepsilon^{bd}(\sigma^0)_{dA}.$$

In applying T to a product of operators, the prescription is to reverse the order of factors without regard to the fermion character of the fields. Thus we have

$$\chi^{*a}(x)(\sigma^\alpha)_{aA}\chi^A(x) \to \varepsilon_{BD}(\hat\sigma^0)^{Da}(\sigma^\alpha)_{aA}(\hat\sigma^0)^{Ab}\varepsilon_{bd}\chi^{*d}(Tx)\chi^B(Tx)$$

$$= \varepsilon_{BD}\begin{pmatrix} \hat\sigma^0 \\ -\hat\sigma^i \end{pmatrix}^{Db} \varepsilon_{bd}\chi^{*d}(Tx)\chi^B(Tx)$$

$$= \chi^{*d}(Tx)\begin{pmatrix} \sigma^0 \\ -\sigma^i \end{pmatrix}_{dB} \chi^B(Tx). \tag{56a}$$

In the first step we inserted the relation mentioned above, in the second we used $(\sigma^{\alpha *})_{Bd} = (\sigma^\alpha)_{dB}$, i.e. the unitarity of σ^α. In a similar manner

$$\phi_A^*(\hat\sigma^\alpha)^{Ab}\phi_b \to \phi_B^*(Tx)\begin{pmatrix} (\hat\sigma^0)^{Ba} \\ (-\hat\sigma^i)^{Ba} \end{pmatrix}\phi_a(Tx). \tag{56b}$$

Thus, v^α and a^α behave as follows under T:

$$v^\alpha(x) \to (-)^{1+\delta_{\alpha 0}} v^\alpha(Tx), \tag{57}$$

$$a^\alpha(x) \to (-)^{1+\delta_{\alpha 0}} a^\alpha(Tx). \tag{58}$$

Note, however, that here again the fields $\Psi(x)$ and $\overline{\Psi(x)}$ are interchanged.

[It is instructive to compare the behaviour of v^α under P, eq. (52), C, and T, eq. (57), with the behaviour of a classical vector current, given by the product of a charge and a four-velocity, under the same transformations.]

Finally, from this discussion we can deduce the behaviour under parity, charge conjugation and time reversal, of vector and axial vector currents which are composed of two *different* fields, viz.

$$v^\alpha(x; i \to k) := \overline{\Psi^{(k)}(x)}\gamma^\alpha\Psi^{(i)}(x), \tag{59}$$

$$a^\alpha(x; i \to k) := \overline{\Psi^{(k)}(x)}\gamma^\alpha\gamma_5\Psi^{(i)}(x). \tag{60}$$

The transformation behaviour of the current operators (59) and (60) with respect to proper, orthochronous Lorentz transformations $\Lambda \in L_+$ is the same as for the operators (51′) and (53′) where particle field i and particle field k were identical, that is

$$v'^\alpha(\Lambda x; i \to k) = \Lambda^\alpha{}_\beta v^\beta(x; i \to k)$$

and similarly for a^α.

Regarding parity we must keep in mind that the fields i and k may have different intrinsic parities, in which case there is an extra minus sign to eqs. (52) and (54). Intrinsic parity of a particle field is always defined relative to the vacuum state which by convention is assigned positive parity. The intrinsic parities of some particles in nature are fixed and given by experiment (examples are the photon field, and the neutral pion for which the intrinsic parity can be measured).

Similarly, if the intrinsic parity of a fermion is fixed then we know from the theory of Dirac fields that the corresponding antiparticle has the opposite parity. The comparison of relative intrinsic parities for two different particle fields i and k is not possible whenever an absolute conservation law forbids the transformation of one particle of type i into one particle of type k, via a parity conserving reaction. For example, the relative parity of proton and neutron cannot be determined from the parities of the vacuum, the π^0 and the photon alone because charge conservation forbids vertices such as $(\mathrm{pn}\pi^0)$ or $(\mathrm{pn}\gamma)$. However, if, by convention, we define the intrinsic parity of the charged pions π^+, π^- to be the same as for the neutral one, experiments such as $\pi^-\mathrm{p} \to \pi^0\mathrm{n}$ or $\pi\mathrm{d} \to \mathrm{nn}$ will show that p and n then have even relative parity (or vice versa, the assignment of the same parity to p and n will fix the relative parity of $\pi^\pm$ and π^0). Similarly, the relative parity of electron and proton, or even of electron and muon, is not fixed and cannot be taken from experiment as long as electron and muon family numbers and baryon numbers are absolutely conserved. Thus, fixing the relative intrinsic parities of particle fields needs a certain number of conventions (beyond the experimental information as in the case of π^0 etc.). We do not enter this discussion in any more detail but just mention that it is possible to assign the same (positive) intrinsic parity to the lepton families (7) and the quark families (15), without running into conflict with experiment. The intrinsic parity of composite states such as (17)–(19), of course, then is determined by the parity of the orbital quark–antiquark and three quark states, respectively. The physical current operators (59), (60) that we shall consider, connect either leptons of the same family with one another, or quarks within the flavour families (15). So, even though in the case of two different fields the relations (52) and (54) must be multiplied by a factor

$$\eta_P^{(k)*}\eta_P^{(i)},$$

where $\eta_P^{(i)} = \mathrm{e}^{\mathrm{i}\pi\alpha(i)}$ with $\alpha(i) = 0$ or 1, we can choose conventions such that this factor is always $+1$.

In a similar way, charge conjugation C and time reversal T, in general, will give additional signs, or phase factors, when applied to the operators (59) and (60):

$$\overline{\Psi^{(k)}(x)}\gamma^\alpha\Psi^{(i)}(x) \underset{C}{\to} -\eta_C^{(k)*}\eta_C^{(i)}\overline{\Psi^{(i)}(x)}\gamma^\alpha\Psi^{(k)}(x), \tag{61a}$$

$$\overline{\Psi^{(k)}(x)}\gamma^\alpha\gamma_5\Psi^{(i)}(x) \underset{C}{\to} \eta_C^{(k)*}\eta_C^{(i)}\overline{\Psi^{(i)}(x)}\gamma^\alpha\gamma_5\Psi^{(k)}(x), \tag{61b}$$

$$\overline{\Psi^{(k)}(x)}\gamma^\alpha\Psi^{(i)}(x) \underset{T}{\to} \eta_T^{(k)*}\eta_T^{(i)}(-)^{1+\delta_{\alpha 0}}\overline{\Psi^{(i)}(x)}\gamma^\alpha\Psi^{(k)}(Tx), \tag{62}$$

the T-transformation of $a^\alpha(i \to k)$ being the same as that of $v^\alpha(i \to k)$. Again, for the physical currents of interest in electroweak interactions the additional phase factors

$$\eta_C^{(k)*}\eta_C^{(i)} \quad \text{and} \quad \eta_T^{(k)*}\eta_T^{(i)} \tag{63}$$

can be made equal to $+1$ by a suitable choice of conventions. If a composite state is an eigenstate of C (or of G-parity $G = C\exp\{\mathrm{i}\pi I_2\}$), its charge conjugation phase is completely determined by the properties of the bound state wave functions (see exercises 1 and 2).

2.2. *Effective vector (V) and axial vector (A) interactions*

On the basis of the empirical information discussed in sec. 1.2 and equipped with the technical tools of the preceding sec. 2.1 we can now write down the most general, effective Lagrangian for four external fermions interacting via vector and axial vector currents. It reads

$$
\begin{aligned}
-\mathscr{L}_{\mathrm{VA}} = \frac{G}{\sqrt{2}} \Big\{ & \big(\overline{\Psi^{(k)}(x)}\gamma^{\alpha}\Psi^{(i)}(x)\big) \\
& \times \Big[C_{\mathrm{V}}\big(\overline{\Psi^{(m)}(x)}\gamma_{\alpha}\Psi^{(n)}(x) + C'_{\mathrm{V}}\big(\overline{\Psi^{(m)}(x)}\gamma_{\alpha}\gamma_5\Psi^{(n)}(x)\big)\Big] \\
& + \big(\overline{\Psi^{(k)}(x)}\gamma^{\alpha}\gamma_5\Psi^{(i)}(x)\big)\Big[C_{\mathrm{A}}\big(\overline{\Psi^{(m)}(x)}\gamma_{\alpha}\gamma_5\Psi^{(n)}(x)\big) \\
& + C'_{\mathrm{A}}\big(\overline{\Psi^{(m)}(x)}\gamma_{\alpha}\Psi^{(n)}(x)\big)\Big] + \text{hermitean conjugate}\Big\},
\end{aligned}
\tag{64}
$$

where $C_{\mathrm{V}},\ldots,C'_{\mathrm{A}}$ are four complex constants. For the sake of convenience a real constant $G/\sqrt{2}$ has been factorized in eq. (64), so that, in fact, C_{V} (if it does not vanish) can be chosen to be a simple phase, or if that phase is not relevant in the interference of the interaction (64) with other interactions, to be $+1$. Even though there is this redundance, we keep C_{V} explicitly in order to preserve the symmetry of eq. (64) in the four types of couplings. Let us rewrite eq. (64), in a more compact form and omitting the space–time argument x which is the same in all current operators,

$$
\begin{aligned}
-\mathscr{L}_{\mathrm{VA}} = \frac{G}{\sqrt{2}} \big\{ & v^{\alpha}(i \to k)\big[C_{\mathrm{V}} v_{\alpha}(n \to m) + C'_{\mathrm{V}} a_{\alpha}(n \to m)\big] \\
& + \big[C^{*}_{\mathrm{V}} v^{\alpha}(m \to n) + C'^{*}_{\mathrm{V}} a^{\alpha}(m \to n)\big] v_{\alpha}(k \to i) \\
& + a^{\alpha}(i \to k)\big[C_{\mathrm{A}} a_{\alpha}(n \to m) + C'_{\mathrm{A}} v_{\alpha}(n \to m)\big] \\
& + \big[C^{*}_{\mathrm{A}} a^{\alpha}(m \to n) + C'^{*}_{\mathrm{A}} v^{\alpha}(m \to n)\big] a_{\alpha}(k \to i)\big\}.
\end{aligned}
\tag{64'}
$$

Here we have used $\gamma^0(\gamma^{\alpha})^{\dagger}\gamma^0 = \gamma^{\alpha}$ and $\gamma^0(\gamma^{\alpha}\gamma_5)^{\dagger}\gamma^0 = \gamma^{\alpha}\gamma_5$. By construction, $\mathscr{L}_{\mathrm{VA}}$ is invariant under all $\Lambda \in \mathrm{L}_{+}^{\uparrow}$. In order to make it conserve electric charge and all other, additively conserved, quantum numbers such as L_{f} and B, the fields i, k, m, n must be chosen such that the net balance of that "charge" is zero. For example, in the case of a charged current (CC) leptonic weak interaction we have

$$
\begin{aligned}
& k \equiv \mathrm{f}, \qquad i \equiv \nu_{\mathrm{f}}, \\
& m \equiv \nu_{\mathrm{f}'}, \qquad n \equiv \mathrm{f}'.
\end{aligned}
\tag{65}
$$

Indeed, the combination

$$
\big(\overline{\Psi^{(\mathrm{f})}(x)}\Gamma^{\alpha}\Psi^{(\nu_{\mathrm{f}})}(x)\big)\big(\overline{\Psi^{(\nu_{\mathrm{f}'})}(x)}\Gamma'_{\alpha}\Psi^{(\mathrm{f}')}(x)\big) + \text{h.c.}
\tag{65'}
$$

of charged leptons (f, f′) and uncharged leptons (ν_{f}, $\nu_{\mathrm{f}'}$) preserves electric charge, as well as the lepton family numbers L_{f} and L'_{f}. In the case of neutral current (NC) interactions, clearly, i and k must be identical, as well as m and n. For example, leptonic NC interactions will have

$$i \equiv k = \mathrm{f} \quad \text{or} \quad i \equiv k = \nu_{\mathrm{f}},$$

and similarly, $m \equiv n = \mathrm{f}'$ or $\nu_{\mathrm{f}'}$.

Let us then study the behaviour of eq. (64′), under P, C and T.

a) *Parity* P: The parity operation leaves invariant the product of two like operators ($v \cdot v$ or $a \cdot a$) but changes the sign of the product of two unlike operators ($v \cdot a$ or $a \cdot v$), up to a common phase factor $\eta_P = \eta_P^{(k)*}\eta_P^{(i)}\eta_P^{(m)*}\eta_P^{(n)}$. As pointed out before this factor can be chosen to be $+1$ in all cases of interest here. Thus the terms with unprimed coefficients C_i are even, the terms multiplied by primed constants C_i' are odd, or, written symbolically,

$$\{C_i, C_i'\} \underset{P}{\rightarrow} \{C_i, -C_i'\}. \tag{66}$$

b) *Charge conjugation* C: From eqs. (61) one sees that under C the terms $v \cdot v$ and $a \cdot a$ are transformed into their hermitean conjugates without extra signs, whilst the products $v \cdot a$ and $a \cdot v$ are transformed into minus their hermitean conjugates. In all cases this holds up to the phase factor $\eta_C = \eta_C^{(k)*}\eta_C^{(i)}\eta_C^{(m)*}\eta_C^{(n)}$ which, however, can be chosen to be $+1$. Thus, the action of C on $\mathscr{L}_{\mathrm{VA}}$ can be summarized by

$$\{C_i, C_i'\} \underset{C}{\rightarrow} \{C_i^*, -C_i'^*\}. \tag{67}$$

c) *Time reversal* T: As is evident from eq. (62) all terms in $\mathscr{L}_{\mathrm{VA}}$ are transformed into their hermitean conjugate, irrespective of whether they are products of like or unlike current densities. This holds again up to a common phase factor $\eta_T = \eta_T^{(k)*}\eta_T^{(i)}\eta_T^{(m)*}\eta_T^{(n)}$. This phase factor can be taken to be $+1$ in all cases of interest. Thus $\mathscr{L}_{\mathrm{VA}}$ transforms according to

$$\{C_i, C_i'\} \underset{T}{\rightarrow} \{C_i^*, C_i'^*\}. \tag{68}$$

We conclude this section with some remarks on these results.

(i) Let us combine the three operators P, C, and T to their product

$$\Theta := PCT. \tag{69}$$

From the results a)–c), eqs. (66)–(68), we see that the interaction (64) is invariant under the combined operation (69),

$$\Theta \mathscr{L}_{\mathrm{VA}} \Theta^{-1} = \eta_P \eta_C \eta_T \mathscr{L}_{\mathrm{VA}} \quad \text{or} \quad \{C_i, C_i'\} \underset{\Theta}{\rightarrow} \eta_P \eta_C \eta_T \{C_i, C_i'\}. \tag{70}$$

This is a special case of the *PCT*-theorem of Lüders and Jost (Lüders 1957, Jost 1957, 1963). Thus, if $\mathscr{L}$ is not invariant under one of the three symmetries P, C or T, it must break at least one more of them: For example, P and C could be violated but T conserved.

Parity invariance is broken if simultaneously some C_i as well as some C_i' are different from zero. On the contrary, if $\mathscr{L}_{\mathrm{VA}}$ contains only unprimed couplings, or primed couplings, it is invariant under parity. [In this case parity could still be broken through interference with some other interaction Lagrangian whose behaviour under P is different from the behaviour of $\mathscr{L}_{\mathrm{VA}}$ eq. (64).]

For time reversal invariance to be broken at least some of the coupling constants must be relatively complex, cf. eq. (68). For violation of C-invariance either of the two previous conditions, or both, must be fulfilled.

(ii) An important special case, relevant for leptonic CC interactions, is one where all C_i and C_i' are relatively real and where $C_i' = -C_i$. In this case $\mathscr{L}_{VA}$ is invariant under time reversal whilst invariance under parity and charge conjugation is violated in a *maximal* way: the P-odd terms in $\mathscr{L}_{VA}$ have the same magnitude as the P-even terms, the C-odd terms have the same magnitude as the C-even terms.

(iii) In a contact interaction of the product form (65′) the order of the fields is not fixed uniquely. For instance the fields $\Psi^{(\nu_f)}$ and $\Psi^{(f')}$ in eq. (65′) can be interchanged by means of Fierz reordering (Fierz, 1936), at the price of expressing the product of two specific covariants in one ordering by a sum over S, P, V, A, and T covariants in other ordering. For example (cf. sec. 6.1.4)

$$\begin{aligned}&\left(\overline{\Psi^{(1)}}\gamma^{\alpha}\Psi^{(2)}\right)\left(\overline{\Psi^{(3)}}\gamma_{\alpha}\Psi^{(4)}\right)\\&\quad=-\left(\overline{\Psi^{(1)}}\mathbb{1}\Psi^{(4)}\right)\left(\overline{\Psi^{(3)}}\mathbb{1}\Psi^{(2)}\right)-\left(\overline{\Psi^{(1)}}\gamma_5\Psi^{(4)}\right)\left(\overline{\Psi^{(3)}}\gamma_5\Psi^{(2)}\right)\\&\qquad+\tfrac{1}{2}\left(\overline{\Psi^{(1)}}\gamma^{\alpha}\Psi^{(4)}\right)\left(\overline{\Psi^{(3)}}\gamma_{\alpha}\Psi^{(2)}\right)+\tfrac{1}{2}\left(\overline{\Psi^{(1)}}\gamma^{\alpha}\gamma_5\Psi^{(4)}\right)\left(\overline{\Psi^{(3)}}\gamma_{\alpha}\gamma_5\Psi^{(2)}\right).\end{aligned} \tag{71}$$

A specific ordering is singled out only when the effective Lagrangian $\mathscr{L}_{VA}$ is the limiting form of an interaction due to the exchange of a heavy vector boson which couples to specific and well-defined currents (59) or (60).

(iv) Finally, we note that the C, P and T transformations can also be applied to interactions containing products of other covariants (30), such as scalars, pseudoscalars or tensors. We return to this in sec. 6.

2.3. Charged current and neutral current V and A interactions due to exchange of heavy vector bosons

Suppose the CC leptonic weak interactions are due to the exchange of charged vector mesons $W^{\pm}$ of mass m_w. Write the complex fields describing these particles as linear combinations of two mass degenerate real fields $A_\alpha^{(1)}$ and $A_\alpha^{(2)}$, in the spherical basis,

$$W_\alpha^{\pm} = \mp\frac{1}{\sqrt{2}}\left(A_\alpha^{(1)} \pm \mathrm{i}A_\alpha^{(2)}\right). \tag{72}$$

If these fields couple exclusively to the left-handed current $\gamma^\alpha\frac{1}{2}(\mathbb{1}-\gamma_5) = \gamma^\alpha P_-$, and if the coupling constant is denoted by g, the exchange of a $W^{\pm}$ between two pairs of leptons gives rise to amplitudes of the form

$$(\nu_f f|T|f'\nu_{f'}) \propto \frac{1}{\sqrt{2}}g\left(\overline{u_f}\gamma^\alpha P_- u_{\nu_f}\right)\frac{-g_{\alpha\beta}+q_\alpha q_\beta/m_w^2}{q^2-m_w^2}\frac{1}{\sqrt{2}}g\left(\overline{u_{\nu_{f'}}}\gamma^\beta P_- u_{f'}\right), \tag{73}$$

where q is the four-momentum transferred in the reaction. In the limit of small momentum transfer, $|q^2| \ll m_w^2$, the W-propagator may be replaced by $g_{\alpha\beta}/m_w^2$. In this approximation the amplitude (73) (as well as the analogous amplitudes with the external particles replaced by antiparticles), may be viewed as being due to the effective contact interaction

$$-\mathscr{L}_{CC}^{eff} = \frac{g^2}{8m_W^2}\left(\overline{\Psi^{(f)}(x)}\gamma^\alpha(\mathbb{1}-\gamma_5)\Psi^{(\nu_f)}(x)\right)\left(\overline{\Psi^{(\nu_{f'})}(x)}\gamma_\alpha(\mathbb{1}-\gamma_5)\Psi^{(f')}(x)\right) + \text{h.c.} \tag{74}$$

This has the form of eq. (64), (64′) with the identification (65) and the following special values of the coupling constants:

$$G/\sqrt{2} = g^2/8m_W^2, \tag{75}$$

$$C_V = C_A = 1, \qquad C'_V = C'_A = -1. \tag{76}$$

It follows from the general discussion in sec. 2.2 that the interaction (74) is invariant under time reversal but that it violates both parity and charge conjugation invariance. The violation of invariance under parity, in fact, is maximal: the interaction (74) produces neutrinos in purely left-handed states, antineutrinos in purely right-handed states. The interaction (74) is generally referred to as the *effective V − A four-fermion interaction*.

Suppose further that the NC weak interactions of leptons (and quarks) are due to amplitudes in which a heavy neutral vector boson Z^0 is exchanged between two pairs of external fermions of like charge. In the simplest case the field Z_α and the photon field A_α are orthogonal linear combinations of a field $A_\alpha^{(3)}$, the neutral partner of $A_\alpha^{(1)}$ and $A_\alpha^{(2)}$, and still another neutral vector field $A_\alpha^{(0)}$, viz.

$$A_\alpha(x) = \frac{1}{\sqrt{g^2+g'^2}}\{g'A_\alpha^{(3)}(x) + gA_\alpha^{(0)}(x)\}, \tag{77a}$$

$$Z_\alpha(x) = \frac{1}{\sqrt{g^2+g'^2}}\{gA_\alpha^{(3)}(x) - g'A_\alpha^{(0)}(x)\}. \tag{77b}$$

There is, of course, some arbitrariness in this ansatz. We have written these equations in anticipation of the results of a local gauge theory of electroweak interactions which is based on the gauge group SU(2)×U(1). The fields $A_\alpha^{(0)}$ and $\{A_\alpha^{(1)}, A_\alpha^{(2)}, A_\alpha^{(3)}\}$ are then gauge fields and, therefore, fall into the adjoint representations of U(1) and SU(2), respectively.

These gauge fields, which are defined with respect to the underlying group, are not observable as such. The physical vector mesons with electric charge ± 1 are defined by eq. (72), whilst the electrically neutral photon and Z^0 boson fields are given by eqs. (77). In terms of the latter the gauge fields $A_\alpha^{(3)}$ and $A_\alpha^{(0)}$ are given by

$$A_\alpha^{(3)}(x) = \frac{1}{\sqrt{g^2+g'^2}}\{gZ_\alpha(x) + g'A_\alpha(x)\}, \tag{78a}$$

$$A_\alpha^{(0)}(x) = \frac{1}{\sqrt{g^2+g'^2}}\{-g'Z_\alpha(x) + gA_\alpha(x)\}. \tag{78b}$$

At this stage and without having developed that theory, it is not possible to guess the detailed form of physical couplings between fermions and the neutral vector bosons except for the following input conditions:

(i) The coupling of the charged leptons and quarks to the photon must be of the form $\overline{\Psi^{(i)}}\gamma_\alpha\Psi^{(i)}A^\alpha$, i.e. must conserve parity; the coupling constant must be the electric charge of the external fermion.

(ii) The neutrinos, being electrically neutral, must not couple to the photon field.

(iii) If only lefthanded neutrinos couple in weak interactions, they must couple to the Z^0 via the V − A current $\overline{\Psi^{(\nu)}}\gamma_\alpha(\mathbb{1}-\gamma_5)\Psi^{(\nu)}$.

Furthermore, if ν_f and f form an isospinlike doublet, eq. (7), with respect to the gauge group, we may guess that the field $A_\alpha^{(3)}$ couples to this doublet via pure V − A operators $\gamma^\alpha P_-$ and via an operator of the type σ_3. g being the coupling constant to $A_\alpha^{(1)}$ and $A_\alpha^{(2)}$, this means that ν_f and f couple to $A_\alpha^{(3)}$ with equal and opposite coupling constants, $\pm\frac{1}{2}g$. Let us first consider the neutrino couplings:

$$\mathscr{L}_{NC}^{(\nu)}=\left\{\tfrac{1}{2}gA_\alpha^{(3)}(x)+\kappa A_\alpha^{(0)}(x)\right\}\overline{\Psi^{(\nu_f)}(x)}\gamma^\alpha P_-\Psi^{(\nu_f)}(x). \tag{79}$$

A glance at eqs. (78) shows that the curly brackets exclude the photon field if and only if $\kappa=-\frac{1}{2}g'$. In this case eq. (79) becomes

$$\mathscr{L}_{NC}^{(\nu)}=\tfrac{1}{2}\sqrt{g^2+g'^2}\,Z_\alpha^0(x)\overline{\Psi^{(\nu_f)}(x)}\gamma^\alpha P_-\Psi^{(\nu_f)}(x). \tag{79'}$$

The couplings of its charged partner f^- to the neutral gauge fields must have the form

$$\mathscr{L}_{\text{neutral}}^{(f)}=\overline{\Psi^{(f)}(x)}\left\{-\tfrac{1}{2}gA_\alpha^{(3)}(x)\gamma^\alpha P_--\tfrac{1}{2}g'A_\alpha^{(0)}(x)(\lambda\gamma^\alpha P_++\gamma^\alpha P_-)\right\}\Psi^{(f)}(x). \tag{80}$$

We have fixed the second term of the coupling to $A_\alpha^{(0)}$ on the basis of the following consideration: $A_\alpha^{(0)}$ being a singlet field, we expect this field to couple to the left-handed current $\overline{\Psi^{(f)}}\gamma^\alpha P_-\Psi^{(f)}$ with the same strength $\kappa=-\frac{1}{2}g'$ as to the neutrino field, cf. eq. (79). The constant λ, finally, must be chosen such as to meet the requirement (i) above. This is guaranteed if and only if $\lambda=2$, as in this case the terms with a $\gamma^\alpha\gamma_5$ coupling to the photon field A_α cancel out. Setting $\lambda=2$, eq. (80) becomes

$$\mathscr{L}_{\text{neutral}}^{(f)}=\frac{1}{\sqrt{g^2+g'^2}}\overline{\Psi^{(f)}(x)}\gamma^\alpha\left\{-gg'(P_-+P_+)A_\alpha(x)\right.$$
$$\left.+\tfrac{1}{2}\left[g'^2(2P_++P_-)-g^2P_-\right]Z_\alpha^0(x)\right\}\Psi^{(f)}(x). \tag{80'}$$

Of course, P_++P_- is unity. Therefore, the elementary charge is to be identified as follows:

$$Q_f=-|e|=gg'/\sqrt{g^2+g'^2}\,. \tag{81}$$

As $e^2 \leq g^2$, one can set

$$e^2/g^2 =: \sin^2\theta_w, \tag{82}$$

$$g' = -g \operatorname{tg}\theta_w. \tag{83}$$

Eq. (82) defines the Weinberg angle θ_w.

With this parametrization the neutral interactions can be written as follows:

$$\begin{aligned}\mathscr{L}^{(f)}_{\text{neutral}} = &-Q_f A_\alpha(x)\overline{\Psi^{(f)}(x)}\gamma^\alpha\Psi^{(f)}(x)\\ &+\frac{g}{\cos\theta_w}Z_\alpha(x)\left\{\overline{\Psi^{(f)}(x)}\left[-\tfrac{1}{2}\gamma^\alpha P_- + \sin^2\theta_w\gamma^\alpha\right]\Psi^{(f)}(x)\right.\\ &\left.+\overline{\Psi^{(\nu_f)}(x)}\tfrac{1}{2}\gamma^\alpha P_-\Psi^{(\nu_f)}(x)\right\},\end{aligned} \tag{84}$$

where $e = |e|$ and $Q_f = -e$.

As in the case of the charged bosons $W^\pm$, the exchange of a Z^0 with squared momentum transfer small as compared to its mass, $|q^2| \ll m_Z^2$, gives rise to an effective current–current interaction of the form

$$-\mathscr{L}^{\text{eff}}_{\text{NC}} = \frac{g^2}{16\cos^2\theta_w}\frac{1}{m_Z^2}K^\dagger_\alpha(x)K^\alpha(x), \tag{85}$$

where

$$\begin{aligned}K^{(f)}_\alpha(x) := &\overline{\Psi^{(f)}(x)}\left[-\gamma_\alpha(1-\gamma_5) + 4\sin^2\theta_w\gamma_\alpha\right]\Psi^{(f)}(x)\\ &+\overline{\Psi^{(\nu_f)}(x)}\gamma_\alpha(1-\gamma_5)\Psi^{(\nu_f)}(x).\end{aligned} \tag{86}$$

The effective coupling strength in eq. (85) can also be written in terms of G, eq. (75):

$$\frac{g^2}{16\cos^2\theta_w}\frac{1}{m_Z^2} = \frac{1}{2}\rho\frac{G}{\sqrt{2}}, \tag{87}$$

with the definition

$$\rho := \frac{m_W^2}{m_Z^2\cos^2\theta_w}. \tag{88}$$

The effective interaction (85) is of V and A character with real coupling constants. Therefore, it is invariant under time reversal but breaks both parity and charge conjugation symmetry.

These semiempirical results constitute the leptonic sector of the Glashow–Salam–Weinberg unified theory of electroweak interactions (Glashow 1963, Salam 1967, Weinberg 1967). They will be derived below in the mathematical framework of local gauge theories.

In anticipation of later results we note at this point that the phenomenological analysis of the data yields $G \simeq 1.1663 \times 10^{-5}\ \text{GeV}^{-2}$, $\sin^2\theta_w = 0.22$, $\rho \simeq 1$. This suffices to estimate the masses of $W^\pm$ and Z^0. From eq. (75) we have

$$m_w = \sqrt{\frac{\pi\alpha}{G\sqrt{2}}}\frac{1}{\sin\theta_w} = 37.3/\sin\theta_w\,\text{GeV}, \tag{89}$$

and from eq. (88)

$$m_Z = \frac{m_W}{\cos\theta_W} = 37.3/\sin\theta_W\cos\theta_W\ \text{GeV}. \tag{90}$$

[These expressions hold up to radiative corrections.] Thus m_W is found to be of the order of 80 GeV, m_Z of the order of 90 GeV, or expressed in Compton wave lengths, $\lambda\!\!\!^{-}(W) \simeq 2.5 \times 10^{-16}$ cm, and $\lambda\!\!\!^{-}(Z^0) \simeq 2.2 \times 10^{-16}$ cm. This gives a measure for the range of weak CC and NC interactions.

2.4. *Difficulties of the effective current–current theory*

The effective current–current interactions (74) and (85), as well as their hadronic counterparts, are very useful in practical applications to weak processes at low and intermediate energies, provided they are treated in lowest order perturbation theory. However, if one applies them to weak scattering processes at high energies (say $\gg 100$ GeV) or if one tries to compute higher order effects, one runs into two kinds of difficulties. A contact interaction of the current–current form (74) or (85) leads to neutrino cross sections which increase linearly with increasing neutrino (laboratory) energy, cf. eq. (23b). This linear increase cannot hold indefinitely and is in contradiction to unitarity. Furthermore, such a theory is not renormalizable, that is higher order diagrams cannot be made finite by means of a *finite* number of renormalization constants. Clearly, the two problems are intimately related. Since we do not know how to handle a nonrenormalizable field theory, we cannot decide whether or not a theory with contact interactions makes mathematical sense, in an exact way. That lowest order diagrams (generalized Born terms) lead to conflict with unitarity at high energies is not new, of course. In the frame of perturbation theory unitarity of S-matrix elements is restored by diagrams of higher orders. These, however, cannot be computed in a theory which is not renormalizable.

In order to obtain a more quantitative feeling of where this conflict with unitarity is to be expected let us work out the following very simple example: Consider the elastic scattering process

$$\nu_e + e^- \to e^- + \nu_e \tag{91}$$

on the basis of the CC contact interaction (74) [and neglecting the NC interaction (85) that contributes to this process, too]. From the formulae in App. B we have

$$d\sigma = \frac{(2\pi)^{10}}{4(pq)}\,\frac{1}{2}\sum|T|^2\delta(p+q-p'-q')\frac{d^3q'}{2q'_0}\,\frac{d^3p'}{2p'_0},$$

where p, p' denote the initial and final neutrino momenta, respectively, q, q' those of the electron before and after the scattering.

Let $s=(p+q)^2$ and $d^3q' = x^2\,dx\,d\Omega^*$, where $x \equiv |\boldsymbol{q}'|$. Integrating over d^3p' one finds

$$\frac{d\sigma}{d\Omega^*} = \frac{(2\pi)^{10}}{16(s-m_e^2)}\int_0^\infty x\,dx\,\frac{\delta\left(\sqrt{x^2+m_e^2}+x-\sqrt{s}\right)}{\sqrt{x^2+m_e^2}}\sum|T|^2.$$

The integral over x gives the integrand at $x_0=(s-m_e^2)/2\sqrt{s}$ times a factor $\sqrt{x_0^2+m_e^2}/\sqrt{s}$ which stems from the derivative of the argument of the δ-distribution,

$$\frac{d\sigma}{d\Omega^*}=\frac{1}{32s}(2\pi)^{10}\sum|T|^2.$$

The squared T-matrix element for the interaction (74), when summed over the spins is

$$\begin{aligned}(2\pi)^{12}\sum|T|^2 &= \tfrac{1}{2}G^2\mathrm{Sp}\{\gamma^\alpha(1-\gamma_5)\not p\gamma^\beta(1-\gamma_5)(\not q'+m_e)\}\\ &\quad\times\mathrm{Sp}\{\gamma_\alpha(1-\gamma_5)(\not q+m_e)\gamma_\beta(1-\gamma_5)\not p'\}\\ &= 128G^2(pq)(p'q')=32G^2(s-m_e^2)^2,\end{aligned}$$

so that

$$\frac{d\sigma}{d\Omega^*}=\frac{G^2}{4\pi^2}\frac{(s-m_e^2)^2}{s}=1.34\times10^{-39}\ \mathrm{cm^2GeV^{-2}}\frac{(s-m_e^2)^2}{s} \tag{92}$$

and the integrated cross section is

$$\sigma=\frac{G^2}{\pi}\frac{(s-m_e^2)^2}{s}. \tag{93}$$

Let us now analyze these results in some detail. The differential cross section in the c.m. system (92) as well as the integrated cross section (93), for $s\gg m_e^2$, increase like s, i.e. like the *square* of the neutrino energy in the c.m. frame. This same cross section (93) can also be evaluated in the laboratory system, where

$$s=(p+q)^2=m_e^2(1+2E_\nu^{\mathrm{lab}}/m_e)\equiv m_e^2(1+2\omega)$$

with $\omega:=E_\nu^{\mathrm{lab}}/m_e$. This gives

$$\sigma=\frac{2G^2m_e^2}{\pi}\frac{2\omega^2}{1+2\omega}=8.8\times10^{-45}\ \mathrm{cm}^2\frac{2\omega^2}{1+2\omega}. \tag{93'}$$

For $\omega\gg1$ this cross section increases *linearly* with E_ν^{lab}. For $E_\nu^{\mathrm{lab}}\simeq400$ GeV it is of the order of 7×10^{-39} cm^2. Returning to the c.m. system, we can write eq. (92) in terms of the standard scattering amplitude f, viz.

$$\frac{d\sigma}{d\Omega^*}=\frac{1}{2}\sum|f|^2,$$

where, according to eq. (II.15), $f=(8\pi^5/\sqrt{s})T$ so that

$$\sum|f|^2=\left(\frac{G}{\pi\sqrt{2}}\frac{s-m_e^2}{\sqrt{s}}\right)^2.$$

This spin-average of the squared amplitude is isotropic and, therefore, behaves like a scalar (i.e. spinless) s partial wave. On the basis of unitarity it must have the general form

$$\left(\sum|f|^2\right)^{1/2} = f_{l=0} = \frac{1}{2\mathrm{i}k}\left(\eta \mathrm{e}^{2\mathrm{i}\varepsilon} - 1\right),$$

cf. eq. (I.13), where ε is a real phase and η is the inelasticity bounded by $0 \le \eta \le 1$; k is the c.m. momentum and is given by $k = (s - m_e^2)/2\sqrt{s}$. Unitarity implies an upper bound on the cross section,

$$\frac{\mathrm{d}\sigma}{\mathrm{d}\Omega^*} \le \frac{1}{8k^2}(\eta + 1)^2 \le \frac{1}{2k^2} = \frac{2s}{\left(s - m_e^2\right)^2}. \tag{94}$$

The calculated differential cross section (92) reaches the unitarity bound (94) at a critical value $s = s_c$ which is so large that m_e^2 can be neglected as compared to s_c, viz.

$$s_c \simeq \frac{2\pi\sqrt{2}}{G} = 7.62 \times 10^5\ \mathrm{GeV}^2. \tag{95}$$

The value (95) corresponds to a c.m. energy of the neutrino of the order of $E_\nu \simeq \sqrt{s}/2 \simeq 440$ GeV. This critical energy is very large indeed. Therefore, at low and intermediate energies the effective current–current interaction (74) and (85) is a very good approximation to the interaction due to exchange of $W^\pm$ and Z^0. So in many practical applications one can neglect the typical effects of W- and Z-propagators. The full unified gauge theories are renormalizable, as was first shown by 't Hooft ('t Hooft 1971), and lead to unitary and calculable S-matrices. The current–current contact interactions then appear as the effective interactions, valid at low and intermediate energies.

3. Elements of local gauge theories based on non-Abelian groups

This section deals with the principles of constructing field theories which, in addition to being Lorentz covariant, are invariant under a group of *local* symmetry transformations. Depending on whether the underlying symmetry group is Abelian or non-Abelian, we talk about *Abelian* or *non-Abelian gauge theories*. Quantum electrodynamics (QED) is an example of a Lorentz covariant theory which, in addition, is invariant under local gauge transformations of the photon field, cf. eq. (III.183), and of the matter fields, cf. eq. (III.180).

In this case, as is evident from eq. (III.180), the group of transformations is a one-parameter continuous group, i.e. it is an Abelian group and has the structure of U(1). We develop the more general non-Abelian case in close analogy to QED. We define generalized vector potentials $A_i^{(k)}(x)$, generalized field tensors $F_{\mu\nu}^{(k)}(x)$, and generalized covariant derivatives, eq. (III.184), of matter fields. These notions and definitions provide the tools for the construction of a rather general class of local gauge theories. We develop these theories in a constructive but still elementary way. Although we try to render the main properties and results as transparent as possible by invoking the geometrical interpretation of the basic elements of the theory, we do not enter the mathematical properties of gauge theories in their full rigour. [These form an important and rich topic by themselves which, however, goes far beyond the scope of this book.]

3.1. Groups of local gauge transformations

Let G be a compact Lie group. We shall always assume that G is simple, or is the direct product of a finite number of simple Lie groups. Examples of interest in particle physics are the unitary unimodular groups SU(n) (i.e. the groups generated, for example, by the unitary matrices with determinant 1 in n complex dimensions), such as

$$\mathrm{U}(1), \qquad \mathrm{SU}(2), \qquad \mathrm{SU}(3),$$

or direct products thereof

$$\mathrm{SU}(2)\times\mathrm{U}(1), \qquad \mathrm{SU}(3)\times\mathrm{SU}(2)\times\mathrm{U}(1), \qquad \text{etc.}$$

The generators of infinitesimal transformations in G are written as T_i in abstract notation, i.e. when no reference to a specific representation is made. They obey the commutators

$$\left[T_i, T_j\right] = i \sum_{k=1}^{N} C_{ijk} T_k, \quad i, j = 1, 2, \ldots, N, \tag{96}$$

where C_{ijk} are the structure constants. As is well known the structure constants C_{ijk} can be chosen totally antisymmetric and fulfill the identity [R9, R24]

$$\sum_l \{C_{ikl}C_{lmn} + C_{kml}C_{lin} + C_{mil}C_{lkn}\} = 0, \tag{97}$$

which follows from the Jacobi identity for T_i, T_k, and T_m. N, finally, is the dimension of the Lie algebra (96) of the group G, $N = \dim(G)$. *⁾ Let us consider some examples:

U(1): Here $N = 1$ and T_1 is the unit element.

SU(2): Here we have $N = 3$ and $C_{ikl} = \varepsilon_{ikl}$, the totally antisymmetric tensor in three dimensions.

A concrete realization of U(1) and SU(2) is obtained by considering the group U(2) of unitary matrices in two complex dimensions. Such matrices $u \in \mathrm{U}(2)$ have the form

$$u = \mathrm{e}^{\mathrm{i}\alpha}\begin{pmatrix} a & b \\ -b^* & a^* \end{pmatrix} \quad \text{with } |a|^2 + |b|^2 = 1,$$

so that $uu^\dagger = \mathbb{1}$. Any such matrix depends on four real parameters and can be written as an exponential series

$$u = \exp\left\{\mathrm{i}\sum_{\mu=0}^{3} \Lambda_\mu h_\mu\right\}, \tag{98}$$

*⁾For SU(n), as is well-known, $N = n^2 - 1$.

in the hermitean matrices

$$h_0 = \begin{pmatrix} 1 & 0 \\ 0 & 1 \end{pmatrix}, \qquad h_i = \tfrac{1}{2}\sigma^{(i)}, \quad i = 1,2,3, \tag{99}$$

Λ_μ being arbitrary real parameters. The commutators of these matrices are

$$[h_0, h_\mu] = 0, \quad \mu = 0,1,2,3, \tag{100a}$$

$$[h_i, h_j] = \mathrm{i} \sum_{k=1}^{3} \varepsilon_{ijk} h_k, \quad i, j = 1,2,3. \tag{100b}$$

As h_0 commutes with h_i, we can write eq. (98) equivalently as follows

$$u = \exp\{\mathrm{i}\Lambda_0 h_0\}\exp\left\{\mathrm{i} \sum_{k=1}^{3} \Lambda_k h_k\right\}. \tag{98'}$$

The first factor of eq. (98′) defines an Abelian subgroup of U(2) and, therefore, forms a U(1) group. As to the second factor we note that $H := \sum_{k=1}^{3} \Lambda_k h_k$ is not only hermitean but also traceless, $\mathrm{Sp}\{H\} = 0$. Therefore, the matrices $\exp\{\mathrm{i}H\}$ are unitary and have determinant 1 (see exercise 3). Thus, the second factor in eq. (98′) defines SU(2), the group of all unitary, unimodular matrices in two complex dimensions. In abstract notation we write the generators of U(2) as T_μ with $\mu = 0, 1, 2, 3$ and obtain the commutators

$$[T_0, T_\mu] = 0, \quad \forall \mu, \tag{101a}$$

$$[T_i, T_j] = i \sum_k \varepsilon_{ijk} T_k. \tag{101b}$$

SU(3): Here we have $N = 8$ and the structure constants are $C_{ijk} = f_{ijk}$ with f_{ijk} as indicated in the following scheme.

ikl	123	147	156	246	257	345	367	458	678
f_{ikl}	1	$\frac{1}{2}$	$-\frac{1}{2}$	$\frac{1}{2}$	$\frac{1}{2}$	$\frac{1}{2}$	$-\frac{1}{2}$	$\frac{1}{2}\sqrt{3}$	$\frac{1}{2}\sqrt{3}$

Those structure constants f_{stu} for which stu is not an even or odd permutation of the indices ikl as listed in this table vanish.

A concrete realization of SU(3) is obtained by considering all unitary, unimodular matrices in 3 complex dimensions. As for SU(2) we write

$$u = \exp\left\{\mathrm{i} \sum_{k=1}^{8} \Lambda_k h_k\right\}, \quad u \in \mathrm{SU}(3), \tag{102}$$

with

$$uu^\dagger = \mathbb{1}, \quad \det(u) = 1,$$
$$h_k = h_k^\dagger, \quad \mathrm{Sp}(h_k) = 0,$$
$$h_k = \tfrac{1}{2}\lambda_k,$$

where λ_k are eight linearly independent matrices whose explicit form can be found in the literature [R12]. Λ_k are arbitrary real parameters, as before. In abstract notation the generators of infinitesimal SU(3) transformations are written as T_k, $k = 1, 2, \ldots, 8$. They obey the Lie algebra

$$\left[T_i, T_j\right] = \mathrm{i} \sum_{k=1}^{8} f_{ijk} T_k, \tag{103}$$

with f_{ijk} as defined above.

In all of these examples we are dealing with *compact* groups. The group parameters Λ_k are then generalized angles of rotation. If one wishes, one can choose these angles such that their domain of variation is either the interval $[0, \pi]$ or the interval $[0, 2\pi]$. For example, SU(2) can be parametrized by means of three Euler angles, cf. eqs. (III.19 and 119), with

$$0 \leq \psi \leq 2\pi, \qquad 0 \leq \theta \leq \pi, \qquad 0 \leq \phi \leq 2\pi.$$

Transformations of the kind of eq. (98) or (102), are called *global* transformations. These must be distinguished from *local* transformations where the group parameters are allowed to be functions of space and time to which we now turn. A group element of G is said to be a *local* transformation or a *gauge transformation* if it has the form

$$g(x) = \exp\left\{\mathrm{i} \sum_{k=1}^{N} \Lambda_k(x) T_k\right\}, \tag{104}$$

where $x \equiv \{x^\mu\}$ is a point in Minkowski space. In contrast to the case of global transformations the group parameters are now taken to be (in general infinitely differentiable) functions $\Lambda_k(x)$ of space and time.

It is easily verified that the local transformations (104), for fixed x, form a group with respect to the group multiplication

$$g_1(x) \cdot g_2(x).$$

Thus the definition (104) provides a copy $G(x)$ of the original Lie group G, for every x in Minkowski space. Consider now a set of M matter fields

$$\phi(x) := \{\phi_n(x)\}, \quad n = 1, 2, \ldots, M, \tag{105}$$

which form an M-dimensional representation of the group G. This representation must be unitary but need not be irreducible. For simplicity we take the fields ϕ_n to be spin-zero boson fields, at least for the moment, but note that in the discussion of their transformations under G the spin character is irrelevant. The transformation properties under G apply equally well to fermion fields or fields with spin 1. Let L_k denote the matrix representatives of the generators T_k in the space of the fields ϕ_n,

$$(L_k)_{ij} = U_{ij}(T_k). \tag{106}$$

The action of a local transformation $g(x)$ on the fields is then given by

$$\phi(x) \underset{g(x)}{\rightarrow} \phi'(x) = \exp\left\{\mathrm{i} \sum_{k=1}^{N} \Lambda_k(x) L_k\right\} \phi(x). \tag{107}$$

Of course, we can write this transformation in representation-free notation by defining the abstract group element $g(x)$ as an exponential series in terms of the generators T_k and the parameter functions $\Lambda_k(x)$,

$$g(x) = \exp\{i\Lambda(x)\} \quad \text{with } \Lambda(x) := \sum_{k=1}^{N} \Lambda_k(x) T_k. \tag{108}$$

In the space of the functions $\phi(x)$ the transformation (108) is represented by the matrix

$$\begin{aligned} U(\Lambda(x)) \equiv U(g(x)) &= \exp\left\{i\sum \Lambda_k(x) U(T_k)\right\} \\ &= \exp\left\{i\sum \Lambda_k(x) L_k\right\}. \end{aligned}$$

Suppose we wish to construct a field theory which describes the equations of motion of the fields $\phi(x)$ and which is invariant under the symmetry transformations of G. The group G defines an internal symmetry of this field theory and the indices n on the fields $\phi_n(x)$, with $n = 1, 2, \ldots, M$, refer to a "charge" space which is given by unitary representations of G. As an example consider the three pion fields $\phi_m(x)$, $m = +1, 0, -1$, which are distinguished by the projection I_3^s of the strong interaction isospin $I^s = 1$. In this example G is the SU(2) of isospin, the representation formed by the pion fields is threedimensional and irreducible. Under a transformation g of G which is specified by three parameters (Euler angles), the pion fields transform according to

$$\phi'_m(x) = \sum_{\mu=1}^{3} D^{(1)}_{m\mu}(\Lambda_1, \Lambda_2, \Lambda_3)\phi_\mu(x).$$

If G is to be a *global* symmetry then the Lagrangian of the pion fields $\phi_m(x)$ can depend only on products of the fields and of their derivatives which are scalars under G. The kinetic energy term, in particular, must have the Lorentz invariant and G-invariant form

$$(\partial_\alpha\phi, \partial^\alpha\phi), \tag{109}$$

where the parentheses (,) imply, symbolically, coupling of the bilinear $\phi_m\phi_{m'}$ to an invariant. [In the present example $(\partial_\alpha\phi, \partial^\alpha\phi) \equiv \sum_m (-)^{1-m} \partial_\alpha\phi_m \, \partial^\alpha\phi_{-m}$.]

If G is to be a *local* symmetry, i.e. if the parameters Λ_i depend on space and time, then a new aspect emerges: As we said before the prescription (104) defines an infinity of copies $G(x)$ of the original abstract group G, one for each point of space–time. Regarding the matter fields ϕ this prescription implies that each point x in Minkowski space is endowed with a local charge space ${}^xH$. As the derivative $\partial_\alpha\phi$ connects the fields in neighbouring points, x and $x + dx$, it relates at the same time the charge space ${}^xH$ to the charge space ${}^{x+dx}H$. A component $\phi_n(x)$ of $\phi(x)$ with respect to a given basis in the charge space ${}^xH$ in x is not simply the same component n with respect to ${}^{x+dx}H$. Therefore, the kinetic energy term (109) cannot be invariant under local gauge transformations and must be replaced by a more general, $G(x)$-invariant form.

This problem can be solved in two steps: First, one derives the transformation that carries $\phi_n(x)$ in the space ${}^xH$ into the same component ϕ_n in ${}^{x+\mathrm{d}x}H$ (parallel transport). Second, one constructs the covariant derivative D_α which replaces ∂_α in eq. (109) and which makes this form locally gauge invariant.

3.2. Vector potentials and their transformation properties

Let us consider a given component $\phi_n(x)$ of the representation formed by the fields $\phi(x)$. The field is taken at a fixed point x of space–time. The given index n, in fact, refers to a basis in the internal symmetry space ${}^xH$, in which case we write ${}^x\phi_n(x)$, or the analogous basis in the symmetry space ${}^{x+\mathrm{d}x}H$ attached to a neighbouring point $(x+\mathrm{d}x)$, in which case we write ${}^{x+\mathrm{d}x}\phi_n(x)$, for the sake of clarity. The two fields are not the same. However, it must be possible to relate them by an infinitesimal local gauge transformation, for which we make the following ansatz

$$ {}^{x+\mathrm{d}x}\phi_n(x) - {}^x\phi_n(x) = -\sum_m U_{nm}(A_\alpha)\cdot \mathrm{d}x^{\alpha}\,{}^x\phi_m(x), \tag{110} $$

with

$$ U_{nm}(A_\alpha) := \mathrm{i}e \sum_{k=1}^{N} A_\alpha^{(k)}(x)(L_k)_{nm}, \tag{111} $$

or, in abstract notation,

$$ A_\alpha(x) := \mathrm{i}e \sum_{k=1}^{N} A_\alpha^{(k)}(x) T_k. \tag{112} $$

Note that the fields $A_\alpha^{(k)}$ carry both an internal symmetry index k and a Lorentz vector index α. The Lorentz vector behaviour is needed in order to obtain a Lorentz invariant form $A_\alpha \mathrm{d}x^\alpha$ in eq. (110). Thus, A_α as defined by eq. (112) has a dual nature: on the one hand it transforms like an ordinary Lorentz vector field, on the other hand, through its dependence on the generators T_k, it is an operator in the internal symmetry space. Eq. (112) defines A_α in abstract form, whilst eq. (111) defines its matrix representation in the space of the matter fields ϕ_n.

Of course, we can apply an arbitrary local gauge transformation $g(x)$ and $g(x+\mathrm{d}x)$ to ${}^x\phi_n$ and ${}^{x+\mathrm{d}x}\phi_n$, respectively. Eq. (110) describes the parallel transport of the field ϕ_n provided it commutes with the local transformation g,

$$ U(g(x+\mathrm{d}x))(\mathbb{1} - U(A_\alpha(x))\mathrm{d}x^\alpha)\,{}^x\phi_n = \left(\mathbb{1} - U(A'_\alpha(x))\mathrm{d}x^\alpha\right)U(g(x))\,{}^x\phi_n. $$

We can derive the transformation behaviour of the vector fields $A_\alpha(x)$, eq. (111), under $g(x)$ from this requirement: As it must hold in all possible representations we can work it out in an abstract, i.e. representation-free form,

$$ g(x+\mathrm{d}x)(\mathbb{1} - A_\alpha(x)\,\mathrm{d}x^\alpha) \stackrel{!}{=} \left(\mathbb{1} - A'_\alpha(x)\,\mathrm{d}x^\alpha\right)g(x). $$

Writing $g(x+\mathrm{d}x) \simeq g(x) + \partial_\alpha g(x)\,\mathrm{d}x^\alpha$ and collecting all terms linear in $\mathrm{d}x^\alpha$, we have

$$\partial_\alpha g(x) - g(x)A_\alpha(x) = -A'_\alpha(x)g(x)$$

or

$$A'_\alpha(x) = g(x)A_\alpha(x)g^{-1}(x) - (\partial_\alpha g(x))g^{-1}(x).$$

Since $\partial_\alpha(g(x)g^{-1}(x)) = (\partial_\alpha g)g^{-1} + g(\partial_\alpha g^{-1}) = 0$, this can also be written as follows:

$$A'_\alpha(x) = g(x)A_\alpha(x)g^{-1}(x) + g(x)\partial_\alpha g^{-1}(x). \tag{113}$$

This fixes the transformation behaviour of the quantity $A_\alpha(x)$, eq. (111), under a local gauge transformation $g(x) \in G(x)$.

Let us analyze in more detail the meaning of eq. (113) and of the fields $A^{(k)}_\alpha(x)$. The transformation $A_\alpha \to A'_\alpha$, eq. (113), contains two elements: the first term is a conjugation, i.e. the familiar transformation behaviour of an operator with respect to global transformations g in G. Indeed, if g does not depend on x, the derivative $\partial_\alpha g^{-1}$ vanishes and $A'_\alpha = gA_\alpha g^{-1}$. The second term is a generalized gauge transformation, as may be seen by considering the example of G being a U(1) group. In this case

$$g(x) = e^{i\Lambda(x)}, \qquad g^{-1}(x) = e^{-i\Lambda(x)},$$

$$g(x)\partial_\alpha g^{-1}(x) = -i\partial_\alpha \Lambda(x),$$

so that eq. (113) reduces to

$$A'_\alpha(x) = A_\alpha(x) - i\partial_\alpha \Lambda(x).$$

Furthermore, the gauge group being U(1), the sum on the r.h.s. of eq. (111) contains only one term, $A_\alpha = ieA^{(1)}_\alpha \mathbb{1}$, so that

$$A^{(1)\prime}_\alpha(x) = A^{(1)}_\alpha(x) - \frac{1}{e}\partial_\alpha \Lambda(x). \tag{114}$$

This is precisely the expression for a gauge transformation in electrodynamics, cf. eq. (III.183). Thus, eq. (113) provides the generalization of the familiar gauge transformations of electrodynamics to the case of non-Abelian local gauge groups.

At the same time this comparison suggests that the vector fields $A^{(k)}_\alpha(x)$, of which there are $N = \dim G$ types, are generalizations of the vector potential of electrodynamics. In this context it is instructive to work out the transformation behaviour of these fields under infinitesimal transformations $g(x)$,

$$g(x) \simeq \mathbb{1} + i\sum \Lambda_k(x)T_k.$$

This gives

$$\begin{aligned} gA_\alpha g^{-1} &\simeq \Big(\mathbb{1} + i\sum_k \Lambda_k T_k\Big) ie\sum_i A^{(i)}_\alpha T_i \Big(\mathbb{1} - i\sum_j \Lambda_j T_j\Big) \\ &\simeq ie\sum_i \Big\{T_i + i\sum_k \Lambda_k [T_k, T_i]\Big\} A^{(i)}_\alpha(x) \\ &= ie\sum_i \Big\{T_i - \sum_{k,l} \Lambda_k C_{kil} T_l\Big\} A^{(i)}_\alpha(x) \end{aligned}$$

and

$$g(x)\partial_\alpha g^{-1}(x) \simeq -\mathrm{i}\sum_j T_j \partial_\alpha \Lambda_j(x).$$

Inserting these formulae into eq. (113) and comparing the coefficients of T_i on either side, one obtains, with $|\Lambda_k| \ll 1$,

$$A_\alpha^{(i)\prime}(x) \simeq A_\alpha^{(i)}(x) - \sum_{kl} C_{ikl}\Lambda_k(x)A_\alpha^{(l)}(x) - \frac{1}{e}\partial_\alpha \Lambda_i(x). \tag{115}$$

For $G=\mathrm{U}(1)$ this reduces to the result (114) above. For $G=\mathrm{SU}(2)$ one has, in an obvious vector notation,

$$\boldsymbol{A}_\alpha'(x) \simeq \boldsymbol{A}_\alpha(x) - \boldsymbol{\Lambda}(x) \wedge \boldsymbol{A}_\alpha(x) - \frac{1}{e}\partial_\alpha \boldsymbol{\Lambda}(x).$$

The constant e which appears in the definition (112), in principle, is arbitrary. As in electrodynamics it plays the role of a coupling constant of the matter fields ϕ to the vector bosons represented by the gauge fields $A_\alpha^{(k)}$. For each irreducible component G_i of $G = G_1 \times G_2 \times \cdots$ there is one such constant e_i which can be chosen arbitrarily (cf. exercise 4). We return to this arbitrariness in more detail below.

We summarize the results of this section: The very definition of parallel transport requires the introduction of a set of vector gauge fields $\{A^{(k)}(x);\ k=1,\dots,N\}$ which form the adjoint representation of the group G. Therefore, once G is given, say

$$G = G_1 \times G_2, \tag{116a}$$

where G_i are simple, then

$$N = N_1 + N_2 \quad \text{with } N_i = \dim(G_i). \tag{116b}$$

With respect to infinitesimal local gauge transformations the fields $A_\alpha^{(k)}$ transform according to eq. (115) which behaviour is the generalization of the familiar gauge transformations of electrodynamics to the non-Abelian case. In the expression for the parallel transport, the gauge fields appear in the form of the operator A_α, eq. (112) whose transformation behaviour under finite $g(x)$ is defined by eq. (113) (equivalent to eq. (115)). In those gauges which preserve the manifest covariance of the theory, A_α is a Lorentz vector. At the same time it is an operator with respect to the group G. Eq. (111) gives its matrix representation in the space of a set of matter fields which form a unitary representation of G.

3.3. *Covariant derivatives*

Our aim is to construct a generalized derivative of the matter fields $D_\alpha\phi_n(x)$ such that bilinears of two such forms can be coupled to an invariant, generalized, kinetic energy $(D_\alpha\phi, D^\alpha\phi)$. For this purpose let us consider the difference ${}^A\phi_n(x+\mathrm{d}x) - {}^A\phi_n(x)$ of the field ϕ_n for two infinitesimally separated arguments but expressed in the *same* symmetry space ${}^AH$, attached to the point $A=x$ in space–time. This infinitesimal difference can be expressed as the sum of a term containing the ordinary derivative and a parallel transport. Let

$$A \equiv x \quad \text{and} \quad B \equiv x + \mathrm{d}x.$$

Then

$$^A\phi_n(B) - {}^A\phi_n(A) = \left[{}^B\phi_n(B) - {}^A\phi_n(A)\right] + \left[{}^A\phi_n(B) - {}^B\phi_n(B)\right],$$

where the first term is given by*)

$$^B\phi_n(B) - {}^A\phi_n(A) \simeq \partial_\alpha {}^A\phi_n(A)\,\mathrm{d}x^\alpha,$$

while the second term is a parallel transport of $^B\phi_n(B)$ to the point $A = B - \mathrm{d}x$. This is obtained from eq. (110) with $\mathrm{d}x$ replaced by $-\mathrm{d}x$, viz.

$$^A\phi_n(B) - {}^B\phi_n(B) = \sum_m U_{nm}(A_\alpha)\,\mathrm{d}x^{\alpha\,B}\phi_m(B).$$

Thus, to first order in $\mathrm{d}x$,

$$\begin{aligned} ^A\phi_n(B) - {}^A\phi_n(A) &\simeq \sum_m \left\{\delta_{nm}\partial_\alpha + U_{nm}(A_\alpha)\right\}\mathrm{d}x^{\alpha\,A}\phi_m(A) \\ &\equiv D_\alpha(A)\,{}^A\phi(A)\,\mathrm{d}x^\alpha, \end{aligned}$$

which defines the covariant derivative $D_\alpha(A)\phi$ with

$$D_\alpha(A) := \mathbb{1}\partial_\alpha + U(A_\alpha). \tag{117}$$

Clearly, these considerations do not depend on the nature of the fields ϕ. In particular, they hold equally well for fermion fields Ψ. For example, specializing again to electrodynamics by taking $G = \mathrm{U}(1)$, we see that the definition (117) reduces to the covariant derivative $\vec{D}_\alpha$ of eq. (III.184).

The generalized derivative $D_\alpha(A)\phi$ is called covariant because under a local gauge transformation $g(x) \in G$ it transforms according to the law

$$D_\alpha(A'(x)) = U(g(x))\,D_\alpha(A(x))\,U^{-1}(g(x)), \tag{118}$$

where $U(g)$ is the matrix representation of g in the space of the fields ϕ. Eq. (118) says that $D_\alpha(A)$ transforms like a tensor operator. Therefore, on the basis of these tensors, it will be easy to form invariants under $G(x)$.

The transformation behaviour (118) follows from the construction given above. It may also be verified by explicit calculation as follows:

$$\begin{aligned} D_\alpha(A')\phi'(x) &= \left(\partial_\alpha + U(A'_\alpha)\right)U(g(x))\phi \\ &= \left(\partial_\alpha U(g)\right)\phi + U(g)\,\partial_\alpha\phi \\ &\quad + U(g)\left[U(A_\alpha)U^{-1}(g) + \left(\partial_\alpha U^{-1}(g)\right)\right]U(g)\phi \\ &= U(g)\left(\partial_\alpha\phi + U(A_\alpha)\right)\phi \\ &\quad + \left\{\left(\partial_\alpha U(g)\right)\phi + U(g)\left[\partial_\alpha\left(U^{-1}(g)U(g)\right)\right.\right. \\ &\qquad\qquad \left.\left. - U^{-1}(g)\left(\partial_\alpha U(g)\right)\right]\phi\right\}. \end{aligned}$$

*) Note that the x-dependence appears in the argument *and* in the basis.

As $U^{-1}U = \mathbb{1}$ is independent of x, the term in curly brackets is zero, while the first is precisely $U(g)D_\alpha(A)\phi$. Thus we obtain

$$D_\alpha(A')\phi'(x) = U(g)D_\alpha(A)\phi,$$

from which eq. (118) follows immediately.

3.4. *Field tensor for vector potentials*

The field tensor $F^{\mu\nu}$ in non-Abelian gauge theories which generalizes the field strength tensor (III.169) of electrodynamics, is obtained, for instance, by studying two successive, infinitesimal, parallel transports (110) from a point x to a point $z = x + \mathrm{d}x + \mathrm{d}y$. This transformation can be effected in either of the two following ways which are not equivalent:

(a) $x \to x + \mathrm{d}x = y \to z = y + \mathrm{d}y = x + \mathrm{d}x + \mathrm{d}y$,

(b) $x \to x + \mathrm{d}y = y' \to z = y' + \mathrm{d}x = x + \mathrm{d}x + \mathrm{d}y$.

Expanding $A(x + \mathrm{d}x)$, $A(x + \mathrm{d}y)$ around the point x, their difference to second order in $\mathrm{d}x\,\mathrm{d}y$ is found to be

$$\begin{aligned}&\big(\mathbb{1} - A_\beta(x+\mathrm{d}x)\,\mathrm{d}y^\beta\big)\big(\mathbb{1} - A_\alpha(x)\,\mathrm{d}x^\alpha\big)\\ &\quad - \big(\mathbb{1} - A_\alpha(x+\mathrm{d}y)\,\mathrm{d}x^\alpha\big)\big(\mathbb{1} - A_\beta(x)\,\mathrm{d}y^\beta\big)\\ &= -\big\{\partial_\alpha A_\beta(x) - \partial_\beta A_\alpha(x) + A_\alpha(x)A_\beta(x) - A_\beta(x)A_\alpha(x)\big\}\,\mathrm{d}x^\alpha\,\mathrm{d}y^\beta\\ &=: -F_{\alpha\beta}(x)\,\mathrm{d}x^\alpha\,\mathrm{d}y^\beta,\end{aligned}$$

where the tensor $F_{\alpha\beta}(x)$ is defined by

$$F_{\alpha\beta}(x) := \partial_\alpha A_\beta(x) - \partial_\beta A_\alpha(x) + \big[A_\alpha(x), A_\beta(x)\big]. \tag{119}$$

$F_{\alpha\beta}(x)$ is a tensor with respect to Lorentz transformations. At the same time it has the properties of a tensor operator with respect to the symmetry transformations g. This is easy to see if we notice that $F_{\alpha\beta}$ can be related to the commutator of the covariant derivatives D_α and D_β. In the space of the matter fields

$$\begin{aligned}\big[D_\alpha(A), D_\beta(A)\big] &= \partial_\alpha U(A_\beta) - \partial_\beta U(A_\alpha) + \big[U(A_\alpha), U(A_\beta)\big]\\ &= U\big(\partial_\alpha A_\beta - \partial_\beta A_\alpha + [A_\alpha, A_\beta]\big) = U\big(F_{\alpha\beta}(x)\big).\end{aligned} \tag{120}$$

Thus the matrix representative of $F_{\alpha\beta}$, in the space of the fields ϕ, transforms like the commutator of D_α and D_β whose transformation behaviour, in turn, is given by eq. (118). This proves the tensor character of $F_{\alpha\beta}$ with respect to the gauge transformations of $G(x)$.

Finally, in analogy to the decomposition (112) of A_α, we can write $F_{\alpha\beta}$ as a linear combination of the generators T_k of G, viz.

$$F_{\alpha\beta}(x) = \mathrm{i}e \sum_{k=1}^{N} T_k F_{\alpha\beta}^{(k)}(x), \tag{121}$$

where $F_{\alpha\beta}^{(k)}(x)$ are ordinary Lorentz tensor fields. Their explicit form is obtained from the definition (119) by inserting the decomposition (112):

$$\mathrm{i}e \sum_{k=1} T_k F_{\alpha\beta}^{(k)}(x) = \mathrm{i}e \sum T_k \left(\partial_\alpha A_\beta^{(k)}(x) - \partial_\beta A_\alpha^{(k)}(x) \right) + (\mathrm{i}e)^2 \sum_{i,j} \left[T_i, T_j \right] A_\alpha^{(i)}(x) A_\beta^{(j)}(x).$$

The commutator in this expression is given by eq. (96). As the generators T_k are linearly independent, we can compare the coefficients of T_k to find

$$F_{\alpha\beta}^{(k)}(x) = f_{\alpha\beta}^{(k)}(x) - e \sum_{i,j} C_{kij} A_\alpha^{(i)}(x) A_\beta^{(j)}(x), \tag{122}$$

with

$$f_{\alpha\beta}^{(k)}(x) := \partial_\alpha A_\beta^{(k)}(x) - \partial_\beta A_\alpha^{(k)}(x). \tag{123}$$

Eq. (122) is the direct generalization of the electromagnetic field-strength tensor (III.169) to the case of non-Abelian gauge theories.

3.5. How to construct locally gauge invariant theories

In the previous sections we have established three types of operators which have a simple transformation behaviour under local gauge transformations: the generalized vector potentials $A_\alpha(x)$, the corresponding field tensors $F_{\alpha\beta}(x)$, and the covariant derivative $D_\alpha(A)$ of matter fields ϕ. As to the latter, we noted previously that the spin content of the field ϕ is irrelevant. Therefore, the covariant derivative of *spinor* fields (or any other field) is given by exactly the same definition (117), where $U(A_\alpha)$ is now the matrix representation of the operator A_α in the space of the spinor fields. These operators are the tools which we need to construct Lagrangians which are invariant under local gauge transformations.

Suppose the theory is to contain a set of boson fields

$$\phi(x) = \{ \phi_n(x); n = 1, \ldots, M \}, \tag{124a}$$

as well as a set of spinor fields

$$\Psi(x) = \{ \Psi_p(x); p = 1, \ldots, P \}. \tag{124b}$$

If we require the Lagrangian to be invariant under *global* transformations $g \in G$ then it must have the form

$$\mathscr{L}_0 = \tfrac{1}{2}(\partial_\alpha \phi, \partial^\alpha \phi) + \tfrac{1}{2}\mathrm{i}(\bar{\Psi}, \gamma^\alpha \overleftrightarrow{\partial}_\alpha \Psi) - (\bar{\Psi}, m\Psi) - (\bar{\Psi}, g\phi\Psi) - V(\phi). \tag{125}$$

The parentheses (X, Y) are meant to indicate that X and Y are coupled to a scalar with respect to G. In the third term of eq. (125), in particular, m is a matrix in the space of the Ψ_p (mass matrix). Likewise, the fourth term represents a G-invariant Yukawa coupling of the fields Ψ and ϕ, while the last term ("potential" term) denotes a group-invariant self-interaction of the fields ϕ as well as possible mass terms for that field. The explicit form of these invariants depends on the nature of the Lie algebra (96) and of the multiplets $\phi(x)$ and $\Psi(x)$. In particular, it may happen that it is not possible to construct a group invariant mass term for the fermion fields. [This case will be encountered in the GSW model.]

On the basis of the results obtained in secs. 3.2 to 3.4 it is easy to construct a new version of the theory (125) which is invariant under *local* gauge transformations $g(x) \in G(x)$ as well: Let $U(A_\alpha)$ and $V(A_\alpha)$ be the matrix representatives of the operator A_α, eq. (112), in the space of the boson fields (124a) and in the space of the fermion fields (124b), respectively. In order to obtain gauge invariant kinetic terms of these matter fields, the ordinary derivatives $\partial_\alpha \phi$, $\partial_\alpha \Psi$ must be replaced by the covariant derivatives

$$D_\alpha \phi = (\mathbb{1}\partial_\alpha + U(A_\alpha))\phi, \tag{126a}$$

$$\vec{D}_\alpha \Psi = (\mathbb{1}\partial_\alpha + V(A_\alpha))\Psi, \tag{126b}$$

$$\bar{\Psi}\overleftarrow{D}_\alpha = \partial_\alpha \bar{\Psi}\mathbb{1} + \bar{\Psi}V^\dagger(A_\alpha) = \partial_\alpha \bar{\Psi}\mathbb{1} - \bar{\Psi}V(A_\alpha). \tag{126c}$$

In addition, a term of the form $(F_{\alpha\beta}, F^{\alpha\beta})$ must be added to the Lagrangian which generalizes the well-known kinetic energy term (III.168) of the Maxwell fields, viz.

$$-\mathscr{L}_A = \frac{c}{4}(F_{\alpha\beta}, F^{\alpha\beta})$$

$$= e^2 \frac{c}{4} \sum_{i,k} \mathrm{Sp}(T_i T_k) F^{(i)}_{\alpha\beta} F^{(k)\alpha\beta}.$$

Now, with

$$\mathrm{Sp}(T_i T_k) = \kappa \delta_{ik} \tag{127}$$

c must be chosen to be

$$c = 1/\kappa e^2$$

for the derivative terms $f^{(i)}_{\alpha\beta} f^{(i)\alpha\beta}$ to obtain the same factor $-\frac{1}{4}$ as in the case of the Maxwell field, eq. (III.168). The mass terms, coupling terms and generalized potentials which do not contain derivatives, remain the same as in the globally invariant version of the theory.

The full Lagrangian describing the interacting matter fields ϕ and Ψ, in interaction with the gauge fields $A^{(i)}_\alpha$, is then given by

$$\mathscr{L} = -\frac{1}{4\kappa e^2}(F_{\alpha\beta}, F^{\alpha\beta}) + \frac{1}{2}(D_\alpha\phi, D^\alpha\phi) + \frac{i}{2}(\bar{\Psi}, \gamma^\alpha \overleftrightarrow{D}_\alpha \psi)$$

$$-(\bar{\Psi}, (m + g\phi)\Psi) - V(\phi). \tag{128}$$

Theories of this kind which are invariant under a non-Abelian group of local gauge transformations have a number of remarkable general properties.

(i) As in the Abelian case of electrodynamics, the Lagrangian cannot contain a mass term $m^2 A^{(i)}_\alpha A^{(i)\alpha}$ for the gauge fields because such a term is not invariant under gauge transformations. Therefore, if the internal symmetry remains unbroken, the gauge fields $A^{(i)}_\alpha$ describe *massless* vector bosons. In turn, if we wish to give some of these bosons finite masses, the symmetry must be broken to some extent.

(ii) In the non-Abelian case the first term in $\mathscr{L}$, eq. (128), contains not only the kinetic energy

$$-\frac{1}{4}\sum_i f_{\alpha\beta}^{(i)} f^{(i)\alpha\beta}$$

of the gauge fields but also coupling terms of the kind

$$\sum_{ijk} C_{ijk} f_{\alpha\beta}^{(i)}(x) A^{(j)\alpha}(x) A^{(k)\beta}(x)$$

and

$$\sum_{i\ldots q} C_{ijk} C_{ipq} A_\alpha^{(j)}(x) A_\beta^{(k)}(x) A^{(p)\alpha}(x) A^{(q)\beta}(x),$$

i.e. cubic and quartic interactions of the gauge fields among themselves.

(iii) Owing to the dependence of $D_\alpha(A)$ on the fields $A_\alpha^{(i)}$ the generalized "kinetic" terms of the boson and fermion fields also yield the couplings of the matter fields to the gauge vector bosons. The gauge bosons are seen to couple to currents of the type $\phi^\dagger U(T_i) i\overleftrightarrow{\partial}_\alpha \phi$ and $\overline{\Psi} V(T_i)\gamma^\alpha \Psi$, respectively. In particular, if T_i is diagonal in the boson multiplet or fermion multiplet, this implies that the physical coupling constants of individual members of this multiplet, i.e. their "charges", are proportional to each other. They are given by the constant e multiplied by the diagonal matrix elements of $U(T_i)$ or $V(T_i)$, respectively. This is a new element of universality which does not occur in Abelian theories.

4. Glashow–Salam–Weinberg model for leptons and quarks

The GSW model whose phenomenology was summarized in sec. 2.3 above, is based on the gauge group

$$G = \mathrm{U}(2), \tag{129}$$

whose algebra reduces to the algebras of SU(2) and U(1). Let T_0 denote the generator of the Abelian factor U(1), $\{T_1, T_2, T_3\}$ the generators of the SU(2) factor. The Lie algebra of these operators is given by eqs. (101). As g has four generators, the theory contains four gauge fields $A_\alpha^{(\mu)}(x)$,

$$A_\alpha(x) = \mathrm{i}e \sum_{\mu=0}^{3} T_\mu A_\alpha^{(\mu)}(x), \tag{130}$$

where $A_\alpha^{(0)}$ is a singlet, whilst the fields $\{A_\alpha^{(i)}, i = 1,2,3\}$ form a triplet with respect to SU(2), i.e. the adjoint representation. The two W-bosons which are charged and which are conjugates of each other, must be linear combinations of two of the triplet fields, cf. eq. (72),

$$W_\alpha^{\pm}(x) \coloneqq \mp\frac{1}{\sqrt{2}}\left(A_\alpha^{(1)}(x) \pm \mathrm{i} A_\alpha^{(2)}(x)\right). \tag{131}$$

The photon field A_α and the Z^0 boson field Z_α must be linear combinations of the third triplet field $A_\alpha^{(3)}$ and the singlet field $A_\alpha^{(0)}$, as indicated in eqs. (77) and (78).

The GSW Lagrangian is constructed on the basis of the following assumptions:

(I) There is no direct coupling between different lepton families, or between quark and lepton families.

(IIa) The photon couplings must conserve parity and must have the form $Q_f\overline{\Psi^{(f)}}\gamma^\alpha\Psi^{(f)}A_\alpha$ with Q_f the electric charge of fermion f.

(IIb) In particular, neutrinos must not couple to the photon.

(III) The neutrinos that couple to weak interaction vertices are fully left-handed. In particular, all CC interaction vertices (f, ν_f, W) are of the form V − A. In other terms, only the left-handed part of the massive fermion field $\Psi^{(f)}$ couples to the W-bosons.

(IV) The theory shall exhibit lepton universality in the sense that the Lagrangians describing the interaction of (μ, ν_μ) and (τ, ν_τ) with the gauge bosons $W^\pm$, Z^0 and γ simply are copies of the interaction Lagrangian for the (e, ν_e) family.

It is customary, for the sake of simplifying the notation to write the particle symbol instead of the field operator, i.e.

$$f(x) \equiv \Psi^{(f)}(x), \qquad q(x) \equiv \Psi^{(q)}(x)$$

for leptons f (f = e, μ, τ) or quarks q. Furthermore, it is useful to define the left-handed and right-handed parts of massive fields, viz.

$$f_L(x) := \tfrac{1}{2}(\mathbb{1} - \gamma_5)f(x) \equiv P_- f(x), \tag{132a}$$

$$f_R(x) := \tfrac{1}{2}(\mathbb{1} + \gamma_5)f(x) \equiv P_+ f(x), \tag{132b}$$

and analogously for the quark fields $q_L(x)$ and $q_R(x)$. As we saw in Chap. III, sec. 3, P_+ projects onto spinors of the first kind, P_- onto spinors of the second kind, cf. eq. (III.54).

Assumption (I) implies that the electroweak interaction Lagrangian can be constructed for each of the lepton families (7) and for each of the quark generations separately. Since the case of the quark doublets is complicated by the mixing of d, s and b states, we start with the simpler case of one lepton family (f, ν_f).

4.1. *GSW Lagrangian for one lepton family*

The matter fields $f(x) \equiv \Psi^{(f)}(x)$ and $\nu(x) \equiv \Psi^{(\nu)}(x)$ [we suppress the index f on ν_f, for simplicity], appear in three forms: $f_L(x)$, $f_R(x)$, and $\nu_L(x)$. This means that the massive, charged lepton f is described by a spinor of the first kind and a spinor of the second kind. The neutrino, by virtue of the dynamic properties of the theory, appears only in the form of a spinor of the second kind.

The simplest possibility of classifying these fields with respect to G is to group them in a triplet

$$\Psi(x) := \begin{pmatrix} \nu_L(x) \\ f_L(x) \\ f_R(x) \end{pmatrix}. \tag{133}$$

As f_R does not couple to $W^{\pm}$, it must be a singlet with respect to the SU(2) factor of G. The pair

$$L(x) := \begin{pmatrix} \nu_L(x) \\ f_L(x) \end{pmatrix} \tag{134}$$

on the other hand, forms a doublet of SU(2). Thus, the triplet (133) is a reducible multiplet of SU(2). In the space of the triplet (133) the generators T_μ are represented by the 3×3 matrices [the factor $\frac{1}{2}$ in $V(T_0)$ is introduced for convenience],

$$V(T_0) = \frac{1}{2}\begin{pmatrix} \lambda_d & 0 & 0 \\ 0 & \lambda_d & 0 \\ 0 & 0 & \lambda_s \end{pmatrix}, \qquad V(T_i) = \frac{1}{2}\begin{pmatrix} \sigma^{(i)} & 0 \\ 0 & 0 \end{pmatrix}. \tag{135}$$

Two of these are diagonal, $V(T_0)$ and $V(T_3)$, and have the eigenvalues

$$\begin{Bmatrix} \frac{1}{2}\lambda_d \\ \frac{1}{2}\lambda_d \\ \frac{1}{2}\lambda_s \end{Bmatrix} \quad \text{and} \quad \begin{Bmatrix} \frac{1}{2} \\ -\frac{1}{2} \\ 0 \end{Bmatrix},$$

respectively. The eigenvalues of $V(T_0)$ which pertain to the doublet partners ν_L and f_L must be the same, while the eigenvalue λ_s for f_R can be different from λ_d. Note that the trace (127) is $\frac{1}{2}$ for the generators T_1, T_2, T_3, and is $\frac{1}{4}(2\lambda_d^2 + \lambda_s^2)$ for T_0.

We analyze first the interaction terms which follow from the term

$$\frac{i}{2}\left(\overline{\Psi(x)}\,, \gamma^\alpha \overleftrightarrow{D}_\alpha \Psi(x)\right)$$

in the Lagrangian (128). With the definitions (117) and (112) these are

$$\begin{aligned} \mathscr{L}_I^{(f)} &= -e\left(\overline{\Psi(x)} \sum_{\mu=0}^{3} V(T_\mu)\gamma^\alpha \Psi(x)\right) A_\alpha^{(\mu)}(x) \\ &= -e\sum_{i=1}^{3}\left(\overline{L(x)}\,\tfrac{1}{2}\sigma^{(i)}\gamma^\alpha L(x)\right)A_\alpha^{(i)}(x) - e\tfrac{1}{2}\lambda_d\left(\overline{L(x)}\,\mathbb{1}\gamma^\alpha L(x)\right)A_\alpha^{(0)}(x) \\ &\quad - e\tfrac{1}{2}\lambda_s\left(\overline{f_R(x)}\,\gamma^\alpha f_R(x)\right)A_\alpha^{(0)}(x). \end{aligned} \tag{136}$$

As it stands, this interaction contains three free parameters: e, λ_d, and λ_s. Our aim is now to work out the restrictions on these parameters that follow from the conditions (IIa) and (IIb), and to identify them with the phenomenological coupling constants α and G, eq. (75).*)

*) This discussion follows closely the analysis of O'Raifeartaigh (1979).

4.1.1. CC weak interactions

Rewriting the couplings to the gauge fields $A_\alpha^{(1)}$ and $A_\alpha^{(2)}$ in terms of the W-fields (131) we have

$$\bar{L}\tfrac{1}{2}\left(\sigma^{(1)}A_\alpha^{(1)}+\sigma^{(2)}A_\alpha^{(2)}\right)\gamma^\alpha L=\frac{1}{\sqrt{2}}\left(\bar{L}s_+\gamma^\alpha L\right)W_\alpha^- - \frac{1}{\sqrt{2}}\left(\bar{L}s_-\gamma^\alpha L\right)W_\alpha^+ ,$$

where $s_\pm=\frac{1}{2}(\sigma^{(1)}\pm\mathrm{i}\sigma^{(2)})$ are the usual step operators in the space of the doublet fields L. Inserting the definitions (134) and (132a) the CC interaction in $\mathscr{L}_\mathrm{I}$ is then found to be

$$\mathscr{L}_{\mathrm{CC}}^{(\mathrm{f})}=-\frac{e}{\sqrt{2}}\overline{\nu_\mathrm{f}(x)}\,\tfrac{1}{2}\gamma^\alpha(\mathbb{1}-\gamma_5)f(x)W_\alpha^{(-)}(x)+\mathrm{h.c.} \tag{137}$$

Comparing this to formulae (73) and (74) of our phenomenological discussion in sec. 2.3 we can identify $-e$ with the constant g, and relate its square to the Fermi constant G, viz.

$$-e\equiv g,\qquad g^2/8m_\mathrm{W}^2=G/\sqrt{2}\,. \tag{138}$$

4.1.2. Neutral couplings

From eq. (136) we see that the couplings of the individual lepton fields to $A_\alpha^{(0)}$ and $A_\alpha^{(3)}$ are as follows:

$$\nu_\mathrm{L}(x):\quad \tfrac{1}{2}g\{A_\alpha^{(3)}+\lambda_\mathrm{d}A_\alpha^{(0)}\}\overline{\nu_\mathrm{L}(x)}\,\gamma^\alpha\nu_\mathrm{L}(x), \tag{139a}$$

$$f_\mathrm{L}(x):\quad \tfrac{1}{2}g\{-A_\alpha^{(3)}+\lambda_\mathrm{d}A_\alpha^{(0)}\}\overline{f_\mathrm{L}(x)}\,\gamma^\alpha f_\mathrm{L}(x), \tag{139b}$$

$$f_\mathrm{R}(x):\quad \tfrac{1}{2}g\lambda_\mathrm{s}A_\alpha^{(0)}(x)\overline{f_\mathrm{R}(x)}\,\gamma^\alpha f_\mathrm{R}(x). \tag{139c}$$

Let us work out the consequences of the assumptions (IIa) and (IIb) on these couplings. It follows from the condition (IIb) that the linear combination of $A_\alpha^{(3)}$ and $A_\alpha^{(0)}$ which appears in the curly brackets of eq. (139a) must be proportional to the Z^0-field. Comparing this to the ansatz (77b) we find that g' of eqs. (77) and λ_d must be related by

$$\lambda_\mathrm{d}=-g'/g, \tag{140}$$

and that the NC coupling of the neutrino, eq. (139a), is indeed given by eq. (79′).

Condition (IIa), in turn, implies that f_L and f_R couple to the photon field, eq. (77a), with the same strength. This means that the photon components of eqs. (139b) and (139c) must be the same. With the transformation (78) we obtain the condition

$$-g'+g\lambda_\mathrm{d}=g\lambda_\mathrm{s},$$

hence

$$\lambda_\mathrm{s}=2\lambda_\mathrm{d}=-2g'/g. \tag{141}$$

Inserting the results (140) and (141) and eqs. (139b) into $\mathscr{L}_\mathrm{I}$, eq. (136), the coupling of f to the photon field is

$$\mathcal{L}_{\gamma}^{(\mathrm{f})} = -\frac{gg'}{\sqrt{g^2+g'^2}}\overline{f(x)}\,\gamma^{\alpha}f(x)A_{\alpha}(x),$$

so that the electric charge of the lepton f is to be identified as indicated in eq. (81). Similarly, one verifies easily that the neutral couplings of the Z^0 to the fields f and ν are indeed the ones of eqs. (80′) and (79′), respectively, or, after introducing the parametrization (82) in terms of the Weinberg angle, by eq. (84).

4.1.3. *Weak isospin, weak hypercharge and the electric charge*

Let us return to the general form (136) of the interaction and let us extract from it the couplings to the photon field by means of eqs. (78)

$$\mathcal{L}_{\gamma}^{(\mathrm{f})} = \frac{gg'}{\sqrt{g^2+g'^2}}\overline{\Psi(x)}\left\{V(T_3)+\frac{g}{g'}V(T_0)\right\}\gamma^{\alpha}\Psi(x)A_{\alpha}(x). \tag{142}$$

From eq. (III.171) we know that the interaction of a charged lepton with the photon field has the general form $-Q\bar{\Psi}\gamma^{\alpha}\Psi A_{\alpha}$. The factor in front of eq. (142) is $-e$. Therefore, the eigenvalues of the diagonal matrix $V(T_3)+V((g/g')T_0)$ are the electric charges of the members of the multiplet (133), in units of the elementary charge e.

Because of the close analogy to the Gell-Mann–Nishijima formula relating the electric charge of a hadron to its strong isospin and hypercharge, the SU(2) factor of G is called the *weak isospin* group, and the operator

$$Y := 2\frac{g}{g'}T_0 \tag{143}$$

is called the *weak hypercharge*. In the space of the triplet we have

$$V(T_3) = \begin{pmatrix} \frac{1}{2} & 0 & 0 \\ 0 & -\frac{1}{2} & 0 \\ 0 & 0 & 0 \end{pmatrix}, \qquad V(Y) = \begin{pmatrix} -1 & 0 & 0 \\ 0 & -1 & 0 \\ 0 & 0 & -2 \end{pmatrix}. \tag{144}$$

Denoting the eigenvalues by t_3 and y, respectively, the electric charge of the member m of the triplet (133) is

$$Q(m)/e = t_3(m) + \tfrac{1}{2}y(m). \tag{145}$$

This gives indeed 0 for the neutrino and -1 for the two parts of the charged lepton field.

4.1.4. *Some remarks and open problems*

We do not write down the kinetic energy and interaction Lagrangian of the vector boson fields $W_{\alpha}^{\pm}$ and Z_{α} because we do not need them for our discussion. It should be clear, however, how to construct these terms from our general discussion in secs. 3.2 to 3.5. We note, in particular, that the electromagnetic properties and interactions of $W^{\pm}$-bosons are completely fixed, including their anomalous magnetic moment [this used to be a problem in the older theories with W-bosons].

The extension to all lepton families is simply effected by taking the sum of $\mathscr{L}_{\mathrm{I}}^{(\mathrm{f})}$, eq. (136), over all leptons,

$$\sum_{\mathrm{f}=\mathrm{e},\,\mu\tau} \mathscr{L}_{\mathrm{I}}^{(\mathrm{f})} = \mathscr{L}_{\mathrm{I}}(\text{leptons}). \tag{146}$$

The generalization to the quark families is slightly more complicated and will be dealt with in the next section.

The unified theory that we developed thus far, is invariant under the entire group (129), $G = U(2)$. As it stands, it is still far from a realistic theory for the electromagnetic and the weak interactions because the weak bosons $W^{\pm}$ and Z^0 remain massless, like the photon. Furthermore, the charged leptons f also remain massless because it is not possible to construct an invariant mass term on the basis of the fields $f_{\mathrm{L}}(x)$ and $f_{\mathrm{R}}(x)$ as they appear in the triplet (133).

Indeed, we know from our discussion in sec. (III.8.4) that a particle which carries a conserved charge, can only have a Dirac mass term. Such a mass term is of the form $\{\overline{f_{\mathrm{L}}(x)}\, f_{\mathrm{R}}(x) + \overline{f_{\mathrm{R}}(x)}\, f_{\mathrm{L}}(x)\}$. However, as f_{L} belongs to a doublet of SU(2) whilst f_{R} belongs to a singlet, this term cannot be invariant.

As a consequence of leptons being massless, all particle currents of this theory are conserved ones. While this conservation law is welcome for the diagonal vector currents, it cannot hold for the axial currents and for the nondiagonal (CC) vector currents. This discussion shows that the symmetry group (129) must be broken very strongly and following a specific pattern: One of the gauge bosons, the photon, must remain massless to all orders, while $W^{\pm}$ and Z^0 must become massive and, in fact, very heavy. Thus G must be broken down to the residual U(1) symmetry of electrodynamics. A way to do this is provided by the mechanism of spontaneous symmetry breaking, at the price of introducing further degrees of freedom into the theory. At the same time this extension allows giving the charged fermions finite masses, so that the model becomes realistic and can be compared to experiment. Before we turn to a discussion of symmetry breaking (sec. 5) we conclude this section by extending our results to the quark families.

4.2. *GSW Lagrangian for the quark families*

For the sake of convenience let us introduce the following notation for the quarks with charge $Q = \frac{2}{3}$ (in units of e),

$$\{\mathrm{u}_f;\, f = 1,2,3\} \quad \text{for u, c, and t} \tag{147a}$$

and similarly for the quarks with charge $Q = -\frac{1}{3}$,

$$\{\mathrm{b}_f;\, f = 1,2,3\} \quad \text{for d, s, and b.} \tag{147b}$$

These are the quark states with the quantum numbers (relevant to strong interactions) that we discussed in sec. 1.1.2. They are the constituents of the physical meson and baryon states.

The weak interactions conserve neither the strong isospin nor the additive quantum numbers introduced in sec. 1.1.2. In fact, they couple to new states d_f, with electric charge $-\frac{1}{3}$, which are related to the states (147b) by a unitary transformation, viz.

$$d_f = \sum_{f'} U_{ff'} b_{f'} \quad \text{with } UU^\dagger = \mathbb{1}. \tag{148}$$

These new states are referred to as the *weak interaction eigenstates*. A way to visualize this relation is by assuming that the weak gauge bosons $W^\pm$ and Z^0 couple to quark currents which contain the fields $u_f(x)$ and $d_f(x)$, but that the strong interaction Lagrangian contains quark mass terms which are not diagonal in the basis of the states d_f, i.e.

$$-\mathscr{L}_{\text{mass}} = \sum_f m(f)\overline{u_f(x)}\, u_f(x) + \sum_{f,f'} M_{ff'}\overline{d_f(x)}\, d_{f'}(x). \tag{149}$$

The mass matrix $M_{ff'}$ can be diagonalized by means of the unitary transformation U,

$$\sum_{a,b} U^\dagger_{fa} M_{ab} U_{bf'} = m_f \delta_{ff'}. \tag{150}$$

The states $b_{f'}$, eq. (147b), are then the *mass eigenstates* whilst the states d_f are the *weak eigenstates*.

In the case of two quark families, U can be taken to be the rotation matrix $D^{(1/2)}(2\psi, 2\theta, 2\phi)$, cf. eq. (III.119). The phases $e^{\pm i\psi}$ and $e^{\pm i\phi}$ are irrelevant for any observable because they can be absorbed into the field operators. After this redefinition of the fields we have

$$\begin{pmatrix} d_1 \\ d_2 \end{pmatrix} = \begin{pmatrix} \cos\theta & \sin\theta \\ -\sin\theta & \cos\theta \end{pmatrix} \begin{pmatrix} b_1 \equiv \mathrm{d} \\ b_2 \equiv \mathrm{s} \end{pmatrix} \tag{151}$$

The remaining mixing angle θ (which is measurable) is called the Cabibbo angle.

In the case of three quark families the most general transformation matrix U can be constructed, for example, by taking the product of three successive, two-dimensional transformations of the same kind as above (see exercise 7). Some of the resulting phases can be absorbed in the field operators, as before. The remainder is a unitary matrix that depends on three real mixing angles θ_1, θ_2, θ_3 and one phase $e^{i\delta}$ (Kobayashi et al. 1973),

$$U = \begin{pmatrix} c_1 & s_1 c_3 & s_1 s_3 \\ -s_1 c_2 & c_1 c_2 c_3 + s_2 s_3 e^{i\delta} & c_1 c_2 s_3 - s_2 c_3 e^{i\delta} \\ -s_1 s_2 & c_1 s_2 c_3 - c_2 s_3 e^{i\delta} & c_1 s_2 s_3 + c_2 c_3 e^{i\delta} \end{pmatrix}, \tag{152}$$

where $c_i := \cos\theta_i$, $s_i := \sin\theta_i$. Clearly, if $\theta_2 = \theta_3 = 0$ this matrix reduces to the previous case (151) with the b-quark decoupling from the (d, s) sector. The real angle θ_1, in particular, is seen to take over the role of the Cabibbo angle of the previous case with only two families of quarks.

In analogy to the case of the leptons, and on the basis of the phenomenological information on hadronic weak interactions, the quark fields $u_f(x)$ and $d_f(x)$ of a given family are classified in the following reducible multiplet of G,

$$\Psi_{qf} = \begin{pmatrix} u_f(x)_{\mathrm{L}} \\ d_f(x)_{\mathrm{L}} \\ u_f(x)_{\mathrm{R}} \\ d_f(x)_{\mathrm{R}} \end{pmatrix}, \tag{153}$$

where

$$L_f := \begin{pmatrix} (u_f)_{\mathrm{L}} \\ (d_f)_{\mathrm{L}} \end{pmatrix}$$

forms a doublet with respect to the SU(2) factor, while $(u_f)_{\mathrm{R}}$ and $(d_f)_{\mathrm{R}}$ are singlets. In order to obtain the correct charges from eq. (145), the weak hypercharges must be chosen as follows:

$$V(y) = \begin{pmatrix} \frac{1}{3} & 0 & 0 & 0 \\ 0 & \frac{1}{3} & 0 & 0 \\ 0 & 0 & \frac{4}{3} & 0 \\ 0 & 0 & 0 & -\frac{2}{3} \end{pmatrix}. \tag{154}$$

This choice guarantees the correct coupling of the quark currents to the photon

$$\begin{aligned} \mathscr{L}_{\gamma}^{(q)} &= \frac{gg'}{\sqrt{g^2+g'^2}} \sum_f \overline{\Psi_{qf}(x)} \left\{ V(T_3) + \tfrac{1}{2} V(Y) \right\} \gamma^\alpha \Psi_{qf}(x) A_\alpha(x) \\ &= -e j^\alpha_{\mathrm{e.m.}}(x) A_\alpha(x) \end{aligned} \tag{155}$$

with

$$j^\alpha_{\mathrm{e.m.}} = \sum_{f=1}^{3} \left\{ \tfrac{2}{3} \overline{u_f} \gamma^\alpha u_f - \tfrac{1}{3} \overline{b_f} \gamma^\alpha b_f \right\}. \tag{156}$$

The weak CC interactions are given by the analogue of eq. (137)

$$\begin{aligned} \mathscr{L}_{\mathrm{CC}}^{(q)} &= \frac{g}{\sqrt{2}} \sum_f \overline{L_f(x)}\, s_+ \gamma^\alpha L_f(x) W_\alpha^-(x) + \mathrm{h.c.} \\ &= \frac{g}{2\sqrt{2}} \sum_f \overline{u_f(x)}\, \gamma^\alpha (\mathbb{1} - \gamma_5) d_f(x) W_\alpha^-(x) + \mathrm{h.c.} \end{aligned} \tag{157}$$

The neutral weak interactions, finally, are found by isolating the coupling to the Z^0 boson in the neutral interaction Lagrangian

$$\sum_f \overline{\Psi_{qf}} \{ g V(T_3) \gamma^\alpha A_\alpha^{(3)} + g' \tfrac{1}{2} V(Y) \gamma^\alpha A_\alpha^{(0)} \} \Psi_{qf}.$$

Using the decomposition (78) this gives

$$\mathscr{L}^{(q)}_{NC} = \sum_f \frac{1}{\sqrt{g^2+g'^2}} \overline{\Psi_{qf}}\{ g^2 V(T_3) - \tfrac{1}{2} g'^2 V(Y)\} \gamma^\alpha \Psi_{qf} Z_\alpha(x)$$

$$= \sqrt{g^2+g'^2} \sum_f \left\{ \overline{\Psi_{qf}} V(T_3) \gamma^\alpha \Psi_{qf} - \frac{g'^2}{g^2+g'^2} \overline{\Psi_{qf}} (V(T_3) + \tfrac{1}{2} V(Y)) \gamma^\alpha \Psi_{qf} \right\} Z_\alpha(x)$$

$$= \frac{g}{4\cos\theta_W} \left\{ \sum_f \left[\overline{u_f(x)}\, \gamma^\alpha (\mathbb{1} - \gamma_5) u_f(x) - \overline{b_f(x)}\, \gamma^\alpha (\mathbb{1} - \gamma_5) b_f(x) \right] - 4\sin^2\theta_W\, j^\alpha_{\text{e.m.}}(x) \right\} \cdot Z_\alpha(x). \tag{158}$$

We have rewritten the neutral current which couples to the Z^0 in such a way that it appears as a linear combination of the neutral partner of the CC current in eq. (157) and of the electromagnetic current. This form stresses the analogy to the leptonic neutral current, eq. (86). It is particularly useful when we wish to compute matrix elements of these currents between physical hadron states: In this case these matrix elements are "dressed" or renormalized by the strong interactions so that the bare vertices $\langle \overline{q(x)}\, \gamma^\alpha q(x) \rangle$ and $\langle \overline{q(x)}\, \gamma^\alpha \gamma_5 q(x) \rangle$ are replaced by vertex functions containing the corresponding covariants and a set of invariant form factors which parametrize these dressing effects.

The neutral current

$$K^{(q)}_\alpha(x) := \sum_f \left[\overline{u_f(x)}\, \gamma_\alpha (\mathbb{1} - \gamma_5) u_f(x) - \overline{b_f(x)}\, \gamma_\alpha (\mathbb{1} - \gamma_5) b_f(x) \right] - 4\sin^2\theta_W\, j^{\text{e.m.}}_\alpha(x), \tag{159}$$

which appears in the interaction (158), contains only terms which are diagonal in flavour, i.e. which are of the form $\bar{u}u$, $\bar{d}d$, $\bar{c}c$, $\bar{s}s$, $\bar{t}t$, and $\bar{b}b$. This is so because the fields (148) appear in diagonal form in the neutral currents K_α and $j^{\text{e.m.}}_\alpha$, viz.

$$\sum_f \overline{d_f} \Gamma_\alpha d_f = \sum_f \sum_{f',f''} U^\dagger_{f'f} U_{ff''} \overline{b_{f'}} \Gamma_\alpha b_{f''} = \sum_f \overline{b_f} \Gamma_\alpha b_f. \tag{160}$$

Thus, the interaction (158) does not contain neutral couplings of the type $\bar{d}\Gamma_\alpha s$ or $\bar{s}\Gamma_\alpha d$ that would change the strangeness (Glashow et al. 1970). NC processes with $\Delta S \neq 0$ can only come about in higher orders of perturbation theory. This result is in accord with the empirical findings, see e.g. eqs. (45).

As in the case of the leptons, the quarks remain massless in this version of the theory which still possesses the full internal symmetry $G = U(2)$. In particular, all vector and axial vector currents are exactly conserved at this stage.

4.3. *Spontaneous symmetry breaking*

The GSW unified theory as developed up to this point is gauge invariant with respect to the group G, eq. (129), of local gauge transformations. Very much like in the case of QED (which is Abelian) gauge invariance is essential for the theory to be renormalizable. At the same time, however, the theory is also invariant under G, considered as a *global* symmetry. As a consequence, the vector bosons of the theory are mass-degenerate, i.e. the W and the Z^0 are massless like the photon.

As this is in conflict with observation it is clear that at least part of the internal symmetry must be broken. More precisely, the symmetry should be broken in such a way that the gauge invariance of the theory is preserved (in view of its renormalizability) but that the mass degeneracy of the multiplets of gauge fields is lifted. This can only be achieved if one succeeds in modifying the underlying Lagrangian, while still preserving its full local gauge symmetry, in such a way that the ground state of the theory, i.e. its physical realization, exhibits less symmetry than the Lagrangian itself. The phenomenon which is well-known from the physics of the condensed matter (superconductivity, ferromagnetism, etc.) is called *spontaneous symmetry breaking* or *hidden symmetry*.

Regarding the electroweak interactions the hope is that eventually this spontaneous breakdown of the symmetry will come about in a dynamical way (as it does in the theory of superconductivity, for instance), from elements which are integral part the theory. As yet, there is no convincing scheme or theoretical proof for this, and symmetry breaking must be introduced by hand through what is called the Higgs mechanism: One adds to the Lagrangian of the GSW model an appropriate set of scalar fields $\phi = \{\phi_1(x), \ldots, \phi_M(x)\}$, so-called Higgs fields, with a self-interaction $V(\phi)$ that is chosen such as to induce spontaneous breakdown of the symmetry.

The phenomenon in itself is very interesting and deserves further detailed study. However, in order not to leave the scope of this book, we concentrate on those results which are relevant for gauge theories. We refer the reader to the excellent reviews in the literature (Bernstein 1974, O'Raifeartaigh 1979) for more detailed presentations.

4.3.1. *Definition of spontaneous symmetry breaking*

Let ϕ be a set of scalar fields which form a representation of the symmetry group G, and let $V(\phi)$ be a potential term, which satisfies the following conditions:

(i) $V(\phi)$ is invariant under the whole group G, $V(U(g)\phi) = V(\phi)$,
(ii) $V(\phi)$ has an absolute minimum at $\phi^0 = \{\phi_1^0, \ldots, \phi_M^0\}$,
(iii) this minimum of $V(\phi)$ is degenerate, i.e. ϕ^0 is not invariant under G.

The degeneracy of the absolute minimum, condition (iii), implies that there is at least one $g \in G$ for which

$$\sum_k U_{ik}(g)\phi_k^0(x) \neq \phi_i^0, \tag{161}$$

If we write $U(g)$ as an exponential series, $\exp\{i\sum_{k=1}^{N}\Lambda_k(x)U(T_k)\}$, the condition (161) says that there is at least one generator T_i of the Lie algebra of G for which

$U(T_i)\phi^0$ is not zero. Therefore, it is useful to form independent linear combinations $S_i = \sum_{k=1}^{N} C_{ik} T_k$ of the generators such that S_i fall in either of the two classes

$$\text{(A)} \quad \{S_1,\ldots,S_P\} \text{ for which } U(S_i)\phi^0 = 0, \tag{162a}$$

$$\text{(B)} \quad \{S_{P+1},\ldots,S_N\} \text{ for which } U(S_k)\phi^0 \neq 0. \tag{162b}$$

Thus, all transformations $g \in G$ which contain only generators of class (A), viz.

$$g = \exp\left\{\mathrm{i} \sum_{k=1}^{P} \Lambda_k S_k\right\}, \tag{163}$$

leave the minimum ϕ^0 invariant. In other terms, these transformations (163) form the *little group* H of ϕ^0 with

$$\dim H = P. \tag{164}$$

Expanding the potential $V(\phi)$ around ϕ^0, we have

$$V(\phi) = V(\phi^0) + \frac{1}{2} \sum_{m,n} M_{mn} \delta\phi_m \delta\phi_n = \mathrm{O}\big((\delta\phi)^3\big),$$

where the matrix

$$M_{mn} := \left.\frac{\partial^2 V}{\partial\phi_m\, \partial\phi_n}\right|_{\phi^0}$$

is positive semidefinite, and where the infinitesimal variation of the field ϕ_n is given by

$$\delta\phi_n = \mathrm{i} \sum_{m=1}^{M} \sum_{k=1}^{N} U_{nm}(S_k)\phi_m^0 \delta\Lambda_k.$$

Let us define the following vectors

$$v_n^{(k)} := \sum_{m=1}^{M} U_{nm}(S_k)\phi_m^0. \tag{165}$$

As the variations $\delta\Lambda_k$ of the group parameters are linearly independent, the condition for $V(\phi)$ to have a minimum in $\phi = \phi^0$ reads

$$\sum_{m,n} v_n^{(i)} M_{nm} v_m^{(k)} = 0.$$

Furthermore, as $M \geq 0$, this is fulfilled provided

$$\sum_m M_{nm} v_m^{(k)} = 0. \tag{166}$$

The matrix M takes the role of a mass matrix for the scalar fields ϕ. A number P of its eigenvectors are identically zero. These are the ones for which $k \in \{1,\ldots,P\}$, cf. eq. (162a), i.e. those which are formed with a generator of the little group H of ϕ^0. For all other values of $k \in \{P+1,\ldots,N\}$, $v^{(k)}$ does not vanish. From this we conclude that M has

$$N_G \coloneqq \dim G - \dim H = N - P \tag{167}$$

eigenvalues which are zero. This result is a consequence of a theorem by Goldstone*): A manifestly Lorentz invariant theory which has a hidden symmetry (i.e. whose internal symmetry is spontaneously broken) contains a set of massless scalar fields, the so-called Goldstone bosons. The number of Goldstone fields N_G is given by eq. (167). It is the difference of the dimension of Lie algebras of the full symmetry group G and of the little group $H(\phi^0)$ of ϕ^0, respectively. Clearly, N_G cannot be zero because, by the very definition of spontaneous symmetry breaking, the class (B), eq. (162b), is not empty. As is clear from our analysis above, N_G, the number of Goldstone fields, does not depend on the representation of the scalar fields. It depends solely on the symmetry group G and on $H(\phi^0)$, i.e. on the residual symmetry of the physical realization of the theory, after it has been spontaneously broken.

There is no restriction on the remaining P eigenvalues of the mass matrix M (corresponding to those $v^{(k)}$, eq. (162a) which vanish identically). These eigenvalues pertain to a set of $(M - N_G)$ massive scalar fields. In contrast to the case of the Goldstone fields their number depends on the representation spanned by the fields ϕ.

The massless Goldstone particles have a simple quasi-geometrical interpretation: By assumption the minimum of $V(\phi)$ in ϕ^0 is degenerate. Thus there exist transformations $g \in G$ for which $\phi'^0 = U(g)\phi^0$ is not identical with ϕ^0, $\phi'^0 \not\equiv \phi^0$. The set of all ϕ'^0 which can be reached by applying all possible group transformations to ϕ^0, form the *group orbit* of ϕ^0. The Goldstone fields are proportional to those vectors $v^{(k)} = U(S_k)\phi^0$ which do not vanish identically. On the other hand, an infinitesimal transformation of ϕ^0 is given by

$$\phi'^0 = U(g \simeq \mathbb{1} + \mathrm{i}\Lambda_k S_k)\phi^0 = \{\mathbb{1} + \mathrm{i}\Lambda_k U(S_k)\}\phi^0, \quad \Lambda_k \ll 1.$$

This shows that $U(S_k)\phi^0$ is a tangent to the orbital of ϕ^0. Thus, a Goldstone field can be understood as an excitation of the system along the orbit of ϕ^0. As this does not lead out of the (degenerate) minimum of $V(\phi)$, a Goldstone excitation can have arbitrarily small frequency. In other words, a Goldstone field describes a massless particle.

4.3.2. *Spontaneous symmetry breakdown in the frame of a local gauge theory*

The discussion of the previous section is incomplete insofar as the scalar fields ϕ cannot be discussed in isolation from the rest of the theory. For instance, the question of whether or not the Goldstone fields correspond to physical, massless particles, cannot be answered without knowing how these fields couple to other fields such as spinor or vector fields. It is particularly interesting to investigate the role of the Goldstone fields in the case where G is not only a global symmetry but is also a

*) See Bernstein (1974) for a complete list of references and a detailed exposition of this theorem and its consequences.

local gauge symmetry of the Lagrangian. In this case the Goldstone fields do not describe observable massless scalars. Instead, as we shall see, they decouple from the other particles in the theory. At the same time some of the gauge bosons which formerly were all massless, acquire finite masses. This happens for as many of them as there are Goldstone fields.

Let us return to the Lagrangian (128) which describes a set of spinor fields Ψ as well as a set of scalar fields ϕ, besides the gauge fields A_α. Let us assume that the potential term $V(\phi)$ is constructed such that it leads to spontaneous breakdown of the symmetry of the Lagrangian (128). Thus, $V(\phi)$ fulfills the conditions (i) to (iii) of sec. 4.3.1, i.e. it has a degenerate minimum at ϕ^0. As ϕ^0 is the position of the absolute minimum of $V(\phi)$ we introduce new scalar fields by subtracting $\phi(x)$ and ϕ^0,

$$\Theta_n(x) := \phi_n(x) - \phi_n^0, \quad n = 1,\dots,M. \tag{168}$$

Rewriting eq. (128) in terms of these new dynamical fields we obtain

$$\begin{aligned}\mathscr{L}' = &-\frac{1}{4\kappa e^2}\left(F_{\alpha\beta}, F^{\alpha\beta}\right) + \tfrac{1}{2}\left(U(A_\alpha)\phi^0, U(A^\alpha)\phi^0\right) + \tfrac{1}{2}\left(D_\alpha\Theta(x), D^\alpha\Theta(x)\right)\\ &+ \mathrm{Re}\left(D_\alpha\Theta(x), U(A^\alpha)\phi^0\right) - V\left(\Theta(x)+\phi^0\right) - g\left(\overline{\Psi(x)}, \phi^0\Psi(x)\right)\\ &+ \frac{\mathrm{i}}{2}\left(\overline{\Psi(x)}, \gamma^\alpha \overleftrightarrow{D}_\alpha \Psi(x)\right) - \left(\overline{\Psi}, (m + g\Theta(x))\Psi\right).\end{aligned} \tag{169}$$

Note that in this Lagrangian the fields ϕ^0 are constants whilst the role of the dynamical scalar fields is taken over by the fields $\Theta(x)$. Although its structure is similar to the structure of $\mathscr{L}$, eq. (128), the modified Lagrangian (169) has two remarkable new properties:

(i) The term $-g(\overline{\Psi}, \phi^0\Psi)$ yields a finite mass for the fermion fields, provided the multiplets Ψ and ϕ are chosen such that $\overline{\Psi}$, ϕ^0,and Ψ can be coupled to an invariant with respect to G. The mass term which is due to the spontaneous symmetry breakdown, is particularly relevant for the GSW model for which it was not possible to construct an invariant mass term $(\overline{\Psi}, m\Psi)$, on the basis of the multiplet assignments (133) and (153).

(ii) Second, and perhaps more importantly, the term

$$\tfrac{1}{2}\left(U(A_\alpha)\phi^0, U(A^\alpha)\phi^0\right) \tag{170}$$

provides mass terms for at least some of the gauge fields without destroying the gauge invariance of the theory. This is what we now wish to analyze in more detail.

In the expression (170) only the generators of class (B), eq. (162b), give nonvanishing contributions, viz.

$$\tfrac{1}{2}\left(U(A_\alpha)\phi^0, U(A^\alpha)\phi^0\right) = \frac{1}{2}\sum_{i,k=P+1}^{N} m_{ik} A_\alpha^{(i)}(x) A^{(k)\alpha}(x), \tag{170$'$}$$

where

$$m_{ik} := e^2\left(U(S_i)\phi^0, U(S_k)\phi^0\right). \tag{171}$$

The quadratic mass matrix has the dimension $N_G \times N_G$, with N_G as given by eq. (167). It has N_G positive eigenvalues. The eigenvectors of this matrix are orthogonal linear combinations of the original fields $A_\alpha^{(i)}$. These new vector fields now describe massive gauge bosons. Thus we obtain the following important result: If G is the original full symmetry, H the residual symmetry (after the symmetry is spontaneously broken), of the locally gauge invariant Lagrangian then a number $N_G = \dim G - \dim H$ of the gauge bosons of the theory acquire finite masses. The remaining $P = \dim H$ gauge bosons which correspond to the generators of the residual symmetry group H remain massless. The number of vector particles which become massive is equal to the number of Goldstone scalar bosons. In fact, one can show that these massless scalars decouple from the rest of the theory. Their role is merely to provide the third independent polarization component that distinguishes the massive vector particle from the massless one (which has only two components).

4.3.3. Application to the GSW theory

The analysis of the preceding section provides a constructive principle of how to give the $W^{\pm}$ and Z^0 finite masses while leaving the photon massless. The original symmetry (129) whose Lie algebra has dimension $N = \dim G = 4$, must be spontaneously broken such that the residual symmetry is $H = U(1)$ with dimension $P = \dim H = 1$. If this is achieved three of the gauge fields become massive, one remains massless. In practice, this means that we add the following Higgs sector to the Lagrangian of the GSW model:

$$\tfrac{1}{2}(D_\alpha \phi, D^\alpha \phi) - V(\phi).$$

The potential $V(\phi)$ shall have a degenerate minimum at $\phi = \phi^0$ such that the residual symmetry is U(1). In fact, it is not necessary to write down an explicit form for the potential $V(\phi)$. Indeed, it turns out that it is enough to arrange the eigenvalues of T_3 (weak isospin) and of Y (weak hypercharge) for the Higgs fields such as to make them decouple from the photon field. This is seen as follows.

Let $U(T_\mu)$ be the representation of the generators T_μ in the space of the Higgs fields ϕ. If ϕ^0 denotes the position of the degenerate, absolute minimum of $V(\phi)$, the mass matrix for the vector fields is given by

$$\tfrac{1}{2}(D_\alpha \phi^0, D^\alpha \phi^0) = \tfrac{1}{2}(U(A_\alpha)\phi^0, U(A^\alpha)\phi^0), \tag{172}$$

where

$$U(A_\alpha)\phi^0 = -\mathrm{i}\left\{ g \sum_{k=1}^{3} U(T_k) A_\alpha^{(k)}(x) + \tfrac{1}{2} g' U(Y) A_\alpha^{(0)}(x) \right\} \phi^0. \tag{173}$$

Here we have used the identification $-e = g$, eq. (138), and the definition (143) for Y in terms of T_0. Replacing the fields $A_\alpha^{(\mu)}$ by the physical charged fields (72) and neutral fields (78) we obtain

$$U(A_\alpha)\phi^0 = -\mathrm{i}\left\{ \frac{1}{\sqrt{2}} g\left[U(T_+)W_\alpha^{(-)}(x) - U(T_-)W_\alpha^{(+)}(x)\right] + \sqrt{g^2+g'^2}\, U\left(\frac{g^2}{g^2+g'^2}T_3 - \frac{1}{2}\frac{g'^2}{g^2+g'^2}Y\right) Z_\alpha^0(x) + \frac{gg'}{\sqrt{g^2+g'^2}} U\left(T_3 + \tfrac{1}{2}Y\right) A_\alpha(x) \right\}\phi^0, \tag{173'}$$

where $T_\pm = T_1 \pm \mathrm{i}T_2$. Let t_3 and y be the eigenvalues of T_3 and Y for the fields $\phi(x)$, or $\theta(x) = \phi(x) - \phi^0(x)$, respectively. Eq. (173′) shows that the electric charge of the field ϕ^0 is proportional to

$$Q_\gamma = t_3 + \tfrac{1}{2}y.$$

The photon field remains massless if and only if $Q_\gamma \phi^0$ vanishes. Thus, if $\phi_m^0 \neq 0$, we impose the condition $y = -2t_3$.*) With this condition we see from eq. (173′) that the coupling to the Z^0 field is proportional to

$$\frac{g^2}{g^2+g'^2}t_3 - \frac{g'^2}{2(g^2+g'^2)}y = t_3.$$

On the other hand, in order to make the Z^0 massive we must require t_3 to be different from zero. Therefore, let us assume that $\phi(x)$ belongs to an irreducible representation of SU(2) with total weak isospin t (different from zero) and projection quantum number t_3,

$$U(\boldsymbol{T}^2)\phi = t(t+1)\phi, \qquad U(T_3)\phi = t_3\phi. \tag{174}$$

Eqs. (172) and (173′) then yield the following expressions for the gauge boson masses:

$$m_\mathrm{W}^2 = \tfrac{1}{2}g^2\tfrac{1}{2}\left(\phi^0, \{U(T_+)U(T_-) + U(T_-)U(T_+)\}\phi^0\right) = \tfrac{1}{2}g^2\left[t(t+1) - t_3^2\right](\phi^0, \phi^0), \tag{175}$$

$$m_\mathrm{Z}^2 = (g^2+g'^2)\left(\phi^0, U(T_3)U(T_3)\phi^0\right) = (g^2+g'^2)t_3^2(\phi^0, \phi^0). \tag{176}$$

In eq. (175) we have made use of the relation

$$\tfrac{1}{2}(T_+T_- + T_-T_+) = \boldsymbol{T}^2 - T_3^2.$$

Eqs. (175) and (176) give at once the mass relation

$$\rho \equiv \frac{m_\mathrm{W}^2}{m_\mathrm{Z}^2\cos^2\theta_\mathrm{w}} = \frac{t(t+1) - t_3^2}{2t_3^2}, \tag{177}$$

where we have used the relation $\cos^2\theta_\mathrm{w} = g^2/(g^2+g'^2)$. Note that eq. (177) is

*) I.e. all components of ϕ^0 vanish except the ones for which $y = 2t_3$.

precisely the quantity ρ as defined in eq. (88). The experimental information that ρ is 1 with a rather small error bar [R2] is compatible with the assignment

$$t = \tfrac{1}{2} \tag{178}$$

for the Higgs fields. This means that these fields form a doublet with respect to SU(2). One of the fields is electrically neutral and has $y = -2t_3 = 1$. This is the one that has a nonvanishing vacuum expectation value ϕ^0 and which gives rise to the vector meson masses (175) and (176). Its partner in the doublet has the quantum numbers ($y = 1$, $t_3 = +\frac{1}{2}$) and electric charge $Q = 1$.

Finally, we can introduce a Yukawa coupling of the fermion fields to the Higgs doublet without destroying the local gauge invariance of the theory,

$$g_f\left\{\left(\overline{L_f}\phi\right)f_\mathrm{R} + \overline{f_\mathrm{R}}\left(\phi, L_f\right)\right\}. \tag{179}$$

This extra term gives rise to a genuine interaction of the fermion with the charged Higgs field. As to the neutral partner of the Higgs doublet, its vacuum expectation value gives rise to a fermion mass term of the form (we consider the example of a lepton doublet),

$$g_f\lambda\left\{\overline{f_\mathrm{R}}f_\mathrm{L} + \overline{f_\mathrm{L}}f_\mathrm{R}\right\} = g_f\lambda\overline{f(x)}\,f(x) \tag{180}$$

with

$$\lambda = \sqrt{(\phi^0, \phi^0)}\,.$$

It is gratifying that in the spontaneously broken version of the theory it is possible to construct fermion mass terms which do not violate the gauge symmetry of the theory. On the other hand, eq. (180) neither predicts the scale of the fermion masses nor does it yield relations between the masses of different lepton families. In fact, for every fermion family, the coupling constant g_f must be adjusted such that eq. (180) yields the correct mass term, viz. $g_f = m_f/(\phi^0, \phi^0)^{1/2}$. At the same time this implies that the fermion coupling to the physical, charged Higgs field, eq. (179), is proportional to m_f and hence very small. This makes it very difficult to subject such an interaction to experimental verification.

Finally, we note that eqs. (175) and (176) predict the W- and Z-masses in terms of the unknown quantity (ϕ^0, ϕ^0). The absolute values of m_W and m_Z, as yet, are derived from the empirical coupling constants α, G and from the Weinberg angle θ_w, as given in eqs. (89) and (90). These expressions are modified somewhat by corrections of higher order. As the theory is renormalizable, these corrections can be calculated in a unique way from perturbation theory. Recently published values are (Wheater et al. 1982)

$$m_\mathrm{W} = 38.48\ \mathrm{GeV}/\sin\theta_\mathrm{w}(M_\mathrm{W}), \tag{181}$$

which gives $m_\mathrm{W} \simeq 83$ GeV, if $\sin^2\theta_\mathrm{w} \simeq 0.215$. The radiative corrections being essentially the same for the Z^0, m_Z is given by eq. (90), i.e. $m_\mathrm{Z} = m_\mathrm{W}/\cos\theta_\mathrm{w} \simeq 94$ GeV. Recent experiments at the proton–antiproton collider facility of CERN have indeed established both the W and Z^0 at about these masses.

4.4. *Summary of CC and NC interactions in the GSW theory*

In this section we collect and summarize the leptonic and hadronic charged current and neutral current interactions as they follow from the results in sects. 2.3, 4.2 and 4.3. For convenience we express all coupling constants in terms of the elementary charge in natural units $e=\sqrt{4\pi\alpha}$. The interaction Lagrangian of leptons and quarks with the gauge bosons γ, W and Z^0 is given by

$$\mathscr{L}_{\text{int}} = -e\left\{ j^{\alpha}_{\text{e.m.}}(x)A_{\alpha}(x) + \frac{1}{2\sqrt{2}\,\sin\theta_{\text{w}}}\left(J^{\alpha}(x)W^{-}_{\alpha}(x)+\text{h.c.}\right)\right.$$
$$\left. + \frac{1}{4\sin\theta_{\text{w}}\cos\theta_{\text{w}}}K^{\alpha}(x)Z^{0}_{\alpha}(x)\right\}. \tag{182}$$

In this expression $j^{\alpha}_{\text{e.m.}}$ is the electromagnetic current, J_{α} is the charged current,

$$J_{\alpha}(x) = \sum_{\text{f}=\text{e},\mu,\tau} \overline{\nu_{\text{f}}}\gamma^{\alpha}(\mathbb{1}-\gamma_5)f(x)$$
$$+\sum_{f=1}^{3}\overline{u_f(x)}\,\gamma^{\alpha}(\mathbb{1}-\gamma_5)d_f(x), \tag{183}$$

where u_f denotes the quarks with electric charge $+\frac{2}{3}$,

$$\{u_f\} = \{\text{u},\text{c},\text{t}\}$$

while d_f stands for the weak eigenstates

$$d_f = \sum_{f'=1}^{3} U_{ff'}b_{f'},$$

given in terms of the strong interaction eigenstates d, s and b with electric charge $-\frac{1}{3}$. We recall that $U_{ff'}$ is the mixing matrix (152). The neutral current K_{α} is given by eqs. (86), (159) and (156), viz.

$$K_{\alpha}(x) = \sum_{\text{f}=\text{e},\mu,\tau}\overline{\nu_{\text{f}}(x)}\,\gamma_{\alpha}(\mathbb{1}-\gamma_5)\nu_{\text{f}}(x)$$
$$+\sum_{\text{f}=\text{e},\mu,\tau}\overline{f(x)}\left\{-\gamma_{\alpha}(\mathbb{1}-\gamma_5)+4\sin^2\theta_{\text{w}}\gamma_{\alpha}\right\}f(x)$$
$$+\sum_{f=1}^{3}\overline{u_f(x)}\left\{\gamma_{\alpha}(\mathbb{1}-\gamma_5)-\tfrac{8}{3}\sin^2\theta_{\text{w}}\gamma_{\alpha}\right\}u_f(x)$$
$$+\sum_{f=1}^{3}\overline{b_f(x)}\left\{-\gamma_{\alpha}(\mathbb{1}-\gamma_5)+\tfrac{4}{3}\sin^2\theta_{\text{w}}\gamma_{\alpha}\right\}b_f(x). \tag{184}$$

Note, in particular, that this current has the general form

$$\overline{\Psi^{(i)}(x)}\Gamma^{(i)}_{\alpha}\Psi^{(i)}(x),$$

where the matrix $\Gamma_\alpha^{(i)}$ depends on the charge and weak isospin of the lepton or quark i:

$$\Gamma_\alpha^{(i)} = 2t_3(i)\gamma_\alpha(\mathbb{1} - \gamma_5) - Q(i)\gamma_\alpha 4\sin^2\theta_\text{w}. \tag{185}$$

The exchange of $W^\pm$ and Z^0 bosons, in processes where the momentum transfer is small as compared to the gauge boson masses, gives rises to the effective four-fermion interaction (cf. eqs. (75) and (87)),

$$-\mathscr{L}_\text{int}^\text{eff} = \frac{G}{\sqrt{2}}\left\{J_\alpha^\dagger(x)J^\alpha(x) + \tfrac{1}{2}\rho K_\alpha^\dagger(x)K^\alpha(x)\right\}, \tag{186}$$

where

$$\frac{G}{\sqrt{2}} = \frac{e^2}{8m_\text{W}^2\sin^2\theta_\text{w}} = \frac{\pi\alpha}{2m_\text{W}^2\sin^2\theta_\text{w}} \tag{187a}$$

and

$$\rho = \frac{m_\text{W}^2}{m_\text{Z}^2\cos^2\theta_\text{w}}. \tag{187b}$$

(Eq. (187a) holds for the uncorrected value of m_W, to be distinguished from the corrected value (181).) These formulae summarize the weak interactions of leptons and quarks in the GSW theory.

It is instructive to recapitulate the number and the nature of the parameters of the theory, none of which is predicted. These are

(i) Coupling constants and gauge boson masses: α, $\sin\theta_\text{w}$, m_W, m_Z. The Fermi constant G is determined in terms of the first three parameters by eq. (187a). m_Z is fixed by eq. (177) if the weak isospin of the Higgs field is given.

(ii) Quark masses and mixing matrix (152): m_u, m_d, m_c, m_s, m_t, m_b, θ_1, θ_2, θ_3, δ.

(iii) Lepton masses: m_e, m_μ, m_τ.

This list does not include the Higgs sector of the theory. Furthermore, if the neutrinos are massive, then there are three more mass values as well as another set of four mixing angles because the weak neutrino states could appear to be mixtures of their mass eigenstates in a way analogous to the quark mixing (148) and (152).

4.5. A comment about fermion multiplets in a unified gauge theory of electroweak interactions

4.5.1. Fermion multiplets in the language of SL(2, ℂ) spinors

As we have noted in the introduction to sec. 4 and in sec. 4.1, the GSW theory is built on the assumptions that CC interactions (mediated by $W^\pm$ bosons) are fully left-handed and that physical neutrinos carry negative helicity $h(\nu) = -1$. As a consequence the basic fermion fields of the theory are van der Waerden spinor fields. This is seen, for instance, in the leptonic field operator (133) which contains two spinor fields of the second kind, one spinor field of the first kind. Thus, the natural language in formulating unified gauge theories that contain the weak interactions is the one of SL(2, ℂ) spinors developed in Chap. III.

For the example of one lepton family, eq. (133), we should introduce a doublet of spinors of the second kind

$$\chi^{(m)A}(x), \quad m=+\tfrac{1}{2},-\tfrac{1}{2},$$

describing the neutrino ($m=+\frac{1}{2}$) and the left-handed part of the charged fermion field ($m=-\frac{1}{2}$), as well as a spinor of the first kind, $\phi_a(x)$, which is a singlet with respect to SU(2) and which describes the right-handed part of the charged fermion field. In terms of these spinors the generalized kinetic term reads

$$\frac{\mathrm{i}}{2}\left(\overline{\Psi(x)}, \gamma^\alpha \overleftrightarrow{D}_\alpha \Psi(x)\right)=\frac{\mathrm{i}}{2}\left\{\sum_m \chi^{(m)*a}(\sigma^\alpha)_{aB}\overleftrightarrow{D}_\alpha \chi^{(m)B}+\phi_A^*(\hat{\sigma}^\alpha)^{Ab}\overleftrightarrow{D}_\alpha\phi_b\right\},$$

which contains, in fact, the genuine kinetic energy and the interaction with the gauge fields. D_α is the covariant derivative (117), as before.

The more standard formulation in terms of Dirac fields that we used in sects. 4.1 to 4.4 above, is completely equivalent, of course. However, the formulation in terms of van der Waerden spinors is very useful for the extension to supersymmetries.

4.5.2. Triangle anomaly and renormalizability of the GSW theory

We mention briefly a theoretical and somewhat technical point which is crucial for the renormalizability of the GSW theory and which sheds some light on the empirical symmetry between the leptons and quark families: the so-called triangle anomaly of the axial current (Adler 1969). One can show that the axial vector part

$$\begin{aligned} a_\alpha^{(0)}(x) &= \sum_{\mathrm{f}=\mathrm{e},\mu,\tau} \overline{\Psi_{\mathrm{f}}(x)}\, \gamma_\alpha \gamma_5 V(Y) \Psi_{\mathrm{f}}(x) \\ &\quad + \sum_{q=1}^{3} \sum_{c=1}^{3} \overline{\Psi_{qc}(x)}\, \gamma_\alpha \gamma_5 V(Y) \Psi_{qc}(x) \end{aligned} \tag{188}$$

of the current that couples to the gauge field $A_\alpha^{(0)}(x)$ has an anomalous term in its divergence $\partial^\alpha a_\alpha^{(0)}(x)$ which is proportional to $S=S_{\text{leptons}}+S_{\text{quarks}}$ with

$$S_{\text{leptons}}=\sum_{\mathrm{e},\mu,\tau}\mathrm{Sp}\{V(T_m T_m Y)\}, \qquad S_{\text{quarks}}=\sum_{q,c}\mathrm{Sp}\{V(T_m T_m Y)\}.$$

m is a component of weak isospin. [It is to be contracted with bilinear and trilinear products of vector fields $A_\alpha^{(m)}(x)$.] The sums in eq. (188) run over the three lepton families and over the three, threefold degenerate quark families. However, as the traces contain the weak isospin, only the *doublets* within the multiplets (133) and (153) contribute to the anomaly. It suffices to consider the case $m=3$. From eqs. (144) and (154) we find

$$S_{\text{leptons}}=\tfrac{1}{4}\cdot 3\times(-2)=-\tfrac{3}{2}, \qquad S_{\text{quarks}}=\tfrac{1}{4}\cdot 3\times 3\cdot\tfrac{2}{3}=+\tfrac{3}{2},$$

so that their sum vanishes. In other words, the anomaly due to the quark fields cancels the anomaly due to the leptons provided the quarks have the additional colour degree of freedom.

The renormalizability of a local gauge theory rests on its internal symmetry, on current conservation and on its specific Ward identities. An anomaly in the divergence of a current that couples to a gauge boson would destroy renormalizability. The minimal model avoids this disaster by arranging its fermion multiplets such as to cancel the anomaly.

5. Simple applications of the GSW model at energies below the vector boson masses

In this section we study three simple examples which illustrate the specific predictions of the GSW theory for reactions due to Z^0 exchange and, in particular, to the interference between the exchange of a virtual Z^0 boson and of a photon. In the terminology introduced earlier these examples concern NC interactions for which the GSW theory gives new and specific predictions. The CC interactions of this model are the same as in the older, effective, theory. They are dealt with, in a more general framework, in subsequent sections.

5.1. Scattering of longitudinally polarized electrons on a nucleus with spin zero

In Chap. IV, sec. 2, we have learnt that the two helicity states of a fast electron, eqs. (IV. 17), are scattered in exactly the same way by a target with spin zero, provided the interaction is invariant under rotations *and* under space reflection. This is indeed the case when we consider the electromagnetic interaction only (as we did in Chap. IV).

The interaction Lagrangian (182) of the GSW model contains neutral interactions which are not invariant under parity. This is due to the simultaneous presence of vector and axial vector couplings of the fermion fields to the Z^0 boson. Thus, very much like for the case of the CC weak interactions, one expects to find manifestations of parity violation in purely weak NC reactions, such as neutrino scattering on leptonic or hadronic targets. In addition, amplitudes due to exchange of a virtual Z^0 can interfere with amplitudes due to exchange of a photon. Z^0-exchange gives rise to effective VV, VA, AV, and AA couplings whilst one-photon exchange gives rise to VV couplings only. As the VA and AV terms are pseudoscalars whilst the VV and AA terms are scalars, the γ–Z^0 interference must also lead to observable, parity violating effects.

Taking into account the interference with the weak NC interactions, the scattering amplitudes f and g for the two helicity states of the incident electron are not equal anymore. They differ precisely by the new parity-odd VA and AV terms. Thus, the *difference* of the differential cross sections $d\sigma_+$ and $d\sigma_-$ for electrons with positive and negative helicity, respectively, is a direct measure of the interference of vector and axial vector amplitudes. This is what we now wish to work out in a simple case: scattering on a nucleus with spin zero, using first Born approximation.

The quantity of interest is the asymmetry in the cross sections

$$A := \frac{d\sigma_+/d\Omega - d\sigma_-/d\Omega}{d\sigma_+/d\Omega + d\sigma_-/d\Omega} = \frac{|f|^2 - |g|^2}{|f|^2 + |g|^2}. \tag{189}$$

In computing the scattering amplitudes due to one-photon and to one-Z^0 exchange we need the nucleonic and nuclear matrix elements of the weak neutral current K_α, eq. (184). This is complicated by the fact that nucleons are composite states and that hadronic matrix elements are modified by the strong interactions. Let us denote the vector currents by v_α, the axial vector currents by a_α. The proton contains two u-quarks and one d-quark. Thus by simply counting the quarks, the matrix element of K_α, eq. (184), between two proton states with momenta p and p' is given by

$$\langle p'|K_\alpha(0)|p\rangle = \left(1-4\sin^2\theta_{\rm w}\right)\langle p'|v_\alpha(0)|p\rangle - \langle p'|a_\alpha(0)|p\rangle. \tag{190a}$$

Similarly, a neutron contains one u-quark and two d-quarks and the matrix element of K_α between two neutron states with momenta n and n' is given by

$$\langle n'|K_\alpha(0)|n\rangle = -\langle n'|v_\alpha(0)|n\rangle + \langle n'|a_\alpha(0)|n\rangle. \tag{190b}$$

The vector matrix elements in eqs. (190a) and (190b) have the same general decomposition in terms of Lorentz invariant form factors as the electromagnetic current, eq. (IV. 46). In particular, if v_α is conserved (this is indeed the case from eq. (184)), the third form factor F_3 must vanish identically. The matrix elements of a_α may be decomposed, in a similar fashion, in terms of axial vector covariants, viz.

$$\langle p'|a_\alpha(0)|p\rangle = \frac{1}{(2\pi)^3}\overline{u(p')}\left\{\gamma_\alpha\gamma_5 F_{\rm A}(q^2) + \frac{1}{2m}q_\alpha\gamma_5 F_{\rm P}(q^2) + \frac{\rm i}{2m}\sigma_{\alpha\beta}q^\beta\gamma_5 F_{\rm T}(q^2)\right\}u(p). \tag{191}$$

[The third term in the curly brackets can be shown to vanish if the strong interactions are invariant under G-parity, i.e. $F_{\rm T}(q^2)\equiv 0$.]

Let us now consider scattering of fast electrons on a nucleus with spin zero. Let k, k' be the four-momenta of the electron before and after the scattering, let p, p' be the corresponding momenta of the nucleus and let

$$q = k - k' = p' - p$$

be the momentum transfer. The one-photon exchange amplitude is given by

$$T_\gamma = \frac{e^2}{(2\pi)^3}\overline{u_{\rm e}(k')}\gamma^\alpha u_{\rm e}(k)\frac{-g_{\alpha\beta}}{q^2}\langle p'|j^\beta_{\rm e.m.}(0)|p\rangle, \tag{192}$$

and the one Z^0-exchange amplitude is

$$T_Z = -\frac{e^2}{(2\pi)^3}\frac{1}{16\sin^2\theta_{\rm w}\cos^2\theta_{\rm w}}\overline{u_{\rm e}(k')}\left\{-\gamma^\alpha\left(1-4\sin^2\theta_{\rm w}\right)+\gamma^\alpha\gamma_5\right\}u_{\rm e}(k) \times\frac{-g_{\alpha\beta}+q_\alpha q_\beta/m_Z^2}{q^2-m_Z^2}\langle p'|K^\beta(0)|p\rangle. \tag{193}$$

The nuclear matrix element of the electromagnetic current has the form

$$\langle p'|j^\beta_{\rm e.m.}(0)|p\rangle = \frac{1}{(2\pi)^3}(p+p')^\beta Z F(q^2), \tag{194}$$

where F is the electric form factor of the nucleus and is normalized to $F(0)=1$.

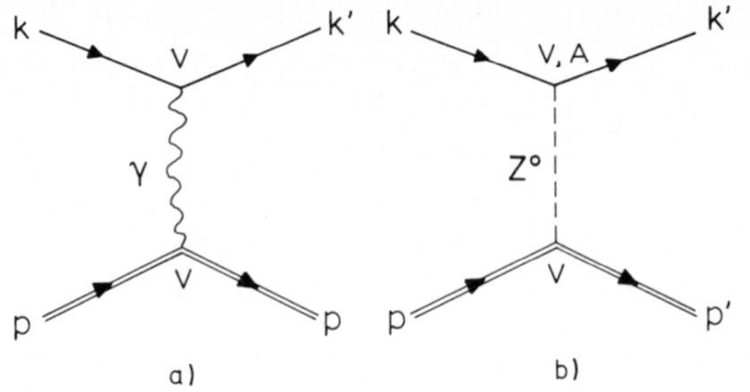

Fig. 6. (a) One-photon exchange in electron-nucleus scattering. (b) Exchange of a Z^0-boson between electron and nucleus.

Regarding the nuclear matrix element of K_α we note the following: Due to angular momentum and parity conservation in the hadronic vertex, the nucleonic axial currents a_α do not contribute to elastic scattering, $\langle p'|a_\alpha|p\rangle = 0$. The vector current of the protons is proportional to the electromagnetic current. Therefore, its *nuclear* matrix element is proportional to the electric form factor $F(q^2)$. If we assume the neutron density to be the same as the proton density, $\rho_{\rm n}(r) = \rho_{\rm p}(r)$, then the matrix element of the vector current due to neutrons is also proportional to that same form factor. In this case we have

$$\langle p'|K_\beta(0)|p\rangle = \frac{1}{(2\pi)^3}(p+p')^\beta\left\{Z(1-4\sin^2\theta_{\rm w}) - N\right\}F(q^2), \tag{195}$$

where N is the neutron number.

It is easy to see that the terms $q_\alpha q_\beta$ of the Z^0-propagator in eq. (193) do not contribute. Thus both the photon and the Z^0 exchange give an effective four-fermion coupling, the first multiplied with the photon propagator $1/q^2$, the second with the factor $1/(q^2 - m_Z^2)$.

The two amplitudes (192) and (193) are depicted in Figs. 6a and b. The leptonic vector coupling in Fig. 6b gives a term which is analogous to the one-photon exchange and whose magnitude depends on the squared momentum transfer. Clearly, at intermediate energies $|q^2|$ is very small as compared to m_Z^2 and this term is then negligible.

Using the following abbreviations:

$$v(q^2) := 1 + \frac{(1-4\sin^2\theta_{\rm w})\left[(1-4\sin^2\theta_{\rm w}) - N/Z\right]}{16\sin^2\theta_{\rm w}\cos^2\theta_{\rm w}}\,\frac{q^2}{q^2 - m_Z^2}, \tag{196a}$$

$$a(q^2) := -\frac{\left[(1-4\sin^2\theta_{\rm w}) - N/Z\right]q^2}{16\sin^2\theta_{\rm w}\cos^2\theta_{\rm w}(q^2 - m_Z^2)}\Big/v(q^2), \tag{196b}$$

and writing the leptonic vectors and axial vectors in terms of $\gamma^\alpha P_+$ and $\gamma^\alpha P_-$, with

$P_{\pm}$ as defined in eq. (III. 54), we then have

$$T_{\gamma}+T_{Z}=-\frac{e^2}{(2\pi)^6}\frac{Ze^2}{q^2}F(q^2)v(q^2)$$
$$\times(p+p')_{\alpha}\overline{u(k')}\gamma^{\alpha}\{(1+a(q^2))P_{+}+(1-a(q^2))P_{-}\}u(k). \tag{197}$$

It is easy to calculate the differential cross section (from the expression (197) and, in fact, on the basis of experience in Chap. IV we guess that we will obtain the Mott cross section (IV. 27), multiplied by $(1\pm a(q^2))^2$ for positive and negative helicity, respectively. For the calculation of the asymmetry (189) eq. (197) is entirely sufficient as it shows that

$$d\sigma_{+}\propto(1+a(q^2))^2 \quad\text{and}\quad d\sigma_{-}\propto(1-a(q^2))^2.$$

Thus we find

$$A=\frac{2a(q^2)}{1+a^2(q^2)}. \tag{198}$$

At intermediate energies of a few hundred MeV q^2 is small as compared to m_Z^2, so that

$$A\simeq 2a\simeq\frac{1-4\sin^2\theta_w-N/Z}{8\sin^2\theta_w\cos^2\theta_w}\frac{q^2}{m_Z^2}, \tag{199}$$

or, with $q^2\simeq-4E^2\sin^2\theta/2$,

$$A\simeq\frac{N/Z-(1-4\sin^2\theta_w)}{2\sin^2\theta_w\cos^2\theta_w}\frac{E^2}{m_Z^2}\sin^2\theta/2. \tag{199'}$$

Assume $E=500$ MeV, $m_Z\simeq 94$ GeV, $\sin^2\theta_w=0.215$, then $A\simeq(N/Z)8.4\times10^{-5}\sin^2\theta/2$. This asymmetry is small but may well be measurable. Obviously, it is of interest to choose a target with a large neutron excess.

As is evident from our derivation this asymmetry is a direct measure for the leptonic axial vector times hadronic vector couplings to the Z^0. Also noteworthy is the fact that the nuclear form factor $F(q^2)$ drops out of the ratio (189). This is a special feature of first Born approximation.

The same asymmetry (189) may also be calculated for the case of scattering on a proton or a neutron, along the same lines as in the example above. In this case there is not only electric (charge) scattering but also magnetic dipole scattering. For small and intermediate scattering angles the asymmetry is dominated by charge scattering, near the backward direction it is given predominantly by M1 scattering.

5.2. Neutrino and antineutrino scattering on electrons

As another example of a clean test of NC interactions we consider the following elastic reactions

$$\nu_{\mu}+e\to e+\nu_{\mu}, \tag{200a}$$

$$\bar{\nu}_\mu + \mathrm{e} \to \mathrm{e} + \bar{\nu}_\mu, \tag{200b}$$

$$\bar{\nu}_\mathrm{e} + \mathrm{e} \to \mathrm{e} + \bar{\nu}_\mathrm{e}. \tag{200c}$$

The first two of these are pure NC reactions. They are observable with the neutrino beams of high-energy accelerators. Reaction (200c) has contributions from both NC and CC interactions. It is observable in experiments at nuclear reactors which produce intense $\bar{\nu}_\mathrm{e}$ beams.

The kinematics is the same as in reactions (21) and (91) that we discussed in secs. 1.2.1 and 2.4, respectively. Denoting the neutrino momenta before and after the scattering by p and p', those of the electron by q and q', we have in the c.m. system

$$s = (p+q)^2 \simeq 4E^{*2}, \tag{201a}$$

$$t = (p-p')^2 = -2E^{*2}(1-z^*) \simeq -\tfrac{1}{2}s(1-z^*), \tag{201b}$$

$$u = (p-q')^2 = 2m_\mathrm{e}^2 - s - t \simeq -\tfrac{1}{2}s(1+z^*). \tag{201c}$$

Here E^* is the neutrino energy, z^* is the cosine of the scattering angle in the c.m. system; the $\simeq$ sign refers to our choosing E^* very much larger than m_e so that m_e can be set equal to zero in eqs. (201). In the laboratory system s is given by (cf. eq. (93′))

$$s = m_\mathrm{e}^2(1+2\omega), \quad \omega := E_\nu^{\mathrm{lab}}/m_\mathrm{e}. \tag{202}$$

So, for $E_\nu^{\mathrm{lab}} \gg m_\mathrm{e}$, $s \simeq 1.022 \times 10^{-3} \cdot E_\nu^{\mathrm{lab}}$ with E_ν^{lab} expressed in GeV. Of relevance is the comparison of s to the W and Z^0 masses:

$$\frac{s}{m_\mathrm{W}^2} \simeq (1.48 \times 10^{-7}\,\mathrm{GeV}^{-1})\,E_\nu^{\mathrm{lab}},$$

$$\frac{s}{m_\mathrm{Z}^2} \simeq (1.16 \times 10^{-7}\,\mathrm{GeV}^{-1})\,E_\nu^{\mathrm{lab}}.$$

The NC contributions to reactions (200) contain a denominator $(t - m_\mathrm{Z}^2)$ due to the Z^0 propagator, the CC contributions contain a denominator $(s - m_\mathrm{W}^2)$ due to the W-propagator. In either case, s and t are very small as compared to m_W^2 and to m_Z^2, for neutrino energies of the order of 10^2–10^3 GeV. Therefore, it is a very good approximation to use the effective contact interaction (186) instead of the full Lagrangian (182).

The differential cross section in the c.m. system is given by the same expression as for reaction (91), viz.

$$\frac{\mathrm{d}\sigma}{\mathrm{d}\Omega^*} = \frac{1}{32 s 4\pi^2}(2\pi)^{12}\sum|T|^2. \tag{203}$$

In calculating reactions (200a) and (200b) let us write the effective contact interaction in the general form

$$-\mathscr{L}_{\mathrm{NC}}^{\mathrm{eff}} = -\frac{G}{2\sqrt{2}}\left(\overline{e(x)}\,(c_\mathrm{V}\gamma_\alpha - c_\mathrm{A}\gamma_\alpha\gamma_5)\,e(x)\right)\left(\overline{\nu_\mu(x)}(\gamma^\alpha - \lambda\gamma^\alpha\gamma_5)\,\nu_\mu(x)\right). \tag{204}$$

In the case of the GSW model, this is identical to the expression (85), so that we must identify the parameters of eq. (204) as follows:

$$c_V = 1 - 4\sin^2\theta_w, \tag{205a}$$

$$c_A = 1, \tag{205b}$$

$$\lambda = 1. \tag{205c}$$

Furthermore, let us assume that the incident ν_μ in reaction (200a) carries helicity h, the incident $\bar{\nu}_\mu$ in (200b) carries helicity $\bar{h}$. In laboratory experiments these neutrinos stem primarily from pion and kaon decays, $\pi \to \mu\nu_\mu$, $K \to \mu\nu_\mu$. As we saw in sec. 1.2.3, (i), these helicities are known to be maximal within about 10%, so that one may assume

$$h = -1 \quad \text{and} \quad \bar{h} = +1. \tag{206}$$

The T-matrix element and the traces in $\Sigma|T|^2$ are worked out along the same lines as in the example of sec. 2.4, except that the incident ν_μ ($\bar{\nu}_\mu$) is polarized along its momentum with polarization h ($\bar{h}$). In calculating the traces this means that we must set

$$u_\nu(p)\overline{u_\nu(p)} = \tfrac{1}{2}(1 + h\gamma_5)\not{p},$$

$$v_\nu(p)\overline{v_\nu(p)} = \tfrac{1}{2}(1 - \bar{h}\gamma_5)\not{p}.$$

These expressions are obtained from eqs. (III. 126) and (III. 127) with $\boldsymbol{n} = h\hat{\boldsymbol{p}}$ (or $\bar{h}\hat{\boldsymbol{p}}$), in the limit $m \to 0$, or, equivalently, from eq. (III. 138), with the identifications $h \equiv 2\lambda$, $\bar{h} = -2\lambda$, respectively.

The differential cross section for reaction (200a) in the c.m. system is found to be

$$\frac{d\sigma}{d\Omega^*}(\nu_\mu e \to e\nu_\mu) = \frac{G^2}{64s \cdot 4\pi^2}\{(s^2 + u^2)(1 + |\lambda|^2 - 2h\,\mathrm{Re}\,\lambda)(|c_V|^2 + |c_A|^2)$$
$$- 2(s^2 - u^2)(h + h|\lambda|^2 - 2\,\mathrm{Re}\,\lambda)\mathrm{Re}(c_V c_A^*)\}. \tag{207}$$

The analogous cross section for reaction (200b) is obtained from eq. (207) by interchanging s and u, and by replacing h with $-\bar{h}$. From these results it is now easy to calculate the integrated elastic cross section. Integrating over $d\Omega^*$ and using eq. (201c) one finds

$$\int (s^2 + u^2)\, d\Omega^* = \frac{16\pi}{3}s^2, \qquad \int (s^2 - u^2)\, d\Omega^* = \frac{8\pi}{3}s^2,$$

so that

$$\sigma(\nu_\mu e) = \frac{G^2 s}{12\pi}\frac{1}{4}\{(1 + |\lambda|^2 - 2h\,\mathrm{Re}\,\lambda)(|c_V|^2 + |c_A|^2)$$
$$- (h + h|\lambda|^2 - 2\,\mathrm{Re}\,\lambda)\mathrm{Re}(c_V c_A^*)\}, \tag{208a}$$

$$\sigma(\bar{\nu}_\mu e) = \frac{G^2 s}{12\pi}\frac{1}{4}\{(1 + |\lambda|^2 + 2\bar{h}\,\mathrm{Re}\,\lambda)(|c_V|^2 + |c_A|^2)$$
$$- (\bar{h} + \bar{h}|\lambda|^2 + 2\,\mathrm{Re}\,\lambda)\mathrm{Re}(c_V c_A^*)\}. \tag{208b}$$

Let us now make the following assumptions:
(i) the incident neutrino ν_μ is fully left-handed, so that $h = -1$ and, correspondingly, $\bar{h} = +1$,
(ii) c_V and c_A are both real.
With these assumptions eqs. (208) simplify to the following expressions:

$$\sigma\begin{pmatrix}\nu_\mu e \\ \overline{\nu}_\mu e\end{pmatrix} = \frac{G^2 s}{12\pi}\frac{|1+\lambda|^2}{4}\left\{c_V^2 + c_A^2 \pm c_V c_A\right\}. \tag{209}$$

The integrated cross sections (209) show the increase with the square of the neutrino energy in the c.m. system (or with the first power of its laboratory energy), which is typical for the contact interaction. (As we noted in sec. 2.4 this increase cannot hold at arbitrarily large energies.)

Suppose that both $\sigma(\nu_\mu e)$ and $\sigma(\overline{\nu}_\mu e)$ are measured and that we know that $\lambda = 1$. We may then use eq. (209) to determine c_V and c_A: Writing $c_V = (1/\sqrt{2})(x+y)$ and $c_A = (1/\sqrt{2})(x-y)$ we see that eq. (209) defines two ellipses whose symmetry axes are rotated by $\pi/4$ with respect to a coordinate frame (c_V, c_A), as shown in Fig. 7. If the two ellipses intersect, c_V and c_A are determined up to a fourfold ambiguity. To this we add the information which comes from a measurement of reaction (200c). The integrated cross section $\sigma(\overline{\nu}_e e)$ is given by an expression analogous to eq. (208b) supplemented by the contribution due to the CC interactions.

Assuming $\bar{h}(\overline{\nu}_e) = +1$, $\lambda = 1$ and assuming that CC couplings are precisely the ones of eq. (74), one finds after some calculation

$$\frac{d\sigma}{d\Omega^*}(\overline{\nu}_e e \to e\overline{\nu}_e) = \frac{G^2}{64 s\pi^2}\left\{s^2(c_V - c_A)^2 + u^2\left((c_V + c_A)^2 + 8(2 - c_V - c_A)\right)\right\}.$$

Integration over $d\Omega^*$ yields the following expression for the total elastic cross

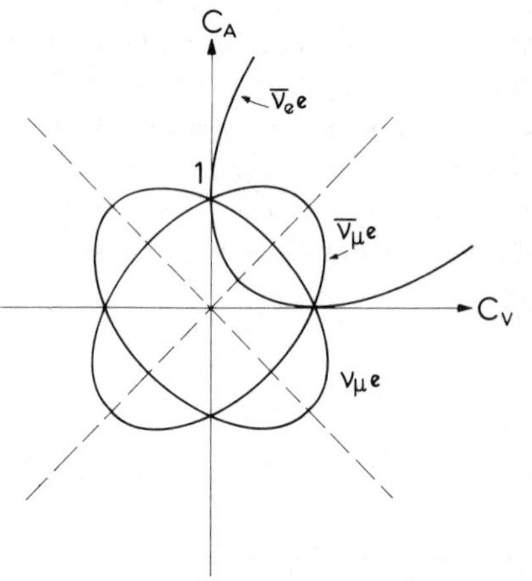

Fig. 7. The elastic neutrino reactions (200) determine three ellipses in the plane of the effective coupling constants c_V and c_A, eq. (204).

section:

$$\sigma(\bar{\nu}_e e) = \frac{G^2 s}{12\pi}\{c_V^2 + c_A^2 - c_V c_A - 2c_V - 2c_A + 4\}. \tag{210}$$

Expressed in terms of the neutrino energy in the laboratory system we have $s = 2m_e E_\nu^{\text{lab}}$. The cross sections (209) have the order of magnitude

$$\sigma_0 = \frac{2m_e G^2}{12\pi} E_\nu^{\text{lab}} = 1.44 \times 10^{-42}\ \text{cm}^2\ \text{GeV}^{-1}\ E_\nu^{\text{lab}}.$$

Suppose that we had found, in these units,

$$\sigma(\nu_\mu e)/\sigma_0 \simeq \sigma(\bar{\nu}_\mu e)/\sigma_0 \simeq 1,$$

$$\sigma(\bar{\nu}_e e)/\sigma_0 \simeq 3.$$

The three ellipses defined by eqs. (209) and (210) would then be the ones shown in Fig. 7. They intersect in the points ($c_V = 0, c_A = 1$) and ($c_V = 1, c_A = 0$), the first of which would be compatible with eqs. (205a) and (205b) and would imply $\sin^2\theta_w = \frac{1}{4}$.

All three reactions have indeed been measured in the laboratory, with error bars of the order of 30 to 50%. The results are found to be consistent with the predictions of the GSW theory but seem to favour a somewhat larger value of $\sin^2\theta_w$. Note that for $\sin^2\theta_w = \frac{1}{4}$, the two cross sections (209) are exactly equal. For $\sin^2\theta_w$ smaller (larger) than $\frac{1}{4}$, $\sigma(\nu_\mu e)$ is larger (smaller) than $\sigma(\bar{\nu}_\mu e)$. The comparison of these two cross sections measures $\sin^2\theta_w$ relative to the value $\frac{1}{4}$, at which they are equal.

One would think that relative phases of parity-even to parity-odd interactions can only be obtained from a spin-momentum correlation or some other pseudoscalar observable. Therefore, one might wonder why and how a measurement of total cross sections can determine the relative strength and relative sign of vector to axial vector NC couplings. The answer to this question can be derived from eqs. (208). If we restrict our considerations to NC couplings only, we may assume c_V to be real, without loss of generality, and take this constant out of the curly brackets in eqs. (208a) and (208b). In fact, by redefining c_A, c_V may be taken to be unity. If indeed, $h = -1$ and $\bar{h} = +1$ the cross sections are proportional to

$$\sigma\begin{pmatrix} \nu_\mu e \\ \bar{\nu}_\mu e \end{pmatrix} \propto |1+\lambda|^2\{1 + |c_A|^2 \pm \text{Re}(c_A)\}. \tag{211}$$

On the other hand, if the incident ν_μ and $\bar{\nu}_\mu$ were unpolarized, i.e. if the beam contained an equal amount of neutrinos of either helicity, we would obtain from eqs. (208)

$$\sigma\begin{pmatrix} \nu_\mu e \\ \bar{\nu}_\mu e \end{pmatrix}_{\text{unpol.}} \propto \{(1+|\lambda|^2)(1+|c_A|^2) \pm 2\,\text{Re}(\lambda)\text{Re}(c_A)\}. \tag{212}$$

Clearly, eq. (212) does not allow to determine $\text{Re}\, c_A$, unless λ is known and is different from zero. Therefore, it contains less information than eq. (211). (The two are the same if and only if $\lambda = 1$.) This analysis shows that in order to extract $\text{Re}\, c_A$, some information on neutrino helicities must be given: In the first case, eq. (211), we have assumed $h(\nu_\mu) = -1$, and therefore $h(\overline{\nu_\mu}) = +1$. In the second, the knowledge of $\lambda \neq 0$ implies that we know the longitudinal polarization of the outgoing neutrino. In either of these cases, we make use of input information on a *pseudoscalar* quantity, the neutrino helicity. (See also exercises 10 and 11.)

Finally, the assumption that both c_V and c_A are real was essential in determining both of them (up to a twofold ambiguity) from the three cross sections discussed above. [It is not sufficient that they only be relatively real because of reaction (200c) for whose calculation we need to know the phases of c_V, c_A relative to the CC couplings.]

5.3. *Angular asymmetry in* $e^+e^- \to \mu^+\mu^-$ *and* $e^+e^- \to \tau^+\tau^-$

In the two previous examples one could not discriminate between the full GSW theory (182), with large but finite masses of the gauge bosons, and the effective contact interaction (186) (corresponding to the former with $m_Z, m_W \to \infty$). This was so because these examples concerned reactions on fixed targets in which case a major fraction of the projectile energy resides in the kinetic energy in the center-of-mass and only a small fraction goes into the momentum variable that appears in the gauge boson propagators. The third example we choose is a colliding beam reaction,

$$e^+ + e^- \to F^+ + F^-, \quad F = \mu \text{ or } \tau, \tag{213}$$

in which the squared momentum of the gauge boson is equal to $s = 4E^{*2}$, i.e. where the virtual boson carries the full energy that is available in the reaction.

To lowest order in e, the amplitude for reaction (213) is given by the sum of the one-photon and the one-Z^0 diagrams of Fig. 8. The kinematics is as indicated in Fig. 8a. For simplicity, we consider the case where s is large as compared to both m_e^2 and m_F^2. (For the more general case with $m_F^2 \neq 0$ see exercise 12.) With

$$s = (p_+ + p_-)^2 = (q_+ + q_-)^2 \simeq 4k^{*2}, \tag{214a}$$

where k^* is the magnitude of the c.m. three-momentum, we can express t and u in

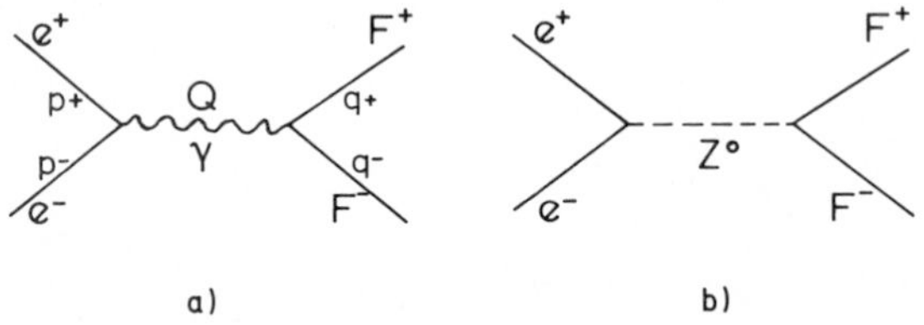

Fig. 8. One-photon and one Z^0-exchange in electron-positron annihilation and pair creation of muons or τ-leptons.

terms of s and of θ^*, the scattering angle in the c.m. system:

$$t=(p_- - q_-)^2 \simeq -\frac{s}{2}(1-\cos\theta^*), \tag{214b}$$

$$u=(p_- - q_+)^2 \simeq -\frac{s}{2}(1+\cos\theta^*). \tag{214c}$$

The incident, colliding beams are taken to be unpolarized. Therefore, as the spins are not discriminated, one might be tempted to conclude from the symmetry of Fig. 8 that the differential cross section is left invariant if we interchange q_+ and q_-. Interchanging the final state momenta means interchanging t and u, or from eqs. (214b) and (214c), means replacing θ^* by $\pi-\theta^*$. Thus, we would conclude that the cross section $\mathrm{d}\sigma/\mathrm{d}\Omega^*$ is a function only of $\cos^2\theta^*$ and is symmetric about $\theta^*=90°$. However, a closer examination of the diagrams, taking into account the helicity transfer at the two vertices, shows that this is not true in general.

We know from the general analysis in secs. 1.2.3 and 1.2.4 that vector and axial vector vertices connect the helicities of massless particles as indicated in Figs. 4a and 4b. Two examples of the helicity transfer at the photon and Z^0 vertices are shown in Figs. 9a and 9b. (Since the incident particles are unpolarized, we must add incoherently the contribution of these two amplitudes and of the ones with all four helicities reversed.) As can be seen from Fig. 9, the interchange of q^+ and q^- effectively means that diagram 9a goes over into diagram 9b, and vice versa. Therefore, the cross section is t–u symmetric only if the two amplitudes have the same relative weight. This happens if the sum of the diagrams, Fig. 9, contains either pure VV or pure AA couplings, but not both. Let us calculate this in more detail.

The differential cross section in the c.m. system (this is the reference system in the laboratory if the colliding beams have equal and opposite momenta), in the limit $s \gg m_e^2, m_F^2$, is given by

$$\frac{\mathrm{d}\sigma}{\mathrm{d}\Omega^*} \simeq \frac{1}{16s}(2\pi)^{10}\frac{1}{4}\sum|T|^2. \tag{215}$$

In writing down the T-matrix element for the one-Z^0 exchange we note that the term $Q_\alpha Q_\beta/m_Z^2$ in the propagator gives a negligibly small contribution: Q_α is equal to $(p_+ + p_-)_\alpha$ and to $(q_+ + q_-)_\alpha$ so that by virtue of the Dirac equations (III. 62) and (III. 63′)

$$\overline{v(p_+)}\,\gamma^\alpha u(p_-)Q_\alpha = 0,$$

$$\overline{v(p_+)}\,\gamma^\alpha\gamma_5 u(p_-)Q_\alpha = -2m_e\overline{v(p_+)}\,\gamma_5 u(p_-),$$

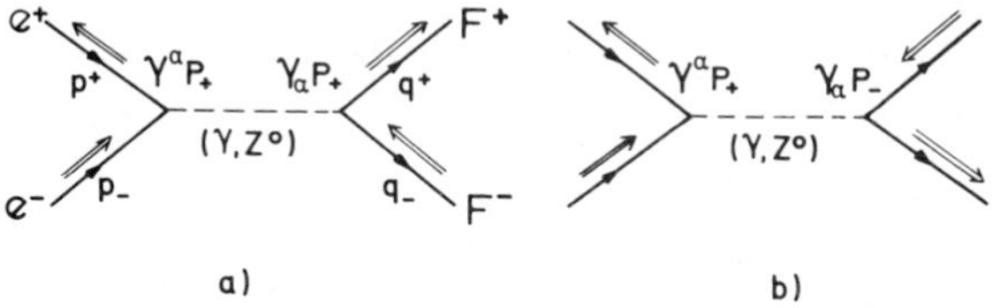

Fig. 9. Two examples for helicity transfers in reaction (213) as described by the diagrams of Fig. 8.

and analogously for Q_β contracted with the (F^+F^-) vertex. Thus, the axial couplings yield a negligible contribution of the order $m_e m_F/m_Z^2$, relative to the term $g_{\alpha\beta}$. With this simplification the amplitudes in Fig. 9 are

$$T_\gamma = \frac{1}{(2\pi)^6}\frac{e^2}{s}\left(\overline{v(p_+)}\,\gamma^\alpha u(p_-)\right)\left(\overline{u(q_-)}\,\gamma_\alpha v(q_+)\right), \tag{216}$$

$$\begin{aligned} T_Z \simeq & \frac{1}{(2\pi)^6}\frac{e^2}{16\sin^2\theta_w\cos^2\theta_w}\frac{1}{s-m_Z^2} \\ & \times\left(\overline{v(p_+)}\,\{-c_V\gamma^\alpha + c_A\gamma^\alpha\gamma_5\}\,u(p_-)\right) \\ & \times\left(\overline{u(q_-)}\,\{-c_V\gamma_\alpha + c_A\gamma_\alpha\gamma_5\}\,v(q_+)\right), \end{aligned} \tag{217}$$

where c_V and c_A are defined by eqs. (205a) and (205b). Using the relation $m_Z^2\cos^2\theta_w = m_W^2$ one can write the factor on the r.h.s of eq. (217) as

$$\frac{e^2}{s}\kappa(s) \quad \text{with } \kappa(s) := \frac{-1}{16 m_W^2\sin^2\theta_w}\frac{s}{1-s/m_Z^2}. \tag{218}$$

Decomposing the V and A vertex factors in terms of γ^α multiplied by helicity projection operators (32), that is

$$\gamma^\alpha = \gamma^\alpha P_+ + \gamma^\alpha P_-, \qquad \gamma^\alpha\gamma_5 = \gamma^\alpha P_+ - \gamma^\alpha P_-,$$

and introducing the short-hand notation

$$A_\pm^\alpha := \overline{v(p_+)}\,\gamma^\alpha P_\pm u(p_-),$$

$$B_\pm^\alpha := \overline{u(q_-)}\,\gamma^\alpha P_\pm v(q_+),$$

the sum of amplitudes (216) and (217) is

$$\begin{aligned} T_\gamma + T_Z = & \frac{1}{(2\pi)^6}\frac{e^2}{s}\Big\{(A_+\cdot B_+)\left[1+\kappa(s)(c_V-c_A)^2\right] \\ & + (A_-\cdot B_-)\left[1+\kappa(s)(c_V+c_A)^2\right] \\ & + \left[(A_+\cdot B_-)+(A_-\cdot B_+)\right]\left[1+\kappa(s)\left(c_V^2-c_A^2\right)\right]\Big\}. \end{aligned} \tag{219}$$

The calculation of $\sum_{\text{spins}}|T_\gamma + T_Z|^2$ is greatly simplified by observing that an amplitude that contains A_+ cannot interfere with an amplitude that contains A_-. Likewise B_+ cannot interfere with B_-. Therefore, in the sum over the spins all interference terms vanish. [Note, however, that this holds only in the limit $m_e = m_F = 0$. See exercise 12]. Furthermore, it is easy to see (from the explicit expression (221a) below) that

$$\sum|(A_+\cdot B_+)|^2 = \sum|(A_-\cdot B_-)|^2, \tag{220a}$$

$$\sum|(A_+\cdot B_-)|^2 = \sum|(A_-\cdot B_+)|^2. \tag{220b}$$

Finally, as the amplitude $(A_+ \cdot B_+)$ corresponds to diagram 9a, the amplitude $(A_+ \cdot B_-)$ to diagram 9b, the expression (220b) is obtained from (220a) by the transformation $t \leftrightarrow u$. Therefore, we only need to calculate the term (220a). Neglecting the masses we have

$$\begin{aligned}\sum|(A_+ \cdot B_+)|^2 &= \tfrac{1}{16}\mathrm{Sp}\{\gamma^\alpha(1+\gamma_5)\not p_-\gamma^\beta(1+\gamma_5)\not p_+\}\\ &\quad\times \mathrm{Sp}\{\gamma_\alpha(1+\gamma_5)\not q_+\gamma_\beta(1+\gamma_5)\not q_-\}\\ &= \tfrac{1}{4}\mathrm{Sp}\{(1-\gamma_5)\gamma^\alpha\not p_-\gamma^\beta\not p_+\}\mathrm{Sp}\{(1-\gamma_5)\gamma_\alpha\not q_+\gamma_\beta\not q_-\}\\ &= 4\{p_-^\alpha p_+^\beta - (p_-p_+)g^{\alpha\beta} + p_+^\alpha p_-^\beta - \mathrm{i}\varepsilon^{\alpha\mu\beta\nu}p_{-\mu}p_{+\nu}\}\\ &\quad\times\{q_{+\alpha}q_{-\beta} - (q_+q_-)g_{\alpha\beta} + q_{-\alpha}q_{+\beta} - \mathrm{i}\varepsilon_{\alpha\sigma\beta\tau}q_+^\sigma q_-^\tau\}\\ &= 16(p_+q_-)(p_-q_+) = 4u^2. \end{aligned} \tag{221a}$$

By the t–u symmetry noted above we conclude

$$\sum|(A_+ \cdot B_-)|^2 = 4t^2. \tag{221b}$$

With the results (221) and keeping in mind the remarks made above, we obtain

$$\begin{aligned}(2\pi)^{12}\sum|T_\gamma + T_Z|^2 = \frac{8e^2}{s^2}\Big\{&u^2\Big[\big(1+\kappa c_V^2\big)^2 + 2\kappa c_A^2 + \kappa^2 c_A^4 + 6\kappa^2 c_A^2 c_V^2\Big]\\ &+t^2\Big[\big(1+\kappa c_V^2\big)^2 - 2\kappa c_A^2 + \kappa^2 c_A^4 - 2\kappa^2 c_A^2 c_V^2\Big]\Big\}.\end{aligned}$$

Finally, inserting eqs. (214) and replacing e^2 by $4\pi\alpha$, the differential cross section (215) is

$$\begin{aligned}\frac{\mathrm{d}\sigma}{\mathrm{d}\Omega^*} \simeq \frac{\alpha^2}{4s}\Big\{&(1+\cos^2\theta^*)\Big[\big(1+\kappa c_V^2\big)^2 + \kappa^2 c_A^2\big(c_A^2 + 2c_V^2\big)\Big]\\ &+4\kappa c_A^2\big(1+2\kappa c_V^2\big)\cos\theta^*\Big\}.\end{aligned} \tag{222}$$

The first term in the curly brackets of eq. (222) is symmetric about the angle 90°, the second is not. The asymmetric term is present only if there is both a VV and an AA interaction in the sum of the diagrams in Fig. 8. The forward–backward asymmetry in the cross section

$$A := \left\{\frac{\mathrm{d}\sigma}{\mathrm{d}\Omega^*}(0) - \frac{\mathrm{d}\sigma}{\mathrm{d}\Omega^*}(\pi)\right\}\Big/\left\{\frac{\mathrm{d}\sigma}{\mathrm{d}\Omega^*}(0) + \frac{\mathrm{d}\sigma}{\mathrm{d}\Omega^*}(\pi)\right\} \tag{223}$$

is found to be

$$A = 2\kappa(s)c_A^2\frac{1+2\kappa c_V^2}{\big(1+\kappa c_V^2\big)^2 + \kappa^2 c_A^2\big(c_A^2 + 2c_V^2\big)} \tag{224}$$

with $\kappa(s)$ as defined in eq. (218).

Let us now estimate the magnitude of the asymmetry as it is predicted from the GSW theory. With $\sin^2\theta_W \simeq 0.22$, $m_W^2\sin^2\theta_W$ as given in eq. (181), and $m_Z^2 \simeq 94$ GeV, we have

$$c_V^2 = 0.0144, \qquad c_A = 1,$$

$$\kappa(s) = -4.22 \times 10^{-5}\ \mathrm{GeV}^{-2} \frac{s}{1 - s/m_Z^2}.$$

At an energy of 18 GeV per beam, $\sqrt{s} = 36$ GeV, the propagator effect is about 17%,

$$\frac{1}{1 - s/m_Z^2} \simeq 1.17 \tag{225}$$

and $\kappa(s) = -0.064$. At this energy both $\kappa^2 c_A^2$ and κc_V^2 are very small so that

$$A(\sqrt{s} = 36\ \mathrm{GeV}) \simeq 2\kappa = -12.8\%.$$

Asymmetries of this sign and magnitude have indeed been found in experiments at DESY. There are additional contributions to the asymmetry (223) from two-photon exchange and from radiative corrections which must be calculated before a quantitative comparison with experiment can be made. Also, as s increases and approaches m_Z^2, the finite width of the Z^0 must be taken into account in the propagator (cf. exercise 14).

The numerical example given above shows that already at $\sqrt{s} = 36$ GeV there is a marked difference between the prediction of the GSW theory with finite m_Z and the effective contact interaction ($m_Z \to \infty$), see eq. (225). At $\sqrt{s} = 50$ GeV the analogous number would be 1.39, at $\sqrt{s} = 66$ GeV it would be about 2.

6. Beyond the minimal model

The GSW theory is a great step forward in our understanding of electroweak interactions because it allows casting the well-known extremely successful theory of quantized electrodynamics and the theory of the weak CC and NC interactions into one unified, renormalizable local gauge theory. Renormalizability, in particular, is a very desirable property of the theory because it makes covariant perturbation theory a reasonable and well-defined approximation method for calculating physical quantities beyond the lowest order diagrams. Nevertheless, as mentioned in the introduction to this Chapter, this model, very likely, is not the coping-stone of a final theory of weak and electromagnetic interactions. It contains very many parameters which are not predicted and whose origin remains unclear. The most prominent and specific properties of the weak interactions are built into the model (see e.g. the discussion in secs. 4.1.1 and 4.1.2) and are not predicted. One of these is parity violation: The fact that QED conserves parity but (bare) CC interactions as well as neutrino induced NC interactions break parity maximally, is introduced into the theory by hand. There is not even a hint at an answer to the question why right-handed neutrino states decouple from the physical world. Furthermore, the accuracy to which some of the empirical information on weak interactions is known, is only limited, as we showed in sec. 1.2, and there is indeed room for sizeable deviations from this minimal picture.

Encouraged by the success of the basic concepts of gauge theories, many authors have proposed enlarged unified models of the elementary interactions which contain the GSW model as a limiting case but exhibit specific deviations from its predictions. Whilst some of their features, such as the existence of further massive weak gauge bosons, can be tested through direct search at future high energy accelerators, others can be established (or disproved) by experiments of high precision at low and intermediate energies. Without going into these generalized models, we provide here the general frame of testing for specific deviations from the simple V – A picture of CC interactions and the minimal neutral interactions of the GSW model. We give some typical examples referring to muon decay and NC interactions in atomic systems.

6.1. *Effective SPVAT interactions*

As we saw earlier, weak interactions have very short ranges, of the order of 10^{-16} cm. For the application to processes at low and intermediate energies this means that the interaction effectively acts like a contact interaction that connects four fermion fields at the same point x in Minkowski space. In sec. 2.2 we discussed the example of effective V and A interactions, cf. eq. (64) and studied their behavior under the discrete symmetries P, C, and T, in some detail. In this section we extend these considerations to the case of scalar, pseudoscalar and tensor interactions which are the only other Lorentz structures that can be formed on the basis of spinor fields.

6.1.1. *Scalars and pseudoscalars*

In analogy to the definitions (59) and (60) let us introduce the following covariants

$$s(x; i \to k) := \overline{\Psi^{(k)}(x)}\Psi^{(i)}(x), \tag{226a}$$

$$p(x; i \to k) := \overline{\Psi^{(k)}(x)}\gamma_5\Psi^{(i)}(x). \tag{226b}$$

When written in terms of spinors of first and second kind, the operators (226) have the structure

$$\left.\begin{matrix} s \\ p \end{matrix}\right\} = \chi^{(k)*a}(x)\phi^{(i)}{}_a(x) \pm \phi^{(k)*}{}_A(x)\chi^{(i)A}(x). \tag{227}$$

With the aid of eqs. (50) it is easy to show that s is a scalar under Lorentz transformations,

$$s(x; i \to k) \underset{P}{\to} s(Px; i \to k), \tag{228a}$$

whilst p is a pseudoscalar,

$$p(x; i \to k) \underset{P}{\to} -p(Px; i \to k). \tag{228b}$$

The transformation of s and p under charge conjugation is derived in the same way as for eqs. (55). One finds

$$s(x; i \to k) \underset{C}{\to} s(x; k \to i), \tag{229a}$$

$$p(x; i \to k) \underset{C}{\to} p(x; k \to i). \tag{229b}$$

The behaviour of s and p under time reversal is derived along the same lines as for the case of vector and axial vector covariants, cf. sec. 2.1. For example,

$$\chi^{(k)*a}(x)\phi^{(i)}{}_a(x) \underset{T}{\to} \varepsilon_{BD}(\hat{\sigma}^0)^{Da}(\sigma^0)_{aF}\phi^{(i)*F}(Tx)\chi^{(k)B}(Tx)$$
$$= \phi^{(i)*}{}_B(Tx)\chi^{(k)B}(Tx),$$

so that

$$s(x; i \to k) \underset{T}{\to} s(Tx; k \to i), \tag{230a}$$

$$p(x; i \to k) \underset{T}{\to} -p(Tx; k \to i). \tag{230b}$$

The transformation properties (228), (229), and (230) hold up to the phase factors discussed in sec. 2.1. We can now formulate the most general Lagrangian built from scalar and pseudoscalar covariants, which is invariant under proper orthochronous Lorentz transformations. Dropping the common argument x in the covariants it reads

$$\begin{aligned} -\mathscr{L}_{SP} = \frac{G}{\sqrt{2}} \{ & s(i \to k)[C_S s(n \to m) + C'_S p(n \to m)] \\ & + [C^*_S s(m \to n) - C'^*_S p(m \to n)] s(k \to i) \\ & - p(i \to k)[C_P p(n \to m) + C'_P s(n \to m)] \\ & - [C^*_P p(m \to n) - C'^*_P s(m \to n)] p(k \to i) \}. \end{aligned} \tag{231}$$

For the sake of convenience, we have taken out a factor $G/\sqrt{2}$ such as to have the constants $C_S, \ldots, C'_P$ dimensionless and to make (231) directly comparable to $\mathscr{L}_{VA}$, eq. (64). The minus sign in front of C'^*_S and of C'^*_P is due to hermitean conjugation,

$$\left(\overline{\Psi^{(k)}}\gamma_5\Psi^{(i)}\right)^\dagger = \overline{\Psi^{(i)}}\gamma^0\gamma_5\gamma^0\Psi^{(k)} = -\overline{\Psi^{(i)}}\gamma_5\Psi^{(k)}$$

whereas the minus sign of the pseudoscalar terms (third and fourth term on the r.h.s) is a matter of convention: It is useful (but not generally adopted) to define pseudoscalar invariants with an extra factor i,

$$\overline{\Psi^{(k)}(x)}\Gamma_P\Psi^{(i)}(x) \quad \text{with } \Gamma_P := \mathrm{i}\gamma_5. \tag{232}$$

The advantage of this choice is evident in calculating traces because

$$\gamma^0\Gamma^\dagger_P\gamma^0 = \Gamma_P$$

and one need not worry about extra signs. Since we shall use this convention in the sequel but do not wish to redefine C_P and C'_P then, we introduce an i^2 already at this point.

It is easy to verify that the behaviour of $\mathscr{L}_{SP}$ with respect to the discrete symmetries P, C, and T, is precisely as indicated in eqs. (66)–(68). In particular, $\mathscr{L}_{SP}$ obeys the PCT-theorem in the same form as $\mathscr{L}_{VA}$, eq. (70).

6.1.2. Tensors and pseudotensors

A third type of covariants that can be defined by forming bilinears of two fermion fields are the following

$$t^{\alpha\beta}(x; i \to k) := \overline{\Psi^{(k)}(x)} \frac{1}{\sqrt{2}} \sigma^{\alpha\beta} \Psi^{(i)}(x), \tag{233a}$$

$$t'^{\alpha\beta}(x; i \to k) := \overline{\Psi^{(k)}(x)} \frac{1}{\sqrt{2}} \sigma^{\alpha\beta} \gamma_5 \Psi^{(i)}(x), \tag{233b}$$

where $\sigma^{\alpha\beta}$ stands for an antisymmetric product of γ-matrices which is generally defined by

$$\sigma^{\alpha\beta} := \frac{i}{2}(\gamma^\alpha\gamma^\beta - \gamma^\beta\gamma^\alpha). \tag{234}$$

[The extra factor $1/\sqrt{2}$ in eqs. (233) is a matter of convention, see below.]

The operator $t^{\alpha\beta}$ transforms like a Lorentz tensor, $t'^{\alpha\beta}$ transforms like a Lorentz pseudotensor, i.e. like a tensor times the determinant of the homogeneous Lorentz transformation. This can be seen most easily if we write these operators in terms of spinor fields ϕ and χ. For this purpose we decompose them as follows

$$t^{\alpha\beta} = \frac{1}{\sqrt{2}}(f^{\alpha\beta} + g^{\alpha\beta}), \qquad t'^{\alpha\beta} = \frac{1}{\sqrt{2}}(f^{\alpha\beta} - g^{\alpha\beta}),$$

with

$$f^{\alpha\beta} := \chi^{(k)*a} \frac{i}{2}\{(\sigma^\alpha)_{aB}(\hat{\sigma}^\beta)^{Bd} - (\sigma^\beta)_{aB}(\hat{\sigma}^\alpha)^{Bd}\}\phi_d^{(i)}, \tag{235a}$$

$$g^{\alpha\beta} := \phi^{(k)*}{}_A \frac{i}{2}\{(\hat{\sigma}^\alpha)^{Ab}(\sigma^\beta)_{bD} - (\hat{\sigma}^\beta)^{Ab}(\sigma^\alpha)_{bD}\}\chi^{(i)D}. \tag{235b}$$

Let us calculate the divergence ∂_β of these latter quantities. Making use of the relation (III.42) and noting that $f^{\alpha\beta}$ vanishes if $\alpha = \beta$, one obtains

$$\partial_\beta f^{\alpha\beta} = \frac{i}{2}\chi^{(k)*}\{2\sigma^\alpha\hat{\sigma}^\beta - 2g^{\alpha\beta}\}\partial_\beta\phi^{(i)} - \frac{i}{2}(\partial_\beta\chi^*)\{2\sigma^\beta\hat{\sigma}^\alpha - 2g^{\alpha\beta}\}\phi^{(i)}.$$

This expression can be simplified by means of the Dirac equations (III.47), giving

$$\partial_\beta f^{\alpha\beta} = m_i\chi^{(k)*}\sigma^\alpha\chi^{(i)} + m_k\phi^{(k)*}\hat{\sigma}^\alpha\phi^{(i)} - i\chi^{(k)*}\overleftrightarrow{\partial}^\alpha\phi^{(i)}.$$

In a similar fashion one shows

$$\partial_\beta g^{\alpha\beta} = m_i\phi^{(k)*}\hat{\sigma}^\alpha\phi^{(i)} + m_k\chi^{(k)*}\sigma^\alpha\chi^{(i)} - i\phi^{(k)*}\overleftrightarrow{\partial}^\alpha\chi^{(i)}.$$

From these equations we obtain

$$\partial_\beta t^{\alpha\beta} = \frac{1}{\sqrt{2}}(m_k + m_i)\{\chi^{(k)*}\sigma^\alpha\chi^{(i)} + \phi^{(k)*}\hat{\sigma}^\alpha\phi^{(i)}\} - \frac{i}{\sqrt{2}}\{\chi^{(k)*}\overleftrightarrow{\partial}^\alpha\phi^{(i)} + \phi^{(k)*}\overleftrightarrow{\partial}^\alpha\chi^{(i)}\}, \tag{236a}$$

$$\partial_\beta t'^{\alpha\beta} = \frac{1}{\sqrt{2}}(m_k - m_i)\{-\chi^{(k)*}\sigma^\alpha\chi^{(i)} + \phi^{(k)*}\hat{\sigma}^\alpha\phi^{(i)}\} - \frac{i}{\sqrt{2}}\{\chi^{(k)*}\overleftrightarrow{\partial}^\alpha\phi^{(i)} - \phi^{(k)*}\overleftrightarrow{\partial}^\alpha\chi^{(i)}\}. \tag{236b}$$

Looking back at our discussion of vector and axial vector operators in sec. 2.1 the behaviour of $t^{\alpha\beta}$ and $t'^{\alpha\beta}$ under (proper and improper) Lorentz transformations is now obvious,

$$t^{\alpha\beta}(\Lambda x)=\Lambda^{\alpha}{}_{\sigma}\Lambda^{\beta}{}_{\tau}t^{\sigma\tau}(x), \tag{237a}$$

$$t'^{\alpha\beta}(\Lambda x)=(\det\Lambda)\Lambda^{\alpha}{}_{\sigma}\Lambda^{\beta}{}_{\tau}t'^{\sigma\tau}(x). \tag{237b}$$

The divergences (236) contain the vector and the axial-vector operators (59) and (60), respectively. This fact may be utilized to derive the behaviour of t and t' under charge conjugation. With respect to C, $v^{\alpha}(x; i\to k)$ is odd. Therefore, from eq. (236a), $t^{\alpha\beta}$ is also odd under C. The axial vector $a^{\alpha}(x; i\to k)$, which appears on the r.h.s. of eq. (236b), is even. However, C also interchanges i and k and yields an extra minus sign from the antisymmetric factor in eq. (236b). Thus $t'^{\alpha\beta}$ is odd, too, and we have

$$t^{\alpha\beta}(x; i\to k)\underset{C}{\to}-t^{\alpha\beta}(x; k\to i), \tag{238a}$$

$$t'^{\alpha\beta}(x; i\to k)\underset{C}{\to}-t'^{\alpha\beta}(x; i\to k). \tag{238b}$$

The behaviour with respect to T, finally, is contained in eqs. (237).

A rather useful relation in calculating matrix elements of tensor and pseudotensor covariants is the following

$$\sigma^{\alpha\beta}\gamma_5=-\frac{\mathrm{i}}{2}\varepsilon^{\alpha\beta\mu\nu}\sigma_{\mu\nu}, \tag{239}$$

where $\varepsilon_{\alpha\beta\mu\nu}$ is the totally antisymmetric tensor in four dimensions, with $\varepsilon_{0123}=+1$.

As for the case of V, A and S, P covariants, the general Lagrangian which is invariant under proper orthochronous Lorentz transformations has the form

$$\begin{aligned}-\mathscr{L}_{\mathrm{T}}=\frac{G}{\sqrt{2}}\Big\{&t^{\alpha\beta}(i\to k)\big[C_{\mathrm{T}}t_{\alpha\beta}(n\to m)+C'_{\mathrm{T}}t'_{\alpha\beta}(n\to m)\big]\\&+\big[C^{*}_{\mathrm{T}}t_{\alpha\beta}(m\to n)-C'^{*}_{\mathrm{T}}t'_{\alpha\beta}(m\to n)\big]t^{\alpha\beta}(k\to i)\Big\}.\end{aligned} \tag{240}$$

In writing the hermitean conjugate of the first term on the r.h.s. of eq. (240) we have made use of the equations

$$\gamma^0(\sigma^{\alpha\beta})^{\dagger}\gamma_0=\sigma^{\alpha\beta},\qquad \gamma^0(\sigma^{\alpha\beta}\gamma_5)^{\dagger}\gamma^0=-\sigma^{\alpha\beta}\gamma_5.$$

Here too, one verifies that the behaviour of $\mathscr{L}_{\mathrm{T}}$ under the discrete symmetries is the one of eqs. (66)–(68), and that $\mathscr{L}_{\mathrm{T}}$ obeys the *PCT*-theorem, eq. (70).

6.1.3. *General four-fermion contact interaction*

With the tools of secs. 2.2, 6.1.1, and 6.1.2 at hand we can now formulate the general, effective interaction Lagrangian connecting four-fermion field operators $\Psi^{(\mathrm{f}_1)}(x)$ to $\Psi^{(\mathrm{f}_4)}(x)$ at the same point of space–time. For this purpose let us introduce the following, somewhat symbolic, notation.

$$\Gamma_{\mathrm{S}} \equiv \mathbb{1}, \qquad \Gamma_{\mathrm{P}} \equiv \mathrm{i}\gamma_5, \tag{241}$$

$$\Gamma_{\mathrm{V}} \equiv \gamma^\alpha, \qquad \Gamma_{\mathrm{A}} \equiv \gamma^\alpha\gamma_5, \tag{242}$$

$$\Gamma_{\mathrm{T}} \equiv \frac{1}{\sqrt{2}}\sigma^{\alpha\beta}, \tag{243}$$

with $\sigma^{\alpha\beta}$ as defined by eq. (234) above. Note, in particular, the factor i in our definition of Γ_{P}. This operator, like all others in eqs. (241)–(243), has the property

$$\gamma^0(\Gamma_i)^\dagger\gamma^0 = \Gamma_i, \quad i = \mathrm{S, P, V, A, T}, \tag{244a}$$

which is particularly convenient in practical calculations. Besides these operators, also the products of $\Gamma_i\gamma_5$ will appear in the Lagrangian. The analogous relation of conjugation for these products is easily derived from eq. (244a), viz.

$$\gamma^0(\Gamma_i\gamma_5)^\dagger\gamma^0 = (\gamma^0\gamma_5\gamma^0)(\gamma^0\Gamma_i^\dagger\gamma^0) = -\gamma_5\Gamma_i.$$

Depending on whether γ_5 commutes or anticommutes with Γ_i, this is equal to minus or plus $\Gamma_i\gamma_5$. Thus

$$\gamma^0(\Gamma_i\gamma_5)^\dagger\gamma^0 = \Gamma_i\gamma_5, \quad \text{for } i = \mathrm{V, A}, \tag{244b}$$

$$\gamma^0(\Gamma_k\gamma_5)^\dagger\gamma^0 = -\Gamma_k\gamma_5, \quad \text{for } k = \mathrm{S, P, T}. \tag{244c}$$

Denoting the field operators by the symbol of the particles that they describe, the most general effective Lagrangian (which does not contain derivative couplings) reads

$$-\mathscr{L} = \frac{G}{\sqrt{2}}\sum_i\left\{\left(\overline{f_1(x)}\,\Gamma_i f_2(x)\right)\left[C_i\left(\overline{f_3(x)}\,\Gamma^i f_4(x)\right) + C_i'\left(\overline{f_3(x)}\,\Gamma^i\gamma_5 f_4(x)\right)\right] + \mathrm{h.c.}\right\}. \tag{245}$$

It is understood that the fields $f_1, \ldots, f_4$ are selected such that $\mathscr{L}$ is electrically neutral and is a scalar with respect to all internal symmetries for which one wishes to impose a conservation law.

The Lagrangian (245) contains ten complex quantities, two for each Lorentz structure, i.e. twenty real constants. Only nineteen of these are physically relevant because an unobservable common phase can always be factored out. If all of these constants are different from zero then $\mathscr{L}$ neither conserves parity, nor charge conjugation, nor time reversal symmetry. If one of these discrete symmetries is conserved the number of coupling constants is reduced by a factor of two, as can be seen from eqs. (66)–(68). For instance, if we impose T-invariance, eq. (68) shows that all C_i and C_i' must be real. From the PCT-theorem (70) we then see that either P and C are both violated, in which case at least some unprimed and some primed coupling constants are different from zero, or both are conserved in which case there can be either unprimed or primed couplings, but not both.

6.1.4. Fierz transformations

Of the Lorentz structures S, ..., T that we considered above, the tensor is somewhat exotic but the scalar and pseudo-scalar are not. Indeed, if all fundamental interactions (except gravitation) are mediated by particles with spin 1 there cannot be an effective interaction of the type (240). The vector particles couple to V and A currents. So where do S and P covariants come in?

In order to see this, let us return to the general contact interaction (245). Once we have ascertained that the theory fulfills the proper conservation laws by suitably combining the quantum numbers of particles f_1 to f_4, we can write the field operators in the ordering

$$(\bar{f}_1 \cdots f_4)(\bar{f}_3 \cdots f_2)$$

as well, without changing any of the invariances of the Lagrangian. However, the specific linear combination of covariants in eq. (245) is not the same combination when written in the reordered form.

Let us write the *same* Lagrangian (245) in a form where the field operators $f_2(x)$ and $f_4(x)$ are interchanged, viz.

$$-\mathscr{L}=\frac{G}{\sqrt{2}}\sum_i\left\{(\bar{f}_1\Gamma_i f_4)\left[D_i(\bar{f}_3\Gamma^i f_2)+D_i'(\bar{f}_3\Gamma^i\gamma_5 f_2)\right]+\text{h.c.}\right\}. \tag{246}$$

This reordering of operators (first studied by M. Fierz, 1936) maps the constants $\{C_i, C_i'\}$ onto the constants $\{D_i, D_i'\}$ according to the linear substitutions

$$\begin{pmatrix} D_\text{S}\\ D_\text{P}\\ D_\text{V}\\ D_\text{A}\\ D_\text{T}\end{pmatrix}=\frac{1}{4}\begin{pmatrix} -1 & 1 & -4 & 4 & -6\\ 1 & -1 & -4 & 4 & 6\\ -1 & -1 & 2 & 2 & 0\\ 1 & 1 & 2 & 2 & 0\\ -1 & 1 & 0 & 0 & 2\end{pmatrix}\begin{pmatrix} C_\text{S}\\ C_\text{P}\\ C_\text{V}\\ C_\text{A}\\ C_\text{T}\end{pmatrix}, \tag{247}$$

$$\begin{pmatrix} D_\text{S}'\\ D_\text{P}'\\ D_\text{V}'\\ D_\text{A}'\\ D_\text{T}'\end{pmatrix}=\frac{1}{4}\begin{pmatrix} -1 & 1 & 4 & -4 & -6\\ 1 & -1 & 4 & -4 & 6\\ 1 & 1 & 2 & 2 & 0\\ -1 & -1 & 2 & 2 & 0\\ -1 & 1 & 0 & 0 & 2\end{pmatrix}\begin{pmatrix} C_\text{S}'\\ C_\text{P}'\\ C_\text{V}'\\ C_\text{A}'\\ C_\text{T}'\end{pmatrix}. \tag{248}$$

[These transformations contain an extra minus sign due to the interchange of the two anticommuting fermion fields f_2 and f_4.] In a short-hand notation we shall also write these as

$$D_i=\sum_k M_{ik}C_k, \qquad D_i'=\sum_k F_{ik}C_k'. \tag{249}$$

The matrices M and F have a number of important properties:

(i) The Fierz transformation from (245) to (246), of course, is the same as from (246) to (245). Therefore, M and F are equal to their own inverse, respectively,

$$M^2=\mathbb{1}, \qquad F^2=\mathbb{1}.$$

(ii) It is remarkable that there is no matrix element connecting V, A and T. Thus, if there are only V and A couplings in the Lagrangian (245), i.e. $C_i = 0$, $C_i' = 0$ for $i = \mathrm{S, P, T}$, then (246) contains VA as well as SP couplings but no tensor couplings, viz.

$$\begin{aligned} D_{\mathrm{S}} &= D_{\mathrm{P}} = -C_{\mathrm{V}} + C_{\mathrm{A}}, \\ D_{\mathrm{V}} &= D_{\mathrm{A}} = \tfrac{1}{2}(C_{\mathrm{V}} + C_{\mathrm{A}}), \\ D_{\mathrm{S}}' &= D_{\mathrm{P}}' = C_{\mathrm{V}}' - C_{\mathrm{A}}', \\ D_{\mathrm{V}}' &= D_{\mathrm{A}}' = \tfrac{1}{2}(C_{\mathrm{V}}' + C_{\mathrm{A}}'), \\ D_{\mathrm{T}} &= D_{\mathrm{T}}' = 0. \end{aligned} \tag{250}$$

(iii) In particular, if in the representation (245) the interaction is of the form "V ± A", by which we mean

$$C_0\big(\bar{f}_1\gamma^\alpha(1 \pm \gamma_5)f_2\big)\big(\bar{f}_3\gamma_\alpha(1 \pm \gamma_5)f_4\big), \tag{251a}$$

then

$$C_{\mathrm{V}} = C_{\mathrm{A}} = C_0, \qquad C_{\mathrm{V}}' = C_{\mathrm{A}}' = \pm C_0,$$

and, from eqs. (250),

$$\begin{aligned} &D_{\mathrm{S}} = D_{\mathrm{P}} = 0, \qquad D_{\mathrm{S}}' = D_{\mathrm{P}}' = 0, \qquad D_{\mathrm{T}} = D_{\mathrm{T}}' = 0, \\ &D_{\mathrm{V}} = D_{\mathrm{A}} = C_0, \qquad D_{\mathrm{V}}' = D_{\mathrm{A}}' = \pm C_0. \end{aligned} \tag{251b}$$

Thus, the reordered Lagrangian (246) has the same form "V ± A" as the original (245), "V + A" is mapped onto "V + A", "V − A" onto "V − A".

If, on the other hand, the original Lagrangian is

$$C_0\big(\bar{f}_1(1 \pm \gamma_5)f_2\big)\big(\bar{f}_3(1 \pm \gamma_5)f_4\big), \tag{252a}$$

that is $C_{\mathrm{S}} = -C_{\mathrm{P}} = C_0$, $C_{\mathrm{S}}' = -C_{\mathrm{P}}' = \pm C_0$ (all others vanishing), then there are no V and A couplings in the reordered form (246), $D_{\mathrm{V}} = D_{\mathrm{A}} = D_{\mathrm{V}}' = D_{\mathrm{A}}' = 0$ and

$$D_{\mathrm{S}} = -D_{\mathrm{P}} = D_{\mathrm{T}} = -\tfrac{1}{2}C_0, \qquad D_{\mathrm{S}}' = -D_{\mathrm{P}}' = D_{\mathrm{T}}' = \mp\tfrac{1}{2}C_0. \tag{252b}$$

[Clearly, these results have to do with the specific helicity selection rules of the covariants, cf. sec. 1.2.3.]

We see from these examples that if the interaction is of V and A type in one ordering, it contains S, P, V and A in the other, but no T, cf. eqs. (250). Only if it is precisely "V − A", or "V + A", are there no S and P terms in the reordered form, cf. eq. (251). Thus, in a theory where all interactions are mediated by spin-1 bosons, the effective contact Lagrangian will not contain tensor couplings, no matter in which order the interaction is written. However, if the theory also contains genuine scalar and/or pseudo-scalar interactions then tensor couplings do occur in a Fierz reordering of the interaction Lagrangian, cf. eqs. (252).

6.1.5. Proof of Fierz reordering relations

This section contains the proof of equations (247) and (248). As this is somewhat technical the reader may wish to skip it in a first reading and go directly to the next section. However, even in passing, it may be worth noting relations (255) and (260) which are invariant linear combinations with respect to reordering.

There is a useful relation for the direct product of two σ-matrices (III.4), contracted over their Lorentz indices, which one derives by verification, (cf. exercise III.8):

$$(\sigma^\alpha)_{aB}(\sigma_\alpha)_{dE} = -2\varepsilon_{ad}\varepsilon_{BE}. \tag{253}$$

As this equation is antisymmetric in B and E, one deduces immediately the relation

$$(\sigma^\alpha)_{aB}(\sigma_\alpha)_{dE} + (\sigma^\alpha)_{aE}(\sigma_\alpha)_{dB} = 0. \tag{253a}$$

It is easy to derive analogous relations for the matrices $\hat{\sigma}^\alpha$ by means of eq. (III.38). In particular, the analogue of (253a) reads

$$(\hat{\sigma}^\alpha)^{Ab}(\hat{\sigma}_\alpha)^{De} + (\hat{\sigma}^\alpha)^{Ae}(\hat{\sigma}_\alpha)^{Db} = 0. \tag{253b}$$

Using the hermiticity of σ_α, by which $(\sigma_\alpha)_{dE} = (\sigma_\alpha^*)_{Ed}$, and multiplying eq. (253) by $\varepsilon^{DE}\varepsilon^{df}$ we obtain, finally,

$$(\sigma^\alpha)_{aB}(\hat{\sigma}_\alpha)^{Df} = 2\delta_a^{\,f}\delta^D{}_B. \tag{254}$$

It is convenient to write the covariants S,...,T in terms of two-component spinors and to combine them such as to project onto spinor fields of first and second kind. Thus, from eq. (227)

$$(s+p)_{ki} = 2\chi^{(k)*a}\phi^{(i)}{}_a,$$

$$(s-p)_{ki} = 2\phi^{(k)*}{}_A\chi^{(i)A},$$

where s and p are defined by eqs. (226) (p still without the factor i). Similarly, from eqs. (51) and (53) we have

$$(v^\alpha + a^\alpha)_{ki} = 2\phi^{(k)*}{}_A(\hat{\sigma}^\alpha)^{Ab}\phi^{(i)}{}_b,$$

$$(v^\alpha - a^\alpha)_{ki} = 2\chi^{(k)*a}(\sigma^\alpha)_{aB}\chi^{(i)B}.$$

Finally, from eqs. (233) and (235), one obtains

$$(t^{\alpha\beta} + t'^{\alpha\beta})_{ki} = \frac{\mathrm{i}}{\sqrt{2}}\chi^{(k)*a}\{(\sigma^\alpha)_{aB}(\hat{\sigma}^\beta)^{Bd} - (\sigma^\beta)_{aB}(\hat{\sigma}^\alpha)^{Bd}\}\phi_d^{(i)},$$

$$(t^{\alpha\beta} - t'^{\alpha\beta})_{ki} = \frac{\mathrm{i}}{\sqrt{2}}\phi^{(k)*}{}_A\{(\hat{\sigma}^\alpha)^{Ab}(\sigma^\beta)_{bD} - (\hat{\sigma}^\beta)^{Ab}(\sigma^\alpha)_{bD}\}\chi^{(i)D}.$$

It is not difficult to show that interference terms between $f^{\alpha\beta}$, eq. (235a), and $g_{\alpha\beta}$, eq. (235b), vanish, i.e. $(f^{\alpha\beta})_{ki}(g_{\alpha\beta})_{nm} = 0$. Therefore, one has the relation

$$(t'^{\alpha\beta})_{ki}(t'_{\alpha\beta})_{nm} = (t^{\alpha\beta})_{ki}(t_{\alpha\beta})_{nm},$$

and, for instance

$$(t^{\alpha\beta} \pm t'^{\alpha\beta})(t_{\alpha\beta} \pm t'_{\alpha\beta}) = 2t^{\alpha\beta}(t_{\alpha\beta} \pm t'_{\alpha\beta}).$$

Consider first the symmetric combination

$$\tfrac{1}{4}(v^\alpha + a^\alpha)_{12}(v_\alpha + a_\alpha)_{34} = (\hat{\sigma}^\alpha)^{Ab}(\hat{\sigma}_\alpha)^{Df}\phi^{(1)*}{}_A\phi^{(2)}{}_b\phi^{(3)*}{}_D\phi^{(4)}{}_f.$$

Commuting the operators 2 and 4, and applying eq. (253b), this is equal to

$$-(\hat{\sigma}^{\alpha})^{Af}(\hat{\sigma}_{\alpha})^{Db}\{-\phi^{(1)*}{}_{A}\phi^{(4)}{}_{f}\phi^{(3)*}{}_{D}\phi^{(2)}{}_{b}\} = \tfrac{1}{4}(v^{\alpha}+a^{\alpha})_{14}(v_{\alpha}+a_{\alpha})_{32}.$$

A similar relation holds for the combination "V − A", so that we have

$$(v^{\alpha} \pm a^{\alpha})_{12}(v_{\alpha} \pm a_{\alpha})_{34} = (v^{\alpha} \pm a^{\alpha})_{14}(v_{\alpha} \pm a_{\alpha})_{32}. \tag{255}$$

The combinations $(v \pm a)(v \mp a)$ are reordered by making use of relation (254)

$$\tfrac{1}{4}(v^{\alpha}+a^{\alpha})_{12}(v_{\alpha}-a_{\alpha})_{34} = (\hat{\sigma}^{\alpha})^{Ab}(\sigma_{\alpha})_{dF}\phi^{(1)*}{}_{A}\phi^{(2)}{}_{b}\chi^{(3)*d}\chi^{(4)F}$$
$$= -2\delta_{a}{}^{b}\delta^{A}{}_{F}\phi^{(1)*}{}_{A}\chi^{(4)F}\chi^{(3)*d}\phi^{(2)}{}_{b} = -\tfrac{1}{2}(s-p)_{14}(s+p)_{32}.$$

The same relation holds with the plus and minus signs exchanged on either side, so that we obtain

$$(v^{\alpha} \pm a^{\alpha})_{12}(v_{\alpha} \mp a_{\alpha})_{34} = -2(s \mp p)_{14}(s \pm p)_{32}. \tag{256}$$

Before we go on, let us apply eqs. (255) and (256) to the transformation of the most general VA Lagrangian from the ordering (12) (34) to the ordering (14) (32):

$$C_{\mathrm{V}}(v^{\alpha})_{12}(v_{\alpha})_{34} + C'_{\mathrm{V}}(v^{\alpha})_{12}(a_{\alpha})_{34} + C'_{\mathrm{A}}(a^{\alpha})_{12}(v_{\alpha})_{34} + C_{\mathrm{A}}(a^{\alpha})_{12}(a_{\alpha})_{34}$$
$$= \tfrac{1}{4}\{(C_{\mathrm{V}}+C'_{\mathrm{V}}+C'_{\mathrm{A}}+C_{\mathrm{A}})(v^{\alpha}+a^{\alpha})_{12}(v_{\alpha}+a_{\alpha})_{34}$$
$$+(C_{\mathrm{V}}-C'_{\mathrm{V}}-C'_{\mathrm{A}}+C_{\mathrm{A}})(v^{\alpha}-a^{\alpha})_{12}(v_{\alpha}-a_{\alpha})_{34}$$
$$+(C_{\mathrm{V}}-C'_{\mathrm{V}}+C'_{\mathrm{A}}-C_{\mathrm{A}})(v^{\alpha}+a^{\alpha})_{12}(v_{\alpha}-a_{\alpha})_{34}$$
$$+(C_{\mathrm{V}}+C'_{\mathrm{V}}-C'_{\mathrm{A}}-C_{\mathrm{A}})(v^{\alpha}-a^{\alpha})_{12}(v_{\alpha}+a_{\alpha})_{34}\},$$

The first two terms on the r.h.s., by eq. (255), go over into the same forms in the ordering (14)(32). The remaining two terms are transformed according to eq. (256) and give scalar and pseudoscalar couplings. Inserting eqs. (255) and (256), and remembering that $p_{ki} = -\mathrm{i}(\overline{f}_k \Gamma_P f_i)$, the comparison with the general form in the ordering (14)(32) yields precisely the eqs. (250), thus establishing the third and fourth columns of eqs. (247) and (248).

In order to obtain the remainder of these equations, we need the transformation behaviour of S, P and T terms. Clearly, eq. (256) can also be read from right to left, with 4 and 2 interchanged. Therefore, in order to cope with the most general SP term, we also need relations for $(s \pm p)_{12}(s \pm p)_{34}$. These are obtained as follows. Using eq. (III.42), one has

$$(\sigma^{\alpha}\hat{\sigma}^{\beta})_{a}{}^{d}\{(\sigma_{\alpha}\hat{\sigma}_{\beta})_{c}{}^{b} - (\sigma_{\beta}\hat{\sigma}_{\alpha})_{c}{}^{b}\} = (\sigma^{\alpha}\hat{\sigma}^{\beta})_{a}{}^{d}\{2g_{\alpha\beta}\delta_{c}{}^{b} - 2(\sigma_{\beta}\hat{\sigma}_{\alpha})_{c}{}^{b}\}.$$

The sums over the Lorentz indices α and β are performed by means of eq. (254), giving eventually

$$= 8\delta_{a}{}^{d}\delta_{c}{}^{b} - 16\delta_{a}{}^{b}\delta_{c}{}^{d}. \tag{257}$$

Multiplying this equation by

$$\chi^{(1)*a}\phi^{(2)}{}_{b}\chi^{(3)*c}\phi^{(4)}{}_{d}$$

or by $-\chi^{(1)*}\phi^{(4)}\chi^{(3)*}\phi^{(2)}$, depending on which pair of indices must be contracted, this yields a relation for $(s+p)_{12}(s+p)_{34}$ in terms of $(s+p)_{14}(s+p)_{32}$ and $t_{14}(t+t')_{32}$. Clearly, an analogous identity holds if σ and $\hat{\sigma}$ are interchanged in (257). Thus we obtain the formulae

$$(s\pm p)_{12}(s\pm p)_{34} = -\tfrac{1}{2}(s\pm p)_{14}(s\pm p)_{32} - \tfrac{1}{2}(t^{\alpha\beta})_{14}\left(t_{\alpha\beta}\pm t'_{\alpha\beta}\right)_{32}. \qquad (258)$$

As one easily verifies, eqs. (258) and eqs. (256) with $(4 \leftrightarrow 2)$, establish the first and second columns of eqs. (247) and (248). [Note also that relation (258) is identical with eq. (252b).]

Regarding the tensor covariants, we now derive two more relations which, when combined with eq. (258), yield the full transformation formula for T couplings. Let us return to eq. (257) to which we add the term $8\delta_a{}^b\delta_c{}^d$,

$$(\sigma^\alpha\hat{\sigma}^\beta)_a{}^d\left\{(\sigma_\alpha\hat{\sigma}_\beta)_c{}^b - (\sigma_\beta\hat{\sigma}_\alpha)_c{}^b\right\} + 8\delta_a{}^b\delta_c{}^d. \qquad (259a)$$

The first term in this expression can be reordered by means of eq. (253b) to $(\sigma^\alpha\hat{\sigma}^\beta)_a{}^b(\sigma_\alpha\hat{\sigma}_\beta)_c{}^d$. The second and third term cancel against each other, as one verifies by means of eq. (254). Therefore, the combination (259a) is equal to

$$(\sigma^\alpha\hat{\sigma}^\beta)_a{}^b\left\{(\sigma_\alpha\hat{\sigma}_\beta)_c{}^d - (\sigma_\beta\hat{\sigma}_\alpha)_c{}^d\right\} + 8\delta_a{}^d\delta_c{}^b. \qquad (259b)$$

Multiplying this identity with $\chi^{(1)*a}\phi^{(2)}{}_b\chi^{(3)*c}\phi^{(4)}{}_d$ (or the combination with $\phi^{(2)}$ and $\phi^{(4)}$ interchanged), one obtains the following relation with the upper sign:

$$\begin{aligned}(t^{\alpha\beta})_{12}\left(t_{\alpha\beta}\pm t'_{\alpha\beta}\right)_{34} &- (s\pm p)_{12}(s\pm p)_{34}\\ &= (t^{\alpha\beta})_{14}\left(t_{\alpha\beta}\pm t'_{\alpha\beta}\right)_{32} - (s\pm p)_{14}(s\pm p)_{32}.\end{aligned} \qquad (260)$$

The corresponding relation with the lower sign is derived in a similar way.

Eqs. (260) and (258) are combined to yield the relations

$$(t^{\alpha\beta})_{12}\left(t_{\alpha\beta}\pm t'_{\alpha\beta}\right)_{34} = \tfrac{1}{2}(t^{\alpha\beta})_{14}\left(t_{\alpha\beta}\pm t'_{\alpha\beta}\right)_{32} - \tfrac{3}{2}(s\pm p)_{14}(s\pm p)_{32} \qquad (261)$$

from which one reads off the fifth columns of eqs. (247) and (248). This completes the proof.

6.2. *Precision tests in muon decay*

Besides the nuclear β-transitions the ordinary, most frequent decay mode of μ^- (or μ^+)

$$\mu^- \to e^-\bar{\nu}_e\nu_\mu \qquad \left(\mu^+ \to e^+\nu_e\bar{\nu}_\mu\right) \qquad (262)$$

provides a very precise source of information on CC weak interactions. It is a purely leptonic process and, therefore, is particularly well suited for testing the standard theory and for identifying specific deviations from the simple "V − A" picture. Restricting the analysis to the electron (or positron) in the final state there are at least nine observables which are measurable realistically and all of which carry some characteristic information on the basic interaction: The rate or lifetime τ_μ, three parameters (ρ, η, δ) determining shapes of spectra, two strength parameters (ξ, ξ')

characterizing spin-momentum correlations, and four parameters $(\alpha, \beta, \alpha', \beta')$ determining the transverse polarization of the electron in the final state. [The spectrum parameter η is linearly dependent on α and β.] The precise definitions are given below.

This analysis disregards the neutrinos as these are not readily observable. There is, however, some information on the character of the neutrinos in the decay (262): If muonic lepton number, instead of being conserved additively, were some kind of parity quantum number, i.e. if instead of $\Sigma_i L_\mu(i)$ the "parity" $\Pi_i(-)^{L_\mu(i)}$ were conserved, then a μ^+ could decay following either of the two branches

$$\mu^+ \to e^+ \nu_e \bar{\nu}_\mu, \tag{263a}$$

$$\mu^+ \to e^+ \bar{\nu}_e \nu_\mu. \tag{263b}$$

In the additive scheme, on the other hand, the mode (263b) is forbidden. An experiment which was designed to identify the $\bar{\nu}_e$ through inverse β-decay, gave the following result for the decay width (Willis et al., 1980):

$$\frac{\Gamma(\mu^+ \to e^+ \bar{\nu}_e \nu_\mu)}{\Gamma(\mu^+ \to \text{all})} = -0.001 \pm 0.061 \text{ (90\% C.L.)}. \tag{264}$$

This important result shows that the decay mode (263b) is absent, at the level of a few percent, and that the additive mode of conserving L_μ is strongly supported (see also exercise 16).

In this section we analyze the differential decay probability of polarized muons in a rather general manner. As a matter of example and in order to demonstrate the sensitivity of the observables to specific types of couplings we discuss the case of S, P, V, and A interactions. This is not quite the most general one as it leaves out possible T couplings. We choose this example because it illustrates well the information content of muon decay while still being manageable as far as practical calculations are concerned. [The general case is found in the literature, see e.g. Scheck (1978).] We also discuss briefly the reaction $\nu_\mu e^- \to \mu^- \nu_e$, so-called inverse muon decay, and show in which respect it yields information which is complementary to the information from the decay.

6.2.1. *Muon decay: kinematics*

Let q be the momentum of the muon, p the momentum of the electron, k_1 and k_2 the momenta of the neutrinos, respectively, and let Q be the sum of k_1 and k_2:

$$q = p + k_1 + k_2 = p + Q,$$

where

$$Q = k_1 + k_2 = q - p. \tag{265}$$

From the general formulae of App. B the differential decay probability, when expressed in the muon's rest frame, is given by

$$\frac{d^3\Gamma}{d^3p} = \frac{(2\pi)^7}{4m_\mu E} \int \frac{d^3k_1}{2E_1} \int \frac{d^3k_2}{2E_2} \delta(Q - k_1 - k_2)|T|^2, \tag{266}$$

where E is the energy of the electron in the final state. The physical domain of E is

$$m_e \le E \le W,$$

with

$$W = \frac{1}{2m_\mu}\left(m_\mu^2 + m_e^2\right). \tag{267}$$

The direction of $\boldsymbol{p}$, the three-momentum of the electron, is understood to be taken with respect to the direction of the muon spin (see Fig. 10 below).

Obviously, the kinematics is symmetric in the momenta of the two neutrinos. As we integrate over them, only those covariants in $|T|^2$ will contribute which are symmetric in k_1 and k_2. Antisymmetric combinations do not contribute. Furthermore, the result of the integration in eq. (266) can only depend on Q. For S, P, V, and A couplings the pertinent integrals are

$$\int \frac{\mathrm{d}^3 k_1}{2k_1^0} \int \frac{\mathrm{d}^3 k_2}{2k_2^0} (k_1 \cdot k_2)\,\delta(Q - k_1 - k_2) = \frac{\pi}{4} Q^2, \tag{268a}$$

$$\int \frac{\mathrm{d}^3 k_1}{2k_1^0} \int \frac{\mathrm{d}^3 k_2}{2k_2^0} \left(k_1^\alpha k_2^\beta - (k_1 \cdot k_2) g^{\alpha\beta} + k_2^\alpha k_1^\beta\right) \delta(Q - k_1 - k_2)$$

$$= \frac{\pi}{6}\left(Q^\alpha Q^\beta - Q^2 g^{\alpha\beta}\right). \tag{268b}$$

(See exercise 17.)

Let $\zeta_\mu = P_\mu \boldsymbol{n}_0$ be the expectation value of the spin of the decaying muon, where $\boldsymbol{n}_0$ is a unit vector and P_μ is the muon polarization. In the muon's rest frame where $q = (m; 0)$, $p = (E; \boldsymbol{p})$ and where the covariant spin vector is $s_0 = (0; \zeta_\mu)$, we choose polar coordinates such that $\boldsymbol{p}$ defines the z-axis and that $\boldsymbol{n}_0$ lies in the xz-plane, see Fig. 10. Let $\boldsymbol{n}_1$ be another unit vector with polar coordinates (ϕ, ψ), as shown in the figure. Below we derive the probability of finding the electron spin pointing in the direction $\boldsymbol{n}_1$; this means that we must calculate the decay amplitude for a polarized electron state whose covariant spin vector (III.124) is

$$s_1 = \left(\frac{1}{m_e}\boldsymbol{p}\cdot\boldsymbol{n}_1;\, \boldsymbol{n}_1 + \frac{\boldsymbol{p}\cdot\boldsymbol{n}_1}{m_e(E + m_e)}\boldsymbol{p}\right).$$

In the muon's rest frame, according to Fig. 10, we have

$$\boldsymbol{n}_0 = (\sin\theta, 0, \cos\theta), \qquad \boldsymbol{n}_1 = (\sin\phi\cos\psi, \sin\phi\sin\psi, \cos\phi),$$

$$(s_0 \cdot p) = -|\boldsymbol{p}|\cos\theta \simeq -E\cos\theta,$$

$$m_e(s_1 \cdot q) = m_\mu |\boldsymbol{p}|\cos\phi \simeq m_\mu E\cos\phi,$$

$$m_e(s_0 \cdot s_1) = -E\cos\theta\cos\phi - m_e \sin\theta\sin\phi\cos\psi \simeq -E\cos\theta\cos\phi. \tag{269}$$

The $\simeq$ sign refers to the case where E is large compared to the rest mass of the electron.

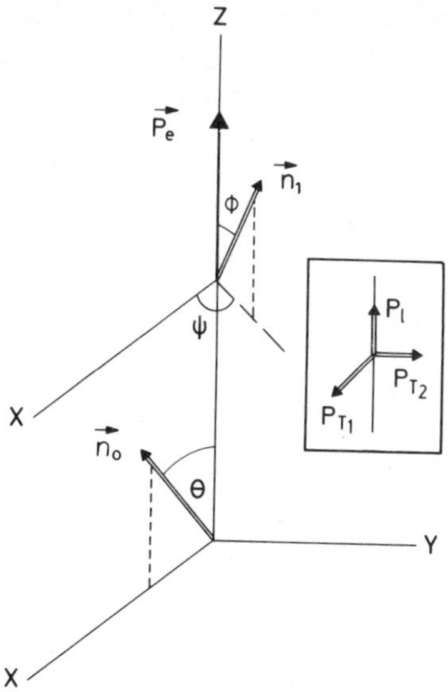

Fig. 10. Definitions of kinematic variables in muon decay. $\boldsymbol{n}_0$ is the direction of the muon's spin expectation value, $\boldsymbol{n}_1$ the direction of the electron's spin expectation value. The insert shows the decomposition of the electron polarization along its momentum (P_ℓ), transverse to the momentum and in the plane spanned by $\boldsymbol{n}_0$ and $\boldsymbol{p}$ (P_{T1}), and normal to that plane (P_{T2}).

6.2.2. *Muon decay: interaction, decay probability and observables*

The general, effective interaction responsible for muon decay has the form given in eq. (245), viz.

$$\mathscr{H} = -\mathscr{L} = \frac{G}{\sqrt{2}} \sum_i \left\{ \overline{e(x)}\, \Gamma_i \nu_e(x) \left[G_i \overline{\nu_\mu(x)}\, \Gamma^i \mu(x) + G_i' \overline{\nu_\mu(x)}\, \Gamma^i \gamma_5 \mu(x) \right] + \text{h.c.} \right\}. \tag{270}$$

For the sake of simplicity, we consider only S, P, V and A interactions, but no tensors, and refer to Scheck (1978) for the complete case.*) Furthermore, on the basis of our experience gained in secs. 6.1.4 and 6.1.5 we rewrite the interaction

$$\begin{aligned} \mathscr{H} = \frac{G}{\sqrt{2}} \{ & h_{11}(s+p)_{\bar{e}\nu_e}(s+p)_{\bar{\nu}_\mu\mu} + h_{12}(s+p)(s-p) \\ & + h_{21}(s-p)(s+p) + h_{22}(s-p)(s-p) \\ & + g_{11}(v^\alpha + a^\alpha)_{\bar{e}\nu_e}(v_\alpha + a_\alpha)_{\bar{\nu}_\mu\mu} + g_{12}(v^\alpha + a^\alpha)(v_\alpha - a_\alpha) \\ & + g_{21}(v^\alpha - a^\alpha)(v_\alpha + a_\alpha) + g_{22}(v^\alpha - a^\alpha)(v_\alpha - a_\alpha) + \text{h.c.} \} \end{aligned} \tag{270'}$$

*) Eq. (3.17) of that reference contains a sign error and should read $c' = -2\,\mathrm{Re}(C_T C_T'^*)$.

(we have written the field symbols only once for each class of covariants). Keeping in mind the definition of Γ_P in eq. (232) and of p in eq. (266b) it is easy to express the h_{ik} in terms of the G_i and G_i' and vice versa:

$$\left.\begin{matrix} h_{11} \\ h_{22} \end{matrix}\right\} = \tfrac{1}{4}\left[(G_S - G_P) \pm (G_S' - G_P')\right], \tag{271a}$$

$$\left.\begin{matrix} h_{12} \\ h_{21} \end{matrix}\right\} = \tfrac{1}{4}\left[(G_S + G_P) \mp (G_S' + G_P')\right]. \tag{271b}$$

In a similar way, the g_{ik} are linear functions of $G_V, \ldots, G_A'$:

$$\left.\begin{matrix} g_{11} \\ g_{22} \end{matrix}\right\} = \tfrac{1}{4}\left[(G_V + G_A) \pm (G_V' + G_A')\right], \tag{271c}$$

$$\left.\begin{matrix} g_{12} \\ g_{21} \end{matrix}\right\} = \tfrac{1}{4}\left[(G_V - G_A) \mp (G_V' - G_A')\right]. \tag{271d}$$

The "V – A" interaction corresponds to

$$\begin{aligned} &g_{22} = 1, \qquad g_{11} = g_{12} = g_{21} = 0, \\ &h_{11} = h_{12} = h_{21} = h_{22} = 0. \end{aligned} \tag{272}$$

In view of the fact that the neutrinos are not detected it is appropriate to apply a Fierz transformation to eq. (270′) to the effect of grouping the (observable) charged leptons in one covariant, and the two (unobserved) neutrinos in the other. In sec. 6.1.5 we have learnt how to do this. Using eqs. (255), (256) and (258) we find at once

$$\begin{aligned} \mathscr{H} = \frac{G}{\sqrt{2}} \Big\{ &-\tfrac{1}{2} h_{11} (s+p)_{\bar{e}\mu} (s+p)_{\bar{\nu}_\mu \nu_e} - 2 g_{21} (s+p)(s-p) \\ &- 2 g_{12} (s-p)(s+p) - \tfrac{1}{2} h_{22} (s-p)(s-p) \\ &+ g_{11} (v^\alpha + a^\alpha)_{\bar{e}\mu} (v_\alpha + a_\alpha)_{\bar{\nu}_\mu \nu_e} - \tfrac{1}{2} h_{21} (v^\alpha + a^\alpha)(v_\alpha - a_\alpha) \\ &- \tfrac{1}{2} h_{12} (v^\alpha - a^\alpha)(v_\alpha + a_\alpha) + g_{22} (v^\alpha - a^\alpha)(v_\alpha - a_\alpha) \\ &- \tfrac{1}{2} h_{11} (t^{\alpha\beta})_{\bar{e}\mu} \left(t_{\alpha\beta} + t'_{\alpha\beta}\right)_{\bar{\nu}_\mu \nu_e} - \tfrac{1}{2} h_{22} t^{\alpha\beta} \left(t_{\alpha\beta} - t'_{\alpha\beta}\right) + \text{h.c.} \Big\}. \end{aligned} \tag{273}$$

The form (270) or (270′) of the Lagrangian is usually referred to as the *charge changing* form. The same Hamiltonian is said to be given in *charge retention* form if it is written as done in eq. (273).

The calculation of the amplitude and of $|T|^2$ is straightforward but rather lengthy. It is simplified somewhat by the observation that due to the symmetric integration over the neutrino momenta, all terms which are antisymmetric in k_1 and k_2 can be skipped. One finds the following result for the decay of a negative muon:

$$\frac{\mathrm{d}^3\Gamma}{\mathrm{d}^3p} \Big/ \left\{\frac{\pi G^2}{8m_\mu E(2\pi)^5}\right\} = aQ^2\left[(pq) - m_e m_\mu (s_0 s_1)\right]$$
$$+ \alpha Q^2\left[m_e m_\mu + (s_1 q)(s_0 p) - (pq)(s_0 s_1)\right] - a' Q^2\left[m_\mu (s_0 p) - m_e (s_1 q)\right]$$
$$+ \tfrac{2}{3} b\left[m_e m_\mu (s_0 s_1) Q^2 - 2 m_\mu m_e (s_1 q)(s_0 p) + (pq) Q^2 + 2(Qq)(Qp)\right]$$
$$+ \tfrac{2}{3}\beta\left[(s_0 s_1)\left(2 m_e^2 m_\mu^2 - \left(m_e^2 + m_\mu^2\right)(pq)\right)\right.$$
$$\left. + \left(m_e^2 + m_\mu^2\right)(s_1 q)(s_0 p) - 3 m_e m_\mu Q^2\right]$$
$$+ \tfrac{2}{3} b'\left[m_e\left(2(Qq)(s_1 q) + Q^2(s_1 q)\right) + m_\mu\left(-2(Qp)(s_0 p) + Q^2(s_0 p)\right)\right]$$
$$+ \tfrac{2}{3} c\left[4(Qq)(Qp) - Q^2(qp) + m_e m_\mu\left(4(s_0 p)(s_1 q) + Q^2(s_0 s_1)\right)\right]$$
$$- \tfrac{2}{3} c'\left[4 m_\mu (Qp)(s_0 p) + 4 m_e (Qq)(s_1 q) + Q^2\left(m_\mu (s_0 p) - m_e (s_1 q)\right)\right]$$
$$- \left(\alpha' Q^2 + \tfrac{2}{3}\beta'\left(m_\mu^2 - m_e^2\right)\right)\varepsilon_{\alpha\beta\sigma\tau} q^\alpha s_0^\beta p^\sigma s_1^\tau. \tag{274}$$

In this expression the real constants $a, \ldots, \beta'$ are given by

$$\left.\begin{matrix} a \\ a' \end{matrix}\right\} = 16\left(|g_{12}|^2 \pm |g_{21}|^2\right) + |h_{22}|^2 \pm |h_{11}|^2, \tag{275a}$$

$$\left.\begin{matrix} b \\ b' \end{matrix}\right\} = 4\left(|g_{11}|^2 \pm |g_{22}|^2\right) + |h_{21}|^2 \pm |h_{12}|^2, \tag{275b}$$

$$\left.\begin{matrix} c \\ c' \end{matrix}\right\} = \tfrac{1}{2}\left(|h_{22}|^2 \pm |h_{11}|^2\right), \tag{275c}$$

$$\left.\begin{matrix} \alpha \\ \alpha' \end{matrix}\right\} = 8\left\{\begin{matrix} \mathrm{Re} \\ \mathrm{Im} \end{matrix}\right\}\left(g_{21} h_{22}^* \pm g_{12} h_{11}^*\right), \tag{275d}$$

$$\left.\begin{matrix} \beta \\ \beta' \end{matrix}\right\} = -4\left\{\begin{matrix} \mathrm{Re} \\ \mathrm{Im} \end{matrix}\right\}\left(g_{22} h_{21}^* \pm g_{11} h_{12}^*\right). \tag{275e}$$

We do not evaluate eq. (274) in its full complexity but rather assume the energy E of the electron to be large as compared to the rest mass. Then m_e may be set equal to zero except where it appears multiplied with s_1, in which case eqs. (269) apply. In this approximation

$$W \simeq \tfrac{1}{2} m_\mu, \qquad Q^2 = (q-p)^2 \simeq m_\mu (m_\mu - 2E),$$
$$\mathrm{d}^3 p = |\boldsymbol{p}|^2 \mathrm{d}|\boldsymbol{p}| \mathrm{d}\Omega \simeq E^2 \mathrm{d}E \mathrm{d}\Omega.$$

It is convenient to replace E with the dimensionless variable

$$x := \frac{E}{W} \simeq \frac{2E}{m_\mu}, \tag{276}$$

whose range of variation is

$$x_0 \leq x \leq 1 \quad \text{with } x_0 = m_e/W,$$

or, in the approximation $m_e = 0$: $0 \leq x \leq 1$.

The differential decay probability for emission of an electron with its energy between x and $x + dx$ at an angle between θ and $\theta + d\theta$ with respect to the muon spin, is then found to be*)

$$\frac{d^2\Gamma(x,\theta,\phi,\psi)}{dx\, d(\cos\theta)}$$

$$\simeq \frac{Am_\mu^5 G^2}{2^{10}\pi^3 6} x^2 \Big\{ [6(1-x) + \tfrac{4}{3}\rho(4x-3)] + \xi\cos\theta[2(x-1) + \tfrac{4}{3}\delta(3-4x)]$$

$$+ \xi'\cos\phi[6(x-1) + 4\delta'(3-4x)] + \xi''\cos\theta\cos\phi[2(1-x) + \tfrac{4}{3}\rho'(4x-3)]$$

$$+ 2\sin\theta\sin\phi\left[\left(3(1-x)\frac{\alpha}{A} + 2\frac{\beta}{A}\right)\cos\psi + \left(3(1-x)\frac{\alpha'}{A} + 2\frac{\beta'}{A}\right)\sin\psi\right]\Big\}. \tag{277}$$

In this expression α, β, α', β' are as given above in eqs. (275d) and (275e). The other parameters are appropriately chosen combinations of $a, \ldots, c'$, viz.

$$A = a + 4b + 6c, \tag{278a}$$

$$\rho = \frac{1}{A}(3b + 6c), \tag{278b}$$

$$\xi = -\frac{1}{A}(3a' + 4b' - 14c'), \tag{278c}$$

$$\delta = \frac{1}{A\xi}(-3b' + 6c'), \tag{278d}$$

$$\xi' = -\frac{1}{A}(a' + 4b' + 6c'), \tag{278e}$$

$$\delta' = -\frac{1}{A\xi'}(b' + 2c'), \tag{278f}$$

$$\xi'' = \frac{1}{A}(3a + 4b - 14c), \tag{278g}$$

$$\rho' = \frac{1}{A\xi''}(3b - 6c). \tag{278h}$$

ρ is called the Michel parameter. It was introduced by L. Michel who gave the first general analysis of muon decay. [As may be seen from eq. (274) there are also spin independent terms in the spectrum which are proportional to m_e and which depend on α and β. They are proportional to $\eta x_0(1-x)$ with $\eta = (1/A)(\alpha - 2\beta)$. As we

*) This holds for a fully polarized muon. If the muon carries partial polarization P_μ, the terms in $\cos\theta$ and $\sin\theta$ are multiplied by P_μ.

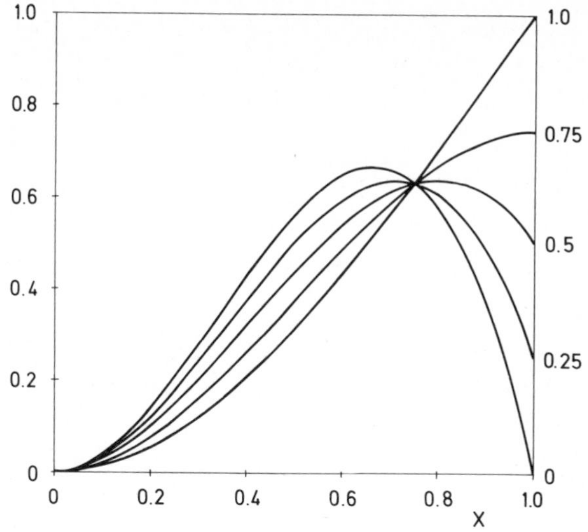

Fig. 11. Michel spectrum, i.e. isotropic part of differential decay probability (277), for various values of ρ. The spectrum is normalized such that it equals ρ at the maximal energy $x=1$.

neglect m_e they do not appear in eq. (277).] Eq. (277) holds for μ^- decay. For μ^+ decay the signs of the terms $\xi \cos\theta$ and $\xi' \cos\phi$ must be reversed.

What are the observables that can be determined in an experiment on muon decay? The parameters that were measured in the past are the following: The spectrum parameters ρ and δ are known to rather high precision (see below). The Michel parameter ρ, for instance, determines the shape of the spectrum for the decay of unpolarized muons, see Fig. 11. ξ is determined by a measurement of the asymmetry of electron emission with respect to the muon spin direction. ξ' determines the longitudinal polarization of the outgoing electron, $\{\alpha, \beta\}$ the transverse component P_{T1}, and $\{\alpha', \beta'\}$ the normal component P_{T2} of the electron polarization, all of which have been measured (for definitions see Fig. 10 and next section). The quantity A, finally, appears in the expression for the total rate. The remaining parameters δ', ξ'', and ρ' have not been measured.

6.2.3. *Analysis and examples*

As an example, let us analyze muon decay in terms of the interaction (270′) and compare the predictions for the lifetime and the decay parameters to their measured values.

The total *rate* is obtained from eq. (277) by integrating over x and θ, and by summing over the spin directions of the outgoing electron. One finds

$$\Gamma \simeq \frac{A m_\mu^5 G^2}{2^8 \pi^3 6} \int_0^1 x^2 \left[6(1-x) + \tfrac{4}{3}\rho(4x-3)\right] \mathrm{d}x = \frac{m_\mu^5 G^2}{192\pi^3} \frac{A}{16} . \tag{279}$$

Note that this result is independent of the value of ρ. In the limit of the exact "V − A" interaction, $g_{22}=1$ whilst all other constants g_{ik} and h_{ik} vanish. From eqs. (275) we then have $a=a'=0$, $b=-b'=4$, $c=c'=0$, $\alpha=\beta=0$, $\alpha'=\beta'=0$, or, from eqs. (278),

$$A=16, \qquad \rho=\tfrac{3}{4}=\delta=\rho', \qquad \delta'=\tfrac{1}{4},$$
$$\xi=1=\xi'=\xi''. \tag{280}$$

ρ and δ are indeed found to be $\frac{3}{4}$ within 0.4% and 1.2%, respectively [R2],

$$\rho=0.752(3), \qquad \delta=0.755(9). \tag{281a}$$

From eqs. (278b) and (278d) this means that $a\simeq 2c$ and $a'\simeq 2c'$ and $\delta'=\frac{1}{4}$, $\rho'=\frac{3}{4}$, or, from eqs. (275a) and (275c), $g_{12}\simeq g_{21}\simeq 0$. Let us assume that, indeed, $g_{12}=0=g_{21}$. Then, from eqs. (275),

$$A=4(a+b)\simeq 16\left(|g_{11}|^2+|g_{22}|^2\right)+4\left(|h_{11}|^2+|h_{22}|^2+|h_{12}|^2+|h_{21}|^2\right),$$

$$1-\xi=\frac{a+b-a'+b'}{a+b}=\frac{8}{A}\left(4|g_{11}|^2+|h_{21}|^2+|h_{11}|^2\right),$$

$$1-\xi'=\frac{a+b+a'+b'}{a+b}=\frac{8}{A}\left(4|g_{11}|^2+|h_{21}|^2+|h_{22}|^2\right),$$

$$\alpha\simeq\alpha'\simeq 0,$$

β,β' as given by eq. (275e).

With these restrictions on the parameters eq. (277) reduces to

$$\frac{\mathrm{d}^2\Gamma}{\mathrm{d}x\,\mathrm{d}(\cos\theta)}\simeq\frac{(a+b)m_\mu^5G^2}{2^8 6\pi^3}$$
$$\times x^2\Big\{(3-2x)(1-\xi'\cos\phi)-(2x-1)\cos\theta(\xi-\xi''\cos\phi)$$
$$+\sin\theta\sin\phi\frac{4}{A}(\beta\cos\psi+\beta'\sin\psi)\Big\}. \tag{282}$$

[If the initial polarization P_μ is not unity, the second and third term are to be multiplied by P_μ.]

The experimental values for ξ and ξ' are

$$\xi=0.975(15), \qquad \xi'=1.01(6). \tag{281b}$$

The value of ξ', in particular, is obtained by measuring the longitudinal polarization of the electron (Corriveau et al. 1981), cf. eq. (25), which is given by

$$P_\ell=\frac{\mathrm{d}\Gamma(x,\theta,\phi=0,\psi=0)-\mathrm{d}\Gamma(x,\theta,\phi=\pi,\psi=0)}{\mathrm{d}\Gamma(x,\theta,0,0)+\mathrm{d}\Gamma(x,\theta,\pi,0)}.$$

With $\rho=\delta=\rho'=\frac{3}{4}$ and $\delta'=\frac{1}{4}$ this is approximately

$$P_\ell\simeq-\xi'+\frac{(2x-1)\cos\theta}{3-2x+\xi(2x-1)\cos\theta}(\xi''-\xi\xi'). \tag{283}$$

β and β', as well as α and α', can be determined, for instance, by measuring the two transverse components of the electron polarization (cf. Fig. 10), viz.

$$P_{\mathrm{T1}}=\frac{\mathrm{d}\Gamma(x,\theta,\pi/2,0)-\mathrm{d}\Gamma(x,\theta,-\pi/2,0)}{\mathrm{d}\Gamma(x,\theta,\pi/2,0)+\mathrm{d}\Gamma(x,\theta,-\pi/2,0)},$$

which, in the situation considered above, is

$$P_{\mathrm{T1}}\simeq\frac{4\beta/A}{3-2x+\xi(2x-1)\cos\theta}\sin\theta. \tag{284}$$

Similarly,

$$P_{\mathrm{T2}}=\frac{\mathrm{d}\Gamma(x,\theta,\pi/2,\pi/2)-\mathrm{d}\Gamma(x,\theta,-\pi/2,\pi/2)}{\mathrm{d}\Gamma(x,\theta,\pi/2,\pi/2)+\mathrm{d}\Gamma(x,\theta,-\pi/2,\pi/2)},$$

which is approximately

$$P_{\mathrm{T2}}\simeq\frac{4\beta'/A}{3-2x+\xi(2x-1)\cos\theta}\sin\theta. \tag{285}$$

P_{T2}, the component perpendicular to the plane spanned by the muon spin and by the electron momentum, is particularly interesting because it can only be different from zero if the interaction (270′) is not invariant under time reversal. Indeed, we see from eqs. (275d) and (275e) that α' and β' are different from zero only if some of the coupling constants are relatively imaginary.

The measurement of P_{T1} and P_{T2} gave values compatible with zero (Corriveau et al. 1983) (with $\alpha=\alpha'=0$),

$$\beta/A=-0.002(16), \qquad \beta'/A=0.007(16). \tag{286}$$

[The measurement of P_{T1} also fixes the parameter η mentioned above.]

In conclusion, the comparison of the experimental results with the theoretical predictions shows that there is good evidence for the "V − A" interaction but that the data do not exclude sizeable contributions of the order of up to 10%, from other types of interaction in the Lagrangian (273).

A case of special interest is the class of left–right symmetric unified gauge theories which are extensions of the GSW theory and which aim at explaining parity violation as a phenomenon typical for low energies. In such theories there is a second charged gauge boson $\mathrm{W_R}$ which couples to V + A currents. In order to obtain the observed "V − A" interaction at low energies the mass m_{R} of $\mathrm{W_R}$ must be significantly larger than the mass m_{L} of its sister boson $\mathrm{W_L}$. Since the two bosons are gauge bosons of the same local gauge theory, they couple to the matter fields with the same coupling constant g. Thus, the effective weak CC Lagrangian is

$$-\mathscr{L}=\frac{g^2}{8}\frac{m_{\mathrm{R}}^2+m_{\mathrm{L}}^2}{m_{\mathrm{R}}^2m_{\mathrm{L}}^2}$$
$$\times\left\{\frac{m_{\mathrm{R}}^2}{m_{\mathrm{R}}^2+m_{\mathrm{L}}^2}(v^\alpha-a^\alpha)(v_\alpha-a_\alpha)+\frac{m_{\mathrm{L}}^2}{m_{\mathrm{R}}^2+m_{\mathrm{L}}^2}(v^\alpha+a^\alpha)(v_\alpha+a_\alpha)\right\}.$$

Setting

$$\frac{G}{\sqrt{2}} = \frac{g^2(m_R^2 + m_L^2)}{8m_L^2 m_R^2},$$

we have

$$g_{11} = \frac{m_L^2}{m_R^2 + m_L^2}, \qquad g_{22} = \frac{m_R^2}{m_R^2 + m_L^2},$$

$g_{12} = 0 = g_{21}$. This is a special case of the analysis given above. The parameter A which determines the rate (279) is $A = 16(m_R^4 + m_L^4)/(m_R^2 + m_L^2)^2$. All parameters except ξ and ξ' have their standard "V − A" values. ξ and ξ' carry the same information on the mass ratio m_R/m_L, viz.

$$\xi = 1 - \frac{2m_L^4}{m_R^4 + m_L^4}.$$

Present data, eqs. (281), to which one must add the available information on $\pi \to \mu\nu_\mu$ and nuclear β-decay (cf. eq. (28)) yield a lower limit on the mass ratio m_R/m_L of about 2.8, i.e. $m_R \gtrsim 230$ GeV. To see the sensitivity of muon decay to the right-handed interactions induced by the existence of W_R, let us assume that ξ is known to be 1 with an error bar of 0.1%. From the result above one would then conclude that $m_R/m_L \gtrsim 6.7$ or $m_R \gtrsim 550$ GeV.

In addition, the two physical W-boson states W_1 and W_2 may be orthogonal mixtures of the states W_R and W_L. This happens if the mass Lagrangian is not diagonal in the basis of the states W_R and W_L. In this case the state mixture leads to additional interaction terms of the type $(v^\alpha \pm a^\alpha)(v_\alpha \mp a_\alpha)$ (Bég et al. 1977). It is not difficult to show that, in our notation, one has

$$g_{12} = g_{21} = -\frac{(m_R^2 - m_L^2)\operatorname{tg}\phi}{(m_R^2 + m_L^2)(1 + \operatorname{tg}^2\phi) - 2(m_R^2 - m_L^2)\operatorname{tg}\phi},$$

where ϕ is the mixing angle. The limits on ϕ come primarily from the Michel parameter ρ and are of the order of $|\phi| \lesssim 0.05$.

6.2.4. Additional remarks

(i) Clearly, the expression (274) which is exact and quite general holds also for the decays

$$\tau^- \to \mu^- \bar{\nu}_\mu \nu_\tau,$$

$$\tau^- \to e^- \bar{\nu}_e \nu_\tau, \tag{287}$$

or for semileptonic decays of bare quarks. In deriving eq. (277) one must check, of course, whether or not the mass of the daughter lepton can be neglected. We have not written out the additional mass dependent terms in eq. (277), for the sake of economy, but it is easy to identify them in eq. (274) and to insert them into eq. (277). In muon decay, eq. (262), they are negligible unless one measures decay electrons with very low energies. The only exception is the rate for which a very accurate experimental value is available and where the mass terms must be taken into account, cf. eqs. (289) and (290) below.

(ii) Radiative corrections to muon decay are important.*⁾ For instance, in the spectrum they amount to about 6% in the determination of the Michel parameter. [The experimental value quoted above was already corrected for this effect.] In the rate, the radiative correction is of the order of 0.4%. One finds, to order αG,

$$\Gamma = \Gamma^{(0)}\left[1 + \frac{\alpha}{2\pi}\left(\tfrac{25}{4} - \pi^2\right) + \frac{3}{5}\frac{m_\mu^2}{m_W^2}\right], \tag{288}$$

where $\Gamma^{(0)}$ is the uncorrected expression.**⁾ In the general case, and including the electron mass terms, it is given by

$$\Gamma^{(0)} = \frac{m_\mu^5 G^2}{192\pi^3}\frac{A}{16}\left\{1 + 4\frac{m_e}{m_\mu}\frac{\alpha - 2\beta}{A} - 8\frac{m_e^2}{m_\mu^2} + O\left(\frac{m_e^3}{m_\mu^3}\right)\right\}. \tag{289}$$

The lifetime is indeed known to very high accuracy

$$\tau_\mu = 2.197\,14(7) \times 10^{-6}\ \text{sec}. \tag{290}$$

In the case of the "V – A" interaction we have $A = 16$, $\alpha = \beta = 0$, so that we can extract G from this datum, viz.

$$G = 1.166\,32(2) \times 10^{-5}\ \text{GeV}^{-2}. \tag{291}$$

Radiative corrections in the electron polarization $P = \{P_{T1}, P_{T2}, P_\ell\}$ are large at low energies (of the order of 10%) but become very small for, say, $x \gtrsim 0.25$.

6.2.5. *Inverse muon decay*

The scattering reaction (so-called inverse muon decay)

$$\nu_\mu e^- \to \mu^- \nu_e \tag{292}$$

provides additional information about the weak CC Lagrangian responsible for muon decay. This reaction was measured at high energies (Jonker et al. 1980, Bergsma et al. 1983) and the cross section was compared to its value as predicted by the pure "V – A" Lagrangian (74) and (186). In the experiment certain integrals of the cross section over the kinematics of the neutrino beam are determined and are compared to the same integrals over the theoretical expressions.

In this section we give a somewhat simplified analysis of reaction (292) by calculating the cross sections at fixed energy and by ignoring these integrations (see however exercise 18.) Let k and k' be the initial and final neutrino momenta, respectively, p the electron momentum and q the muon momentum. Neglecting the

*⁾For a summary see Heil et al. (1984), Scheck (1978).

**⁾The last term on the r.h.s. of eq. (288) is a propagator effect taking into account the finiteness of the W-mass. It is, however, negligibly small.

electron and muon masses, the cross section in the c.m. system reads

$$\frac{d\sigma}{d\Omega^*} \simeq \frac{1}{64\pi^2 s} \tfrac{1}{2}(2\pi)^{12} \sum |T|^2.$$

ν_μ is the incident particle, the muon is the outgoing particle which is detected, so the invariant squared momentum transfer is $t = (k-q)^2 \simeq -(s/2)(1-\cos\theta^*)$. The invariant differential cross section is

$$\frac{d\sigma}{dt} = \frac{d\sigma}{d\Omega^*} 2\pi \frac{d(\cos\theta^*)}{dt} = \frac{1}{32\pi s^2}(2\pi)^{12} \sum |T|^2.$$

The T-matrix element for the interaction (270′) and the spin summations over $|T|^2$ are calculated along the standard lines. If h is the helicity of the incoming ν_μ, the cross section is found to be

$$\frac{d\sigma}{dt} = \frac{G^2}{8\pi s^2}\Big\{ s^2\big(|g_{11}|^2(1+h) + |g_{22}|^2(1-h)\big) + (s+t)^2\big(|g_{12}|^2(1-h) + |g_{21}|^2(1+h)\big) + t^2\big(|h_{11}|^2 + |h_{21}|^2\big)(1-h) + t^2\big(|h_{22}|^2 + |h_{12}|^2\big)(1+h)\Big\}. \tag{293}$$

The muon neutrino in the laboratory stems from pion and kaon decays and, therefore, h is -1, to a very good approximation. If one compares the result (293) to the prediction of the standard interaction (186) one must remember that, from eq. (279), the quantity G^2 in the "V − A" case is replaced by $G^2 A/16$ in the case of the more general interaction (270′). Therefore,

$$\frac{d\sigma}{dt} \Big/ \left(\frac{d\sigma}{dt}\right)_{V-A} = \frac{16}{A}\left\{ |g_{22}|^2 + \frac{(s+t)^2}{s^2}|g_{12}|^2 + \frac{t^2}{s^2}\big(|h_{11}|^2 + |h_{21}|^2\big)\right\}, \tag{294}$$

with A as given in sec. 6.2.3. Remember that we had concluded $g_{21} \simeq g_{12} \simeq 0$. Reaction (292) therefore gives additional limits on the scalar couplings h_{11} and h_{21}. Experimentally, the ratio (294), after integration over the experimental spectrum, is indeed found to be 1 within about 18%.

6.3. *Neutral currents in muonic atoms*

Muonic atoms provide a promising, but still largely unexplored, system for testing parity violating effects due to weak neutral currents. There are three main reasons why precision tests of NC interactions in muonic atoms are superior, except for problems of purely experimental nature, to tests in electronic atoms or even in electron scattering:

(i) On atomic scales the range of the weak NC interaction is practically zero, so that the effective lepton–nucleon interaction acts like a contact interaction. Due to its larger mass the bound muon moves closer to the nucleus than a bound electron, and, therefore, weak NC effects in muonic atoms are enhanced by some power of the mass ratio m_μ/m_e over such effects in electronic atoms.

(ii) The muon, from pion decay, is fully polarized when it is captured in the Coulomb field of the nucleus. The amount of depolarization (through hyperfine interactions) during the cascade can be measured as well as calculated; the residual polarization in the lowest states of the cascade is finite and is well under control. This is important if one wishes to identify the NC effects by means of some spin-momentum correlation such as the asymmetry of a cascade X-ray with respect to the muon spin.

(iii) The muonic energies and wave functions are calculable to any desired accuracy and the analysis of measured effects in terms of the elementary muon–nucleon NC interaction is well defined. This is in contrast to the case of electronic atoms where the analysis of measured effects (dichroism and optical rotation) is beset with uncertainties about the atomic configuration.

The elementary effective Lagrangian is given by the neutral current part of eq. (186) (with $\rho = 1$),

$$-\mathscr{L}_{\mathrm{NC}}^{\mathrm{eff}} = \frac{G}{2\sqrt{2}} K_{\alpha}^{\dagger}(x) K^{\alpha}(x), \tag{295}$$

where the cross product of the muonic neutral current and of the nucleonic current, cf. eq. (184), is relevant here. The Lagrangian contains Lorentz *scalar* pieces which stem from the product of leptonic vector V_{ℓ} and the nucleonic vector V_{n}, as well as from the product of leptonic axial vector A_{ℓ} and nucleonic axial vector A_{n}. These terms will hardly be observable, as they have to compete with the electromagnetic interaction which is of the type $V_{\ell}V_{\mathrm{n}}$. It also contains Lorentz *pseudo*scalar pieces which stem from the couplings $V_{\ell}A_{\mathrm{n}}$ and $A_{\ell}V_{\mathrm{n}}$. These latter terms are parity-odd relative to the electromagnetic interaction and give rise to observable, parity violating effects in the cascade of a muonic atom.

As an example, let us consider the $n = 2$ system in a light muonic atom. Without the weak NC interaction, the 2p-state decays very quickly into the 1s-state through emission of an X-ray photon with the characteristics of an E1-field. The 2s-state is metastable and it will disintegrate to the ground state primarily by emitting two photons. As another possibility it can emit a single photon with the quantum numbers of an M1-field. This latter possibility is due to the relativistic components in the s-states, (remember that a state with $\kappa = -1$, $j = \frac{1}{2}$ contains a "large" component with $l = 0$ and a small component with $l = 1$). Clearly, as a light muonic atom is only moderately relativistic, this M1-transition will be very weak.

If we add the static, Lorentz pseudoscalar NC interaction to the electromagnetic one, the atomic states are no more eigenstates of parity. In particular, the 2s-state will receive a small admixture from the 2p-state, and vice versa. As a consequence, what was a weak (2s–1s) M1-transition before will now have an admixture of the strong (2p–1s)E1-transition, see Fig. 12a. The admixture coefficient η is proportional to the matrix element of the parity-odd interaction between the 2s and 2p states divided by an energy denominator which contains the difference of the real energies and the sum of the radiative widths of the 2s and 2p states, respectively, viz.

$$\eta \simeq \frac{\langle 2\mathrm{p}|H^{\mathrm{NC}}|2\mathrm{s}\rangle}{E_{2\mathrm{s}} - E_{2\mathrm{p}} + (\mathrm{i}/2)(\Gamma_{2\mathrm{s}} + \Gamma_{2\mathrm{p}})}. \tag{296}$$

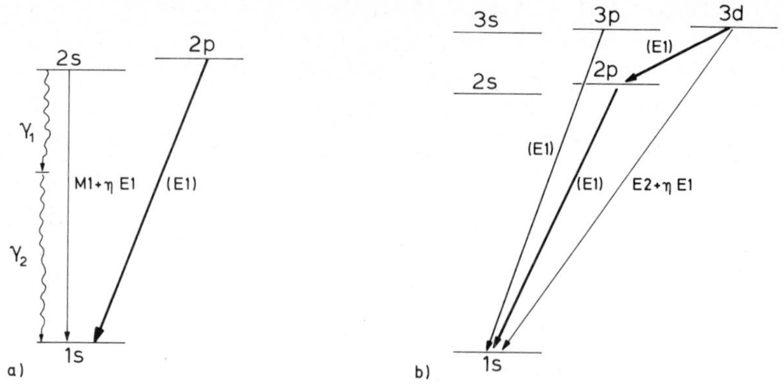

Fig. 12. Observable, parity violating effects in the muonic atom cascade as caused by NC weak interactions. (a): The 2p–2s state mixing causes a small admixture of E1 to the direct M1-transition from 2s- to 1s-states. (b) The 3d–3p state mixing gives rise to a small admixture of E1 to the dominant (3d–1s) E2-transition.

Here H^{NC} is an effective Hamiltonian which stems from the $V_\ell A_n$ and $A_\ell V_n$ couplings in eq. (295). An admixture of this kind is detectable by measuring the angular distribution of the photon (with the characteristics of the M1-, E1-mixture), with respect to the orientation of the muon's spin.

Another example is provided by the $n = 3$ system in muonic atoms, cf. Fig. 12b): The 3d-state decays preferentially via E1-emission to the 2p-state (and then on to the 1s-state via another E1-transition). However, it also has a small but nonvanishing probability of decaying directly to the 1s-state through emission of an E2 X-ray. The neutral current interaction will cause a small admixture of the 3p- to the 3d-state (and vice versa), so that the weak E2 transition receives a small contribution from the strong (3p–1s) E1 transition. Here again this admixture will show up in the angular distribution of the photon relative to some spin orientation. [In fact, this need not necessarily be the muon spin. It could as well be the spin of the nucleus provided the atom is formed in a polarized target.]

These two examples, as well as many others which have been examined (Missimer et al. 1984), have in common that one studies a weak transition T_0 to which H^{NC} admixes some other, strong transition T_1, such that the parity violating effect in the resulting transition amplitude

$$T \simeq T_0(1 + \eta T_1/T_0) \tag{297}$$

is enhanced by the factor T_1/T_0. [Clearly, T_0 must not be suppressed too much because otherwise the transition will have no measurable yield.] Let us briefly analyze H^{NC} as it follows from the basic interaction (186). The motion of the nucleons in the nucleus is essentially nonrelativistic. Furthermore, in first approximation the bound nucleons have the same weak coupling constants as when they are free. In the nonrelativistic reduction of the nucleonic vector current only the time component $\langle p'|v_0(0)|p\rangle$ survives, the space components $\langle p'|\boldsymbol{v}(0)|p\rangle$ being of higher order in (v/c). This is easy to see if one examines these matrix elements in the standard representation (III.56) or (III.59) and (III.68), in the limit $v/c \to 0$. Therefore, the interaction term due to the $A_\ell V_\mathrm{n}$-couplings becomes

$$H_1 \simeq \frac{G}{\sqrt{2}} \left\{ C_{1\mathrm{p}} \sum_{i=1}^{Z} \delta(\boldsymbol{r}_i - \boldsymbol{r}_\mu) + C_{1\mathrm{n}} \sum_{k=1}^{N} \delta(\boldsymbol{r}_k - \boldsymbol{r}_\mu) \right\} \gamma_5, \tag{298}$$

where the first sum runs over the protons, the second over the neutrons in the nucleus, and where γ_5 acts on the relativistic bound state wave functions of the muon. The parameters in eq. (298) are easy to identify from eqs. (186), (184), and (190):*)

$$C_{1\mathrm{p}} = \tfrac{1}{2}\left(1 - 4\sin^2\theta_\mathrm{w}\right),$$

$$C_{1\mathrm{n}} = -\tfrac{1}{2}.$$

Similarly, it is easy to see that the nonrelativistic limit of the nucleonic axial current $\langle p'|a_\alpha(0)|p\rangle$ vanishes if $\alpha = 0$, and gives a matrix element of the spin operator $\sigma^{(i)}$ for $\alpha = i$. Thus the interaction due to the $V_\ell A_\mathrm{n}$ couplings is approximately

$$H_2 \simeq \frac{G}{\sqrt{2}} \left\{ C_{2\mathrm{p}} \sum_{i=1}^{Z} \boldsymbol{\sigma}_i \delta(\boldsymbol{r}_i - \boldsymbol{r}_\mu) + C_{2\mathrm{n}} \sum_{k=1}^{N} \boldsymbol{\sigma}_k \delta(\boldsymbol{r}_k - \boldsymbol{r}_\mu) \right\} \cdot \boldsymbol{\alpha}, \tag{299}$$

where the matrices $\boldsymbol{\alpha}$ are defined in eqs. (III.59) and act on the muonic wave functions. The constants in eq. (299) are found from eqs. (190) and (191), viz.

$$C_{2\mathrm{p}} = -\tfrac{1}{2}F_\mathrm{A}(0)\left(1 - 4\sin^2\theta_\mathrm{w}\right),$$

$$C_{2\mathrm{n}} = \tfrac{1}{2}F_\mathrm{A}(0)\left(1 - 4\sin^2\theta_\mathrm{w}\right).$$

Without going into the detailed analysis of nuclear matrix elements of H_1, eq. (298), and H_2, eq. (299), we can make the following general remarks. The expectation value of H_1 in the nuclear ground state is proportional to a linear combination of proton and neutron densities,

$$ZC_{1\mathrm{p}}\rho_\mathrm{p}(\boldsymbol{r}_\mu) + NC_{1\mathrm{n}}\rho_\mathrm{n}(\boldsymbol{r}_\mu)$$

with $C_{1\mathrm{p}} \simeq 0.06$ (for $\sin^2\theta_\mathrm{w} \simeq 0.22$) and $C_{1\mathrm{n}} = -0.50$. In a static situation where the nucleus remains in the ground state, the nucleons contribute coherently, each with its own weak neutral coupling, $(1 - 4\sin^2\theta_\mathrm{w})$ for protons, 1 for neutrons.

In the same situation, the nuclear matrix element of H_2 either vanishes, when the nucleus has spin $J = 0$. Or else, if $J \geq \frac{1}{2}$, is proportional to $\langle \boldsymbol{\sigma} \rangle$, the matrix element of the spin operator of the unpaired valence nucleon. Except for the possibility of collective state admixtures, this matrix element is a typical one-particle matrix element. Therefore, in the static situation, one expects the interaction H_1 to predominate over the term H_2. This is indeed the result of detailed investigations of this

*)Here we are using the fact that the "charge" form factor of the vector current at zero momentum transfer is not renormalized by strong interactions if CVC holds. See the discussion of CVC in sec. 7.2 below.

problem. Therefore, experiments with muonic atoms test primarily the $A_\ell \cdot V_n$ couplings. The couplings of the type $V_\ell A_n$ which appear in H_2, eq. (299), should be accessible, too, but will be more difficult to isolate.

In order to give a feeling for the order of magnitude, we quote the example of muonic thulium $^{169}_{69}\mathrm{Tm}_\mu$: The forward–backward asymmetry of X-rays from the 3d–1s transition in this atom (with respect to the muonic or nuclear spin direction), due to parity violating NC interactions is calculated to be -1.2×10^{-5}.

Finally, we note and emphasize that the muonic atom is perhaps the best possibility of testing muon–electron universality of the weak NC interactions.

7. Hadronic charged current interactions

In this section we return to another facet of weak interactions that we mentioned in the introduction to this chapter: Assuming the Lorentz structure and selection rules of hadronic weak interactions to be known, the weak currents can be regarded as still another class of probes for the structure of hadronic targets. In close analogy to the role of the electromagnetic current in electron scattering on hadrons, the weak vector and axial vector currents give access to internal properties of mesons and baryons, mainly through the characteristics of their weak decay modes. Nucleons and nuclei, in particular, can also be probed in muon capture and in exclusive and inclusive neutrino scattering.

While muon capture is touched upon briefly in sec. 7.2, we concentrate primarily on *semileptonic* decays of mesons and nucleons for which we choose some illustrative and instructive examples.

7.1. Semileptonic interactions, structure of hadronic charged current

Our starting point is the effective current × current interaction (186) with $J_\alpha(x)$, the weak charged current, as given by eq. (183). In the product of this current with its hermitean conjugate there appear terms of the form

$$-\mathscr{L}^{\mathrm{CC}}_{\mathrm{semileptonic}} = \frac{G}{\sqrt{2}} \sum_{\mathrm{f,f'}} \left\{ \left(\overline{f(x)}\, \gamma^\alpha (1-\gamma_5) \nu_{\mathrm{f}}(x) \right) \left(\overline{u_{\mathrm{f'}}(x)}\, \gamma_\alpha (1-\gamma_5) d_{\mathrm{f'}} \right) + \mathrm{h.c.} \right\}, \tag{300}$$

which couple two lepton fields to two quark fields at the same point of space–time. Eq. (300), as well as its analogue with the neutral current K_α, eq. (184), defines and describes the so-called *semileptonic processes*. After the purely leptonic weak interactions this is a second, simple class of interactions: A semileptonic process involves one hadronic matrix element of the charged current multiplied by a leptonic vertex which is assumed to be known and which, therefore, is perfectly calculable. In other words, semileptonic processes give information on matrix elements of the hadronic pieces of J_α, viz.

$$\langle \mathrm{B}| \sum_{f=1}^{3} \overline{u_f(0)\, \gamma_\alpha(1-\gamma_5) d_f(0)} |\mathrm{A}\rangle. \tag{301}$$

Here A is a one-boson or one-baryon state, whilst B can be the vacuum, another one-boson or one-fermion state, etc., depending on the selection rules which apply to the matrix elements (301). Examples we shall encounter are

$$\langle 0|J_\alpha(0)|\pi^+\rangle, \quad \langle 0|J_\alpha(0)|\mathrm{K}^+\rangle, \quad \langle \pi^0|J_\alpha(0)|\pi^+\rangle,$$
$$\langle \gamma|J_\alpha(0)|\pi^+\rangle, \quad \langle \mathrm{p}|J_\alpha(0)|\mathrm{n}\rangle, \quad \text{etc.} \tag{301'}$$

All effective coupling constants in eq. (300) are known combinations of G and the elements of the mixing matrix U, eq. (152). What we do not know are the effects of the *strong* interactions on the matrix elements (301) which originate in the nature of hadrons being bound states of quarks and in hadronic corrections of higher order. These effects are parameterized in the frame of Lorentz scalar form factors which then are purely hadronic objects, i.e. in some general sense are "structure functions" of the hadronic system under investigation.

The charged current (183) contains nine different terms

$$\sum_{f=1}^{3} \sum_{f'=1}^{3} \overline{u_f(x)\, \Gamma^\alpha U_{ff'} b_{f'}(x)}, \quad \Gamma^\alpha \equiv \gamma^\alpha(1-\gamma_5),$$

which have different selection rules as far as the internal quantum numbers of hadronic states are concerned. In table 1 we list the pieces $\bar{\mathrm{u}}\mathrm{d}$ and $\bar{\mathrm{u}}\mathrm{s}$ which are relevant for the discussion in this section, together with their strong isospin, spin/parity, and strangeness properties. Unlike the case of the electromagnetic current, cf. table IV.1, the $\bar{\mathrm{u}}\mathrm{d}$-current is not diagonal under charge conjugation: Indeed, from the discussion in sec. 2.1 we know that

$$\overline{u(x)}\, \gamma^\alpha d(x) \underset{C}{\rightarrow} -\overline{d(x)}\, \gamma^\alpha u(x),$$

$$\overline{u(x)}\, \gamma^\alpha\gamma_5 d(x) \underset{C}{\rightarrow} +\overline{d(x)}\, \gamma^\alpha\gamma_5 u(x).$$

However, if we apply a rotation $\exp\{\mathrm{i}\pi I_2\}$ in isospin space to the fields $u(x)$ and $d(x)$, the currents on the r.h.s. are mapped back onto the originals on the l.h.s.. From eq. (III.23) we have

$$\mathrm{e}^{\mathrm{i}\pi I_2} u(x) = d(x), \qquad \mathrm{e}^{\mathrm{i}\pi I_2} d(x) = -u(x).$$

Thus, if one applies the combined operation of G-parity to the $\bar{\mathrm{u}}\mathrm{d}$-currents $G = \exp\{\mathrm{i}\pi I_2\}C$, the vector current is seen to be even, the axial current to be odd.

In the last column of table 1, finally, we list a few meson states that have the same quantum numbers as the currents listed in the first column. It is worth nothing that the $\bar{\mathrm{u}}\mathrm{d}$ vector current has the same quantum numbers as the isovector part of the electromagnetic current as shown in the second line of table IV.1.

Table 1
Quantum numbers of hadronic weak currents in the u, d, s-sector of the Lagrangian.

Current	Strong Isospin I	I_3	Spin/Parity J^π	ΔS	G-parity G	Analogue meson states
$\bar{u}\gamma^\alpha d\cos\theta_1$	1	1	$\begin{Bmatrix} 0^+ \\ 1- \end{Bmatrix}$	0	+	$\rho(770)$
$\bar{u}\gamma^\alpha\gamma_5 d\cos\theta_1$	1	1	$\begin{Bmatrix} 0^- \\ 1+ \end{Bmatrix}$	0	−	$\pi(139)$, $A_1(1275)$
$\bar{u}\gamma^\alpha s\sin\theta_1\cos\theta_3$	$\frac{1}{2}$	$\frac{1}{2}$	$\begin{Bmatrix} 0^+ \\ 1^- \end{Bmatrix}$	1		K*(892)
$\bar{u}\gamma^\alpha\gamma_5 s\sin\theta_1\cos\theta_3$	$\frac{1}{2}$	$\frac{1}{2}$	$\begin{Bmatrix} 0^- \\ 1^+ \end{Bmatrix}$	1		K(494), $Q_1(1280)$

7.2. *Pion beta decay and conserved vector current* (CVC)

We noted above the close similarity of the ūd vector current, first line of table 1, and of $j_\alpha^{(1)}$, the isovector part of the electromagnetic current, second line of table IV.1. In a world of free u- and d-quarks these currents are, respectively,

$$j_\alpha^{(1)}(x)=\tfrac{1}{2}\left\{\overline{u(x)}\,\gamma_\alpha u(x)-\overline{d(x)}\,\gamma_\alpha d(x)\right\}=\tfrac{1}{2}\overline{N(x)}\,\gamma_\alpha\tau_3 N(x), \tag{302a}$$

where $N\equiv\binom{u}{d}$, and

$$v_\alpha^{(1+i2)}(x)=\overline{u(x)}\,\gamma_\alpha d(x)=\tfrac{1}{2}\overline{N(x)}\,\gamma_\alpha(\tau_1+i\tau_2)N(x), \tag{302b}$$

where eq. (302a) follows from eq. (155). It is obvious that these currents are components of one and the same triplet of isovector operators. If the masses of u and d quarks are exactly equal then v_α is conserved,

$$\partial^\alpha v_\alpha^{(1+i2)}(x)=0. \tag{303}$$

The hypothesis of the conserved vector current (CVC) states that these properties hold for the strangeness-conserving vector current, even when it is "dressed" by the strong interactions, viz.
(i) $v_\alpha^{(i\pm i2)}$ is conserved,
(ii) $v_\alpha^{(1+i2)}$ and $v_\alpha^{(1-i2)}$ are isospin components of a triplet of operators whose third member is $j_\alpha^{(1)}$, the isovector part of the electromagnetic current.

As an example let us consider pion β-decay

$$\pi^-(q)\to\pi^0(q')+e(p)+\bar{\nu}_e(k), \tag{304}$$

where we have written the momenta in parentheses. The T-matrix element reads

$$T(\pi^-\to\pi^0 e^-\bar{\nu}_e)=\frac{G}{\sqrt{2}}\cos\theta_1\frac{1}{(2\pi)^3}\overline{u_e(p)}\,\gamma^\alpha(1-\gamma_5)v_\nu(k)$$
$$\times\langle\pi^0(q')|v_\alpha^{(1+i2)}(0)|\pi^-(q)\rangle. \tag{305}$$

A general form factor decomposition of the pionic matrix element is

$$\langle \pi^0(q')|v_\alpha^{(1+i2)}(0)|\pi^-(q)\rangle = \frac{1}{(2\pi)^3}\{(q_\alpha + q'_\alpha)f_+(Q^2) + (q_\alpha - q'_\alpha)f_-(Q^2)\}, \tag{306}$$

with $Q^2 := (q-q')^2 = m^2 + m_0^2 - 2m\sqrt{m_0^2+\kappa^2}$ and where κ is the magnitude of the 3-momentum of π^0 in the rest system of the decaying π^-. m denotes the mass of the charged pion, m_0 the mass of π^0. If v_α is conserved, we have the condition

$$(m^2 - m_0^2)f_+(Q^2) + Q^2 f_-(Q^2) = 0.$$

The contribution of f_- to T, eq. (305), is multiplied by $(q-q') = (p+k)$ and, therefore, by a factor m_e which is small compared to the term $(q+q')f_+$. In the limit $m = m_0$ (exact isospin symmetry), one has $f_-(Q^2) \equiv 0$. Neglecting the second form factor one obtains the following expression for the decay width (see below)

$$\Gamma(\pi^- \to \pi^0 e^- \bar{\nu}_e) \simeq \frac{G^2\cos^2\theta_1 \Delta^5}{60\pi^3}\left(1 - \frac{3}{2}\frac{\Delta}{m}\right)K(\varepsilon)f_+^2(0), \tag{307}$$

where $\Delta = m - m_0 \simeq 4.6$ MeV, $\varepsilon \equiv m_e^2/\Delta^2$; $K(\varepsilon)$ is a correction factor given by

$$K(\varepsilon) = \sqrt{1-\varepsilon}\,(1 - \tfrac{9}{2}\varepsilon + 4\varepsilon^2) + \tfrac{15}{2}\varepsilon^2 \ln\left(\frac{1+\sqrt{1-\varepsilon}}{\sqrt{\varepsilon}}\right) \simeq 1 - 5\varepsilon.$$

We have replaced $f_+(Q^2)$ by its value at $Q^2 = 0$ because the domain of variation is very small in this decay, $m_e^2 \le Q \le \Delta^2$.

The value of $f_+(0)$ is fixed by the second part of the CVC hypothesis. In order to see this let us note that $I_+ := \int d^3x\, v_0^{(1+i2)}(x)$ is a generator of infinitesimal transformations in isospin space, provided isospin is an exact global symmetry. Thus

$$\langle \pi^0|\int d^3x\, v_0^{(1+i2)}(x)|\pi^-\rangle = \langle \pi^0|I_+|\pi^-\rangle = \sqrt{2}\, 2E_q \delta(\boldsymbol{q} - \boldsymbol{q}').$$

On the other hand, making use of the translation formula (IV.42), the left-hand side is equal to

$$\int d^3x\, e^{ix(q'-q)}\langle \pi^0|v_0^{(1+i2)}(0)|\pi^-\rangle = (2\pi)^3\delta(\boldsymbol{q} - \boldsymbol{q}')\frac{1}{(2\pi)^3}2E_q f_+(0),$$

which gives the result $f_+(0) = \sqrt{2}$. Note that in this derivation we have used exact isospin symmetry in the pion states and, consequently (except for the kinematics), we have set $m = m_0$.

The absolute square of the matrix element, summed over the spin is worked out to be

$$(2\pi)^{12}\sum|T|^2 = 4G^2\cos^2\theta_1 \times\{f_+^2(2(ap)(ak) - a^2(kp)) + f_-^2 m_e^2(kp) + 2f_+f_- m_e^2(ak)\}, \tag{308}$$

where we have set $a = q + q' = 2q - Q$. The terms which contain the factor m_e^2 are negligibly small, of order m_e^2/m_π^2, as compared to the others. For the calculation of the total width we use the analogue of eq. (266) above, viz.

$$\frac{d^3\Gamma}{d^3q'} = \frac{1}{4mE_{q'}(2\pi)^5} \int \frac{d^3p}{2E_p} \int \frac{d^3k}{2E_k} \delta(Q-p-k)(2\pi)^{12} \sum |T|^2. \tag{309}$$

The integrations over $\boldsymbol{p}$ and $\boldsymbol{k}$ are performed in the same way as the integration over the neutrino momenta in the case of muon decay, except that here we must keep the mass of the electron. Explicit calculation (exercise 19) shows that eq. (268b) is replaced by

$$\int \frac{d^3p}{2E_p} \int \frac{d^3k}{2E_k} \left(p^\alpha k^\beta - (pk) g^{\alpha\beta} + k^\alpha p^\beta \right) \delta(Q-p-k)$$

$$= \frac{\pi}{6} \frac{(Q^2 - m_e^2)^2}{(Q^2)^2} \left\{ \left(1 + 2\frac{m_e^2}{Q^2}\right) Q^\alpha Q^\beta - \left(1 + \frac{m_e^2}{2Q^2}\right) Q^2 g^{\alpha\beta} \right\}. \tag{310}$$

It is not possible to expand this expression in terms of m_e^2 because m_e^2 is not small compared to Q^2. In calculating

$$\Gamma = \int \boldsymbol{q}'^2 d|\boldsymbol{q}'| \frac{1}{E_{q'}} \cdots$$

it is useful to transform both $|\boldsymbol{q}'|$ and $E_{q'}$ to the variable Q^2. From

$$E_{q'} = \frac{1}{2m}(m^2 + m_0^2 - Q^2)$$

we have

$$\frac{1}{E_{q'}} \boldsymbol{q}'^2 d|\boldsymbol{q}'| = |\boldsymbol{q}'| dE_{q'} = -\frac{1}{2m} |\boldsymbol{q}'| d(Q^2)$$

with

$$|\boldsymbol{q}'| = \frac{1}{2m} \sqrt{(m^2 + m_0^2 - Q^2)^2 - 4m^2 m_0^2}$$

$$= \frac{1}{2m} \sqrt{(\Delta^2 - Q^2)(\Delta^2 - Q^2 + 4mm_0)}$$

$$= \frac{1}{m} \sqrt{mm_0} \sqrt{(\Delta^2 - Q^2)\left(1 + \frac{\Delta^2 - Q^2}{4mm_0}\right)}.$$

The following relations are needed when eq. (310) is contracted with $a_\alpha a_\beta$, as prescribed by eq. (308),

$$(aQ) = \Delta(m + m_0), \qquad a^2 = 2(m^2 + m_0^2) - Q^2.$$

Finally, it is convenient to introduce the dimensionless integration variable

$$z := Q^2/\Delta^2.$$

Neglecting small terms of the order of Δ^2/m^2 in the integrand one then has the following expression for the total width:

$$\Gamma \simeq \frac{G^2\cos^2\theta_1 f_+^2(0)}{24\pi^3 m}\,\frac{(m+m_0)^2}{4m^2}\sqrt{mm_0}\,\Delta^5\int_\varepsilon^1 \mathrm{d}z\sqrt{1-z}\,\frac{(z-\varepsilon)^2}{z^2}\left\{1-\frac{\varepsilon}{2}-z+2\frac{\varepsilon}{z}\right\}.$$

The mass factors can be approximated as follows: With $m_0 = m - \Delta$ one has

$$\frac{1}{4m^3}(m+m_0)^2\sqrt{mm_0} \simeq 1 - \frac{3}{2}\frac{\Delta}{m}.$$

The integral over z can be performed by elementary means [R25], giving $\int_\varepsilon^1 \mathrm{d}z \cdots = \frac{2}{5}K(\varepsilon)$ with $K(\varepsilon)$ as given above. Thus, eq. (307) is proven.

The experimental branching ratio of pion beta decay

$$\frac{\Gamma(\pi^- \to \pi^0 \mathrm{e}^- \bar{\nu}_\mathrm{e})}{\Gamma(\pi \to \mu\nu_\mu)}$$

is found to be $1.04(4)\times 10^{-8}$, which is in a very good agreement with the theoretical prediction of $1.035(5)\times 10^{-8}$ (which includes radiative corrections), and provides the most precise test of the CVC hypothesis at this time. The applications and predictions of CVC in nucleonic currents are dealt with in sec. 7.4 below.

7.3. *The strangeness conserving axial current*

7.3.1. *Pion decays $\pi\ell 2$ and $\tau \to \pi\nu$ decay*

The decay modes

$$\pi^- \to \mu^- \bar{\nu}_\mu, \qquad \pi^- \to \mathrm{e}^- \bar{\nu}_\mathrm{e}, \tag{311a}$$

$$\tau^+ \to \pi^+ \bar{\nu}_\tau, \tag{311b}$$

all involve a leptonic factor times a matrix element of the hadronic weak current between a one-pion state and the vacuum. In reactions (311a) it is $\langle 0|J_\alpha^{(1+\mathrm{i}2)}(0)|\pi^-(q)\rangle$, in (311b) it is $\langle \pi^+(q)|J_\alpha^{(1+\mathrm{i}2)}(0)|0\rangle$. The discussion in sec. 1 showed that only the axial part of the current can contribute to this matrix element, cf. eq. (6). Furthermore, knowing the behaviour of the axial current under time reversal and charge conjugation one shows that F_π is pure imaginary (exercise 20), thus

$$\langle 0|a_\alpha^{(1+\mathrm{i}2)}(0)|\pi^-(q)\rangle = \frac{\mathrm{i}}{(2\pi)^{3/2}} f_\pi q_\alpha. \tag{312}$$

The amplitude for the decays (311a) then reads

$$T_{\pi^-\to \mathrm{f}^-\bar{\nu}_\mathrm{f}} = \frac{\mathrm{i}}{(2\pi)^{9/2}}\frac{f_\pi G\cos\theta_1}{\sqrt{2}}\overline{u_\mathrm{f}(p)}\,\not{q}(1-\gamma_5)v_\nu(k)$$

$$= \frac{\mathrm{i}}{(2\pi)^{9/2}}\frac{f_\pi G\cos\theta_1}{\sqrt{2}} m_\mathrm{f}\overline{u_\mathrm{f}(p)}\,(1-\gamma_5)v_\nu(k),$$

with f = e or μ.

It is not difficult to calculate the decay width from this amplitude. One finds

$$\Gamma(\pi \to \mathrm{f}\nu_{\mathrm{f}}) = \frac{G^2\cos^2\theta_1 f_\pi^2 m_\pi}{8\pi} m_{\mathrm{f}}^2 \left(1 - m_{\mathrm{f}}^2/m_\pi^2\right)^2, \tag{313}$$

from which one obtains the ratio of $\pi \to e\nu_e$ to $\pi \to \mu\nu_\mu$ that we anticipated in eq. (38) (see also the discussion there). Knowing that the mode $\pi \to \mu\nu_\mu$ amounts to practically 100% of the pion decays, one deduces from the experimental lifetime $\tau_\pi = 2.6030(23) \times 10^{-8}$ sec and from $|\cos\theta_1| = 0.974$ a value for f_π:

$$f_\pi \simeq 0.944 m_{\pi^+} \simeq 0.132 \text{ GeV}. \tag{314}$$

In a similar way the decay (311b) is described by the amplitude

$$T_{\tau^+\to\pi^+\bar{\nu}_\tau} = \frac{\mathrm{i}}{(2\pi)^{9/2}} \frac{f_\pi G\cos\theta_1}{\sqrt{2}} \overline{v_\nu(k)}\, \not{q}(1-\gamma_5) v_\tau(p)$$

$$= \frac{\mathrm{i}}{(2\pi)^{9/2}} \frac{f_\pi G\cos\theta_1}{\sqrt{2}} m_\tau \overline{v_\nu(k)}\,(1+\gamma_5) v_\tau(p). \tag{315}$$

In this case it is useful to keep the correlation term between the spin of the τ and the pion momentum. If the pion is emitted at an angle θ with respect to the expectation value of s_τ, the differential decay probability is found to be, from eq. (315),

$$\frac{\mathrm{d}\Gamma(\tau^+ \to \pi^+\bar{\nu}_\tau)}{\mathrm{d}(\cos\theta)} = \Gamma(\tau \to \pi\nu_\tau)\tfrac{1}{2}(1 - \cos\theta), \tag{316}$$

where

$$\Gamma(\tau \to \pi\nu_\tau) = \frac{G^2\cos^2\theta_1 f_\pi^2 m_\tau^3}{16\pi}\left(1 - m_\pi^2/m_\tau^2\right)^2. \tag{317}$$

The decay mode has a branching ratio of about 10% and is found in good agreement with the prediction (317), (see exercise 21).

There are two different definitions of f_π, eq. (312), in the literature on particle physics. The first is the one we have adopted in eq. (312), where a *physical* matrix element of the isospin raising operator $a_\alpha^{(1+\mathrm{i}2)}$ between a one-pion state and the vacuum is decomposed in terms of covariants. We can write this current in terms of Cartesian components, viz. $a_\alpha^{(1+\mathrm{i}2)} = a_\alpha^{(1)} + \mathrm{i}a_\alpha^{(2)}$. If these were given in terms of free quark fields they would read

$$a_\alpha^{(i)} = \overline{N(x)}\,\gamma_\alpha\gamma_5 \frac{\tau^{(i)}}{2} N(x).$$

Obviously, the π^- state can be written in Cartesian coordinates, too,

$$|\pi^-\rangle = \frac{1}{\sqrt{2}}\{|\pi_1\rangle - \mathrm{i}|\pi_2\rangle\},$$

so that

$$\langle 0|a_\alpha^{(1+i2)}|\pi^-\rangle = \frac{1}{\sqrt{2}}\{\langle 0|a_\alpha^{(1)}|\pi_1\rangle + \langle 0|a_\alpha^{(2)}|\pi_2\rangle\}.$$

This implies that the Cartesian matrix elements are

$$\langle 0|a_\alpha^{(j)}(0)|\pi_k(q)\rangle = \frac{\mathrm{i}}{(2\pi)^{3/2}}\delta_{jk}\frac{1}{\sqrt{2}}f_\pi q_\alpha. \tag{318}$$

On many occasions it is convenient to work with Cartesian, rather than spherical coordinates. For example, in studying commutators of current densities $v_\alpha^{(i)}(x)$ and $a_\alpha^{(j)}(x)$ one usually makes use of the Cartesian notation which treats the three isospin components in a symmetric (in fact, cyclic) way. For this reason some authors prefer to use eq. (318) as the defining equation for the pion decay constant, but without the factor $1/\sqrt{2}$ on the r.h.s. Thus, another definition (not used in this book) is

$$\bar{f}_\pi = \frac{1}{\sqrt{2}}f_\pi \simeq 0.667 m_{\pi^+} \simeq 0.0933 \text{ GeV}. \tag{319}$$

Additional remark. Unfortunately this is not the only point where confusion may arise. In accord with standard phase conventions for the rotation group it would seem appropriate to express a π^+-state in the form

$$|\pi^+\rangle = -\frac{1}{\sqrt{2}}\{|\pi_1\rangle + \mathrm{i}|\pi_2\rangle\}, \tag{320}$$

with the characteristic minus sign of the spherical basis. With this convention the isospin rotation contained in the definition of G-parity is indeed $\mathrm{e}^{\mathrm{i}\pi I_2}$ whose matrix representation in the space of a unitary irreducible multiplet is $(-)^{I-\mu}\delta_{\mu-m}$. This rotation (which effects the transition from cogredience to contragredience) commutes with the generators of the strong isospin, as it should. Unfortunately, since the early days of analyzing pion–nucleon scattering it has become customary to define the $|\pi^+\rangle$ state with a plus sign, in disaccord with the spherical basis. In this latter convention the rotation factor in G must be chosen to be $\mathrm{e}^{\mathrm{i}\pi I_1}$.

There is no problem in adopting the spherical basis (320) provided it is done consistently throughout the calculation. Attention to phases must be paid, however, in handling matrix elements of isospin raising and lowering currents $j_\alpha^{(1\pm i2)} = j_\alpha^{(1)} \pm \mathrm{i}\, j_\alpha^{(2)}$ which neither are normalized nor are written in the spherical basis.

7.3.2. *Axial current and pion field:* PCAC

It is remarkable that the axial currents $a_\alpha^{(3)}(x)$ and $a_\alpha^{(1\pm i2)}(x)$ carry the internal quantum numbers of the pion states π^0, and $\pi^\pm$, respectively. Their divergences $\partial^\alpha a_\alpha^{(\lambda)}$, in particular, define pseudoscalar fields which have the same behaviour with respect to Lorentz transformations as the pion fields. For example, the pion-to-vacuum matrix element of the divergence of the isospin raising component can be calculated from eq. (312). Using the translation formula (IV.42) we have

$$\partial_x^\alpha \langle 0| a_\alpha^{(1+i2)}(x)|\pi^-(q)\rangle = \frac{\mathrm{i}}{(2\pi)^{3/2}} f_\pi q_\alpha \partial^\alpha \mathrm{e}^{-\mathrm{i}qx}$$

$$= \frac{1}{(2\pi)^{3/2}} f_\pi q^2 \mathrm{e}^{-\mathrm{i}qx} = \frac{1}{(2\pi)^{3/2}} f_\pi m_\pi^2 \mathrm{e}^{-\mathrm{i}qx}. \qquad (321)$$

This equation shows clearly that the axial current cannot be conserved. If it were then either f_π would have to vanish, in which case pions would not decay, or m_π^2 would have to be zero. [The second alternative is very interesting because it shows that in models with a conserved axial current pions may appear as massless Goldstone bosons.]

The divergence $\partial^\alpha a_\alpha^{(\lambda)}(x)$ is a pseudoscalar field operator which has all properties of a pion field $\phi^{(\lambda)}(x)$, except for the normalization. Indeed, an interpolating pion field must be normalized such that a one-pion matrix element is

$$\langle 0|\phi_\pi(x)|\pi(q)\rangle = \frac{1}{(2\pi)^{3/2}} \mathrm{e}^{-\mathrm{i}qx}.$$

Thus (expressed in Cartesian coordinates) we can define

$$\phi_\pi^{(i)}(x) := \frac{\sqrt{2}}{f_\pi m_\pi^2} \partial^\alpha a_\alpha^{(i)}(x) \qquad (322)$$

and use this operator as interpolating field for the pion. As it stands, this definition is just one possible choice and contains no more than the statements that the divergence carries the internal quantum numbers of the pion and that it is correctly normalized.

By the assumption of PCAC (partial conservation of the axial current) it is understood that typical one-particle matrix elements of the kind

$$\langle \mathrm{B}; q'|(\Box + m_\pi^2)\phi_\pi(x)|\mathrm{A}; q\rangle \qquad (323)$$

are smooth functions of the invariant, squared momentum transfer $Q^2 = (q-q')^2$, in the interval $-m_\pi^2 < Q^2 \lesssim m_\pi^2$. In other words, it is assumed that a vertex function $\langle \mathrm{B}|\phi_\pi|\mathrm{A}\rangle$ is dominated by the pion pole $1/(Q^2 - m_\pi^2)$ whose residue is a slowly varying function of Q^2.

In the next section we shall discuss two examples of applying PCAC to pion–nucleon vertices. Generally speaking, PCAC is a useful approximation whenever one has to estimate simple matrix elements of weak axial currents, or combinations of such currents with other currents. It is also useful when one wishes to take external pions off their mass shell, replace their interpolating field by the divergence $\partial^\alpha a_\alpha(x)$ and extrapolate to $q^2 = 0$ (so-called *soft-pion* method). In practice, this approximation is usually found to be a rather good one, in those examples where it can actually be tested. This, presumably, is a reflection of the observation that the physical world of hadrons is not very far from an idealized world in which the axial current is strictly conserved, like the vector current, and in which pions are massless Goldstone bosons of a spontaneously broken symmetry. This limit is called the limit of *chiral symmetry*.

7.4. CVC *and* PCAC *as applied to nucleonic currents*

7.4.1. *Nucleonic vector currents and* CVC

The weak charged vector current, taken between a neutron state of momentum q and a proton state of momentum q' has the general decomposition into covariants,

$$\langle \mathrm{p}; q' | v_\alpha^{(1+\mathrm{i}2)}(0) | \mathrm{n}; q\rangle = \frac{1}{(2\pi)^3} \overline{u_\mathrm{p}(q')} \left\{ \gamma_\alpha F_\mathrm{V}(Q^2) + \mathrm{i}\sigma_{\alpha\beta} \frac{Q^\beta}{2m_\mathrm{N}} F_\mathrm{M}(Q^2) + \frac{Q_\alpha}{2m_\mathrm{N}} F_3(Q^2) \right\} u_\mathrm{n}(q), \tag{324}$$

with $Q = q' - q$ (see eq. (IV.46)). The CVC hypothesis, when applied to this specific case, has two consequences:

(i) Very much like in the case of the electromagnetic current, the conservation condition (303) implies that $F_3(Q^2)$ vanishes identically.

(ii) The isotriplet character of $v_\alpha^{(1\pm\mathrm{i}2)}$ and $j_\alpha^{(1)}$ implies that the form factors F_V and F_M are identical with the nucleonic isovector form factors (IV.66b), up to a factor 2,

$$F_\mathrm{V}(Q^2) = F_1^{(\mathrm{p})}(Q^2) - F_1^{(\mathrm{n})}(Q^2), \tag{325a}$$

$$F_\mathrm{M}(Q^2) = F_2^{(\mathrm{p})}(Q^2) - F_2^{(\mathrm{n})}(Q^2) \tag{325b}$$

(see also exercise 22). In particular, CVC makes the important prediction

$$F_\mathrm{V}(0) = 1, \tag{326}$$

which implies that superallowed nuclear Fermi transitions have the strength $G\cos\theta_1$, without any renormalization by the strong interactions. Furthermore, eq. (325b) implies

$$F_\mathrm{M}(0) = \mu_\mathrm{an}^\mathrm{p} - \mu_\mathrm{an}^\mathrm{n} \simeq 3.706, \tag{327}$$

so that the "magnetic" terms in eq. (324) are fixed at $Q^2 = 0$. This latter prediction is the basis of a famous test of CVC in the isotriplet of 1^+-states in ^{12}B(g.s.), ^{12}N(g.s.), and ^{12}C*, which decay to the state ^{12}C(0^+ g.s.) through β^-, β^+ and M1-γ-decay, respectively.

Finally, CVC predicts that the dependence of the form factors (325) on Q^2 is the same as that of the isovector, electric and magnetic form factors.

7.4.2. *Nucleonic axial vector currents and* PCAC

Let us consider the vertex pnπ, Fig. 13, with the pion on or off its mass shell, and carrying the momentum $\boldsymbol{Q} = q' - q$.

The pion's source term j_π is defined by the action of the Klein–Gordon operator on the field $\phi_\pi(x)$:

$$j_\pi(x) := (\Box + m_\pi^2)\phi_\pi(x). \tag{328a}$$

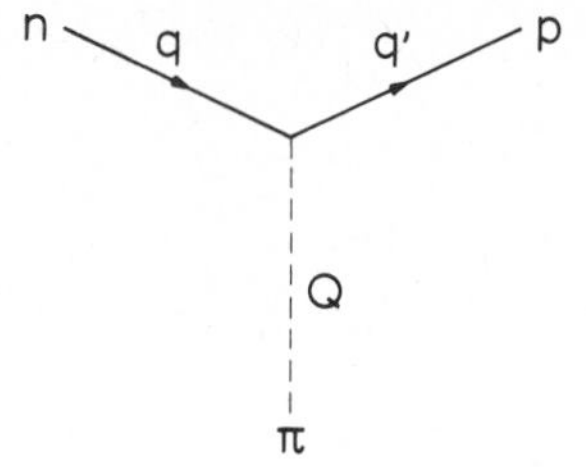

Fig. 13. Strong interaction vertex proton–neutron–pion.

$j_\pi(x)$ is a complicated operator function which contains the couplings of the pion to the other particles in the theory. Owing to this definition, a matrix element of the type of eq. (323) is

$$\langle \mathrm{B}; q'|(\Box + m_\pi^2)\phi_\pi(x)|\mathrm{A}; q\rangle = (-Q^2 + m_\pi^2)\langle \mathrm{B}; q'|\phi_\pi(x)|\mathrm{A}; q\rangle,$$

so that we have

$$\langle \mathrm{B}; q'|\phi_\pi(0)|\mathrm{A}; q\rangle = \frac{1}{m_\pi^2 - Q^2}\langle \mathrm{B}; q'|j_\pi(0)|\mathrm{A}; q\rangle. \tag{328b}$$

For the example of Fig. 13, a decomposition in Lorentz covariants is

$$\langle \mathrm{p}; q'|j_\pi(0)|\mathrm{n}; q\rangle = \frac{\mathrm{i}\sqrt{2}}{(2\pi)^3} g(Q^2)\overline{u_{\mathrm{p}}(q')}\,\gamma_5 u_{\mathrm{n}}(q). \tag{329}$$

Here $g(Q^2)$ is a Lorentz scalar form factor; at the point $Q^2 = m_\pi^2$ it is the conventional pion–nucleon coupling constant

$$g(Q^2 = m_\pi^2) \equiv g_{\pi\mathrm{NN}} \quad \text{with } g_{\pi\mathrm{NN}}^2/4\pi = 14.64. \tag{330}$$

[The relation to $f_{\pi\mathrm{NN}}$, eq. (I.123), is $f_{\pi\mathrm{NN}}^2 = (g_{\pi\mathrm{NN}}^2/4\pi)(m_\pi/2m_\mathrm{N})^2$.] Making use of eqs. (322) and (328b) we can write

$$\begin{aligned}\langle \mathrm{p}; q'|j_\pi(0)|n; \mathrm{q}\rangle &= (m_\pi^2 - Q^2)\langle \mathrm{p}; q'|\partial^\alpha a_\alpha^{(1+\mathrm{i}2)}(0)|\mathrm{n}; q\rangle \\ &= (m_\pi^2 - Q^2)\frac{\mathrm{i}Q^\alpha}{f_\pi m_\pi^2}\langle \mathrm{p}; q'|a_\alpha^{(1+\mathrm{i}2)}(0)|\mathrm{n}; q\rangle.\end{aligned}$$

On the r.h.s. of this equation we insert the form factor decomposition (191) (with $F_\mathrm{T} \equiv 0$), and obtain

$$= \frac{\mathrm{i}}{(2\pi)^3}\frac{m_\pi^2 - Q^2}{f_\pi m_\pi^2}\overline{u_{\mathrm{p}}(q')}\left\{\not{Q}\gamma_5 F_\mathrm{A}(Q^2) + \frac{Q^2}{2m_\mathrm{N}}\gamma_5 F_\mathrm{P}(Q^2)\right\}u_\mathrm{n}(q).$$

Upon comparison with eq. (329), and making use of the Dirac equations by means of which

$$\overline{u_{\mathrm{p}}(q')}\,\not{Q}\gamma_5 u_\mathrm{n}(q) = (m_\mathrm{p} + m_\mathrm{n})\overline{u_\mathrm{p}(q')}\,\gamma_5 u_\mathrm{n}(q) \simeq 2m_\mathrm{N}\bar{u}_\mathrm{p}\gamma_5 u_n,$$

we obtain the equation

$$\sqrt{2}\, g(Q^2) = \frac{m_\pi^2 - Q^2}{f_\pi m_\pi^2}\left\{2m_N F_A(Q^2) + \frac{Q^2}{2m_N} F_P(Q^2)\right\}. \tag{331}$$

As such and without further knowledge about the form factors g, F_A, and F_P, this equation is barely more than an identity because at least one of its two sides must be off-shell: $g(Q^2)$ is physical, i.e. measurable at $Q^2 = m_\pi^2$, whilst $F_A(Q^2)$ and $F_P(Q^2)$ are measurable in the domain $Q^2 \leq 0$ which does not contain the point m_π^2. It is precisely at this stage that the assumption of PCAC becomes effective: According to this assumption the extrapolation of the form factors away from their physical domain should be as smooth as possible.

Example 1: Goldberger–Treiman relation. As an application let us evaluate eq. (331) at $Q^2 = 0$, and let us assume that $g(Q^2)$ is essentially constant when extrapolated from the physical point m_π^2 to the point zero, $g(0) \simeq g(m_\pi^2)$. In this approximation we obtain the relation

$$F_\pi \simeq \frac{\sqrt{2}\, m_N F_A(0)}{g_{\pi NN}} \tag{332}$$

(so-called Goldberger–Treiman relation), which now relates measurable quantities. $F_A(0)$ is determined in the decay $n \to pe^-\bar{\nu}_e$ with the result $F_A(0) = 1.255(6)$. The r.h.s. of eq. (332) then gives $0.880\ m_{\pi^+} = 0.123$ GeV which is indeed rather close to the measured value of f_π, eq. (314).

Example 2: Muon capture in nuclei. As a second application consider capture of a muon from the 1s-state in a muonic atom via the reaction $\mu^- p \to n\nu_\mu$. For the sake of simplicity let us assume that the muon (momentum p) and the proton (momentum q) on which it is captured, are in a relative s-state and move with small, nonrelativistic velocities, i.e. $p \simeq (m_\mu, \mathbf{0})$, $q \simeq (m_N, \mathbf{0})$. The energy of the neutron in the final state is then approximately

$$E_{q'} \simeq m_N\left\{1 + m_\mu^2/2m_N^2\right\},$$

and the squared momentum transfer is $Q^2 = (q' - q)^2 = 2m_N(m_N - 2E_q') \simeq -m_\mu^2$. In the theory of muon capture it is customary to redefine the form factor F_P as follows

$$G_P := \frac{m_\mu}{2m_N} F_P\left(-m_\mu^2\right).$$

At $Q^2 = -m_\mu^2$ and assuming the extrapolations

$$F_A\left(-m_\mu^2\right) \simeq F_A(0), \qquad g\left(-m_\mu^2\right) \simeq g\left(m_\pi^2\right),$$

eq. (331) leads to the approximate equality

$$m_\mu G_P \simeq 2m_N F_A(0) - \frac{m_\pi^2\sqrt{2}}{m_\pi^2 + m_\mu^2} g_{\pi NN} f_\pi,$$

or, by making use of the relation (332),

$$\frac{G_\mathrm{P}}{F_\mathrm{A}(0)} \simeq \frac{2m_\mu m_\mathrm{N}}{m_\pi^2 + m_\mu^2}. \tag{333}$$

Thus, PCAC predicts $G_\mathrm{P} \simeq 6.5F_\mathrm{A}(0) = 8.13$, in good agreement with the results of muon capture experiments.

Example 3: Adler–Weisberger relation. One can apply PCAC and the method of soft pions to pion–nucleon scattering at threshold. With an additional input from current algebra, stating, in essence, that the equal time commutator of an axial vector charge $Q_\mathrm{A}^{(i)}(x^0) := \int \mathrm{d}^3x\, a_0^{(i)}(x)$ with another axial vector current is proportional to a vector current, viz.

$$\left[Q_\mathrm{A}^{(i)}(0), a_\alpha^{(k)}(0)\right] = \mathrm{i}\varepsilon_{ikl} v_\alpha^{(l)}(0),$$

one derives the following relation between $F_\mathrm{A}(0)$, $g_{\pi\mathrm{NN}}$ and the isovector pion–nucleon scattering length, eq. (II.33b):*)

$$F_\mathrm{A}^2(0) \simeq 6\frac{g_{\pi\mathrm{NN}}^2}{4\pi}\left(\frac{m_\pi}{2m_\mathrm{N}}\right)^2 \frac{1}{m_\pi(1 + m_\pi/m_\mathrm{N})(a_1 - a_3)}. \tag{334}$$

This relation predicts that $(a_3 - a_1)$ is negative; from the values of $g_{\pi\mathrm{NN}}$ and F_A quoted above one finds $(a_3 - a_1)m_\pi \simeq -0.27$, in striking agreement with the experimental value (II.78a).

7.5. *Another example: pion radiative decay*

A further example which illustrates the properties of hadronic weak currents in a transparent and instructive manner is the decay

$$\pi^+ \to \mathrm{e}^+ \nu_\mathrm{e} \gamma \qquad (\pi^- \to \mathrm{e}^- \bar{\nu}_\mathrm{e} \gamma). \tag{335}$$

This decay mode has a branching ratio of $5.6(7) \times 10^{-8}$, and, therefore, is almost as rare as pion beta decay. Although the analogous muonic mode

$$\pi^+ \to \mu^+ \nu_\mu \gamma \tag{336}$$

is much more frequent (branching ratio $1.24(25) \times 10^{-4}$), it is less interesting than the electronic mode (335) because, as we shall see, it is completely dominated by internal bremsstrahlung (that is, by processes in which the photon is shaken off by the external charged particles). In the decay mode (335), to the contrary, there are sizeable contributions from genuine structure terms which are comparable to the contributions from the bremsstrahlung diagrams.

It is convenient to analyze this process in several steps, as follows.

7.5.1. *Internal bremsstrahlung*

The diagrams describing internal bremsstrahlung are shown in Fig. 14. In the diagram 14a the photon is emitted by the outgoing positron, whilst in diagram 14b it

*) We skip the derivation of eq. (334) and refer to the original literature (Adler 1965, Weisberger 1966, Weinberg 1966).

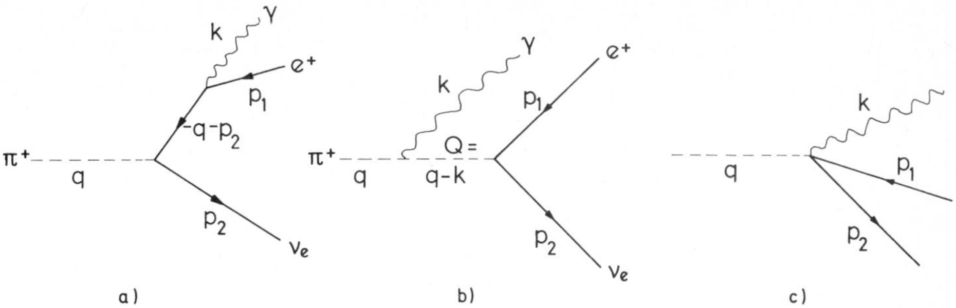

Fig. 14. Internal Bremsstrahlung in the decay $\pi^+ \to e^+ \nu_e \gamma$. In diagram 9a) it is the positron, in diagram (b) it is the pion that radiates. Diagram (c) is the contact term. Only the sum of the three graphs is gauge invariant.

is emitted by the incoming pion. In either case the process contains the decay amplitude $\pi^+ \to e^+ \nu_e$ (on or off-shell) and an ordinary bremsstrahlung process. Following standard rules, the diagram of Fig. 14a corresponds to the amplitude

$$T_{IB}^{(e)} = -\frac{ief_\pi G}{(2\pi)^6\sqrt{2}} \varepsilon_\alpha \overline{u_\nu(p_2)}\, \not{q}(1-\gamma_5) \frac{-(\not{p}_1 + \not{k}) + m_e}{2(p_1 k)} \gamma^\alpha v_e(p_1), \tag{337a}$$

where q is the momentum of the decaying pion; p_1, p_2 and k are the momenta of the positron, neutrino, and photon, respectively, whilst ε_α is the photon's polarization vector. The energy denominator in the propagator is $(p_1 + k)^2 - m_e^2 = 2(p_1 k)$.

The amplitude 14b (the photon being emitted by the pion) reads

$$T_{IB}^{(\pi)} = \frac{ief_\pi G}{(2\pi)^6\sqrt{2}} \varepsilon_\alpha \frac{(q+Q)^\alpha Q^\beta}{(q-k)^2 - m_\pi^2} l_\beta, \tag{337b}$$

where

$$l_\beta := \overline{u_\nu(p_2)}\, \gamma_\beta(1-\gamma_5) v_e(p_1) \tag{338}$$

is the leptonic vertex and where $Q := q - k$ is the momentum of the pion in the intermediate state.

Eq. (337b) deserves some more explanation and a word of caution. The pion in the intermediate state is not on its mass shell. Therefore, at the weak vertex we have to insert $f_\pi(Q^2)$ at a squared momentum transfer off the pion's mass shell. Furthermore, the electromagnetic vertex describes the transition from an external pion to an offshell pion state and, therefore, must be described by a general form factor decomposition of the form of eq. (306), viz.

$$\begin{aligned} &f_+(Q^2, k^2)(q+Q)^\alpha + f_-(Q^2, k^2)(q-Q)^\alpha \\ &= f_+(Q^2, k^2)(2q-k)^\alpha + f_-(Q^2, k^2)k^\alpha, \end{aligned} \tag{339a}$$

where $f_+(Q^2 = m_\pi^2, k^2)$ is the pion's electric form factor, (with $f_+(0) = 1$), whilst f_- is a form factor that would not contribute if Q^2 were equal to m_π^2. Indeed one can show (see remarks at the end of this section) that f_+ and f_- are linearly dependent and are related by

$$(m_\pi^2 - Q^2)f_+(Q^2, k^2) + k^2 f_-(Q^2, k^2) = m_\pi^2 - Q^2. \tag{339b}$$

This relation (taken at $k^2 \neq 0$) is instructive as it says that f_- is proportional to the squared r.m.s. radius of the pion,

$$\lim_{Q^2 \to m_\pi^2} \left(\frac{1}{m_\pi^2 - Q^2} f_-(Q^2, k^2) \right) = \frac{1}{k^2}\left(1 - f_+(m_\pi^2, k^2)\right) = -\tfrac{1}{6}\langle r^2 \rangle + \mathrm{O}(k^2). \tag{339c}$$

In the amplitude (337b) we have replaced $f_\pi(Q^2)$ by its on-shell value $f_\pi(m_\pi^2)$, and have set $f_+(Q^2, k^2) \equiv 1$. In terms of physics this means that we have taken the pion to be a pointlike particle which possesses no internal structure. This is legitimate provided we lump the extra terms into what we shall call the structure terms proper (see below).

The sum of the amplitudes (337a) and (337b) is not gauge invariant for, if we replace ε_α by k_α, we obtain

$$\begin{aligned} k_\alpha \overline{u_\nu(p_2)} \Bigg\{ & \frac{1}{2(p_1 k)} \not{q}(1 - \gamma_5)(m_\mathrm{e} - \not{p}_1 - \not{k})\gamma^\alpha \\ & + \frac{1}{2(qk)}(2q - k)^\alpha(\not{q} - \not{k})(1 - \gamma_5) \Bigg\} v_\mathrm{e}(p_1) \\ = & \overline{u_\nu}\{(-\not{q} + \not{q} - \not{k})(1 - \gamma_5)\} v_\mathrm{e} = -\overline{u_\nu}\not{k}(1 - \gamma_5) v_\mathrm{e}. \end{aligned}$$

In this equation we substituted $-\not{p}_1\not{k} = \not{k}\not{p}_1 - 2(p_1 k)$, applied the Dirac equation and used the mass shell condition $k^2 = 0$. This calculation shows that gauge invariance can be restored by adding the following amplitude to eqs. (337a) and (337b):

$$T_{\mathrm{IB}}^{(\pi\mathrm{e})} = -\frac{\mathrm{i} e f_\pi G}{(2\pi)^6 \sqrt{2}} \varepsilon_\alpha g^{\alpha\beta} l_\beta. \tag{337c}$$

Indeed, the sum of $T_{\mathrm{IB}}^{(\mathrm{e})}$, $T_{\mathrm{IB}}^{(\pi)}$, and $T_{\mathrm{IB}}^{(\pi\mathrm{e})}$ is now gauge invariant.

There is no arbitrariness in this procedure. This additional term, needed to make internal bremsstrahlung gauge invariant by itself, is fixed uniquely to the order $(k)^0$ in an expansion in terms of the photon momentum. Physically it represents a contact term (see Fig. 14c) where the photon and the lepton pair emerge at the same point of space–time. It arises naturally in models in which the pions remain structureless such as the nonrelativistic quark model (Scheck et al. 1973), see also exercise 23. Finally, having ascertained that the contributions from bremsstrahlung are gauge invariant, the structure terms proper must also be gauge invariant by themselves. This is a useful restriction in formulating form factor decompositions of those structure terms. With the abbreviations

$$s := (qk), \qquad \kappa := -\frac{eG}{(2\pi)^6 \sqrt{2}}, \tag{340}$$

the sum of the three amplitudes (337) is

$$T_{\mathrm{IB}} = \kappa \mathrm{i} f_\pi \varepsilon_\alpha \left\{ \left[g^{\alpha\beta} + \frac{1}{s} q^\alpha (q^\beta - k^\beta) \right] l_\beta \right.$$

$$\left. + \frac{1}{2(p_1 k)} \overline{u_\nu(p_2)} \not{q} (1-\gamma_5)(m_e - \not{p}_1 - \not{k}) \gamma^\alpha v_e(p_1) \right\}, \tag{341}$$

with l_β as defined by eq. (338).

7.5.2. Structure terms

Let us now turn to the structure terms which, by definition, describe all contributions to the decay (335) which are not contained in the amplitude (341). In the amplitude (341) the pion was taken to be structureless and no other intermediate states than those of Figs. 14a and 14b were admitted.

In the conversion of a pion into a photon via the weak hadronic currents both the vector and the axial vector current can contribute. This is so because the photon contains either G-parity, plus and minus (cf. Table IV.1). The pion has $G(\pi) = -1$. Therefore, in the matrix element $\langle \gamma | v_\alpha(0) | \pi \rangle$ it is the iso*scalar* part of the photon that contributes (cf. Table 1),

$$\langle \gamma(I=0, G=-) | v_\beta(I=1, G=+) | \pi(I=1, G=-) \rangle. \tag{342a}$$

Similarly, in the matrix element $\langle \gamma | a_\alpha(0) | \pi \rangle$ it is the iso*vector* piece that contributes,

$$\langle \gamma(I=1, G=+) | a_\beta(I=1, G=-) | \pi(I=1, G=-) \rangle. \tag{342b}$$

In decomposing the matrix elements (342) in terms of Lorentz covariants, we have to keep in mind the following points:

(i) Both expressions (342) must be proportional to ε_α, the polarization vector of the photon. ε_α is an axial vector.

(ii) The decomposition of either matrix element, (342a) or (342b), must be gauge invariant on its own.

(iii) If $\varepsilon_\alpha M^{\alpha\beta}$ is to be a vector (an axial vector), $M^{\alpha\beta}$ must be a pseudotensor (genuine tensor) constructed on the basis of $g^{\mu\nu}$, $\varepsilon^{\mu\nu\sigma\tau}$, and the momenta k and q.

The only pseudotensor we can form is $\varepsilon^{\alpha\beta\sigma\tau} k_\sigma q_\tau$. It is automatically gauge invariant since its contraction with k_α vanishes. Thus, the matrix element (342a) must be proportional to

$$\varepsilon_\alpha(k) \varepsilon^{\alpha\beta\sigma\tau} k_\sigma q_\tau .$$

Regarding proper tensors the only possible forms are $g^{\alpha\beta}$, $q^\alpha k^\beta$, $k^\alpha q^\beta$, $k^\alpha k^\beta$ and $q^\alpha q^\beta$, the first two of which can be combined to a gauge invariant form,

$$(qk) g^{\alpha\beta} - q^\alpha k^\beta .$$

The third and fourth give zero upon contraction with ε_α, whilst the fifth is not gauge invariant.

In conclusion, the structure dependent contributions to the process (335) can be written in the following effective form:

$$T_S = \kappa \frac{1}{m_\pi} \varepsilon_\alpha \{ F(s) \varepsilon^{\alpha\beta\sigma\tau} k_\sigma q_\tau + \mathrm{i} a(s)(s g^{\alpha\beta} - q^\alpha k^\beta) \} l_\beta . \tag{343}$$

Here $F(s=(qk))$ is an invariant form factor that describes the vector structure (342a), $a(s)$ is the form factor that describes the axial vector structure (342b); s and κ are defined by eqs. (340). We have taken out a factor $1/m_\pi$ in order to make $F(s)$ and $a(s)$ dimensionless. [We note, but do not prove here, that from time reversal and charge conjugation invariance F and a are real.]

The hypothesis of CVC implies that the vector form factor $F(s)$ is related to the amplitude for the decay of the neutral pion into two photons,

$$\pi^0 \to \gamma\gamma . \tag{344}$$

This is seen as follows. In analyzing the properties of the matrix element $\langle \gamma_1 \gamma_2 | \pi^0 \rangle$ with respect to internal quantum numbers we realize that the isospins and G-parities contained in the photon states must occur in either of the two combinations

$$\begin{aligned} &(\text{photon 1:} \quad I^G = 1^+, \text{photon 2:} \quad I^G = 0^-), \\ \text{or} \\ &(\text{photon 1:} \quad I^G = 0^-, \text{photon 2:} \quad I^G = 1^+), \end{aligned} \tag{345}$$

in order to match the quantum numbers $I^G = 1^-$ of the pion. However, due to their boson nature, the two photons are indistinguishable and, therefore, these two possibilities are identical. In other words, the decay (344) is characterized by one single amplitude. Furthermore, the structure of the matrix element (342a) is exactly as indicated in eq. (345). With the CVC relation between the electromagnetic current (the source of the photon field) and the weak vector current v_α, it follows that the form factor $F(s)$ must be proportional to the amplitude characteristic for π^0-decay.

The exact relationship is this: Let k, k' be the momenta of the two photons, respectively, ε^α, ε'^β their polarization vectors, and let q be the momentum of the pion. For the same reasons as for eq. (342a), the amplitude for the decay (344) must have the form

$$T_{\pi^0 \to \gamma\gamma} = \frac{e^2}{(2\pi)^{9/2}} \frac{1}{m_\pi} 2 a_0 \varepsilon_{\alpha\beta\sigma\tau} \varepsilon^\alpha \varepsilon'^\beta k'^\sigma k^\tau . \tag{346a}$$

[We have, arbitrarily, inserted a factor 2 because of the two identical possibilities (345). The factor $1/m_\pi$ is introduced in order to make a_0 dimensionless.]

The decay width is easily worked out to be

$$\Gamma(\pi^0 \to \gamma\gamma) = \alpha^2 \pi a_0^2 m_\pi . \tag{346b}$$

From the measured value $\Gamma(\pi^0 \to \gamma\gamma) = 7.85(54)$ one finds

$$|a_0| = \frac{\sqrt{\Gamma(\pi^0 \to \gamma\gamma)}}{\alpha \sqrt{\pi m_\pi}} = 1.865(64) \times 10^{-2} . \tag{347}$$

Using the Wigner–Eckart theorem we have

$$\langle\gamma(I^G=0^-)|j_\alpha^{(1)}(0)|\pi^0\rangle = \frac{1}{\sqrt{3}}(0\|j_\alpha\|1), \tag{348a}$$

$$\langle\gamma(I^G=0^-)|\frac{1}{\sqrt{2}}v_\alpha^{(1-i2)}(0)|\pi^0\rangle = -\frac{1}{\sqrt{3}}(0\|j_\alpha\|1), \tag{348b}$$

where the factor on the r.h.s. is the reduced (isospin) matrix element of the triplet current. The same matrix elements, when expressed in terms of the form factors $F(s)$ and a_0, read

$$\begin{aligned}\langle\gamma(0^-)|j_\alpha^{(1)}(0)|\pi^0\rangle &= \frac{e}{(2\pi)^6}\frac{1}{m_\pi}2a_0\varepsilon_{\alpha\beta\sigma\tau}\varepsilon^\beta k'^\sigma k^\tau \\ &= \frac{e}{(2\pi)^6}\frac{1}{m_\pi}2a_0\varepsilon_{\alpha\beta\sigma\tau}\varepsilon^\beta q^\sigma k^\tau,\end{aligned}$$

$$\langle\gamma(0^-)|v_\alpha^{(1-i2)}(0)|\pi^+\rangle = \frac{e}{(2\pi)^6}\frac{1}{m_\pi}F(s)\varepsilon_{\alpha\beta\sigma\tau}\varepsilon^\beta k^\sigma q^\tau.$$

Upon comparison with the isospin decomposition (348), and noting that in the decay $\pi^0\to\gamma\gamma$ the variable s has the value $s=(qk)=\frac{1}{2}q^2=\frac{1}{2}m_\pi^2$, one obtains the CVC relation

$$F(s=\tfrac{1}{2}m_\pi^2) = a_0\sqrt{2}\,. \tag{349}$$

One important aspect of this result is that pion radiative decay now depends on only one unknown form factor, $a(s)$, defined in eq. (343). Thus, assuming CVC, the data can be analyzed in terms of the ratio

$$R \coloneqq a(s)/F(s). \tag{350}$$

[Past experiments were all analyzed in this fashion. It seems possible, however, to determine both $a(s)$ and $F(s)$ from the double-differential decay probability, eqs. (355) below, in future precision measurements. This would provide still another, direct, test of CVC, via eq. (349).]

The axial form factor is an interesting quantity for testing theoretical models of chiral symmetry. We do not go into these more theoretical speculations and we just note that, at the very least, it must contain the form factor effect that we discussed in connection with eq. (337b).*) Indeed, from eq. (339c) we see that we must have

$$a = \bar{a} + 2m_\pi f_\pi \left.\frac{\partial f_+}{\partial k^2}\right|_{k^2=0} = \bar{a} + \frac{1}{3}m_\pi f_\pi\langle r_\pi^2\rangle, \tag{351}$$

where $\bar{a}$ describes other structure terms contributing to the axial matrix element. We

*) The difference between $f_\pi^2(Q^2)$ and $f_\pi(q^2=m_\pi^2)$ is negligibly small since $Q^2-q^2=-2s$, with $s=(qk)$ varying between 0 and $m_\pi^2/2$. So when one expands

$$f_\pi(Q^2) \simeq f_\pi(q^2) - 2s\,\partial f_\pi/\partial Q^2|_{Q^2=m_\pi^2},$$

the derivative is multiplied by a small number.

note that the second term in eq. (351), with $\langle r_\pi^2 \rangle = 0.46\ \mathrm{fm}^2$, would give $\frac{1}{3} m_\pi f_\pi \langle r_\pi^2 \rangle \simeq 7.24 \times 10^{-2}$. The curious observation is that the data seem to favour $a \simeq 0$. This would imply a rather strong cancellation in eq. (351). An analysis in the framework of quantum chromodynamics based on sum rules (Nasrallah et al. 1982) does indeed give a close to zero.

7.5.3. *The differential decay spectrum*

The calculation of the differential decay probability from the amplitudes (341) and (343) is rather tedious and we skip its details. It is convenient to choose the energies of the positron (E_e) and of the photon (E_γ) as the independent variables, rather than one energy and the opening angle θ between the photon and the positron. The standard formulae of App. B give, in the pion's frame,

$$\frac{\mathrm{d}^2 \Gamma}{\mathrm{d}(\cos\theta)\,\mathrm{d}E_\gamma} = \frac{(2\pi)^9}{8m_\pi} \int \mathrm{d}E_\mathrm{e} \sum |T|^2 \frac{|\boldsymbol{p}_1| E_\gamma}{E_\nu} \delta(E_\nu + E_\gamma + E_\mathrm{e} - m_\pi).$$

Neglecting the electron mass, we have $E_\mathrm{e} \simeq |\boldsymbol{p}_1|$ and, from energy and momentum conservation,

$$2E_\mathrm{e} E_\gamma (1 - \cos\theta) - 2m_\pi (E_\mathrm{e} + E_\gamma) + m_\pi^2 = 0. \tag{352}$$

With $m_\mathrm{e} \simeq 0$ the maximal energy of the positron as well as that of the photon is $E_{\max} = m_\pi/2$. It is convenient to introduce the dimensionless variables

$$x := \frac{E_\gamma}{E_{\max}} \simeq \frac{2E_\gamma}{m_\pi}, \qquad y := \frac{E_\mathrm{e}}{E_{\max}} \simeq \frac{2E_\mathrm{e}}{m_\pi}, \tag{353}$$

and to transform the differential decay rate accordingly. The Jacobian is

$$\frac{\partial(E_\gamma, \cos\theta)}{\partial(x, y)} = \frac{x-1}{xy^2} m_\pi .$$

Working out the traces and transforming to the variables x and y, one finds eventually:

$$\frac{\mathrm{d}^2 \Gamma(\pi \to \mathrm{e}\nu\gamma)}{\mathrm{d}x\,\mathrm{d}y} = \frac{\alpha}{2\pi} \Gamma(\pi \to \mathrm{e}\nu) \{ W_{\mathrm{IB}}(x, y) + W_{\mathrm{int}}(x, y) + W_{\mathrm{S}}(x, y) \}, \tag{354}$$

where $\Gamma(\pi \to \mathrm{e}\nu)$ is the decay width (313), with $\mathrm{f} \equiv \mathrm{e}$, W_{IB} is due to the bremsstrahlung diagrams, W_S to the structure terms, whilst W_{int} contains the interference terms, viz.

$$W_{\mathrm{IB}}(x, y) = \frac{1-y}{x^2} \frac{(x-1)^2 + 1}{x+y-1}, \tag{355a}$$

$$W_{\mathrm{int}}(x, y) = \frac{m_\pi}{f_\pi} \frac{1-y}{x} \left\{ (F+a)(1-x) - (F-a)\left(2 - y + \frac{(1-y)^2}{x+y-1} \right) \right\}, \tag{355b}$$

$$W_\mathrm{S}(x, y) = \left(\frac{m_\pi}{f_\pi} \right)^2 \left(\frac{m_\pi}{2m_\mathrm{e}} \right)^2 (1-x) \{ (F+a)^2 (x+y-1)^2 + (F-a)^2 (1-y)^2 \}. \tag{355c}$$

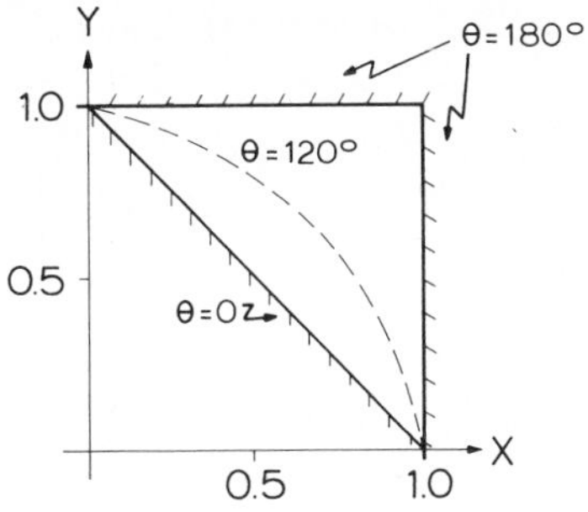

Fig. 15. The triangle indicates the domain of allowed photon and positron energies in the decay $\pi^+ \to e^+ \nu_e \gamma$. x is the reduced photon energy $x = 2E_\gamma/m_\pi$, y is the reduced positron energy $y = 2E_e/m_\pi$. θ is the opening angle between photon and positron momentum.

Let us now discuss these formulae and add a few remarks on this decay. With $m_e \simeq 0$ the kinematically allowed region for x and y is the triangle indicated in Fig. 15. According to eq. (352), which now reads

$$\tfrac{1}{2}(1-\cos\theta)xy - (x+y) + 1 = 0, \tag{352'}$$

the outer boundaries correspond to $\theta = 180°$, the lower side of the triangle corresponds to $\theta = 0°$. The dashed line corresponds to an intermediate opening angle ($\theta = 120°$).

If in an actual experiment one integrates the rates over the upper part of the triangle, say from some x_0 to 1 and from y_0 to 1 with e.g. $x_0 \simeq y_0 \simeq 0.7$, then it is not difficult to see that W_{IB} and W_S give contributions of comparable magnitude, whilst W_{int} contributes much less. Actually, within W_S and under identical experimental conditions, the term with the factor $(F+a)^2$ would contribute about ten times more than the term with $(F-a)^2$. Therefore, it is primarily the combination $(F+a)^2$ which is determined by experiment.

As one sees from eqs. (354) and (355a) the contribution from bremsstrahlung has the same suppression factor m_e^2 which was characteristic for the decay $\pi \to e\nu$ and whose origin we discussed in sec. 1.2.4 in terms of the helicity transfer at V and A vertices. Clearly, the helicity selection rules do not depend on whether the pion (or positron) is on or off its mass shell. In the structure term proper, eq. (355c), there is no such suppression factor (the factor m_e^2 cancels out), because the helicity of the photon can always compensate for the helicity mismatch at the leptonic vertex, without any conflict with the conservation of total angular momentum.

The extra factor $m_\pi^2/2m_e^2$ in W_S compensates for the smallness of $F = a_0\sqrt{2}$, with a_0 as given in eq. (347). For example, at $x = y = 0.7$ we find $W_{IB} \simeq 1.67$ while $W_S \simeq 1 \times 10^3 (F+a)^2 \simeq 0.70$ (assuming $a = 0$), so that the structure terms are of the same order of magnitude as the bremsstrahlung. On the basis of these results we can now comment on the muonic decay mode (336). The structure terms are proportional to $\Gamma(\pi \to \mu\nu_\mu) m_\pi^2/2m_\mu^2$, and are of the same order of magnitude as in the electronic mode (335). However, the bremsstrahlung terms are larger by a factor

$(m_\mu/m_e)^2 \simeq 4.3 \times 10^4$. The phase space available for the branch (336) is smaller than for the branch (335) (see exercise 24), and there are some additional mass terms in the decay probabilities. Nevertheless, the estimate is essentially correct: The branching ratio for the decay (336) is about four orders of magnitude larger than for the decay (335). In the muonic mode internal bremsstrahlung predominates by about 10^3 over the structure terms.

Additional theoretical remarks. Both the vector form factor F and the axial vector form factor concern intrinsic properties of the pion and, therefore, are purely hadronic properties. One can show, using PCAC, that the amplitude a_0 for the decay $\pi^0 \to \gamma\gamma$ is given by (Adler 1968)

$$a_0 = \frac{g_{\pi \mathrm{NN}}}{4\pi^2 F_{\mathrm{A}}(0)} \frac{m_\pi}{m_{\mathrm{N}}} S, \tag{356}$$

where S is a pure number that depends on the charges Q_{f} of the u- and d-quarks: $S = \Sigma_i a_i Q_i^2$. The a_i are the coefficients of the individual quark terms in $a_\alpha^{(3)}(x)$, the isospin partner of $a_\alpha^{(1\pm \mathrm{i}2)}(x)$, viz.

$$a_\alpha^{(3)}(x) = \sum_{i=\mathrm{u,d}} a_i \sum_{c=1}^{3} \overline{q_{ic}(x)}\, \gamma_\alpha \gamma_5 q_{ic}(x) = \sum_c (\bar{u}_c \bar{d}_c) \gamma_\alpha \gamma_5 \frac{\tau_3}{2} \begin{pmatrix} u_c \\ d_c \end{pmatrix}$$

(c being the colour index), so that $a_{\mathrm{u}} = \frac{1}{2}$, $a_{\mathrm{d}} = -\frac{1}{2}$, $a_{\mathrm{s}} = 0$, and $S = 3\{\frac{1}{2}(\frac{2}{3})^2 - \frac{1}{2}(\frac{1}{3})^2\} = \frac{1}{2}$. The prediction (356) gives $a_0 = 1.96 \times 10^{-2}$, in good agreement with the experimental number (347). This quantitative agreement rests on the threefold degeneracy in the colour degree of freedom and is an important argument in favour of QCD.

Regarding the axial form factor $a(s)$, eq. (343), one can show that the natural scale which determines the orders of magnitude is the ratio of $f_\pi m_\pi$ to the square of the mass of the ρ-meson $\rho(769)$,

$$f_\pi m_\pi / m_\rho^2 \simeq 3.1 \times 10^{-2}. \tag{357}$$

This is plausible from eq. (351) if one assumes the pion's charge form factor to be dominated by the ρ-meson, because in this case $\langle r_\pi^2 \rangle$ is proportional to m_ρ^{-2}.

The remaining structure term $\bar{a}$, eq. (351), has the same scale (357) in all models where it is expressed in terms of commutators of currents which are saturated with ρ- and A1-vector meson states.

Finally we wish to comment on eqs. (339). If the pion in the intermediate state is not on its mass shell, the matrix element of the electromagnetic current $\langle Q | j_\alpha | q \rangle$ is to be replaced by the expression

$$\begin{aligned} &\frac{\mathrm{i}}{(2\pi)^{3/2}} \int \mathrm{d}x\, \mathrm{e}^{\mathrm{i}Qx} (\Box + m_\pi^2) \langle 0 | T\phi_\pi(x)\, j_\alpha(0) | q \rangle \\ &= \frac{1}{(2\pi)^3} \{ f_+(Q^2, k^2)(q+Q)_\alpha + f_-(Q^2, k^2)(q-Q)_\alpha \}. \end{aligned} \tag{358}$$

Making use of translational invariance the integral on the l.h.s. can also be written as follows:

$$\int \mathrm{d}x\, \mathrm{e}^{-ikx}(\Box + m_\pi^2)\langle 0|T\phi_\pi(0)\, j_\alpha(-x)|q\rangle,$$

with $k = q - Q$. Transforming x to $-x$, then multiplying with k^α gives the divergence of the integrand,

$$\begin{aligned}
&\partial^\alpha\langle 0|T\phi_\pi(0)\, j_\alpha(x)|q\rangle \\
&= \langle 0|(\partial^0\theta(x^0))\, j_0(x^0, \boldsymbol{x})\phi_\pi(0) + (\partial^0\theta(-x^0))\phi_\pi(0)\, j_0(x^0, \boldsymbol{x})|q\rangle \\
&= \langle 0|[\, j_0(0, \boldsymbol{x}), \phi_\pi(0)]\,|q\rangle\delta(x^0)
\end{aligned}$$

(noting that the divergence of j_α vanishes). The commutator of j_0 with ϕ_π is

$$[\, j_0(0, \boldsymbol{x}), \phi_\pi(0)] = -\phi_\pi(0)\delta(\boldsymbol{x}).$$

Therefore, contracting eq. (358) with k^α we obtain for the l.h.s.

$$-\frac{1}{(2\pi)^{3/2}}\int \mathrm{d}x\, \mathrm{e}^{iQx}(\Box + m_\pi^2)\langle 0|\phi_\pi(x)|q\rangle\delta(x^0)\delta(\boldsymbol{x}).$$

Integrating the $\Box$-operator by parts this gives, finally,

$$\frac{1}{(2\pi)^3}(m_\pi^2 - Q^2),$$

which is equal to the r.h.s. of eq. (358) contracted with k^α,

$$\frac{1}{(2\pi)^3}\{f_+(m_\pi^2 - Q^2) + f_- k^2\}.$$

This completes the proof of eq. (339b).

8. New perspectives and open problems

There are many open questions in the present theory of the weak interactions some of which were already mentioned in sec. 4.4 and in the introduction to sec. 6. In this section we touch on some topics where present and future experimentation is likely to yield further insight and, perhaps, some clues for further progress in our understanding of weak interactions.

In sec. 6 we discussed some possibilities for precision tests of the Lorentz structure of weak CC and NC interactions. In the present section we deal with two topics which have some, though model dependent, relation to this problem, but which are also of great interest in their own right: The question of neutrino masses and the nature of the leptonic family numbers.

The present upper limits on possible neutrino masses are not very stringent. The best mass limit for $\bar{\nu}_e$ comes from the β-decay of the triton,

$${}^3\mathrm{H} \to {}^3\mathrm{He} + \mathrm{e}^- + \bar{\nu}_e,$$

where one measures the spectrum of the electron near its kinematical endpoint. Although the issue is not completely settled (cf. sec. 1.1.1), the present upper limit is of the order of 40 eV. For the case of the muon neutrino ν_μ the best limit comes from a precision measurement of the muon momentum $|\boldsymbol{p}|$ in pion decay at rest,

$$\pi^+ \to \mu^+ \nu_\mu .$$

Using the muon and pion masses as input, information on $m(\nu_\mu)$ is deduced from the kinematic relation

$$m_{\nu_\mu}{}^2 = m_\pi^2 + m_\mu^2 - 2m_\pi \sqrt{p^2 + m_\mu^2} . \tag{359}$$

This method yields an upper limit of about 430 keV. Regarding the τ-neutrino a mass limit is obtained from the kinematic analysis of the leptonic decays of the τ-lepton. It is of the order of 100 MeV.

In view of this state of affairs, one must be prepared to investigate two extreme, in fact complementary, possibilities: If neutrinos are at all massive then either

(i) some of the masses are "large", in the sense that $m(\nu_\mu)$ and $m(\nu_\tau)$ could be of the same order of magnitude as m_e and/or m_μ, respectively, or

(ii) some or all masses are different from zero but small, say $\lesssim 1\text{eV}$, and at least one mass *difference* is small as compared to a typical experimental energy resolution.

In the first case there is hope that it might be possible to find direct evidence for a nonvanishing mass in the kinematics of a leptonic or semileptonic process. The chances for this are even better, as we shall see, if there is neutrino state mixing. In the second case, and if the leptonic family numbers are not conserved, oscillations between different neutrino states should be observable. If these occur, and if the state mixing involves no more than two or three neutrino states, then it should be possible to extract information on masses and mixing matrix elements.

It is plausible that neutrinos which have finite, nondegenerate masses, will occur as mixed states in the weak interactions. In other terms, the weak eigenstates "ν_e", "ν_μ", "ν_τ" which couple to CC vertices (i.e. vertices of the type $(\mathrm{f}\nu_\mathrm{f}\mathrm{W})$) may not be identical with the mass eigenstates "n_1", "n_2", "n_3". For the case of the three lepton families, we would have

$$(\nu_\mathrm{f})_L = \sum_{i=1}^{3} U_{\mathrm{f}i}(\alpha_1, \alpha_2, \alpha_3, \varepsilon)(n_i)_\mathrm{L}, \quad \mathrm{f} = \mathrm{e}, \mu, \tau \tag{360}$$

where $U_{\mathrm{f}i}$ is the leptonic analogue of the quark mixing matrix (152). Like the quark mixing matrix it depends on three real angles and one phase. More precisely, this is true if the physical neutrinos are purely left-handed. If, in addition, also right-handed neutrino states couple to other particles, then these can be mixed states, too, viz.

$$(\nu_\mathrm{f})_\mathrm{R} = \sum_{i=1}^{3} V_{\mathrm{f}i}(n_i)_\mathrm{R}, \tag{361}$$

with a mixing matrix a priori independent of $U_{\mathrm{f}i}$, eq. (360).

We emphasize that the ansatz (360) should not be analyzed in isolation and disregarding the mass sector of the theory since it is the structure of the mass

Lagrangian that determines the mixing matrix elements. In particular, the limit of all masses going to zero must be studied with some caution.

8.1. "Heavy" neutrinos

8.1.1. Neutrino masses from two-body decays

Relation (359) shows that the momentum of the charged lepton in the final state depends on the square of the neutrino mass. Therefore, the value of this momentum is not very sensitive to masses in a range of masses small as compared to the mass of the parent particle. On the other hand the two-body decay of a stopped particle has the advantage that the charged lepton in the final state is monochromatic and, therefore, provides a clear and unique signature for the number and masses of companion neutrino states. Indeed, suppose that in the decay of a pseudoscalar meson P

$$P^+ \to f^+ \nu_f, \tag{362}$$

the weak interaction state "ν_f" is a superposition of different mass eigenstates n_i with masses m_i, as indicated in eq. (360). For every mass state with m_i in the interval

$$0 \le m_i < m_P - m_f \tag{363}$$

there is a monochromatic state of the charged lepton f^+ with momentum

$$|\boldsymbol{p}^{(i)}| = \tfrac{1}{2} m_P \sqrt{\left(1 - r_f^2\right)^2 + \left(1 - r_i^2\right)^2 - \left(1 + 2 r_i^2 r_f^2\right)}\,, \tag{364}$$

where we have introduced the mass ratios

$$r_f := m_f / m_P, \qquad r_i := m_i / m_P. \tag{365}$$

In other words, the decay (362) in reality consists of several branches

$$P \to f(\boldsymbol{p}^{(i)}) + n_i(m_i).$$

The corresponding partial decay width is calculated as follows. According to App. B we have

$$\Gamma = \frac{(2\pi)^7 \pi}{2 m_P^2} |\boldsymbol{p}^{(i)}| \sum |T|^2.$$

The T-matrix element being

$$T = \frac{i}{(2\pi)^{9/2}} \frac{\kappa_P G}{\sqrt{2}} U_{fi} \overline{u_i(k)}\, \not{q} (1 - \gamma_5) v_f(p),$$

where $\kappa_P = f_\pi \cos\theta_1$ in the case of pions, $\kappa_P = f_K \sin\theta_1 \cos\theta_3$ in the case of kaons (f_K being defined in analogy to eq. (312)), and where $q = p + k$. Making use of the Dirac equations (III.63) and (III.62′), this gives

$$T = -\frac{i}{(2\pi)^{9/2}} \frac{\kappa_P G}{\sqrt{2}} U_{fi} \overline{u_i(k)} \left\{ m_f - m_i + (m_f + m_i)\gamma_5 \right\} v_f(p),$$

and therefore

$$\sum |T|^2 = \frac{4\kappa_P^2 G^2}{(2\pi)^9} |U_{fi}|^2 \{ (m_f^2 + m_i^2)(pk) + 2m_i^2 m_f^2 \},$$

with $(pk) = \frac{1}{2}(m_P^2 - m_f^2 - m_i^2)$. The partial decay width is then found to be

$$\Gamma(P \to f^{(i)}\nu_i) = \frac{\kappa_P^2 |U_{fi}|^2 G^2 |\boldsymbol{p}^{(i)}| m_P^2}{4\pi} \{ r_f^2 + r_i^2 - (r_f^2 - r_i^2)^2 \}, \tag{366}$$

where $|\boldsymbol{p}^{(i)}|$ is given by eq. (364). [As one verifies easily, eq. (366) with $r_i = 0$, $\kappa_P = f_\pi \cos\theta_1$ and $U = 1$ reduces to eq. (313) for the decay $\pi \to f\nu_f$. Note also the complete symmetry of eq. (366) in i and f.]

Let us discuss the result (366) in a little more detail. Possible reactions that can be investigated in the laboratory are

$$\pi^+ \to \mu^+ \nu_\mu, \tag{367a}$$

$$\pi^+ \to e^+ \nu_e, \tag{367b}$$

$$K^+ \to \mu^+ \nu_\mu, \tag{367c}$$

$$K^+ \to e^+ \nu_e. \tag{367d}$$

A good starting point is the idea that the predominant states in ν_e and ν_μ, respectively, have rather small masses and that one (or several) heavy mass eigenstates are mixed into them. This would imply that both $|U_{e1}|$ and $|U_{\mu 2}|$ are large as compared to $|U_{e3}|$ or $|U_{\mu 3}|$ etc. The strengths of the heavy neutrino branches relative to the dominant light neutrino branch is determined by $|U_{e3}/U_{e1}|^2$ or $|U_{\mu 3}/U_{\mu 2}|^2$, respectively (which is a small number), and by the mass factor in curly brackets on the r.h.s. of eq. (366). This mass factor indicates to which extent the decay is inhibited by the helicity selection rule. For this reason, the electronic modes (367b) and (367d) are more sensitive to heavy neutrinos than the muonic modes: the dominant mode is suppressed because both e and n_1 are light, the heavy neutrino mode is not because n_3 is heavy. As an example consider the decay (367b) with m_1, $m_2 \ll m_e$, but $m_3 = 80$ MeV. Then $|\boldsymbol{p}^{(1)}| \simeq \frac{1}{2} m_\pi$, $|\boldsymbol{p}^{(3)}| \simeq 0.67 \frac{1}{2} m_\pi$ and from eq. (366)

$$\frac{\Gamma(\pi \to e\nu_3)}{\Gamma(\pi \to e\nu_1)} = \frac{|U_{e3}|^2}{|U_{e1}|^2} 1.65 \times 10^4.$$

Therefore, this mode allows detecting even very small admixtures of heavy neutrinos. Indeed, present limits are of the order of $|U_{e3}|^2 < 10^{-5}$. In fact, information on possible heavy neutrinos can be obtained both from a direct search for monochromatic charged leptons in the decays (367), and from a comparison of measured e/μ branching ratios

$$\frac{\Gamma(\pi \to e\nu_e)}{\Gamma(\pi \to \mu\nu_\mu)}, \qquad \frac{\Gamma(K \to e\nu_e)}{\Gamma(K \to \mu\nu_\mu)}$$

to the expected results (Shrock 1981). The different reactions (367) are complementary insofar as they scan different mass regions, owing to their different kinematics.

8.1.2. Neutrino masses from three-body decays

From a kinematic point of view decays of elementary systems into three bodies, one of which is a massive neutrino, are more favourable than two-body decays because the neutrino can be produced with arbitrarily small velocity. In this nonrelativistic limit the energy of the neutrino is a function of its mass, and not of the square of the mass. Therefore in the appropriate kinematic situation one expects measurable effects which are *linear* in the mass of the neutrino. Examples of such decays are

$$^{3}\mathrm{H} \to {}^{3}\mathrm{He} + \mathrm{e}^{-} + \bar{\nu}_{\mathrm{e}}, \qquad (368\mathrm{a})$$

$$(\mathrm{e}(Z, A))_{\text{K-shell}} \to (Z-1, A) + \gamma + \nu_{\mathrm{e}}, \qquad (368\mathrm{b})$$

$$\pi^{+} \to \mu^{+} + \gamma + \nu_{\mu}, \qquad (369\mathrm{a})$$

$$(\mu^{-}\,{}^{6}\mathrm{Li})_{\text{s-state}} \to {}^{3}\mathrm{H} + {}^{3}\mathrm{H} + \nu_{\mu}. \qquad (369\mathrm{b})$$

Similarly, if we suspect m_{ν_μ} to be much larger than m_{ν_e}, and m_{ν_τ} to be much larger than m_{ν_μ}, the decays

$$\mu^{+} \to \mathrm{e}^{+} + \nu_{\mathrm{e}} + \bar{\nu}_{\mu}, \qquad (370\mathrm{a})$$

$$\tau^{+} \to \mu^{+} + \nu_{\mu} + \bar{\nu}_{\tau}, \qquad (370\mathrm{b})$$

$$\tau^{+} \to \mathrm{e}^{+} + \nu_{\mathrm{e}} + \bar{\nu}_{\tau}, \qquad (370\mathrm{c})$$

may also be suitable for detection of these masses. The difference being that in the first group of decays, eqs. (368) and (369), there are two particles which can be detected, in the second group, eqs. (370), only one particle can be studied.

Here we analyze primarily the first group (368) and (369). For the sake of simplicity we neglect neutrino state mixing, i.e. we take the matrix (360) to be (approximately) diagonal. It will be easy, however, to extend the results to the more general situation with nontrivial mixing.

Let us consider the decay of a particle, or atomic system, of mass M into three particles with masses m_1, m_2, m_3, respectively, in a frame of reference where the decaying system is at rest. Suppose that 1 and 2 are particles that can be detected in an experimental arrangement, and that 3 is a neutrino whose mass we wish to determine. We determine first the *maximal* energy of a particle 1. At this kinematic point the three momenta of the particles are collinear, as shown in Fig. 16, with $\boldsymbol{p}_1 = -k\boldsymbol{n}$, $\boldsymbol{p}_2 = xk\boldsymbol{n}$, and $\boldsymbol{p}_3 = (1-x)k\boldsymbol{n}$. x is a number between 0 and 1 that one determines as follows. For E_1 to be maximal, $k = k(x)$ must be maximal. From energy conservation we have the condition

$$M = E_1 + E_2 + E_3 = \sqrt{m_1^2 + k^2} + \sqrt{m_2^2 + x^2 k^2} + \sqrt{m_3^2 + (1-x)^2 k^2}$$
$$=: F(x, k(x)). \qquad (371)$$

The maximum of the function $k(x)$ is found from the equation for its derivative

$$\frac{\mathrm{d}k}{\mathrm{d}x} = -\frac{\partial F}{\partial x}\Big/\frac{\partial F}{\partial k} = -kE_1 \frac{xE_3 - (1-x)E_2}{E_2E_3 + x^2E_1E_3 + (1-x)^2E_1E_2} = 0,$$

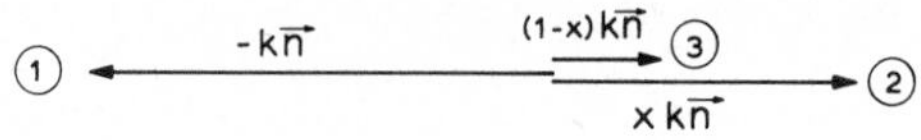

Fig. 16. Collinear decay of a particle where decay particle 1 has maximal energy.

which yields the condition

$$xE_3 = (1-x)E_2. \tag{372a}$$

By squaring this equation, one finds $xm_3 = (1-x)m_2$ and, finally,

$$x = \frac{m_2}{m_2 + m_3}. \tag{372b}$$

Inserting eq. (372a) into eq. (371), one has $M - E_1 = E_2/x = (m_2 + m_3)E_2/m_2$, the square of which gives the desired result

$$E_1^{\max} = \frac{1}{2M}\left\{M^2 + m_1^2 - (m_2+m_3)^2\right\}. \tag{373a}$$

The energy of particle 2 at this same kinematic point is easily calculated, viz.

$$E_2|_{\text{at } E_1^{\max}} = \frac{m_2}{2M(m_2+m_3)}\left\{M^2 - m_1^2 + (m_2+m_3)^2\right\}. \tag{373b}$$

As in the last section it is convenient to introduce the mass ratios m_i/M; similarly, it is useful to introduce dimensionless energy variables x_i by dividing each energy E_i by $M/2$, its maximal value in case the other two particles are massless:

$$r_i := m_i/M, \qquad x_i := 2E_i/M. \tag{374}$$

Relation (371) then reads

$$x_1 + x_2 + x_3 = 2, \tag{371'}$$

while eqs. (373a) and (373b) become

$$x_1^{\max} = 1 + r_1^2 - (r_2 + r_3)^2, \tag{375a}$$

$$x_2|_{\text{at } x_1^{\max}} = \frac{1}{1 + r_3/r_2}\left\{1 - r_1^2 + (r_2+r_3)^2\right\}. \tag{375b}$$

It is a simple matter to determine the kinematic point where particle 1 has its *minimal* energy $E_1 = m_1$ or $x_1 = 2r_1$. One finds

$$x_1^{\min} = 2r_1, \tag{376a}$$

$$x_2|_{\text{at } x_1^{\min}} = 1 - r_1 + \frac{1}{1-r_1}\left(r_2^2 - r_3^2\right). \tag{376b}$$

Clearly, there is a cyclic symmetry in these equations so that all other extrema can be obtained from them. Fig. 17 shows the boundary of the kinematically allowed domain of energies in the (x_1, x_2)-plane (for the arbitrarily chosen example $r_1 = \frac{1}{16}$, $r_2 = \frac{1}{2}$, $r_3 = \frac{1}{8}$). Eqs. (375) and (376) give the coordinates of the points B_1 and A_1,

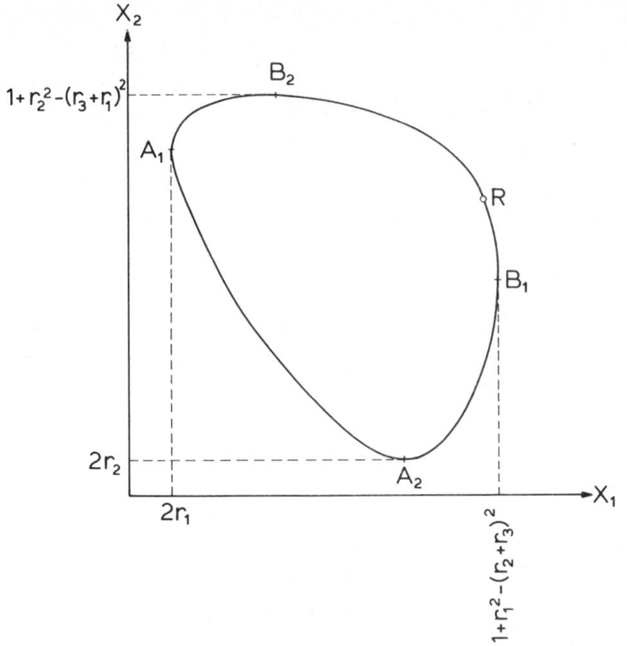

Fig. 17. Boundary of the kinematic domain in the plane spanned by the energies of two particles which emanate from a three-body decay. Point R is the projection of the kinematical point where the third particle (not visible in this plane) is produced at rest.

respectively. The analogous extrema B_2 and A_2 for the energy variable x_2 are obtained from eqs. (375) and (376), respectively, by the permutation 123 → 231.

Of particular interest is the point R at which the neutrino (particle 3) comes to rest. It is in the neighbourhood of this point that we expect effects linear in m_3. Its coordinates x_3, x_1 are found from eqs. (376) by the permutation 123 → 312, viz.

$$x_3 = 2r_3,$$

$$x_1 = 1 - r_3 + \frac{1}{1-r_3}\left(r_1^2 - r_2^2\right), \tag{377a}$$

while its coordinate x_2 is found by means of eq. (371′),

$$x_2 = 1 - r_3 - \frac{1}{1-r_3}\left(r_1^2 - r_2^2\right). \tag{377b}$$

Generally speaking, if we compare the shape of Fig. 17 to what it would be for a massless neutrino, $m_3 = 0$, we see that the coordinates of the points B_1, R, and B_2 depend on terms which shift *linearly* with m_3. For example, the endpoint energy of particle 1 is shifted as follows:

$$\Delta E_1^{\max} := E_1^{\max}(m_3 = 0) - E_1^{\max}(m_3) = m_3 \frac{m_2}{M}\left(1 + \frac{m_3}{2m_2}\right). \tag{378}$$

The larger the ratio m_2/M, the larger the shift. The reader may wish to discuss the specific examples (368), (369) for himself. We quote only one result for the decay (369a), $\pi \to \mu\nu_\mu\gamma$. Here $M = m_{\pi^+}$, $m_1 = 0$, $m_2 = m_\mu$. Because the photon is massless the points A_1 and B_2 coincide, whilst R is rather close to point B_1. The shift of the photon's endpoint energy, eq. (378), is approximately

$$\Delta E_\gamma^{\max} \simeq m_{\nu_\mu} \frac{m_\mu}{m_\pi} = 0.757 m_{\nu_\mu}, \tag{379}$$

a rather large effect indeed.

Unfortunately, a mass determination by simply establishing the exact boundary of the kinematic domain of Fig. 17 is very difficult because the differential rates which fill this figure become very small as one approaches the boundary from the inside. In practice, one will only be able to determine partially integrated rates, by integrating over finite portions of the allowed domain. This raises the question to which extent the differential *rates* are affected by the assumption that the neutrino mass is different from zero. The answer is this: it can be shown that if the CC coupling at the leptonic vertex is exactly V − A, the changes in the differential rates are *quadratic* in m_ν (Missimer et al. 1981). As m_ν^2 appears scaled with M^2, or the square of the energy release, this effect is generally negligible. In other terms, the effect of a nonvanishing neutrino mass, to first approximation, is a purely kinematic one. The differential rate is cut off by the shrinking boundary of the available phase space.

Regarding the possibility of neutrino state mixing, we note that state mixing of several neutrinos $n_i, n_j, \ldots$ with $m_i < m_j < \ldots$ means superimposing several figures of the type shown in Fig. 17. The rates of individual decay channels are added with relative weights $|U_{fi}|^2$. To first approximation the method will be sensitive only to the lightest, most dominant neutrino component.

In the decays of eqs. (370), which are of the type

$$f^+ \to f'^+ \nu_{f'} \bar{\nu}_f, \tag{380}$$

the effects due to possible neutrino masses are even more difficult to identify because only one particle in the final state can readily be detected. For simplicity, let us assume that m_{ν_f} is much larger than $m_{\nu_{f'}}$ so that, in fact, the latter may be neglected. For the decays (370) this is certainly a very good approximation because in the case of muon decay we know already that $m_{\nu_e} \ll m_e \ll m_\mu$, whilst in the case of the τ-decays m_{ν_e} and m_{ν_μ} are certainly negligible as compared to m_τ. In this approximation we have

$$M \equiv m_f, \qquad m_1 \equiv m_{f'}, \qquad m_2 \equiv m_{\nu_{f'}} \simeq 0, \qquad m_3 \equiv m_{\nu_f} \neq 0.$$

The maximum energy of the charged lepton in the final state is given by eq. (373a),

$$E_1^{\max} \equiv W \simeq \frac{1}{2m_f}\left\{m_f^2 + m_{f'}^2 - m_{\nu_f}^2\right\},$$

which should be compared to eq. (267). As in our discussion of muon decay with vanishing neutrino masses, let us introduce the dimensionless variable, cf. eq. (276),

$$y := E_1/W.$$

The decay spectrum for a polarized initial state, but summing over the polarization of f′, is calculated as in sec. 6.2.2. In particular, the integrals (268) are evaluated in close analogy to eq. (310). One finds for the "V – A" interaction

$$\frac{1}{\Gamma}\frac{\mathrm{d}^2\Gamma}{\mathrm{d}x\,\mathrm{d}(\cos\theta)} = A_{\mathrm{F}}(y)\sqrt{y^2-y_0^2}\left\{3y-2y^2-y_0^2+\frac{6\sigma^2 y}{1-y+3\sigma}\right.$$
$$\left.-\xi\cos\theta\sqrt{y^2-y_0^2}\left[2y-1-\frac{m_{\mathrm{f}'}}{m_{\mathrm{f}}}y_0+\frac{2\sigma(1-y)}{1-y+3\sigma}\right]\right\}, \tag{381}$$

where $y_0 = m_{\mathrm{f}'}/W$, as before, and

$$\sigma := m_{\nu_\mathrm{f}}^2/2m_\mathrm{f}W. \tag{382}$$

The function $A_\mathrm{F}(y)$ stems from the integration over the phase space of the two neutrinos and is given by

$$A_\mathrm{F}(y) = \frac{(1-y)^2(1-y+3\sigma)}{(1-y+\sigma)^3}. \tag{383}$$

The differential decay rate (381) has two new features as compared to the case $m_2 = m_3 = 0$:

(i) The isotropic part of the spectrum vanishes at the endpoint $y = 1$, due to the phase space factor (383).

(ii) The isotropic part of the spectrum has its maximum approximately at the kinematic point (377a) where the sensitivity to $m_3 \equiv m_{\nu_\mathrm{f}}$ is greatest, i.e. at

$$E_1 \simeq \frac{m_\mathrm{f}}{2}\left\{1-\frac{m_{\nu_\mathrm{f}}}{m_\mathrm{f}}+\frac{m_\mathrm{f}'^2/m_\mathrm{f}^2}{1-m_{\nu_\mathrm{f}}/m_\mathrm{f}}\right\}. \tag{384}$$

The position of this maximum is indeed a linear function of m_{ν_f}. However, for the cases of practical interest (and taking into account the radiative corrections) this shift is only measurable if $m_{\nu_\mathrm{f}}/m_\mathrm{f}$ is not smaller than, say, 0.05.*) With the present limit on m_{ν_μ}, this excludes muon decay as a realistic possibility. Regarding the τ-neutrino this method is applicable only if $m_{\nu_\tau} \gtrsim 100$ MeV.

8.2. *Neutrino oscillations*

In the preceding sections we considered the case of large neutrino masses, i.e. a situation where at least ν_μ and ν_τ have masses of the same order of magnitude as the electron and the muon mass. Although this possibility cannot be excluded on the basis of extant, direct experimental information, it is perhaps in conflict with conclusions from cosmology (Steigmann 1979). The cosmology of the "big bang" provides indirect limits on possible masses to the extent that there should be no stable neutrino within a mass interval from about 50 eV to about 1 GeV. For unstable neutrinos the mass limits depend strongly on their lifetime and, in fact, are

*) For details see Missimer et al. (1981).

then much less tight. The cosmological bounds rest on a number of assumptions which, however plausible, remain to be tested. Therefore, the direct search for large masses, in laboratory experiments, is of great importance not only as a fundamental question of lepton physics on its own but also as input to astrophysics and cosmology.

In this section we consider the complementary situation of "small" neutrino masses mentioned in the introduction and discuss some experimental possibilities for determining finite masses and mass differences. We base our analysis on the following assumptions:

(i) The weak eigenstates ν_e, ν_μ, ν_τ are nontrivial superpositions of mass eigenstates n_1, n_2, n_3, cf. eq. (360).

(ii) One or several of the mass differences $|m_i - m_j|$ are small as compared to the typical resolution in energy of an actual experiment.

(iii) In particular, the mass differences are small also in comparison with the momentum with which the neutrinos are produced in a given decay process.

For the sake of simplicity we consider at first the case of two states. Let ν_{f1} and ν_{f2} be the weak eigenstates, and let

$$|\nu_{f1}\rangle_0 = |n_1\rangle\cos\alpha_1 + |n_2\rangle\sin\alpha_1, \tag{385a}$$

$$|\nu_{f2}\rangle_0 = -|n_1\rangle\sin\alpha_1 + |n_2\rangle\cos\alpha_1. \tag{385b}$$

be the mixed states (360) with $\alpha_2 = \alpha_3 = 0$. We assume further that initially the state $|\nu_{f2}|$ was produced with a given momentum k, e.g. in a two-body decay $P \to f_2 + \nu_{f2}$ in the system where particle P was at rest. Since n_1 and n_2 are the mass eigenstates the state ν_{f2} has the time evolution

$$|\nu_{f2}\rangle_t = -|n_1\rangle e^{-iE_1 t}\sin\alpha_1 + |n_2\rangle e^{-iE_2 t}\cos\alpha_1, \tag{386}$$

where

$$E_i = \sqrt{m_i^2 + \boldsymbol{k}_i^2} \simeq |\boldsymbol{k}| + \tfrac{1}{2}m_i^2/|\boldsymbol{k}|. \tag{387}$$

Note that here we make use of assumption (iii), $m_i^2 \ll \boldsymbol{k}^2$, and, consequently, we neglect the difference in momentum in the different mass eigenchannels. At the same time it is understood that the momentum resolution in detecting the charged partner f_2 in the two-body decay of particle P is not sufficient to distinguish the two channels. If these conditions are met the state mixture (386) leads to observable oscillations between the states ν_{f2} and ν_{f1}.

In order to see this we recall that the only way of detecting neutrinos is by having them induce another weak reaction. Whatever the experimental arrangement, it will always measure either the probability of finding the initial neutrino after a time t,

$$P(\nu_{f2} \to \nu_{f2}; t), \tag{388a}$$

or of finding another neutrino state ν_{f1}, viz.

$$P(\nu_{f2} \to \nu_{f1}; t). \tag{388b}$$

It is easy to calculate these probabilities from the overlap of the state (386) with the initial states (385), viz.

$$P(\nu_{f2} \to \nu_{f2};\, t) = |_0\langle \nu_{f2} | \nu_{f2} \rangle_t|^2$$
$$= 1 - 2\sin^2\alpha_1 \cos^2\alpha_1 \{1 - \cos(E_2 - E_1)t\}, \tag{389a}$$

$$P(\nu_{f2} \to \nu_{f1};\, t) = |_0\langle \nu_{f1} | \nu_{f2} \rangle_t|^2$$
$$= 2\sin^2\alpha_1 \cos^2\alpha_1 \{1 - \cos(E_2 - E_1)t\}. \tag{389b}$$

In these formulae $(E_2 - E_1)$ can be replaced by $|E_2 - E_1|$, where

$$|E_2 - E_1| \simeq \frac{|m_2 - m_1|}{2|\boldsymbol{k}|} \simeq \frac{|m_2 - m_1|}{2E_\nu}.$$

In practice, the time of flight t is measured by installing the neutrino detectors at a distance L from their source and, if possible, by varying that distance. As v/c is practically 1, $L \simeq t$ in natural units. The oscillation pattern in eqs. (389) is then determined by the quantity

$$\frac{|E_2 - E_1|}{2\pi} \cdot t = \frac{L}{L_{12}},$$

where the oscillation length L_{12} is defined by

$$L_{12} := \frac{2\pi}{|E_1 - E_2|} \simeq \frac{4\pi E_\nu}{|m_1^2 - m_2^2|}. \tag{390}$$

Regarding the units, it is customary to express the length in meters, the neutrino energy E_ν in MeV, the difference of squared masses in $(\text{eV})^2$, so that

$$L_{12}[\text{m}] = \kappa \frac{E_\nu[\text{MeV}]}{|m_1^2 - m_2^2|[\text{eV}]^2} \tag{391a}$$

with

$$\kappa = \hbar c(\text{in MeV} \cdot \text{m}) 4\pi \times 10^{12} = 4\pi \cdot 0.19733 \simeq 2.48. \tag{391b}$$

With these definitions we have

$$P(\nu_{f2} \to \nu_{f2};\, L) = 1 - \sin^2(2\alpha)\sin^2(\pi L/L_{12}), \tag{392a}$$

$$P(\nu_{f2} \to \nu_{f1};\, L) = \sin^2(2\alpha)\sin^2(\pi L/L_{12}). \tag{392b}$$

In a discussion of these results we have to distinguish three limiting situations:

(i) The greatest sensitivity to the phenomenon of neutrino oscillations is obtained if $L \simeq \frac{1}{2}L_{12}$, i.e. if the experimental conditions are chosen such that $E_\nu/L \simeq |m_1^2 - m_2^2|$. For example, an experiment with $E_\nu = 100$ MeV and $L = 20$ m would be most sensitive to a difference of $|m_1^2 - m_2^2| \simeq 5\ \text{eV}^2$.

(ii) If $L/L_{12} \gg 1$, that is if $E_\nu/L \ll |m_1^2 - m_2^2|$, then the interference term $\sin^2(\pi L/L_{12})$ oscillates very rapidly with L, so that only an average effect of neutrino mixing will be visible,

$$\langle P(\nu_{f2} \to \nu_{f2}) \rangle = 1 - \tfrac{1}{2}\sin^2(2\alpha).$$

This could be the case, for instance, for neutrinos coming from weak processes in the sun.

(iii) Finally, if $L/L_{12} \ll 1$, that is if $E_\nu/L \gg |m_1^2 - m_2^2|$, there is practically no observable effect at all.

The case of three (or even more) mass eigenstates mixed into the weak interaction states is a little more complicated. Instead of going into all details we just mention one case which may be relevant for astrophysics. In the case of three basis states, eq. (360), and with a mixing of the form of eq. (152), the average probability, for a beam that was initially ν_e, of finding again ν_e after the beam has travelled for a distance $L \gg L_{ij}$, is given by

$$\langle P(\nu_e \to \nu_e)\rangle = 1 - \tfrac{1}{2}\sin^2(2\alpha_1) - \tfrac{1}{2}\sin^4\alpha_1 \sin^2(2\alpha_3). \tag{393}$$

The minimum of this expression is reached for $\sin^2\alpha_1 = \frac{2}{3}$, i.e. $\sin^2(2\alpha_1) = \frac{8}{9}$, and $\sin^2(2\alpha_3) = 1$, in which case $\langle P(\nu_e \to \nu_e)\rangle = \frac{1}{3}$. This result may be relevant for the so-called solar neutrino problem: In measurements of the ν_e flux from the sun, by means of the inverse β-decay $\nu_e + {}^{37}\mathrm{Cl} \to {}^{37}\mathrm{Ar} + e^-$, one expects $(7.6 \pm 3.3) \times 10^{-36}$ neutrino captures per second per $^{37}\mathrm{Cl}$ nucleus (Bahcall et al. 1972, 1978). This number is calculated on the basis of a standard theory of the sun and assuming the neutrinos to be massless and not oscillating. In the experiment by Davies and collaborators one finds $(1.76 \pm 0.3) \times 10^{-36}$ captures/sec per $^{37}\mathrm{Cl}$.

Finally, we mention that the nondiagonal probabilities $\langle P(\nu_e \to \nu_\mu)\rangle$ and $\langle P(\nu_e \to \nu_\tau)\rangle$ depend also on ε, the phase indicating CP violation. Here there is another, though remote possibility of detecting violation of time reversal invariance in the leptonic world.

8.3. Processes which change lepton family numbers

In this section we return to the question of lepton family numbers that we discussed briefly in the introductory section 1.1.1. We sketch the various options that are conceivable in breaking the conservation of the family numbers (11) and/or of total lepton number (12), and we mention some of the most important consequences that one should be able to test in experiments. We then discuss one class of processes which provide an illustrative example of family number violation.

As this field is still very much open and our ignorance is great, our discussion is neither exhaustive nor conclusive.

8.3.1. Internal quantum numbers of neutrinos

A basic problem of lepton physics is the question whether neutrinos carry any additively conserved quantum numbers at all, and if they do, what the nature and dynamical origin of those quantum numbers is. There is very good evidence from astrophysics that neutrinos have no electromagnetic attributes such as electric charge or magnetic moment. So the discussion concentrates upon the lepton and lepton family numbers, eqs. (11) and (12).

The simple neutrino mixing scheme (360) is based on an assumed analogy of the leptons to the quark families. It implies that weak CC interactions do conserve total lepton number exactly but can change individual family numbers such that only their sum (12) is conserved. The neutrino states ν_f (weak eigenstates) or n_i (mass eigenstates) belong to the eigenvalue +1 of L, eq. (12). The weak Lagrangian then possesses a global U(1)-symmetry. The neutrino states n_i, being massive and carrying nonvanishing eigenvalues of this symmetry, are Dirac fermions as defined in Chap. III. This option can be tested through the mass measurements discussed above, secs. 8.1 and 8.2. It may also be testable in processes of the type $\mu \to e\gamma$ provided the basic CC couplings are not strictly "V − A". [Without interference between V − A and V + A currents the rates, though finite, are hopelessly small.]

Another possibility is that neutrinos are massive and are selfconjugate, i.e. Majorana particles (cf. our discussion in sec. (III.8.4)). In this case they cannot carry total lepton number L. This option opens up a number of very interesting consequences some of which can be tested in experiment. The most direct signal would be provided by positive evidence for processes with $\Delta L \neq 0$ such as *neutrinoless double β-decay*. Examples which are being investigated are the processes

$$\begin{aligned} &{}_{52}\mathrm{Te} \to {}_{54}Xe + 2e^- \quad (\text{isotopes } A = 128 \text{ and } 130), \\ &{}^{82}_{34}\mathrm{Se} \to {}^{82}_{36}\mathrm{Kr} + 2e^-, \\ &{}^{76}_{32}\mathrm{Ge} \to {}^{76}_{34}\mathrm{Se} + 2e^-. \end{aligned} \tag{394}$$

The decays can be thought of as two-step processes involving the sequence $n \to p + e^- + \nu_M$ and $\nu_M + n \to p + e^-$ with ν_M being a Majorana neutrino. As yet no positive evidence for any of these reactions was found.

There are immediate consequences, although somewhat indirect, for $\mu \to e\gamma$ and related processes. Regarding lepton decays such as μ-decay (262), there are consequences in principle but, with present limits on neutrino masses, they are too small to be measurable with any kind of precision.

The central problem in this discussion is the leptonic mass sector about which we know very little. It is the mass matrix which fixes the Dirac and Majorana nature of neutrinos and which determines the mixing matrices in the weak interaction eigenstates. We emphasize again that mixing angles should not be discussed without considering the mass sector. In particular, as should be clear from the discussion in sec. (III.8.4), the difference between the Majorana and Dirac cases fades away when the masses go to zero.

8.3.2. $\mu \to e\gamma$ and related processes

In the frame of a discussion about lepton family numbers the processes

$$\mu^\pm \to e^\pm \gamma, \tag{395a}$$

$$\mu^\pm \to e^\pm e^\mp e^\pm \tag{395b}$$

$$\mu^-(Z, A) \to (Z, A)^* e^-, \tag{395c}$$

are of particular interest because they are amenable to experimental investigations of extraordinary sensitivity. The present upper limits for the branching ratios

$$R_{\mu\to e\gamma} := \frac{\Gamma(\mu\to e\gamma)}{\Gamma(\mu\to \text{all})}, \tag{396a}$$

$$R_{\mu\to 3e} := \frac{\Gamma(\mu\to e\bar{e}e)}{\Gamma(\mu\to \text{all})}, \tag{396b}$$

$$R_{\mu^-\to e^-}(Z,A) := \frac{\Gamma(\mu^-(Z,A)\to(Z,A)^* e^-)}{\Gamma(\mu^-(Z,A)\to \text{all})}, \tag{396c}$$

are of the order of 10^{-10}. It is likely that these will be pushed down to 10^{-11} or 10^{-12} in the near future.

The processes (395), if they exist, are processes of second and higher order. As their detailed analysis is technically complicated we restrict the discussion to qualitative considerations here and refer the reader to the literature for the technical details. The first decay mode (395a) is a purely electroweak one and, independently of any model, depends on an effective ($\mu e\gamma$)-vertex as shown in Fig. 18(a). This vertex is a matrix element of the electromagnetic current $j_\alpha^{\text{e.m.}}$ between interacting lepton states, "dressed" by the weak interactions, viz. (for $\mu^-\to e^-\gamma$)

$$\begin{aligned}\langle e(p)|j_\alpha^{\text{e.m.}}(0)|\mu(q)\rangle &= \frac{1}{(2\pi)^3}\overline{u_e(p)}\left\{G_1(k^2)\left[\gamma_\alpha - \frac{1}{k^2}(m_\mu - m_e)k_\alpha\right]\right.\\ &\quad + G_2(k^2)\left[\gamma_\alpha + \frac{1}{k^2}(m_\mu + m_e)k_\alpha\right]\gamma_5\\ &\quad \left. + \frac{i}{m_\mu}\left[F_1(k^2) + F_2(k^2)\gamma_5\right]\sigma_{\alpha\beta}k^\beta\right\}u_\mu(q). \end{aligned}\tag{397}$$

The Lorentz covariants are combined such as to take account of current conservation, $\partial^\alpha j_\alpha^{\text{e.m.}}(x)=0$. The decomposition (397) contains both parity-even and -odd terms, allowing for an arbitrary amount of parity violation from the weak interactions. The current operator j_α is hermitean. If the weak interactions are invariant under time reversal, then one can show that the form factors F_i and G_i, as defined by eq. (397), are real. Finally, for photons on the mass shell, $k^2=0$, eq. (397) remains

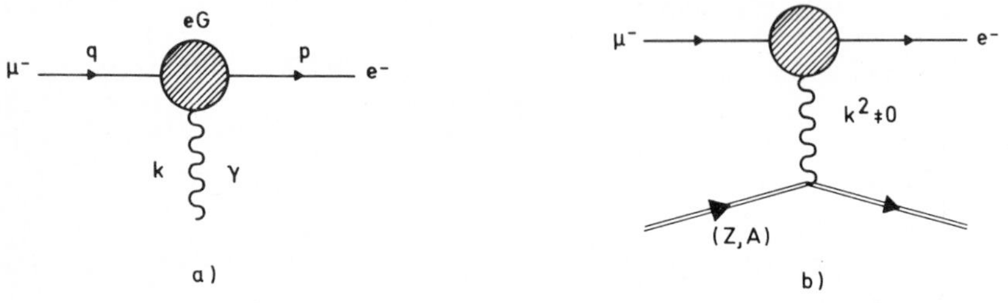

Fig. 18. (a) Effective $\mu e\gamma$-vertex. (b) Photon amplitude in $\mu\to e$ conversion on nuclei.

finite only if G_1 and G_2 vanish at least like k^2,

$$G_1(k^2), G_2(k^2) \sim k^2 \quad \text{for } k^2 \to 0. \tag{398}$$

It is easy to work out the decay rate for process (395a). In the limit $m_e^2 \ll m_\mu^2$ one finds (exercise 27)

$$\Gamma(\mu \to e\gamma) = \tfrac{1}{2}\alpha m_\mu \{ |F_1|^2 + |F_2|^2 \}. \tag{399}$$

In the processes (395b) and (395c) the contribution of Fig. 18(b), eq. (397), with $k^2 \neq 0$ constitutes what is called the *photonic* amplitude. The virtual photon then couples to the e^+e^--pair, or to the nucleus, respectively, via an ordinary electromagnetic vertex. In addition to this there can also be contributions from box diagrams in which two gauge bosons are exchanged between the four external fermion legs, with various possibilities for the intermediate states. The sum of these constitutes what is called the *weak* amplitude. The two amplitudes (weak and photonic) are usually of comparable magnitude.

Specific models which provide nonvanishing rates for the process (395) can be classified as follows.

(i) *Lepton mixing models:* These models start from the observation that both the neutral weak eigenstates as well as the charged weak eigenstates may be nontrivial mixtures of a set of mass eigenstates. Furthermore, in case the weak interactions contain left- and right-handed couplings, the left-handed and right-handed fields could be characterized by independent mixing schemes, cf. Eqs. (360) and (361).

In the standard GSW model, supplemented by the neutral mixing matrix (360), but no more, the amplitude for $\mu \to e\gamma$ is found to be proportional to m_i^2/m_W^2, the ratio of a typical squared neutrino mass to the square of the W-mass. For example, in a model with two families one would find typically

$$R_{\mu \to e\gamma} \simeq \frac{75\alpha}{128\pi} \sin^2(2\alpha_1) \frac{|m_1^2 - m_2^2|^2}{m_W^4}, \tag{400}$$

which can be at most 10^{-24} and hence remains far below any measurable level. This strong, dynamic suppression comes about in the same way as the suppression of neutral current interactions with $\Delta S \neq 0$, cf. eq. (160) in sec. 4.2, in the quark sector: Let U_{ik}^- be the mixing matrix of the charged, left-handed leptons. The amplitude for $\mu \to e\gamma$ is proportional to $\Sigma_f (U_{\mu f}^-)^* (U^-)_{ef}$, where the sum runs over all left-handed doublets $\in$ SU(2). If the sum is complete, i.e. if there is no charged, left-handed lepton in a singlet of SU(2), than this amplitude vanishes due to the unitarity of the matrix U^-.

The situation changes dramatically if the model is extended by allowing for more complicated lepton representations, for a larger Higgs sector and/or for interference terms between left-handed and right-handed CC couplings (such as are assumed in left–right symmetric models). In the latter case, for example, the amplitude for $\mu \to e\gamma$ is proportional to

$$\frac{m_i}{m_\mu} \operatorname{tg}\phi,$$

where m_i is a typical neutrino mass, ϕ is the mixing angle of W_L and W_R (see sec. 6.2.3). Even with the existing rather low limits on ϕ, the rates for the processes (395) could come very close to the present upper limits.

(ii) *Horizontal symmetries:* There is, of course, always the possibility of introducing direct interfamily couplings

$$g'\{\overline{e(x)}\,(a\gamma^{\alpha}+b\gamma^{\alpha}\gamma_5)\mu(x)X_{\alpha}(x)+\text{h.c.}\}$$

by means of some new very heavy vector bosons. Assuming coupling constants of the same order of magnitude as in the GSW model, i.e. $g' \sim e$, we would conclude from the present limit for the branching ratio (396a)

$$\frac{\alpha}{\pi}\left(\frac{m_W}{m_X}\right)^4 \lesssim 10^{-10},$$

or $m_X \gtrsim 70\ m_W \simeq 6$ TeV.

These superheavy bosons could be the gauge bosons of some new local gauge theory G_H whose particle multiplets are spread over different lepton and quark families of the standard electroweak gauge theory SU(2)×U(1). While the $W^{\pm}$ and Z^0, in the patterns of eqs. (7) and (15), couple only "vertically" within each family, the gauge bosons of G_H would mediate "horizontally" between the families.

In conclusion, the class of processes (395), as well as the related rare kaon decays, bears on some fundamental problems of weak interactions: In the first alternative discussed above one is testing the leptonic mass sector, the multiplet structure of the theory, as well as certain aspects of the Lorentz structure of weak CC interactions. In particular, there is a clear relationship to the other topics and problems that we discussed in sec. 6 and in secs. 8.1 and 8.2 above: The Lorentz structure of leptonic CC couplings and the question of neutrino masses. In the second alternative—should neutrinos turn out to remain massless and should there be no detectable deviation from the minimal model—the rare decays open up a first window onto physics in the region of energies of 10^{12} to 10^{13} eV, a virgin soil still open for our imagination.

References

Abela, R., G. Backenstoss, W., Kunold, L.M. Simons and R. Metzner, 1983, *Nucl. Phys.* A395, 413.

Abers, E. and B. Lee, 1973, *Phys. Rep.* 9C, 2.

Adler, S.L., 1965, *Phys. Rev.* 140B, 736.

Adler, S.L., 1969, *Phys. Rev.* 177, 2426.

Akhmanov, V.V., I.I. Gurevich, Yu.P. Dobretsov, L.A. Makar'ina, A.P. Mishakova, B.A. Nikol'skiĭ, B.V. Sokolov, L.V. Surkova and V.D. Shestakov 1967, *Yad. Fiz.* 6, 316 (engl. transl. Sov. J. of Physics 6(1968) 230).

Bahcall, J., 1978, *Rev. Mod. Phys.* 50, 881.

Bahcall, J. and R. Sears, 1972, *Ann. Rev. Astron. and Astrophys.* 10, 25.

Becher, P., M. Böhm and H. Joos, 1981, Eichtheorien der starken und elektroschwachen Wechselwirkung (Teubner, Stuttgart).

Bég, M.A.B., R.V. Budny, R. Mohapatra and A. Sirlin, 1977, *Phys. Rev. Lett.* 38, 1252.

Bergsma, F. et al., 1983, *Phys. Lett.* 122B, 465.

Bernstein, J. 1974, *Rev. Mod. Phys.* 46, 7.
Bilenky, S. and B. Pontecorvo, 1978, *Phys. Rep.* C41, 225.
Corriveau, F., J. Egger, W. Fetscher, H.J. Gerber, K.F. Johnson, H. Kaspar, H.J. Mahler, M. Salzmann and F. Scheck, 1981, *Phys Rev.* D24, 2004; 1983, *Phys. Lett.* 129B, 260.
Fierz, M., 1937, *Z. Physik* 104, 553.
Glashow, S., 1961, *Nucl. Phys.* 22, 579.
Glashow, S.L., J. Iliopoulos and L. Maiani, 1970, *Phys. Rev.* D2, 1285.
Heil, A. and F. Scheck, 1984, The Physics of the Lepton Families, *Phys. Reports* (to be published).
Jonker, M. et al., 1979, *Phys. Lett.* 86B, 229; 1983, *Z. Physik* 17, 211.
Jonker, M. et al., 1980, *Phys. Lett.* 93B, 203.
Jost, R., 1957, *Helv. Phys. Acta* 30, 409.
Jost, R., 1963, *Helv. Phys. Acta* 36, 77.
Kobayashi, M. and K. Maskawa, 1973, *Prog. Theor. Phys.* 49, 652.
Koks, F.W. and J. van Klinken, 1976, *Nucl. Phys.* A272, 61.
Lüders, G., 1957, *Ann. Phys.* (N.Y.) 2, 1.
Missimer, J., R. Tegen and F. Scheck, 1981, *Nucl. Phys.* B188, 29.
Missimer, J. and L. Simons, 1984, The weak neutral current of the muon, to be published.
Nasrallah N., N. Papadopoulos and K. Schilcher, 1982, *Phys. Lett.* 113B, 61.
O'Raifeartaigh, L., 1979, *Rep. Progr. Phys.* 42, 159.
Roesch, L.Ph., V.L. Telegdi, P. Truttmann, A. Zehnder, L. Grenacs and L. Palffy, 1982, *Helv. Phys. Acta.* 55, 74.
Salam, A., 1968, in "Elementary Particle Theory", ed. N. Svartholm (Almqvist and Wiksells, Stockholm).
Scheck, F., 1978, *Phys. Rep.* 44, 187.
Scheck, F. and A. Wullschleger, 1973, *Nucl. Phys.* B67, 504.
Shrock, R.E., 1981, *Phys. Rev.* D24, 1232.
Steigmann, G., 1979, *Ann. Rev. Nucl. Part. Sci.* 29, 313.
't Hooft, G., 1971, *Nucl. Phys.* B33, 173 and B35, 167.
Weinberg, S., 1966, *Phys. Rev. Lett.* 17, 17.
Weinberg, S., 1967, *Phys. Rev. Lett.* 19, 1264.
Weisberger, W.I., 1966, *Phys. Rev.* 143, 1302.
Wheater, J.F. and C.H. Llewellyn-Smith, 1982, *Nucl. Phys.* B208, 27.
Willis, S.E., V.W. Hughes, P. Némethy, R.L. Burman, D.R.F. Cochran, J.S. Frank, R.P. Redwine, J. Duclos, H. Kaspar, C.K. Hargrove, and U. Moser, 1980, *Phys. Rev. Lett.* 44, 522, Erratum 45, 1370.

Exercises

1. A quark and an antiquark of the same flavour form a bound state $(q\bar{q})^J_{l,S}$ with relative orbital angular momentum l and total spin S, coupled to angular momentum J. Show that this is an eigenstate of P and of C and give the corresponding eigenvalues. Apply the results to π^0, η, ρ^0, ω, ϕ, and A_2^0 mesons. What is the wave function of ω if ϕ contains only strange quarks? *Hint*: $(j_1 m_1 j_2 m_2 | JM) = (-)^{j_1+j_2-J} (j_2 m_2 j_1 m_1 | JM)$.

2. A meson is said to have natural parity if $P = (-)^J$. Show that in this case the bound states of exercise 1 necessarily have $P = C$.

3. Show that the matrices $\exp\{iH\}$ where $H = \sum_{i=1}^{3} \alpha_i \sigma^{(i)}$ and α_i real are unitary and have determinant 1. *Hint*: Diagonalize first the matrix H.

4. Show that in a local gauge theory built on the Lie algebra of $G = \mathrm{SU(P)} \times \mathrm{SU(Q)}$ there are two constants e_{P} and e_{Q} which can be chosen independently.

5. From eq. (113) it is clear that A_α can be "gauged to zero", i.e. be transformed to $A'_\alpha \equiv 0$, if and only if there exists a $g(x)$ for which

$$-\left(\partial_\alpha g^{-1}(x)\right) g(x) = A_\alpha(x).$$

This is a differential equation for $g(x)$ with a given inhomogeneity $A_\alpha(x)$. A condition of integrability for this equation is $(\partial_\alpha\partial_\beta - \partial_\beta\partial_\alpha)g^{-1}(x) = 0$. Work this out and show that $A'_\alpha \equiv 0$ can be obtained if and only if the field tensor $F^{\alpha\beta}$ vanishes identically. What is the analogy to electrodynamics?

6. Work out the globally symmetric Lagrangian (125) as well as the locally invariant version (128) for the case $G = \mathrm{SO}(3)$ and with real boson and fermion multiplets.

7. Derive the matrix (152) that describes state mixing of the quarks d, s, b. *Hints*: The matrix U can be written as a product of three unitary matrices, $U_1(2)U_2(1)U_1(3)$, where $U_i(k) \equiv U_i(\psi_k, \theta_k, \phi_k)$ leaves invariant the component i. Some of the resulting phases can be absorbed in the fields.

8. Write out explicitly the generalized kinetic Lagrangian for the gauge boson fields in the GSW theory. In particular, isolate the couplings of $W^\pm$ to the photon field and compare to what one would have obtained from a Klein–Gordon equation for the W-field supplemented by minimal substitution.

9. Starting from eqs. (172) and (173) and the assignment $y = 2t_3$ for the neutral Higgs field, construct the mass matrix of the vector bosons in the basis of the fields $A_\alpha^{(\mu)}$. Diagonalize this matrix.

10. Consider neutrino–electron scattering as in sec. 5.2. In the case dealt with in eq. (212) calculate the longitudinal polarization of the outgoing neutrino.

11. Suppose you had data on $(\nu_e e)$ scattering (integrated elastic cross section). Analyze the cross section in the context of eqs. (209) and (210).

12. Consider the cross section

$$\frac{d\sigma}{d\Omega^*}(e^+e^- \to \tau^+\tau^-)$$

for $m_\tau \neq 0$. B_+ and B_- can now interfere. Calculate $\Sigma(A_+B_-)(A_+B_-)^*$ and the cross section noting that in eqs. (214) t and u are replaced by $(t - m_\tau^2)$ and $(u - m_\tau^2)$, respectively.

13. Suppose the Z^0 had only vector couplings to electrons and only axial vector couplings to muons. What asymmetry would one predict?

14. How would you modify eq. (217) for s approaching m_Z^2? If the Z^0 boson has a width of 2 GeV, estimate the cross section at resonance and compare to the purely electromagnetic cross section at the same energy.

15. Knowing the behaviour of $v_\alpha(x)$ and $a_\alpha(x)$ under the discrete symmetries P, C, and T, sec. 2.1, derive the transformation properties of s and p, eqs. (226), by studying divergences of v_α and a_α.

16. Predict the isotropic spectra of ν_e and of $\bar{\nu}_\mu$ in positive muon decay, for the case of "V – A", eq. (280). Discuss the relevance of this result for reaction (263b) and the experiment that led to the result (264).

17. Calculate the integrals (268). *Hint*: Go to a system where $Q = (Q^0, \mathbf{0})$.

18. In the reaction (292) introduce the variable $y = E_\mu^{\mathrm{lab}}/E_\nu^{\mathrm{lab}}$, transform the cross section (293) to $d\sigma/dy$, and integrate from some $y_{\min}$ to 1.

19. Calculate the integrals (268) for one massive and one massless particle. *Hints*: The integral (268a) can be done in a frame where $\mathbf{Q} = 0$. The integral (268b) can be written as $A(Q^2)Q^\alpha Q^\beta - B(Q^2)Q^2 g^{\alpha\beta}$. Isolate and calculate $A(Q^2)$ and $B(Q^2)$.

20. Using the behaviour of the axial current $a_\alpha(x)$ under T and C, show that F_π, as defined by eq. (6), is purely imaginary.

21. Compare the result (317) for the process $\tau \to \pi\nu_\tau$, divided by the sum of the rates for $\tau \to \mu\nu_\mu\nu_\tau$ and $\tau \to e\nu_e\nu_\tau$ to results in the Data Tables [R2]. In turn, use the leptonic branching ratios to estimate the lifetime of τ.

22. Prove eqs. (325) by making use of the Wigner–Eckart theorem.

23. Suppose the pionic axial current were simply

$$a_\alpha^{(i)} = -\frac{1}{\sqrt{2}} f_\pi \partial_\alpha\left(\phi_\pi^{(i)}(x)\right),$$

so that semileptonic interaction were

$$L = -f_\pi \partial_\alpha\left(\phi_\pi^{(1)}(x) - \mathrm{i}\phi_\pi^{(2)}(x)\right)\sum_f \overline{\nu_f(x)}\, \gamma_\alpha(1-\gamma_5) f(x) + \text{h.c.}$$

Introduce the coupling to the photon through minimal substitution and study the process $\pi^+ \to e^+\nu_e\gamma$.

24. Apply the results of sec. 8.1.2 to the special case $\pi \to \mu^+\nu_\mu\gamma$. Perform the integral over the surface of the allowed region of phase space and compare to $\pi \to e\nu\gamma$.

25. Consider the β^+-decay from ^{14}O(g.s., 0^+) to ^{14}N(0^+), whose masses differ by $\Delta E = 2.32$ MeV. As these states are members of an isotriplet one can calculate the decay amplitude in the same way as for pion β-decay. Calculate the decay width

$$\Gamma \simeq (2\pi)^7 \int \frac{\mathrm{d}^3 p_e}{2E_e} \int \frac{\mathrm{d}^3 p_\nu}{2E_\nu} \delta(E_e + E_\nu - \Delta E) \sum |T|^2.$$

Show, in particular, that

$$\Gamma \simeq \frac{1}{\pi^3} G^2 \cos^2\theta_1 \int_{m_e}^{\Delta E} \mathrm{d}E_e\, P(E_e),$$

where $P(E) = Ep(\Delta E - p)^2$ and $p = (E - m_e^2)^{1/2}$. The integral can be performed analytically giving

$$\Gamma = \frac{1}{\pi^3} G^2 \cos^2\theta_1 m_e^5 F(\eta),$$

with

$$\eta := \sqrt{\Delta E^2/m_e} - 1,$$

$$F(\eta) = -\tfrac{1}{4}\eta - \tfrac{1}{12}\eta^3 + \tfrac{1}{30}\eta^5 + \tfrac{1}{4}\sqrt{1+\eta^2}\, \ln\left(\eta + \sqrt{\eta^2+1}\right).$$

If an experiment gave the result $\tau F(\eta) \ln 2 = 3075(10)$ sec and if G were the same as in muon decay, what would be $\cos\theta_1$?

26. Draw the simplest Feynman diagram for the process $\pi^0 \to e^+e^-$ and estimate $\Gamma(\pi^0 \to e^+e^-)$ knowing that $\tau(\pi^0 \to 2\gamma) \simeq 10^{-16}$ sec. Show that the unitarity condition for the decay amplitude yields a lower bound for $\Gamma(\pi^0 \to e^+e^-)$.

27. Work out the decay rate (399).

BOOKS, MONOGRAPHS AND GENERAL REVIEWS OF DATA

[R1] M. Abramowitz and I.A. Stegun, Handbook of Mathematical Functions (Dover Publ., New York, 1965).

[R2] Review of Particle Properties, Phys. Lett. 111B (1982) 1.

[R3] J.J. Sakurai, Advanced Quantum Mechanics (Addison-Wesley, Reading, 1967).

[R4] U. Fano and G. Racah, Irreducible Tensorial Sets (Academic Press, New York, 1959).

[R5] I.S. Gradshteyn and I.M. Ryzhik, Table of Integrals, Series, and Products (Academic Press, New York, 1965).

[R6] A.R. Edmonds, Angular Momentum in Quantum Mechanics (Princeton University Press, 1957).

[R7] A. de Shalit and I. Talmi, Nuclear Shell Theory (Academic Press, New York, 1963).

[R8] W. Rühl, The Lorentz Group and Harmonic Analysis (W.A. Benjamin, New York, 1970).

[R9] M. Hamermesh, Group Theory (Addison-Wesley, Reading, 1962).

[R10] R. Omnès, Introduction à l'Etude des Particules Elémentaires (Ediscience, Paris, 1970).

[R11] L.J. Schiff, Quantum Mechanics (McGraw-Hill, New York, 1955).

[R12] S. Gasiorowicz, Elementary Particle Physics (John Wiley & Sons, New York, 1966).

[R13] J.J.J. Kokkedee, The Quark Model (W.A. Benjamin, New York, 1969).

[R14] M.E. Rose, Multipole Fields (John Wiley & Sons, New York, 1955).

[R15] J.D. Jackson, Classical Electrodynamics (John Wiley & Sons, New York, 1975).

[R16] M.E. Rose, Elementary Theory of Angular Momentum (John Wiley & Sons, New York, 1957).

[R17] W. Heitler, The Quantum Theory of Radiation (Oxford University Press, 1953).

[R18] G.N. Watson, Theory of Bessel Functions (Cambridge University Press, 1958).

[R19] H. Überall, Electron Scattering from Complex Nuclei (Academic Press, New York, 1971).

[R20] M.E. Rose, Relativistic Electron Theory (John Wiley & Sons, New York, 1961).

[R21] C.S. Wu and V. Hughes, editors, Muon Physics, Vols. I–III (Academic Press, New York, 1977).

[R22] E. Artin, Einführung in die Theorie der Gammafunktion, Hamb. math. Einzelschriften 11 (Leipzig, 1931).

[R23] L.D. Landau and E.M. Lifshitz, Theoretical Physics, Vol. IV b (Pergamon Press, New York, 1975).
[R24] G. Racah, Group Theory and Spectroscopy, Springer Tracts in Modern Physics, Vol. 37 (Springer, Berlin, 1964).
[R25] G. Källén, Elementary Particle Physics (Addison-Wesley, Reading, 1964).
[R26] C. Itzykson and J.-B. Zuber, Quantum Field Theory (McGraw-Hill, New York, 1980).

Appendix A

LORENTZ TRANSFORMATIONS

In this Appendix we collect some properties of homogeneous Lorentz transformations, selected mainly on the basis of what we need in the main text and for the purpose of defining our notation and conventions. The reader who wishes to study the representations of the Lorentz and Poincaré groups in greater depth should consult the excellent and specialized treatises on this subject.

Contravariant vectors are denoted as follows:

$$a^{\mu}=(a^{0},\boldsymbol{a}).$$

The corresponding *covariant* vector, i.e. the vector which is contragredient to a^{μ} with respect to Lorentz transformations, is then given by

$$a_{\mu}=g_{\mu\nu}a^{\nu}=(a^{0},-\boldsymbol{a}),$$

where $g_{\mu\nu}$ is the metric tensor,

$$g_{\mu\nu}=\begin{pmatrix}1&0&0&0\\0&-1&0&0\\0&0&-1&0\\0&0&0&-1\end{pmatrix}. \tag{A.1}$$

(Summation over identical upper and lower indices is implied.)

General homogeneous Lorentz transformations are written as

$$x^{\mu}\to x'^{\mu}=\Lambda^{\mu}{}_{\nu}x^{\nu}, \tag{A.2}$$

where x' and x represent the same point in Minkowski space: x^{ν} are the coordinates with respect to the original frame of reference, x'^{μ} are the coordinates with respect to the new, transformed frame. The invariance of the scalar product

$$(xy)=x^{\mu}g_{\mu\nu}y^{\nu}=x^{\mu}y_{\mu}=x_{\mu}y^{\mu}=x^{0}y^{0}-\boldsymbol{xy} \tag{A.3}$$

implies the condition

$$g_{\mu\nu}\Lambda^{\mu}{}_{\alpha}\Lambda^{\nu}{}_{\beta}=g_{\alpha\beta}, \tag{A.4}$$

from which one deduces two basic properties of homogeneous Lorentz transformations: (i) By taking the determinant of eq. (A.4), noting that Λ is real and that $\det g=-1$, one finds

$$\det\Lambda=\pm1. \tag{A.5}$$

(ii) Taking eq. (A.4) at $\alpha=\beta=0$ yields

$$\left(\Lambda^{0}{}_{0}\right)^{2}=1+\sum_{i=1}^{3}\left(\Lambda^{i}{}_{0}\right)^{2} \tag{A.6}$$

and hence

$$\Lambda^0{}_0 \geq +1 \quad \text{or} \quad \Lambda^0{}_0 \leq -1. \tag{A.7}$$

Lorentz transformations with $\det \Lambda = +1$ are called *proper* transformations and are designated by a suffix +. Transformations which have the property $\Lambda^0{}_0 \geq +1$ are called *orthochronous* and are identified by an arrow pointing upward. Analogously, the set of transformations with $\det \Lambda = -1$ or with $\Lambda^0{}_0 \leq -1$ is designated by the suffix −, or an arrow pointing downward, respectively.

Eqs. (A.5) and (A.7) allow to identify four connected pieces of the real Lorentz group:

a) $L_+^\uparrow$: the group of all proper, orthochronous Lorentz transformations. This set contains the unit element $\mathbb{1}$ and is a subgroup. Any element of $L_+^\uparrow$ can be deformed continuously into the identity.

b) $L_-^\uparrow$: This set is characterized by $\Lambda^0{}_0 \geq +1$, $\det \Lambda = -1$. It contains the parity transformation P, eq. (III.36a).

c) $L_-^\downarrow$: This set is characterized by $\Lambda^0{}_0 \leq -1$, $\det \Lambda = -1$. It contains the time reflection T, eq. (III.36b).

d) $L_+^\downarrow$: This set has $\Lambda^0{}_0 \leq -1$ but $\det \Lambda = +1$. It is easy to see that it contains the product PT.

Note that the set $L_-^\uparrow$ is obtained from $L_+^\uparrow$ by multiplication with P, $L_-^\downarrow$ is obtained by multiplication with T, and $L_+^\downarrow$ is obtained by multiplication with PT. Note also that the elements $\mathbb{1}$, P, T, and PT form a discrete subgroup of the Lorentz group.

Rotations in three-dimensional space have the form

$$\mathscr{R} = \begin{pmatrix} 1 & 0 \\ 0 & R \end{pmatrix}, \tag{A.8}$$

where $R \in SO(3)$ is an ordinary rotation matrix. As $\det \mathscr{R} = +1$, and $\mathscr{R}^0{}_0 = 1$, the rotations belong to $L_+^\uparrow$. Similarly, special Lorentz transformations, eqs. (III.15) and (III.16) (so-called "boosts"), also belong to $L_+^\uparrow$. For example, the transformation that transforms from the rest system of a particle with mass m to the frame with respect to which the particle moves with the three-momentum $\boldsymbol{p}$, is given by

$$L^\mu{}_\nu(p) = \begin{pmatrix} \dfrac{E_p}{m} & \dfrac{p^{\nu=j}}{m} \\ \dfrac{p^{\mu=i}}{m} & \delta_i{}^j + \dfrac{p^i p^j}{m(E_p+m)} \end{pmatrix}. \tag{A.9}$$

This can be proven from eq. (III.16) using $\cosh\lambda = E_p/m$ and $\sinh\lambda = |\boldsymbol{p}|/m$, carrying out the matrix multiplication $Y = HXH^\dagger$, and solving for y^μ in terms of x^ν. One verifies $L(p)(m,0) = (E_p, \boldsymbol{p})$ and, more generally,

$$L(p)(0,\boldsymbol{a}) = \left(\frac{1}{m}\boldsymbol{p}\cdot\boldsymbol{a},\ \boldsymbol{a} + \frac{(\boldsymbol{p}\cdot\boldsymbol{a})}{m(E_p+m)}\boldsymbol{p}\right). \tag{A.10}$$

Theorem A1. Any $\Lambda \in L_+^\uparrow$ can be written as the product of a special Lorentz transformation ("boost") $L(\boldsymbol{v})$ and a rotation $\mathscr{R}$ in three-dimensional space. This decomposition is unique.

Proof. Choose $v^i := \Lambda^0{}_i/\Lambda^0{}_0$. This is an acceptable velocity because it does not exceed the velocity of light.

$$\boldsymbol{v}^2 = \frac{1}{(\Lambda^0{}_0)^2} \sum_{i=1}^{3} (\Lambda^0{}_i)^2 \leq 1,$$

owing to eq. (A.6). Furthermore we take

$$L^0{}_0(-\boldsymbol{v}) = \cosh\lambda = \frac{1}{\sqrt{1-\boldsymbol{v}^2}} = \Lambda^0{}_0,$$

$$L^0{}_i(-\boldsymbol{v}) = -\hat{\boldsymbol{v}}_i \sinh\lambda = -\frac{\boldsymbol{v}_i}{\sqrt{1-\boldsymbol{v}^2}} = -\Lambda^0{}_i,$$

and $L^i{}_0(-\boldsymbol{v}) = L^0{}_i(-\boldsymbol{v})$. Define now the matrix $\mathscr{R} := \Lambda L(-\boldsymbol{v})$. It is easy to show that this is a rotation: Indeed

$$\mathscr{R}^0{}_0 = (\Lambda^0{}_0)^2 - \sum_{i=1}^{3} (\Lambda^0{}_i)^2 = 1.$$

Therefore we obtain the decomposition

$$\mathscr{R}L(\boldsymbol{v}) = \Lambda L(-\boldsymbol{v}) L(\boldsymbol{v}) = \Lambda.$$

It remains to show that R and $\boldsymbol{v}$ are unique. Suppose we had found two different matrices R, $\bar{R} \in \mathrm{SO}(3)$ and two different velocities $\boldsymbol{v}$, $\bar{\boldsymbol{v}}$, such that $\Lambda = \mathscr{R}L(\boldsymbol{v}) = \bar{\mathscr{R}}L(\bar{\boldsymbol{v}})$. Consider then

$$\mathscr{R}^{-1}\Lambda L(-\boldsymbol{v}) = \mathbb{1} = \mathscr{R}^{-1}\bar{\mathscr{R}}L(\bar{\boldsymbol{v}})L(-\boldsymbol{v}).$$

The component $\mu = \nu = 0$ of this last equation is

$$1 = \{1 - \bar{\boldsymbol{v}}\cdot\boldsymbol{v}\}/\sqrt{1-v^2}\sqrt{1-\bar{v}^2}.$$

This equation holds only if $\boldsymbol{v} = \bar{\boldsymbol{v}}$. In this case, however, we must also have $\bar{\mathscr{R}} = \mathscr{R}$ or $\bar{R} = R$. This completes the proof.

Theorem (A1) shows that $\Lambda \in \mathrm{L}_+^\uparrow$ depends on six real parameters: three Euler angles characterizing the rotation, three velocity components characterizing the boost.

Derivatives with respect to the contravariant variable $x^\mu = (x^0, \boldsymbol{x})$ are denoted by

$$\partial_\mu = \frac{\partial}{\partial x^\mu} = \left(\frac{\partial}{\partial x^0}, \nabla\right). \tag{A.11}$$

Note that the spatial part of eq. (A.11) is the ordinary gradient operator in three dimensions. The divergence of a vector field $A^\mu(x)$, therefore, is given by

$$\partial_\mu A^\mu(x) = \frac{\partial}{\partial x^0} A^0(x) + \nabla \cdot \boldsymbol{A}(x). \tag{A.12}$$

Similarly derivatives with respect to the covariant variable x_μ are given by

$$\partial^\mu \equiv \frac{\partial}{\partial x_\mu} = \left(\frac{\partial}{\partial x^0}, -\nabla \right).$$

The four-dimensional Laplace operator, finally, is given by

$$\square \equiv \partial_\nu \partial^\mu = \frac{\partial^2}{(\partial x^0)^2} - \Delta. \tag{A.13}$$

Theorem A2. There are precisely two, linearly indendent, Lorentz invariant distributions $\Delta_i(z; m)$ which obey the Klein–Gordon equation for mass m. These are

$$\Delta_0(z; m) := -\frac{\mathrm{i}}{(2\pi)^3} \int \frac{\mathrm{d}^3 k}{2\omega_k} (\mathrm{e}^{-ikz} - \mathrm{e}^{ikz}), \tag{A.14}$$

$$\Delta_1(z; m) := \frac{1}{(2\pi)^3} \int \frac{\mathrm{d}^3 k}{2\omega_k} (\mathrm{e}^{-ikz} + \mathrm{e}^{ikz}), \tag{A.15}$$

with $\omega_k \equiv k^0 = \sqrt{\boldsymbol{k}^2 + m^2}$. These distributions have the following properties:

(i) $\{\square + m^2\}\Delta_i(z; m) = 0, \quad i = 0, 1,$ (A.16)

(ii) $\Delta_0(z^0 = 0, \boldsymbol{z}; m) = 0,$ (A.17a)

(iii) $\left.\frac{\partial}{\partial z^0} \Delta_0(z; m)\right|_{z^0=0} = -\delta(\boldsymbol{z}),$ (A.17b)

(iv) $\Delta_0(-z; m) = -\Delta_0(z; m), \qquad \Delta_1(-z; m) = \Delta_1(z; m),$ (A.18)

(v) $\Delta_0(z; m) = 0 \quad \text{for } z^2 < 0.$ (A.19)

Note, however, that $\Delta_1(z; m)$ does not vanish for spacelike argument, $z^2 < 0$.

Appendix B

S-MATRIX, CROSS SECTIONS, DECAY PROBABILITIES

Write the scattering matrix (somewhat symbolically) as

$$S_{\mathrm{fi}} = \delta_{\mathrm{fi}} + R_{\mathrm{fi}}, \tag{B.1}$$

where f and i are asymptotic free states. δ_{fi} means "no scattering" and R_{fi} is the reaction matrix proper. R_{fi} necessarily contains a δ-distribution expressing conservation of total energy and momentum. Besides this distribution it is convenient to take out a factor $\mathrm{i}(2\pi)^4$ and to define the T-matrix by (see sec. II.1.3),

$$R_{\mathrm{fi}} = \mathrm{i}(2\pi)^4 \delta(P_{\mathrm{f}} - P_{\mathrm{i}}) T_{\mathrm{fi}}. \tag{B.2}$$

The differential cross section for the reaction

$$\mathrm{a} + \mathrm{b} \to 1 + 2 + \cdots + N$$

(all particles being described asymptotically by plane waves), is given by the general expression

$$\mathrm{d}\sigma_{\mathrm{fi}}(\mathrm{a} + \mathrm{b} \to 1 + \cdots + N) = \frac{(2\pi)^{10} \delta(P_{\mathrm{f}} - P_{\mathrm{i}})}{2E_{\mathrm{a}} 2E_{\mathrm{b}} |\boldsymbol{v}_{\mathrm{ab}}|} |T_{\mathrm{fi}}|^2 \prod_{n=1}^{N} \frac{\mathrm{d}^3 p^{(n)}}{2E_N}. \tag{B.3}$$

In this expression

$$E_{\mathrm{a}} = \sqrt{\boldsymbol{p}^{(\mathrm{a})2} + m_{\mathrm{a}}^2}, \qquad E_{\mathrm{b}} = \sqrt{\boldsymbol{p}^{(\mathrm{b})2} + m_{\mathrm{b}}^2}, \qquad E_n = \sqrt{\boldsymbol{p}^{(n)2} + m_n^2};$$

$$P_{\mathrm{i}} = p^{(\mathrm{a})} + p^{(\mathrm{b})}, \qquad P_{\mathrm{f}} = \sum_{n=1}^{N} p^{(n)},$$

and $|\boldsymbol{v}_{\mathrm{ab}}|$ is the relative velocity of the incoming particles, $\mathrm{d}^3 p^{(n)}/2E_n$ is the Lorentz invariant volume element in the phase space of particle number n. This term as well as the incoming flux factor in the first denominator is in accord with our covariant normalization $\langle \boldsymbol{p}' | \boldsymbol{p} \rangle = 2E_p \delta(\boldsymbol{p}' - \boldsymbol{p})$. Formula (B.3) holds in all systems of reference where the momenta $\boldsymbol{p}^{(\mathrm{a})}$ and $\boldsymbol{p}^{(\mathrm{b})}$ are collinear (e.g. the laboratory and center-of-mass systems). It can be extended to any system by replacing the flux factor (so-called Møller factor) with the invariant on the r.h.s. of the following equation:

$$E_{\mathrm{a}} E_{\mathrm{b}} |\boldsymbol{v}_{\mathrm{ab}}| = \sqrt{\left(p^{(\mathrm{a})} \cdot p^{(\mathrm{b})}\right)^2 - p^{(\mathrm{a})2} p^{(\mathrm{b})2}}. \tag{B.4}$$

(One should verify that this equation does indeed hold if the 3-momenta of particles a and b are collinear.)

The observable cross sections are obtained from eq. (B.3) by integration over those momentum variables in the final state which are not observed. Similarly, depending on whether or not particles a and b have nonvanishing spin and are polarized, the appropriate average over spin projections must be taken. If the spin orientations in the final state are not discriminated, eq. (B.3) must be summed over them.

In a similar fashion the differential decay rate of a particle a, with mass m_a and momentum q, into a final state with N particles,

$$a \to 1 + 2 + \cdots + N,$$

is given by

$$d\Gamma_{fi} = (2\pi)^4 \delta(p^{(1)} + p^{(2)} + \cdots + p^{(N)} - q) \frac{(2\pi)^3}{2E_q} |T_{fi}|^2 \sum_{n=1}^{N} \frac{d^3 p^{(n)}}{2E_n}. \tag{B.5}$$

Again, depending on what shall be observed, integration over some of the momentum variables and, possibly, sums over spin projections in the final state must be performed. If the decaying particle has nonvanishing spin and if the spin orientation is not known, the formula must be averaged over all spin projections.

From eq. (B.5) one sees that the squared decay amplitude has the dimension $[|T_{fi}|^2] = E^{2(3-N)}$. Thus, in a two-body decay the dimension is (energy)^2, whereas in a three-body decay T is dimensionless. This can be useful in checking calculations.

Let us consider a few examples:

(i) *Differential cross section for $a + b \to 1 + 2$*. The example of elastic scattering of two massive particles (masses m and M, respectively) is worked out in detail in sec. IV.4.2. Examples of neutrino reactions are treated in secs. V.2.4 and V.6.2.5.

(ii) *Two-body decay*. In the rest system of the decaying particle the two particles in the final state have the momenta

$$p^{(1)} = \{E_1, \boldsymbol{\kappa}\}, \qquad p^{(2)} = \{E_2, -\boldsymbol{\kappa}\},$$

with $E_1 + E_2 = m_a$, $E_i = (m_i^2 + \kappa^2)^{1/2}$, and $\kappa := |\boldsymbol{\kappa}|$. Integrating over the 3-momentum of particle 2 one obtains from eq. (B.5)

$$d^3\Gamma = \frac{(2\pi)^7}{8m_a E_1 E_2} |T(a \to 1 + 2)|^2 \delta(E_1 + E_2 - m_a)\, d^3 p^{(1)},$$

where $d^3 p^{(1)}$ can be expressed in polar coordinates, $d^3 p^{(1)} = \kappa^2\, d\kappa\, d\Omega$. Noting that $\kappa\, d\kappa = E_1\, dE_1$ one can convert the integration over κ into an integration over E_1. For this we need the derivative of the argument of the δ-distribution with respect to E_1, viz.

$$\frac{d}{dE_1}\{E_1 + E_2 - m_a\} = 1 + \frac{dE_2}{d\kappa} \cdot \frac{d\kappa}{dE_1} = \frac{E_1 + E_2}{E_2} = \frac{m_a}{E_2}.$$

One obtains

$$d^2\Gamma = \frac{(2\pi)^7 \kappa}{8m_a^2} |T(a \to 1 + 2)|^2\, d\Omega.$$

The decay probability is independent of the azimuth ϕ. Integrating over this angle we have

$$d\Gamma = \frac{(2\pi)^8 \kappa}{8m_a^2} |T(a \to 1 + 2)|^2\, d(\cos\theta), \tag{B.6}$$

where θ is the opening angle between the spin expectation value of the decaying particle a and the momentum of particle 1. Eq. (V.316) for the decay of a polarized τ into a pion and a neutrino provides an example for this case. If the decaying particle is spinless, or if it has spin but is unpolarized, $|T|^2$ is isotropic. Integrating over $\mathrm{d}(\cos\theta)$ one obtains the total decay rate

$$\Gamma = \frac{(2\pi)^8 \kappa}{4m_\mathrm{a}^2} |T(\mathrm{a} \to 1+2)|^2. \tag{B.7}$$

(iii) *Three-body decays*. Here we distinguish several situations: If two of the particles in the final state are not observed (cf. the example of $\mu \to e\nu\bar{\nu}$) one proceeds as described in sec. V.6.2.1 and obtains a differential decay rate

$$\mathrm{d}^2\Gamma/\mathrm{d}E\,\mathrm{d}(\cos\theta),$$

where θ is the opening angle between the spin of the decaying particle and the momentum of the observed particle in the final state. In other situations one may proceed as follows. Integrate first eq. (B.5) over $\mathrm{d}^3p^{(3)}$ to obtain

$$\mathrm{d}^6\Gamma = \frac{(2\pi)^7}{16m_\mathrm{a}} \frac{\kappa_1\kappa_2}{E_3} |T(\mathrm{a} \to 123)|^2 \delta(E_1+E_2+E_3-m_\mathrm{a})\,\mathrm{d}E_1\,\mathrm{d}E_2\,\mathrm{d}\Omega_1\,\mathrm{d}\Omega_2,$$

where $E_3 = [m_3^2 + (p^{(1)}+p^{(2)})^2]^{1/2}$ and $\kappa_i := |\boldsymbol{p}^{(i)}|$. Then integrate over $\mathrm{d}\Omega_1$ for particle 1 which is emitted isotropically, take $\boldsymbol{p}^{(1)}$ as the 3-axis and make use of the axial symmetry around this direction, viz.

$$\mathrm{d}^3\Gamma = \frac{(2\pi)^9}{8m_\mathrm{a}} |T(\mathrm{a} \to 123)|^2 \frac{\kappa_1\kappa_2}{E_3} \delta(E_1+E_2+E_3-m_\mathrm{a})\,\mathrm{d}E_1\,\mathrm{d}E_2\,\mathrm{d}(\cos\theta), \tag{B.8}$$

where now

$$E_3 = \left(m_3^2 + \kappa_1^2 + \kappa_2^2 + 2\kappa_1\kappa_2\cos\theta_2\right)^{1/2}.$$

This formula may be transformed to the variables E_1, E_2, E_3 by means of the Jacobian

$$\frac{\partial(E_1, E_2, \cos\theta_2)}{\partial(E_1, E_2, E_3)} = \frac{E_3}{\kappa_1\kappa_2}$$

and finally to the variables $s := E_2 + E_3$ and $t := E_2 - E_3$, giving

$$\mathrm{d}^2\Gamma = \frac{(2\pi)^9}{16m_\mathrm{a}} |T(\mathrm{a} \to 123)|^2\,\mathrm{d}E_1\,\mathrm{d}t, \tag{B.9}$$

from which the total rate is obtained by integration over the kinematic range of E_1 and of t.

Note that T contains a factor $(2\pi)^{-3/2}$ for each external particle. So $|T(\mathrm{a} \to 1+2)|^2$ produces a factor $(2\pi)^{-9}$, $|T(\mathrm{a} \to 123)|^2$ produces a factor $(2\pi)^{-12}$.

Appendix C

SOME FEYNMAN RULES FOR QUANTUM ELECTRODYNAMICS OF SPIN-1/2 PARTICLES $f^{\pm}$

The rules hold for the matrix R, as defined by eq. (B.1). The T-matrix is obtained upon comparison with the defining equation (B.2).

(i) *Diagrams*. One draws all connected diagrams of the process under consideration, at the order n in the coupling constant that one wishes to calculate. External and internal fermion lines are provided with arrows which point in the direction of the flow of *negative* charge. The momenta of internal lines are chosen such as to follow the arrow. All factors prescribed by the following rules must be written down from *right* to *left* following the direction of the arrows.

(ii) *External lines*. For each external, incoming f^- write a spinor in momentum space $u_f(p)$, for each incoming f^+ write $\overline{v_f(p)}$. Similarly, for an outgoing f^- write $\overline{u_f(p)}$, for an outgoing f^+ write $v_f(p)$. For an incoming or outgoing photon write a polarization vector $\varepsilon_\alpha(\boldsymbol{k}, \lambda)$ with the index α to be contracted with γ^α at the fermion vertex to which it couples. In addition, each external particle obtains a factor $(2\pi)^{-3/2}$.

(iii) *Vertices*. Each vertex $(\mathrm{ff}\gamma)$ has a factor $e\gamma^\alpha$ and a δ-distribution for energy–momentum conservation at that vertex.

(iv) *Internal fermion lines*. An internal fermion line is represented by a propagator

$$\frac{\not{p} + m_f}{p^2 - m_f^2 + i\varepsilon},$$

where the direction of p is chosen in accordance with rule (i).

(v) *Internal photon lines*. An internal photon line with momentum k connects two vertices $(\mathrm{ff}\gamma)$ characterized by the Lorentz indices α and β (cf. rule (iii)) and yields a factor

$$-\frac{g^{\alpha\beta}}{k^2 + i\varepsilon}.$$

(vi) *Integrations*. All internal momenta must be integrated over. In all cases this yields a δ-distribution $\delta(P_i - P_f)$ for conservation of total energy–momentum. In orders of e which are higher than the lowest nontrivial order this rule also gives rise to some nontrivial integrations over internal momenta. Such integrals can turn out to be divergent and must then be analyzed in the framework of renormalization.

(vii) *Factors*. R_{fi} has a factor $(-)^P$ where P is the permutation of the fermions in the final state, as well as a factor $(-)^L$ if L is the number of closed fermion loops. In addition, R_{fi} obtains the following factors:

$$i^{n+f_i+b_i}(2\pi)^{4(n-f_i-b_i)},$$

where n is the order of perturbation theory, f_{i} the number of internal fermion lines, b_{i} the number of internal photon lines.

(viii) *Closed fermion loops*. Closed loops which couple to an *odd* number of photon lines give a vanishing amplitude. This is a consequence of C-invariance of QED.

(ix) *External potentials*. An external potential is an approximation for the interaction with a very heavy particle which therefore can absorb or provide an arbitrary amount of 3-momentum. Therefore, for an external potential one has to write a δ-distribution only for energy conservation, whilst the vertex factor $e\gamma^{\alpha}$ must be replaced with

$$\delta^{\alpha 0}\frac{Ze}{(2\pi)^3}\frac{v(\boldsymbol{k})}{k^2},$$

where Ze is the total charge that creates the potential, $v(\boldsymbol{k})$ is the form factor of the corresponding charge distribution,

$$v(\boldsymbol{k})=\int \mathrm{d}^3x\,\mathrm{e}^{-\mathrm{i}\boldsymbol{k}\cdot\boldsymbol{x}}\rho(\boldsymbol{x}),$$

with $\int\rho(\boldsymbol{x})\,\mathrm{d}^3x=1$.

Appendix D

TRACES

The following formulae are all derived from eqs. (III.51) to (III.53):

$$\mathrm{Sp}\,\mathbb{1} = 4, \qquad \mathrm{Sp}\,\gamma^\alpha = \mathrm{Sp}\,\gamma_5 = 0.$$

The trace of a product with an *odd* number of factors vanishes. For products with an even number of γ-matrices the following relations are useful:

$$\mathrm{Sp}\{\gamma^\alpha\gamma^\beta\} = 4g^{\alpha\beta},$$

$$\mathrm{Sp}\{\gamma^\alpha\gamma^\beta\gamma_5\} = 0,$$

$$\mathrm{Sp}\{\gamma^\alpha\gamma^\beta\gamma^\sigma\gamma^\tau\} = 4(g^{\alpha\beta}g^{\sigma\tau} - g^{\alpha\sigma}g^{\beta\tau} + g^{\alpha\tau}g^{\beta\sigma}\},$$

$$\mathrm{Sp}\{\gamma^\alpha\gamma^\beta\gamma^\sigma\gamma^\tau\gamma_5\} = 4i\varepsilon^{\alpha\beta\sigma\tau},$$

$$\begin{aligned}\mathrm{Sp}\{\gamma^\alpha\gamma^\beta\gamma^\mu\gamma^\nu\gamma^\sigma\gamma^\tau\} = {} & g^{\alpha\beta}\mathrm{Sp}\{\gamma^\mu\gamma^\nu\gamma^\sigma\gamma^\tau\} - g^{\alpha\mu}\mathrm{Sp}\{\gamma^\beta\gamma^\nu\gamma^\sigma\gamma^\tau\}\\ & + g^{\alpha\nu}\mathrm{Sp}\{\gamma^\beta\gamma^\mu\gamma^\sigma\gamma^\tau\} - g^{\alpha\sigma}\mathrm{Sp}\{\gamma^\beta\gamma^\mu\gamma^\nu\gamma^\tau\}\\ & + g^{\alpha\tau}\mathrm{Sp}\{\gamma^\beta\gamma^\mu\gamma^\nu\gamma^\sigma\}.\end{aligned}$$

In many cases the Lorentz indices of some of the γ-matrices in a product have to be contracted. The following formulae are then useful:

$$\gamma_\alpha\gamma^\alpha = 4, \qquad \gamma_\alpha \not{a}\gamma^\alpha = -2\not{a},$$

$$\gamma_\alpha \not{a}\not{b}\,\gamma^\alpha = 4ab, \qquad \gamma_\alpha \not{a}\not{b}\not{c}\gamma^\alpha = -2\not{c}\not{b}\not{a},$$

$$\gamma_\alpha \not{a}\not{b}\not{c}\not{d}\gamma^\alpha = 2(\not{d}\not{a}\not{b}\not{c} + \not{c}\not{b}\not{a}\not{d}).$$

Note that in our conventions $\gamma_5 = i\gamma^0\gamma^1\gamma^2\gamma^3$, $\varepsilon_{0123} = +1$. Note also the relation

$$\varepsilon^{\alpha\beta\mu\nu}\varepsilon_{\alpha\beta\sigma\tau} = -2\{\delta^\mu{}_\sigma\delta^\nu{}_\tau - \delta^\mu{}_\tau\delta^\nu{}_\sigma\}.$$

DIRAC EQUATION WITH CENTRAL FIELDS

The Hamiltonian form (III.60a) of the Dirac equation is well adapted for a discussion of interactions with external fields. For the case of an external, spherically symmetric potential $V(r)$ and for stationary states ($\propto \mathrm{e}^{-\mathrm{i}Et}$) it reads

$$E\Psi(\boldsymbol{r}) = \{-\mathrm{i}\boldsymbol{\alpha}\cdot\nabla + V(r) + m\beta\}\Psi(\boldsymbol{r}), \tag{E.1}$$

with α and β as given by eqs. (III.59). Using the vector identities

$$\begin{aligned}\nabla &= \hat{\boldsymbol{r}}(\hat{\boldsymbol{r}}\cdot\nabla) - \hat{\boldsymbol{r}}\wedge(\hat{\boldsymbol{r}}\wedge\nabla)\\ &= \hat{\boldsymbol{r}}(\hat{\boldsymbol{r}}\cdot\nabla) - \frac{\mathrm{i}}{r}\hat{\boldsymbol{r}}\wedge\boldsymbol{l}\end{aligned}$$

one has

$$\boldsymbol{\alpha}\cdot\nabla = \boldsymbol{\alpha}\cdot\hat{\boldsymbol{r}}\frac{\partial}{\partial r} - \frac{\mathrm{i}}{r}\boldsymbol{\alpha}\cdot(\hat{\boldsymbol{r}}\wedge\boldsymbol{l}) = \gamma_5\boldsymbol{S}\cdot\hat{\boldsymbol{r}}\left\{\frac{\partial}{\partial r} - \frac{1}{r}\boldsymbol{S}\cdot\boldsymbol{l}\right\},$$

where γ_5 is given by eq. (III.56), whilst the matrix S stands for

$$\boldsymbol{S} := \begin{pmatrix}\boldsymbol{\sigma} & 0\\ 0 & \boldsymbol{\sigma}\end{pmatrix}.$$

Finally, upon introduction of Dirac's angular momentum operator

$$K := \beta(\boldsymbol{S}\cdot\boldsymbol{l} + \mathbb{1}) \equiv \begin{pmatrix}K^{(0)} & 0\\ 0 & -K^{(0)}\end{pmatrix} \tag{E.2}$$

with $K^{(0)} = \boldsymbol{\sigma}\cdot\boldsymbol{l} + \mathbb{1}$, eq. (E.1) takes the form

$$\left\{-\mathrm{i}\gamma_5\boldsymbol{S}\cdot\hat{\boldsymbol{r}}\left(\frac{\partial}{\partial r} + \frac{1}{r} - \beta\frac{K}{r}\right) + V(r) + \beta m\right\}\Psi = E\Psi =: H\Psi. \tag{E.3}$$

One verifies by explicit calculation that K commutes with H, $[H, K] = 0$, but that H neither commutes with the orbital angular momentum nor with the spin. The operator K contains the entire dependence on angular momenta so that eq. (E.3) lends itself to separation in radial and angular coordinates. To see this, we note first that $K^{(0)}$ can be written as

$$K^{(0)} = \boldsymbol{\sigma}\cdot\boldsymbol{l} + 1 = 2\boldsymbol{s}\cdot\boldsymbol{l} + 1 = \boldsymbol{j}^2 - \boldsymbol{l}^2 - \boldsymbol{s}^2 + 1.$$

Its eigenfunctions are the coupled states

$$\varphi_{jlm} = \sum_{m_l m_s}(lm_l, \tfrac{1}{2}m_s|jm)Y_{lm_l}\chi_{m_s}.$$

Denote the eigenvalues of $K^{(0)}$ by $-\kappa$, i.e.

$$K^{(0)}\varphi_{jlm} = -\kappa\varphi_{jlm} \quad \text{with} \quad \kappa = -j(j+1) + l(l+1) - \tfrac{1}{4}.$$

Note also that $(K^{(0)})^2 = \mathbb{1} + \boldsymbol{\sigma}\cdot\boldsymbol{l} + \boldsymbol{l}^2 = \boldsymbol{j}^2 - \boldsymbol{s}^2 + \mathbb{1}$, from which one deduces $\kappa^2 = (j+\frac{1}{2})^2$. From these formulae one sees that

$$\begin{aligned}
&\text{for } \kappa > 0: \quad l = \kappa, \\
&\text{for } \kappa < 0: \quad l = -\kappa - 1, \\
&\text{in all cases} \quad j = |\kappa| - \tfrac{1}{2}.
\end{aligned} \tag{E.4}$$

Therefore, the eigenfunctions of total angular momentum $\boldsymbol{j}$ can be written in the compact notation $\varphi_{jlm} \equiv \varphi_{\kappa m}$, the modulus of κ giving the value of j, the sign giving the value of $l = j \pm \frac{1}{2}$, according to the rules (E.4). As $K^{(0)}\varphi_{\kappa m} = -\kappa\varphi_{\kappa m}$, the eigenvalues and eigenfunctions of K, eq. (E.2), are

$$K\begin{pmatrix} \varphi_{\kappa m} \\ \varphi_{-\kappa m} \end{pmatrix} = -\kappa\begin{pmatrix} \varphi_{\kappa m} \\ \varphi_{-\kappa m} \end{pmatrix}.$$

For the eigenfunctions Ψ of eq. (E.1) or eq. (E.3) one makes the ansatz

$$\Psi_{\kappa m}(r, \hat{\boldsymbol{r}}) = \begin{pmatrix} g_\kappa(r)\varphi_{\kappa m}(\hat{\boldsymbol{r}}) \\ \mathrm{i} f_\kappa(r)\varphi_{-\kappa m}(\hat{\boldsymbol{r}}) \end{pmatrix}, \tag{E.5}$$

the factor i being introduced for convenience so that the resulting differential equations for the radial functions f and g become real. As a last step one verifies by explicit calculation that $(\boldsymbol{\sigma}\cdot\hat{\boldsymbol{r}})\varphi_{\kappa m} = -\varphi_{-\kappa m}$. With these tools at hand one deduces from eq. (E.3) the following system of differential equations:

$$\begin{aligned}
f_\kappa' &= \frac{\kappa - 1}{r} f_\kappa - (E - V(r) - m) g_\kappa, \\
g_\kappa' &= -\frac{\kappa + 1}{r} g_\kappa + (E - V(r) + m) f_\kappa.
\end{aligned} \tag{E.6}$$

Clearly, this result does not depend on the specific representation (III.60a) of the Dirac equation we started from. For example in the representation (III.52) we would obtain

$$\Psi_{\kappa m}(\boldsymbol{r}) = \frac{1}{\sqrt{2}}\begin{pmatrix} g_\kappa\varphi_{\kappa m} + \mathrm{i} f_\kappa\varphi_{-\kappa m} \\ g_\kappa\varphi_{\kappa m} - \mathrm{i} f_\kappa\varphi_{-\kappa m} \end{pmatrix}$$

(cf. eq. (IV.107)).

Eq. (E.5) shows very clearly that the central field solutions are not eigenfunctions of orbital angular momentum: For example for $\kappa = -1$, the upper component has $l = 0$, the lower has $\bar{l} = 1$. Thus, the relativistic analogue of an s-state has a component proportional to a p-state, cf. the discussion of the M1-transition 2s → 1s in sec. V.6.3.

SUBJECT INDEX